THE YOUNG BIRDER'S GUIDE TO BIRDS OF NORTH AMERICA

Bill Thompson III

Illustrations by Julie Zickefoose and Michael DiGiorgio

Houghton Mifflin Company
Boston New York

For information about permission to
reproduce selections from this book, write to
Permissions, Houghton Mifflin Company,
215 Park Avenue South,
New York, New York 10003.

www.hmhbooks.com

Library of Congress Cataloging-in-Publication Data

Thompson, Bill, date.
The young birder's guide to birds of North America / Bill Thompson III;
illustrations by Julie Zickefoose and Michael DiGiorgio.
p. cm.
Includes index.
ISBN 978-0-547-44021-7
1. Bird watching—North America—Juvenile literature. I. Zickefoose,
Julie, ill II. DiGiorgio, Michael, ill. III. Title.
QL681.T454 2012
598.072'347—dc23 2011046149

Printed in China

SCP 10 9 8 7 6 5 4 3

Dedication

To Phoebe and Liam
with my love and fondest wishes for
a lifetime of bird watching.

And to Julie, who gave me the joy
of being a parent.

And to to the memory of my dad,
William H. Thompson, Jr.
(1932–2011)

Photo Credits

Glenn Bartley: 139, 173, 255 right, 260
Cliff Beittel: 62 left, 90 left, 92, 93, 96, 98 left, 102, 113, 115 left, 123, 124 left, 125 left, 130 left, 133 left, 134, 151, 152,153, 158, 166, 168, 189 left, 196 right, 198 left, 211, 228 left, 229, 240, 242, 244, 261, 265, 268 left, 306 right, 315 right, 323 right, 330 left
David Dvorak, Jr.: 111, 137 right, 223 left, 253 left
Brian Henry: 87, 161, 171 right, 241, 254 left, 254 right, 268 right, 273 right, 278 right, 321 left
Jerry Liguori: 101 left, 112 left, 114 left
Maslowski Productions: 54, 68 right, 76 left, 100 left, 131 right, 163, 164, 176, 177 left, 177 right, 184 right, 199, 214, 218, 223 right, 226, 234, 251, 252 left, 277 right, 280 left, 289 right, 302 right, 308 left, 312 right, 313 left, 315 left, 320 right, 328 right, 338
Garth McElroy: 252 right, 277 left, 278 left, 279 left, 282, 283 left, 283 right
Charles Melton: 47 right, 61, 64 left, 64 right, 67 left, 67 right, 75 left, 79 right, 84, 88, 90 right, 91 bottom, 97, 99 right, 101 right, 121, 124 right, 132 right, 136, 138 left, 140 right, 160, 167, 175, 178 left, 178 right, 179 left, 179 right, 180 right, 181 left, 181 right, 182 top right, 182 bottom right, 183 left, 183 right, 187, 191, 195, 206 left, 206 right, 210, 232, 233, 246, 256 left, 257, 262, 270, 274 left, 276, 286, 288 left, 290 left, 290 right, 291, 292 right, 293 left, 293 right, 294, 295 right, 305, 309 left, 309 right, 311, 313 right, 327 right, 329 left, 329 right, 331, 335 left, 335 right, 340 left, 340 right, 341 left, 341 right
Gary Meszaros: 162 right, 174, 253 right
Arthur Morris/Birds As Art: 16, 46, 47 left, 52, 58, 77 left, 78, 83, 86 right, 89, 91 top, 98 right, 99 left, 100 right, 106 right, 108, 110 right, 119, 122, 126 left, 127, 129, 130 right, 131 left, 132 left, 137 left, 143 right, 146, 148 left, 148 right, 149 right, 162 left, 272, 279 right, 307 left, 310 left
Marie Read: 49, 51, 53, 56 left, 57, 59 left, 63 left, 65, 72 left, 72 right, 73 left, 73 right, 82 right, 85 right, 94 left, 94 right, 106 left, 107, 112, 116, 120, 141, 142, 153 right, 154, 156, 172, 186 left, 186 right, 188, 190 left, 190 right, 192 left, 194 right, 203, 209, 212, 215, 217, 222, 227 left, 227 right, 230, 237, 238, 249, 259 left, 266, 275, 285 left, 289 left, 292 left, 298, 301 right, 302 left, 305 left, 306 left, 308 right, 310 right, 317 left, 321 right, 322, 326 left, 328 left, 334 left, 337 left, 337 right
© robertmccaw.com: 50, 59 right, 56 right, 55 left, 55 right, 60 left, 60 right, 62 right, 63 right, 66 left, 66 right, 68 left, 68 left, 69, 70 left, 70 right, 71 left, 71 right, 74, 75 right, 77 right, 79 left, 80, 81 left, 81 right, 82 left, 86 left, 95, 110 left, 114 right, 115 right, 117, 118, 125 right, 126 right, 128, 133 right, 135, 138 right, 143 left, 144 left, 144 right, 145, 147 left, 147 right, 149 left, 159, 169, 170 left, 170 right, 180 left, 182 left, 193, 194 left, 196 left, 197 left, 197 right, 201, 202, 204, 207, 208, 213, 216, 219, 220, 221, 224, 225, 231, 235, 236, 247, 245, 248, 250, 255 left, 256 right, 258, 263, 264, 267 left, 267 right, 269 left, 271 left, 271 right, 273 left, 274 left, 274 right, 280 right, 281 left, 284, 285 right, 287, 288 right, 295 left, 296, 297, 299, 300, 301 left, 303 right, 307 right, 314 left, 316 right, 318 left, 319 left, 319 right, 320 left, 323 left, 324 left, 324 right, 325 left, 325 right, 326 right, 327 left, 330 right, 332 left, 332 right, 333 left, 333 right, 334 right, 336 left, 336 right, 339 left, 339 right, 342, 343 left, 343 right

Barth Schorre: 269 right, 281 right
Bill Thompson III: 1, 3, 9, 15, 22, 27, 37, 45, 316 left
VIREO/Glenn Bartley: 157
VIREO/Roger Brown: 165
VIREO/R. Curtis: 304
VIREO/R. Day: 105 left
VIREO/W. Greene: 171 left
VIREO/Greg Lasley: 318 right
VIREO/Arthur Morris: 155, 189 right, 303 left
VIREO/Brian Small: 150, 239 right, 314 right, 317 right
VIREO/Kevin Smith: 76 right, 185
VIREO/Bob Steele: 239 left, 312 left
VIREO/Michael Stubblefield: 228 right
VIREO/Frederick Kent Truslow: 192 right
VIREO/Doug Wechsler: 200
VIREO/J. Wedge: 243, 259 right
VIREO/Brian Wheeler: 103, 104 left, 104 right, 105 right, 109 left, 109 right
Lisa White: 41
Julie Zickefoose: 32–33, 184 left

CONTENTS

WHAT IS BIRDING?

Birding, or bird watching (the two terms mean essentially the same thing), is just what it sounds like. The activity of identifying a bird involves seeing a bird, looking at it closely through binoculars, and then trying to match what you see to a picture in a field guide (like the one you are currently reading).

Birds are divided into all sorts of categories as determined by bird scientists, who are called ornithologists. The two categories that are most important to the bird watcher are bird families and bird species. A bird species is defined as a distinct group of identical or very similar individuals that can successfully produce offspring. A bird family is a group of related bird species. The Downy Woodpecker is a bird species. The Downy Woodpecker is a member of the woodpecker family, which includes 23 different related species in North America.

Some birders enjoy correctly identifying the birds they see. "*That* bird is an Eastern Wood-Pewee! I know that bird as sure as I know my annoying little sister!" Other birders just want to see the birds and enjoy them without worrying about putting a name on each bird.

Birds, because they have wings and tend to use them, can show up anywhere at any time. Every little scrap of habitat can contain birds. When your family stops to gas up the car on a trip, there will be birds. Walking home from school, you'll hear birds in the trees and see them flying overhead. Birds are everywhere. Bird watchers know this and take advantage of the chances to see them and enjoy them.

Mrs. Huck's fifth grade class helped create some of the content for this book.

GETTING STARTED IN BIRD WATCHING

Most birders start out looking at birds in the backyard—maybe those species visiting the bird feeders or birdbath. After becoming familiar with most of the birds in the backyard, a bird watcher might venture out to look for birds in a local park or along a nearby stream or lake.

Once you get interested in birds, don't be surprised if you find that you want to see more and different birds. This usually means that you'll need to travel a bit. And there's a whole world of birding adventure out there just waiting for you. More than 800 bird species are regularly found in the United States and Canada. Worldwide, there are more than 10,000 bird species. So if you start birding today, it's going to be a long time before you run out of birds to see and enjoy.

Birding is a hobby that is inexpensive and easy to do. You can watch birds anywhere and at almost any time. For many of us who enjoy watching birds, we can trace our interest to a single encounter that sparked our imagination. This is our "spark" bird. My spark bird was a Snowy Owl that drifted into the giant oak tree in front of our house in Pella, Iowa, on a cold November day. I was six years old and was helping my parents rake the leaves off the front yard when a large white bird caught my eye. I was stunned at the bird's size, its clean white plumage, and its mysterious appearance. I knew it was an owl, but what kind of owl? And weren't owls creatures of the night? I ran inside the house for our very basic field guide. There it was: Snowy Owl! I devoured the short description: *A huge, mostly white bird of the Arctic tundra. Ventures southward in some years when food shortages and severe weather force it to move.* I spent the next few days carrying that guide around our large backyard, looking at birds and then trying to find them in the guide.

Once you see your spark bird, or even if you have yet to see it, the best way to get good at birding is to watch birds whenever you can. Even better, join up with a friend or a local bird club. You will see more birds and learn more about them if you go birding with others. It's simple: more eyes, ears, and brains mean you see more birds and have more fun.

BUT IS BIRDING COOL?

Birding is one of North America's fastest-growing and most popular hobbies—there are as many as 44 million bird watchers in the United States. Back in 1978, when my family began publishing *Bird Watcher's Digest* from our living room, bird watching was still considered a little bit odd. Today things are different. Birding is no longer considered a hobby for little old ladies and absent-minded professors in funny hats. There are former presidents, rock stars, movie stars, and millions of regular folks like you and me who enjoy watching birds. Everyone knows at least one bird watcher among his or her friends and family. Once your friends realize that you know something about birds, they'll ask you all kinds of questions and will share bird sightings with you. Don't let anyone else's perception of birding bother you. If you like to watch birds, then, as one famous company advises, *Just do it*.

WHY WATCH BIRDS?

There's no question that birds themselves are cool. Birds have inspired human beings for thousands of years. Why? There are several reasons. Birds can fly whenever they want, wherever they want to go. We humans figured out how to fly only about 100 years ago. Birds are amazing and beautiful creatures—they have brilliant plumage and may change their colors seasonally. Many birds are master musicians, singing beautiful, complex songs. They possess impressive physical abilities—hovering, flying at high speeds, and surviving extreme weather and the rigors of long migration flights. In short, we admire them because they inspire us. This makes us want to know them better and bring them closer to us. We accomplish this by attracting them to our backyards and gardens, by going to find birds where they hang out, and by using optics to see them more clearly.

BASIC GEAR

If you're just starting out as a birder, you may need to acquire the two basic tools of bird watching—binoculars and a field guide. The binoculars (often referred to as "binocs" by bird watchers) help you see the bird better. The field guide helps you identify what it is you're seeing.

Binoculars

Binoculars are like two miniature connected telescopes that enlarge distant birds so that we can see them well enough to identify them.

You may be able to borrow optics from a friend or family member, but if your interest in birding takes off, you'll certainly want to have your own binoculars to use anytime you wish. (Try dropping hints to your parents: "Gee, I sure hope the Birthday Elves will bring me some good binoculars for birding!") Fortunately, a decent pair of get-you-started binoculars costs less than $100. And some really nice binoculars (even used ones) cost just a bit more.

The magnification powers that are commonly used for bird watching are 6x, 7x, 8x, and 10x. Power is always the first number listed in a binoculars description, as in 8x40. The second number refers to the size of the objective lens (the big end) of the binoculars. The bigger the second number, the brighter the view presented to your eye. In general, for bird-watching binoculars the first number should be between 6 and 10, and the second number should be between 30 and 45.

Try to find binoculars that are easy to use. Make sure they are comfortable to hold (not too large or heavy), fit your eye spacing, and focus easily, giving you a clear image. Every set of eyes is different, so don't settle for binoculars that don't feel right. Over time, you will become superskilled at using your binocs, and that's when they will feel like an extension of your own eyes. You won't even notice that you're using them.

Field Guide

When choosing a field guide, it's helpful to know what type of birding you'll be doing and where you plan to do it. If nearly all of your bird watching will be done at home, you might want to get a basic field guide to the backyard birds of your region, or at least a field guide that limits its scope to your half of the continent. Many field guides are offered in eastern (east of the Rocky Mountains) and western (west of the Great Plains) versions. These geographically limited formats include only those birds that are commonly found in that part of the continent, rather than continent-wide guides, which include more than 800 North American bird species. Choose a field guide that is appropriate for you, and you'll save a lot of searching time. And that's time that can be better spent looking at birds! This Field Guide is intended to be a good starter book as you begin watching birds. It contains 300 of the most common and often encountered birds of North America. Other field guides can be found at bookstores, in nature centers, and on the Internet. See the Resources section at the end of this book for suggestions.

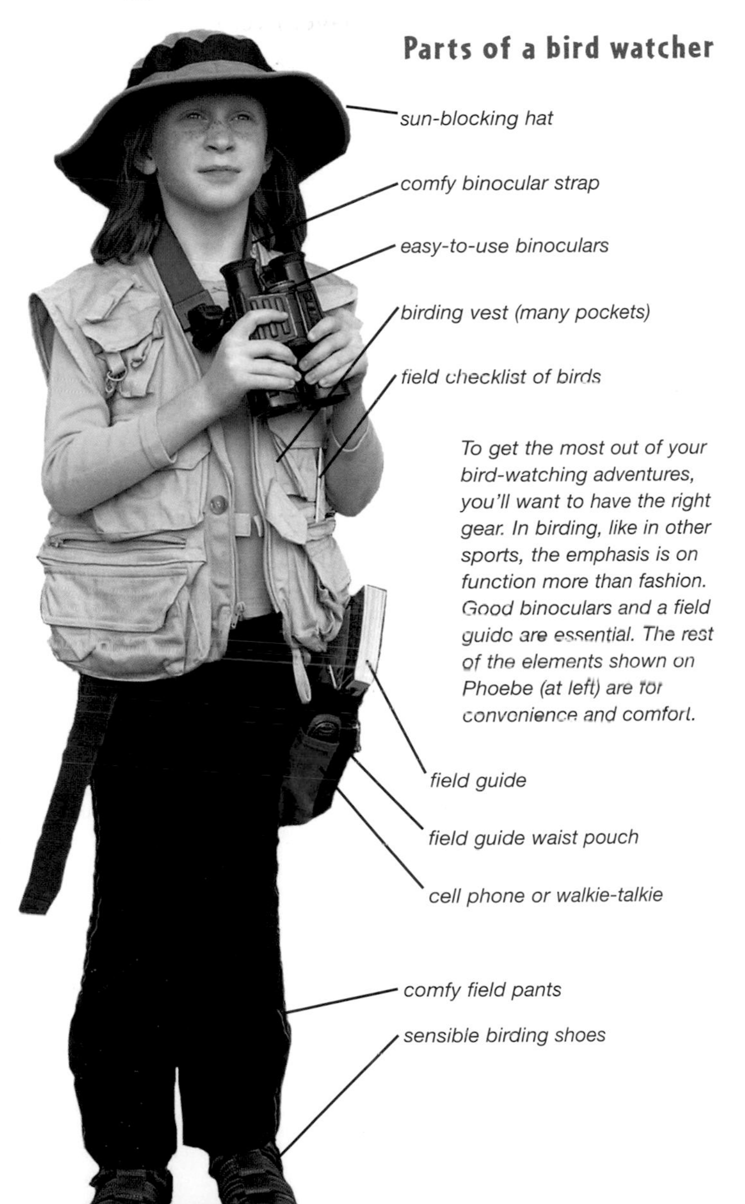

IDENTIFICATION BASICS

Identifying birds is at the very heart of bird watching. Each bird encountered is like a little puzzle or mystery to solve, because, while birds of a single species all share a certain set of physical traits, no two individual birds, like no two individual humans, are exactly alike. You solve the mystery of a bird's identity by gathering clues, just like a detective.

Most of the clues we birders use are called **field marks.** Field marks are most often physical things we can see—visual clues such as a head crest, white bars on the wing (called wing bars), a forked tail, patches of color, spots on a breast, rings around eyes (yes, called eye-rings), long legs,

Parts of a bird

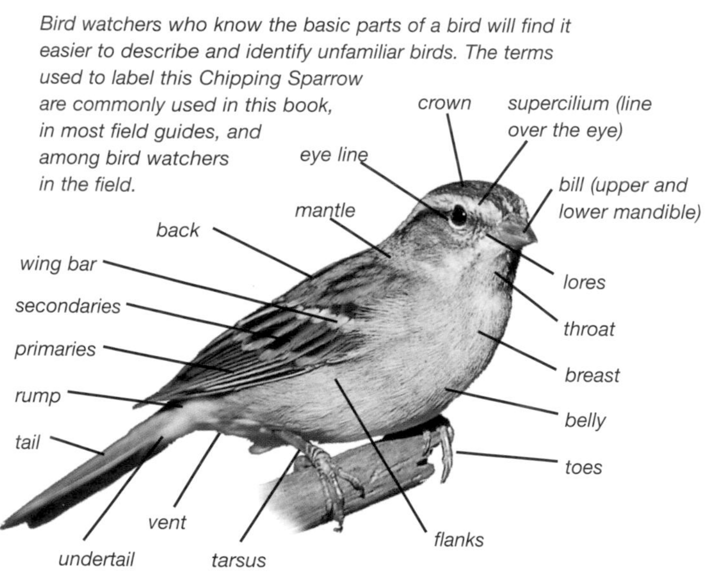

Bird watchers who know the basic parts of a bird will find it easier to describe and identify unfamiliar birds. The terms used to label this Chipping Sparrow are commonly used in this book, in most field guides, and among bird watchers in the field.

and a curved bill, for example. But field marks also include behavior, such as hovering in flight, probing with a bill, pecking on a tree (woodpecker!), and flitting about actively. And field marks include sounds, too—songs, calls given in flight, chip notes, and even the whistle of wings. When added up, these clues should lead you to a correct answer: the bird's identity.

New birders are wise to start with the obvious visual field marks of a bird. You'll want to collect these field-mark clues in a logical way: Start at the top of the bird (by the head and bill) and work your way down and back. Most North American birds can be identified by field marks above the bird's shoulders: on the bill, head, and neck, and near the bend of the wing.

Tip: Resist the urge to take a quick glance at a strange bird, note one field mark, then drop your binocs and grab the field guide. Birds can fly and are prone to sudden decisions. Field guides are books and cannot fly. Watch the bird as long as it lets you, *and then* reach for the field guide.

Some birds may appear to be completely plain and will require a longer look. But plainness itself can be a field mark! Don't give up. The clues are there, waiting for you to notice them.

Remember: Look carefully at your mystery bird, gather the clues, and then refer to your guide to solve the mystery.

The challenge of identifying birds is one of the best parts of bird watching. It can seem difficult and frustrating, but there are a few hints that will make it easier and more enjoyable.

The steps for identifying a bird are the same no matter where you are, no matter what bird you are watching. Here are some basic steps to follow.

Size: The first thing to notice is the size of the bird. Size will narrow down your choices a lot. To start with, think of birds as falling into the categories of small, medium, and large. Try associating those categories with objects you are familiar with: a cell phone, a soda can, a loaf of bread. It won't be long before your judgment of a bird's size is automatic. In most field guides, the size given is for the bird's length, measured from the tip of its bill to the end of its tail.

Here is an important point about judging size: A bird hunched over on the ground picking up seeds appears shorter and fatter than the same bird perched on a tree limb. Birds startled by a sudden noise or the appearance of a predator will stretch their necks, making them look considerably longer than when they are relaxed. Also, birds look thin when they are holding their feathers close to their bodies, and fat and round when they have fluffed them out, as they do in very cold weather. The key to judging size is to watch the bird for several minutes and to look at both its length and its bulk.

Look at the Bird: Start with a general impression: What is *the* most noticeable thing about this bird? The answer to this question is a basic description of the bird's shape, size, and appearance. For example: This is a big, tall, thin bird with long legs.

Sometimes the general impression is enough; it is certainly a good starting point. Begin at the front (the head or bill) of the bird and work backward. The key to identification is often in the pattern of the head. Does the bird have stripes on the head? A line over the eye? Is there a noticeable color on the face or head?

Pay particular attention to the bill. Bill shape and size often indicate the family to which a bird belongs. A family is made up of bird species that are closely related and that share many characteristics. For example: All sparrows have short, thick bills. Warblers have short, thin bills. Thrashers and mockingbirds have long, thin bills, usually down-curved.

After you have looked at the head, check the back, wings, and underparts. Ask yourself: Is the back darker or lighter than the head or the belly? Does it have streaks? Are the underparts plain, or are there streaks or spots?

Wings often provide a key identification feature: the presence or absence of wing bars. Wing bars are contrasting, usually pale, lines across the wings. Many groups of birds—warblers and sparrows, for example—are divided, for ID (identification) purposes, into those that have wing bars and those that do not.

Last, look at the tail. Is it long or short, rounded or forked, darker or lighter than the back? Are there white panels or spots? Is it all one color?

Does the bird bob or wag its tail persistently? Is the tail held cocked up or angled down?

All this sounds like a lot to remember, but after a few tries these questions will become automatic. As in most things, there is no substitute for practice. Try these steps several times on familiar birds. Remember, the key is to look for the most obvious clues to a bird's identification.

Range: Expect the expected. A bird's range (where it is regularly found) can be a valuable clue to its identity. If you live in Florida and identify a bird at your feeder that, according to the field guide map, occurs only in Oregon, your identification may be incorrect. Reconsider the other clues you have and try again. Birds occasionally show up a long way from where they are supposed to be, but it isn't common.

Look at the Book: Now it is time to use your field guide to identify the bird you have been watching. At first, the number of choices can be confusing. For new bird watchers, one of the best ways to find a species in a field guide is simply to start at the beginning and work through to the end. Don't worry, it won't be long before you are instinctively turning to certain sections. Small brown birds with thick bills will have you checking the sparrows, and chunky gray-on-brown birds with long tails and bobbing heads will have you checking the pigeons and doves.

THREE TIPS FOR USING YOUR FIELD GUIDE

1. The first several times you go through your field guide, be sure to go all the way through, even if you think you have found the bird. Also check to see if there are similar species with which the bird might be confused. A common mistake many people make is to settle on the first bird in the book that looks something like the bird they have seen.

2. The bird may not be identical to the picture in the book. Birds, like people, are variable. If almost everything matches and there are no very similar species, then you have almost certainly got the right bird.

3. If the possibilities have been narrowed down to three or four birds but you're still not sure, check another book. Every field guide has information that others lack.

MISIDENTIFYING BIRDS

Everyone does it. Beginners do it frequently, but even experts make mistakes. Misidentifying birds is part of the learning process. Don't worry about it. The occasional mistake will make you a better birder in the long run.

FIELD SKILLS

Getting better at bird watching is simply a matter of practicing as much as possible. Once you get the hang of it, you may find, as I do, that you're never *not* watching birds (and this can cause a slight problem if it's done during school, while riding your bike, in the middle of playing soccer, or at other inappropriate times). The point is to look at birds every chance you get.

FINDING THE BIRD IN YOUR BINOCS

Practicing with your binoculars is very important. Many new birders get frustrated at trying to find birds through their binoculars. Here's how to avoid this common problem: Lock your eyes on the bird, as if you are engaging in a staring contest. Don't move your eyes or your head. Then slowly bring the binoculars up until they are

Binoculars magnify our view of birds so we can see them better.

aligned with your eyes. Did it work? If not, try practicing this same technique on a stationary object, such as a building or a tree. Trust me, this motion becomes automatic very quickly.

PROPER SPACING AND FOCUS

Getting the most out of your birding binoculars requires a few minor adjustments the first time you use them. Start by adjusting the distance between the two barrels of the binoculars so that it is the right width for your eyes. Too far apart or too close together, and you will see black edges in your field of view. If you have the spacing right, your view should be a perfect circle.

Next, find something to focus on, such as a dark tree branch against the blue sky, a street sign, or an overhead wire. Focus, with both eyes open, by turning the central focus wheel.

Now it's time for the fine focusing. First, close your right eye, and then, using only your left eye, adjust the focus wheel until your chosen object is in focus. Next, you'll use the diopter—a secondary focus wheel, usually located on the right eyepiece. The diopter compensates for the strength difference between your two eyes, giving you a finely focused image. Now, close your left eye, open your right eye, and adjust the diopter to bring your view into sharp focus.

Now open both eyes and see if your focus is crystal clear. If the image is not clearly focused, repeat these steps, making small adjustments using one focus wheel or the other. Once you have your binoculars focused properly, you'll use only the central focus wheel. The diopter setting will stay put.

How can you tell if your focus is correct? First of all, with the diopter adjusted, the view through your binoculars should appear almost three-dimensional. It should really pop with clarity. Your eyes should not have to work very hard when using your binoculars. If your eyes feel tired after a bit of birding with your binoculars, check the focus again. If the problem does not go away, you may need to have a professional check your binoculars for alignment.

Tip: If you see black every time you raise the binoculars to your eyes, your eyes may be too close to the lens. Roll or twist out the eyecups until you stop seeing black. (If you wear glasses, set the eyecups flat against the binoculars.)

OTHER OPTICS TIPS

Clean binoculars make for happy birding. Protect your optics at all times, and clean them frequently. Use a soft brush (to get rid of dust and dirt), and then clean the surfaces with a soft cloth and lens fluid (available at most pharmacies or from your eye doctor).

Always wear your binoculars with a neck strap or a binocular harness. This is the safest and most convenient way to carry your binocs when birding. Don't carry them by the strap—that just puts them in danger of banging on things or being dropped. And don't carry them in your pocket, because you won't be able to get them up to your eyes fast enough.

A spotting scope is a fantastic tool for watching distant birds, such as ducks and shorebirds, and for scanning large expanses of habitat. You can certainly watch birds without a scope, and the cost and effort involved in using one can be considerable. Before you buy a spotting scope, try out several models owned by birding friends and see which one you prefer.

GET FIELD-GUIDE FAMILIAR

Take some time when you're not bird watching to look through your field guide and get familiar with how it is organized. Most guides are organized taxonomically, with closely related bird species grouped together. But this won't seem logical until you are familiar with the taxonomic order of birds. Getting to know where the hawks, ducks, warblers, and sparrows are in your field guide will be very helpful to you when you need to find a bird quickly. Besides, it's fun to flip through a field guide and dream about seeing some of the amazing birds it contains.

Here's a tip that veteran birders often forget: Use the guide's index! Some guides have color-coded page edges to help you find each bird family. Others just have an index at the back, telling you the page number for each entry. Flipping through the pages to find your bird may seem smart, but I guarantee you that using the index is faster.

JOIN A CLUB AND GO BIRDING A LOT

Getting out in the field with more experienced bird watchers is the fastest way to improve your skills. Don't be afraid to ask questions ("How did you know that was an Indigo Bunting? How could you tell that tiny speck in the sky was a Red-tailed Hawk?"). Don't worry if you feel overwhelmed by the amount of new information. All new bird watchers experience this. When it happens, relax and take some time simply to watch the birds and your fellow birders.

I joined a bird club when I was 10 years old, and I'm still a member today. Most of my best field trips and some of the most memorable birds I've ever seen came to me because I was part of an active bird club. Find a club near you (most clubs have a website) and get out on a club field trip. (Try using Google to search for the name of the nearest bird club.) Introduce yourself as a new bird watcher, and, because birders are the friendliest folks on the planet, you'll soon have a batch of new birding pals eager to help you see more birds and have more fun.

NOTES AND MEMORY DEVICES

All bird watchers can benefit from taking notes about birds. These may be simple ID tips added to your field guide or field sketches and longer notes in a separate notebook or field journal. The point is to help you remember what you've seen and learned.

Another way to remember is to use memory

Joining a bird club for field trips is both fun and helpful.

devices. My birding mentor, Pat Murphy, had a memory device for almost all the birds she taught me. I learned to remember the difference between Downy and Hairy Woodpeckers as this: *Downy is dinky, Hairy is huge.* It also works for bird songs. *Pleased to meetcha, Miss Beecher* is how I remember the song of the Chestnut-sided Warbler, and *who cooks for you-all!* is the call of the Barred Owl. You may prefer to come up with your own memory devices. But whatever you do, write them down so you won't forget.

DON'T FORGET TO LISTEN

When you're out watching birds, don't forget to use your ears, too. You'll frequently hear birds before you see them. (And sometimes you'll hear them and never see them.) Even if you don't immediately recognize the song or call, you can track down the bird by its sound and then let your eyes take over. As your birding enjoyment grows, you'll probably want to take advantage of the many sound recordings of birds that are available.

Your school or public library will have some bird sounds CDs, and there are dozens of others available at nature centers, at wild-bird stores, and on the Internet. I find my ears are better prepared for the sounds of spring migration when I've spent a bit of time in early spring listening to (and relearning) the songs I have on CD and on my iPod.

DRESS RIGHT

Nothing can spoil a fine day of birding like being underdressed for cold or wet weather. Veteran birders know to dress in layers so they can adjust their outerwear to fit the outside temperatures. But you can't put on more clothes if you don't have them with you! Wear comfortable, supportive shoes and all the appropriate outerwear when you head out to see birds, and you'll avoid the pitfall of being underdressed. It's hard to hold binoculars steady when you're shivering!

TAKE IT EASY

Finally, when you're bird watching, have a relaxed attitude and expect to have a good time. Don't get upset if you can't identify every single bird you see. Nobody can! I know lots of supercompetitive birders who have a hard time having fun when the birding is slow or when they miss out on a good bird. Sometimes, the harder you try, the harder it is to have fun. Remember, it's not always about seeing the most birds; it's about enjoying the birds you see.

BIRDING MANNERS

Out in the field with other bird watchers, it's a good idea to remember a few basic manners. These are mostly commonsense rules, but they're helpful to know.

Keep your voice down. Seems basic enough, but excitement at seeing a new bird can sometimes make us exclaim, *"Hey! Look at that Bald Eagle!"* And this makes all the other birds fly away.

Treat others as you'd like to be treated. If a fellow birder makes a bad call (and we all do this from time to time), don't tease him about it. This is the Golden Rule of Birding. I have personally seen some of the world's greatest bird watchers make bad calls—it happens. And when it happens to you, your fellow birders will treat you as you have treated them.

Stay with the group. If you are birding with a group of bird watchers and you wander ahead of them, you may see some birds they don't, but you may also scare off those same birds, which will upset your pals who didn't get to see them.

Share the scope. Because not every bird watcher has his or her own spotting scope, there is normally some sharing of the scope when a cooperative bird is found. When it's your turn at the scope, get a good look, but don't hog it. Once everyone else has gotten a look, you can take another turn.

Help beginners. If you stick with birding long enough, you'll soon be in a position to help new bird watchers. Most bird watchers were helped out by a birding mentor or by someone more experienced when we started watching birds. When it's your turn to be the mentor, repay the favor.

Pish in moderation. *Pish* is the sound we birders make to lure curious birds into view. To pish, softly say the word *spish* or *pish* through your clenched teeth. Many songbirds hear this sound, and think it's a chickadee scolding a predator, so they pop up for a look-see. Some bird watchers prefer to make a soft squeaking noise instead. This can also work. Still others prefer to use a recording of bird songs or sounds to lure birds into view. In recent years, the iPod, smartphones, and other digital sound devices have become popular with birders because they can play hundreds of bird songs at the touch of a button. All of these methods of bird luring can be effective, but we need to use them in moderation. Overplaying a song can wear out a territorial warbler or other songbird, who will try to locate the invisible rival singing in his territory. We owe it to the birds we love so much to respect their privacy. So please pish or play songs in moderation. Your fellow birders can help you know when to say when.

BIRDING BY HABITAT

All birds have specific habitat preferences. Some are specialists: A large woodpecker, such as a Pileated, needs old woods with big trees. Others are generalists: Song Sparrows may be found in city parks, suburban gardens, and brushy field edges in farmland.

When you're in the field bird watching, a great way to prepare yourself for the species you may encounter is to think like a bird. Look at the landscape and all the various habitats it contains and think about where you would go if you were a certain bird. Combine this with your previous experiences encountering bird species in particular habitats, and you are officially "birding by habitat."

Birds' habitat preferences can change seasonally. For example, Eastern Bluebirds prefer open grassy fields and meadows in spring and summer, but in late fall and winter, they move to wooded habitat for shelter and to find food. Many of the migratory birds, including many of our most colorful songbirds, spend spring and summer with us, then leave our midst altogether to spend the winter in the warm, insect-rich tropics.

The illustration on pages 30–31 shows some of the birds to expect on a summer day birding along a country road in the eastern half of North America. Try birding by habitat the next time you are out bird watching. It's easy to do, and you can't help but become a better birder by doing it.

1
2
3

15
11
12
14
10
13
4
5
6
7
9
8

KEY TO ILLUSTRATION:

1. SYCAMORE TREE

- Baltimore and Orchard Orioles
- Yellow-throated Warbler, Northern Parula
- Wood Duck
- Eastern Screech-Owl (in natural cavities)

2. PINES AND CONIFERS

- Owl roosts, nests
- Heron nests
- Raptor nests
- Pine Warbler, Black-throated Green Warbler
- Chickadees and nuthatches

3. BEAVER SWAMP

- Swallow nests
- Ducks
- Spotted or Solitary Sandpiper
- Woodpeckers (on dead trees)
- Herons
- Flycatchers

4. BEAVER POND

- Ducks
- Swallows
- Herons
- Grebes
- Shorebirds

5. CATTAIL MARSH

- Rails
- Blackbirds
- Marsh Wren
- Herons, bitterns

6. TUSSOCK (SEDGE) MARSH

- Swamp Sparrow
- Sedge Wren

7. CUT BANK OF STREAM

- Nesting Belted Kingfisher
- Nesting Bank Swallow
- Rough-winged Swallow
- Spotted Sandpiper

8. GRAVEL ROAD/ROADSIDE

- Eastern Bluebird
- American Goldfinch
- Killdeer
- Ruby-throated Hummingbird
- Indigo Bunting
- Common Yellowthroat

9. BRUSHY OLD FIELD

- Common Yellowthroat
- Prairie Warbler
- Blue-winged Warbler
- Song Sparrow
- Brown Thrasher
- Yellow-breasted Chat
- Northern Cardinal

10. FARMYARD

- Barn Swallow nests
- Cliff Swallow nests
- Barn Owl nests
- American Kestrel nests
- House Sparrow
- European Starling
- Brown-headed Cowbird
- Rock Pigeon

11. HARDWOOD FOREST

- Scarlet and Summer Tanagers
- Warblers
- Vireos
- Thrushes
- Flycatchers
- Cuckoos
- Whip-poor-will, Chuck-will's-widow
- Woodpeckers
- Chickadees
- Nuthatches
- Titmice

12. HAY MEADOW

- Eastern Meadowlark
- Dickcissel
- Blackbirds
- Field Sparrow
- Grasshopper Sparrow
- Eastern Bluebird
- Eastern Kingbird
- Bobolink

13. POWER LINES

- Hawks (perching)
- Eastern Bluebird
- Indigo Bunting
- Eastern Meadowlark
- Mourning Dove
- Eastern Kingbird
- Blackbirds

14. PASTURE

- Killdeer
- Eastern Meadowlark
- Brown-headed Cowbird
- European Starling
- Eastern Bluebird
- Prairie Warbler
- Field Sparrow

15. SKY

- Soaring raptors (hawks and vultures)
- Swallows

BE GREEN: TEN THINGS YOU CAN DO FOR BIRDS

1. Create bird-friendly habitat. There are countless ways to create habitat for birds in your backyard. Perhaps the easiest is to let things go wild in one part of your property. Chances are the plants that grow in your wild area will be natural sources of food for the birds. A more focused approach involves providing birds with the four things they need: food, water, shelter, and a place to nest.

2. No chemicals! It's widely known that seemingly safe lawn and garden insecticides and herbicides can be harmful to birds. Many of these chemicals target the pests that are food sources for birds, so any birds eating treated insects or seeds are also ingesting toxic chemicals. Avoid or at least minimize the use of toxic lawn and garden chemicals.

3. Recycle your trash. Each plastic, glass, aluminum, or tin item you recycle is one less piece of trash cluttering up the planet and one less ugly and hazardous item that we (and the birds) have to deal with in the environment. Recycling also saves money, eases pressure on habitat, and reduces pollution created by the production of first-generation materials such as glass, tin, plastic, and aluminum.

4. Keep your feeders and nest boxes clean. A once-a-month scrub cleaning of bird feeders will go a long way toward reducing disease transmission. Use a solution of one part bleach to nine parts hot water. Keeping your nest boxes clean is equally important. Clean out old nesting material several weeks after the nesting season is over. If the inside is really fouled with droppings, clean it out with the same bleach solution described above. Replace the old nesting material with a fresh handful of dried grasses to give the birds some insulation if they use the box for fall and winter roosting.

But how do you know when the nesting season is over? Read on . . .

Monitoring your nest boxes regularly is good for the birds (and for you, too)!

5. Monitor your nest boxes. Cavity-nesting birds face almost constant competition from non-native species that want to use these same cavities (hollow trees, old woodpecker holes, and nest boxes) for nesting. By checking your nest boxes regularly, you can discourage these introduced species and keep your nest boxes available for native species that need a place to nest or roost. Chickadees, titmice, nuthatches, woodpeckers, Tree and Violet-green Swallows, Purple Martins, wrens, and bluebirds are among the species that commonly use backyard nest boxes.

6. Participate in bird counts. There are dozens of local, national, and even international bird counts in which bird watchers can play a part. The National Audubon Society's annual Christmas Bird Count is one of the longest-running counts. The Cornell Laboratory of Ornithology conducts Project FeederWatch and the Great Backyard Bird Count as well as several other specific annual counts. See Resources, on page 344, for contact information.

7. Reduce window kills. Mylar strips, crop netting, branches, screens, and hawk silhouettes have been suggested as foils to keep birds from flying headlong into your windows. Placing these items outside, in front of the problem panes, breaks up the windows' reflections of the surrounding habitat so that the windows do not fool birds into flying into them.

8. Keep cats indoors. Even the most slothful, couch-potato cats can catch birds if given a chance. It's been estimated that housecats kill many millions of birds each year—deaths that could be avoided if these pets were kept indoors. For more information, write to Cats Indoors! Campaign, American Bird Conservancy, 1250 24th Street, NW, Suite 400, Washington, DC 20037; or go to www.abcbirds.org/cats.

9. Support conservation initiatives. Every day there are a thousand battles we bird watchers can fight on behalf of birds. The key is picking your spots so that you can make the most effective impact. Not all conservation initiatives are created equal, so be sure you're fully informed about the issues. In most cases, if bird habitat is preserved or created, it's a good thing. After you've created healthy habitat for birds in your own backyard, you may wish to contact the American Bird Conservancy or The Nature Conservancy to see how else you can help.

10. Make a new bird watcher today. Why not take a friend along on your next bird-watching trip, to the next bird-club meeting, or on a tour of your bird-friendly backyard? The more bird watchers we have today, the more good we can do for the birds tomorrow.

HOW TO USE THIS GUIDE

This is not the only field guide you will ever need. In fact, I hope it's the first of many bird books you'll want to have. The *Young Birder's Guide* is aimed at giving new bird watchers an easy first step into the world of birds.

The 300 species contained in these pages were chosen because they are birds that either are commonly seen or that every young birder should get to see in his or her lifetime. They are organized in general taxonomic order and follow the species names and Latin names currently agreed upon by most ornithologists.

In order to make this book small enough to be easily carried and used, I've limited it to slightly more than 300 species. I guarantee that while out birding, you will encounter birds that are not in this book. This is the first clue that you need to get a more comprehensive field guide covering *all* of the birds of your region, or even *all* of the birds of North America.

Each bird's page has these main parts:

- A species profile
- The species' common name and Latin name, along with body length for size reference
- One or two photo images of the typical plumages
- One black-and-white drawing of the bird doing something interesting
- Range map showing seasonal distribution

And each species profile has these sections:

- Look For (field marks for identification)
- Listen For (the bird's song and other sounds)
- Find It (its habitat preferences and seasonal occurrence)
- Remember (an additional ID tip, often comparing it to similar species)
- Wow! (an interesting extra fact about this bird)

To get the most out of this book, take it with you whenever you go birding. Flip through it when you are at home or when you can't be outside watching birds. Though we're all told never to write in a book, please write notes in this one. Record your bird sightings at the bottom of each bird's page. Make it your own book. If we've done our work properly, you'll soon outgrow *The Young Birder's Guide* and will move on to larger, more complete field guides covering all the birds of North America.

THE BEST BIRDING TIPS

KEEPING YOUR BIRD LIST

Most bird watchers enjoy keeping a list of their sightings. This can take the form of a written list, notations inside your field guide next to each species' account, or a special journal meant for just such a purpose. There are even software programs available to help you keep your list on your computer. In birding, the most common list is the life list. A life list is a list of all the birds you've seen at least once in your life. Let's say you noticed a bright black and orange bird in your backyard willow tree one morning and then used your field guide to identify it as a male Baltimore Oriole. If it's a species you'd never seen before, now you can put it on your life list. List keeping can be done at any level of involvement, so keep the list or lists that you enjoy. Many birders keep lists for fun—all the birds seen in their yard in one day, for example. Others keep lists for their county, their state or province, or for a complete year.

At the bottom of each bird's page in **THE YOUNG BIRDER'S GUIDE**, you'll find a check box and space for a short note. You can use this to start your life list.

TEN TIPS FOR BEGINNING BIRD WATCHERS

1. Get a decent pair of binoculars, one that is easy for you to use and hold steady.

2. Find a field guide to the birds of your region. (Many guides cover only eastern or western North America.) Guides that cover all the birds of North America contain many species that are uncommon or entirely absent from your area. You can always upgrade to a continent-wide guide later.

3. Set up a basic feeding station in your yard or garden.

4. Start with your backyard birds. They are easiest to see, and you can become familiar with them fairly quickly.

5. Practice your identification skills. Starting with a common bird species, note the most obvious visual features of the bird (color, size, shape, patterns in the plumage). These features are known as field marks and will be helpful clues to the bird's identity.

6. Notice the bird's behavior. Many birds can be identified by their behavior—woodpeckers peck

on wood, kingfishers hunt for small fish, swallows are known for their graceful flight.

7. Listen to the bird's sounds. Bird song is a vital component to birding. Learning bird songs and sounds takes a bit of practice, but many birds make it pretty easy for us. For example, chickadees and Whip-poor-wills (among others) call out their names. The Resources section of this book contains a list of tools to help you learn bird songs.

8. Look at the bird, not at the book. When you see an unfamiliar bird, avoid the temptation to glance at the bird and then grab the guide. Instead, watch the bird carefully for as long as it is present. Your field guide will be with you long after the bird has gone, so take advantage of every moment to watch an unfamiliar bird while you can.

9. Take notes. No one can be expected to remember every field mark and description of a bird. But you can help your memory and accelerate your learning by taking notes on the birds you see. These notes can be written in a small pocket notebook or even in the margins of your field guide.

10. Venture beyond the backyard and find other bird watchers in your area. The bird watching

you'll experience beyond your backyard will be enriching, especially if it leads not only to new birds but also to new birding friends. Ask a local nature center or wildlife refuge about bird clubs in your region. Your state ornithological organization or natural resources division may also be helpful. Being with others who share your interest in bird watching can greatly enhance your enjoyment of this wonderful hobby.

This Carolina Chickadee calls out chick-a-dee-dee-dee.

GREATER WHITE-FRONTED GOOSE

Anser albifrons **Length: 28"**

Look for: Named for the white forehead patch on its face, the Greater White-fronted Goose is a large bird that looks slender in flight. Field marks that set it apart from our other geese include the boldly patterned black-and-white tail with a white tip, a speckled dark-and-light belly, bright orange legs, and a pale pink bill.

Listen for: White-fronts make a high-pitched series of rapid, tooting cackles, nothing like the honking of a Canada Goose. Flocks are very vocal.

Remember: You might confuse this species with the similar Greylag Goose, a domesticated fowl. A single bird in a farmyard or city park is probably not a wild Greater White-fronted Goose.

The splotchy black, gray, and white pattern of this bird's belly has earned it the nickname Specklebelly.

▲ Greater White-fronted Geese are often mistaken for the Greylag Goose (in back) and are often seen with the Canada Goose (right).

Find it: A large population uses the Central Flyway to migrate between the Arctic tundra breeding grounds and the coastal Gulf of Mexico wintering grounds. White-fronts forage in large open fields and shallow marshes.

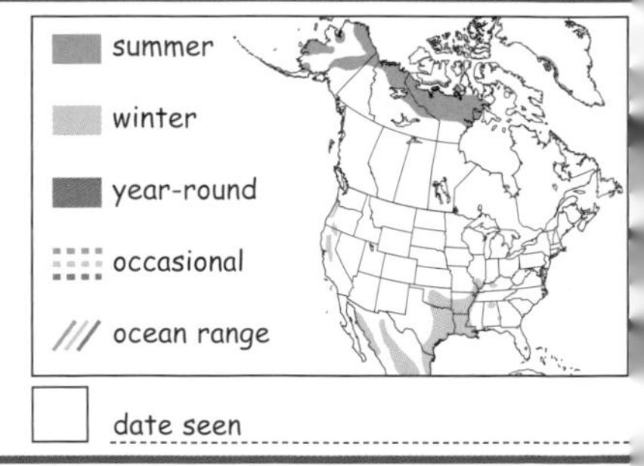

☐ date seen ____________

SNOW GOOSE

Chen caerulescens **Length: 28"–31"**

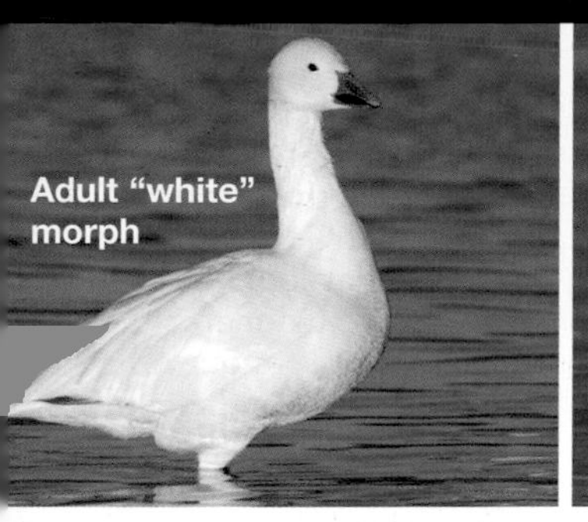

Look for: Named for its snow-white plumage, the Snow Goose comes in two colors (or morphs), white and dark, though white Snow Geese are far more common. The typical white morph has an all-white body with black wingtips, pink legs, and a large pink bill. The dark morph, or Blue Goose, has a dark gray body and white head.

▲ *For a long time, the Blue Goose (top) was thought to be a separate species from the Snow Goose.*

Listen for: A high-pitched *howk-howk* is the most common call, often heard from large flying flocks. Foraging birds utter piglike grunts.

Remember: In the Midwest and West, the similar but smaller Ross's Goose may be found with Snow Geese. The Ross's has a smaller pink bill and looks like a miniature Snow Goose.

WOW!
Snow Geese have enjoyed a population boom in recent years, and this is putting a strain on the very sensitive tundra habitat where they breed.

Find it: Snow Geese nest on the tundra of the far north. Migrating flocks fly very high in long, wavering lines (rather than the V formations that some other geese prefer), which has earned this species the nickname Waveys.

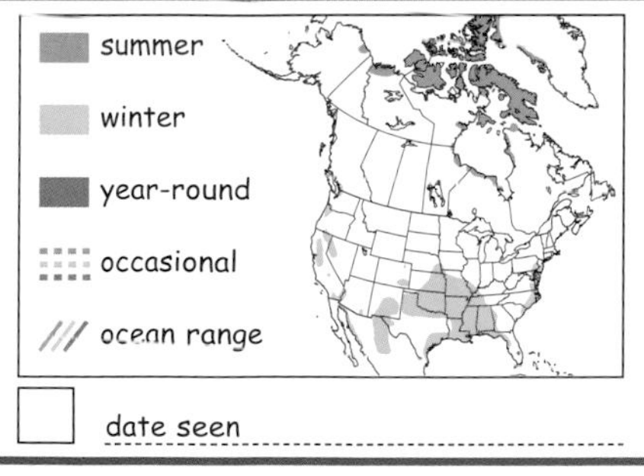

ROSS'S GOOSE

Chen rossii **Length: 23"**

Look for: The Ross's Goose is like the Mini-Me of the Snow Goose, more petite in every way: smaller overall with a smaller, rounder head; shorter neck; and a shorter bill than the Snow Goose. The bill of the Ross's Goose is all pinkish, lacking the black "smile" patch of the Snow Goose.

Listen for: Nasal, high-pitched yelps. Not as vocal as the Snow Goose, but call is similar, pitched higher.

Remember: When scanning through a large flock of Snow Geese, look for the smaller overall size and rounder shape of a Ross's Goose. The difference in head and bill size and shape is a good field mark to check on small white geese.

WOW!

Ross's Geese sometimes occur in the darker-bodied "blue" form, just as Snow Geese do. But this was not discovered until the 1970s!

▼ *A careful observer can pick out a Ross's Goose or two from most large flocks of Snow Geese.*

Find it: Often found in mixed flocks with Snow Geese in winter in fields, marshes, and bays. Nests in summer on the tundra in the far north. Ross's Geese are more commonly found in wintering flocks along the Mississippi Flyway and points west.

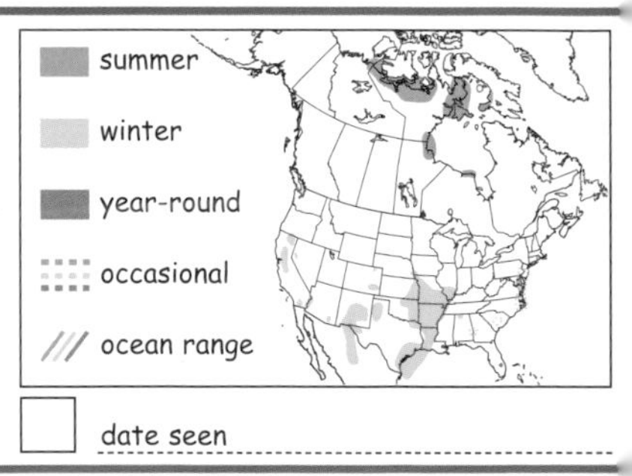

Look for: Almost everything about this small, short-necked goose is dark: head, bill, body, wings, legs. The overall dark head and body contrast with a white lower belly and upper tail, giving flying birds a dark-in-front, white-in-back appearance. If you can get close to a Brant, look for a partial white necklace.

Listen for: A low, rolling honk that sounds almost like a goat or sheep.

Remember: The Brant lacks the bright white cheek patch of the Canada Goose. Brant are also much smaller and shorter necked than Canada Geese.

WOW!
Migrating Brant may fly nonstop from their high-Arctic breeding grounds to the Atlantic Ocean, a distance of more than 1,800 miles.

◀ *Eelgrass is like spaghetti to Brant—they depend on it for much of their diet.*

Find it: Brant breed on the tundra. This small, dark goose is almost never seen away from coastal waters, where huge flocks of Brant can be seen flying in disorganized, dark, wavering lines. Foraging flocks use shallow bays and tidal mud flats.

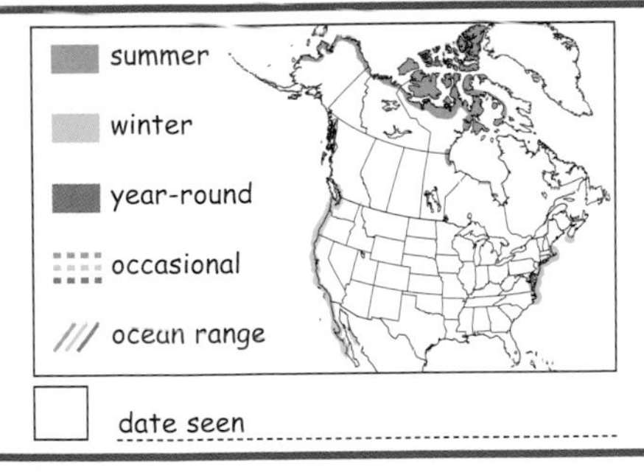

☐ date seen ____________

CANADA GOOSE

Branta canadensis **Length: 24"–45"**

Look for: This familiar large goose, with its black head and neck and white "chinstrap," was once a symbol of wild North America. The dark gray-brown body contrasts with a white rump, the back third ending in a black tail. A black bill and black legs complete the Canada's typical plumage.

Listen for: A deep *honk* or *ha-ronk*, rising in tone, which is considered by many to be the classic goose sound. Flocks in flight are noisy. Flocks and families on the ground make a variety of grunts, gabbles, and hissing sounds.

Remember: Flocks of Canada Geese flying overhead in a V formation may contain other species, including Snow Geese, White-fronted Geese, and other waterfowl.

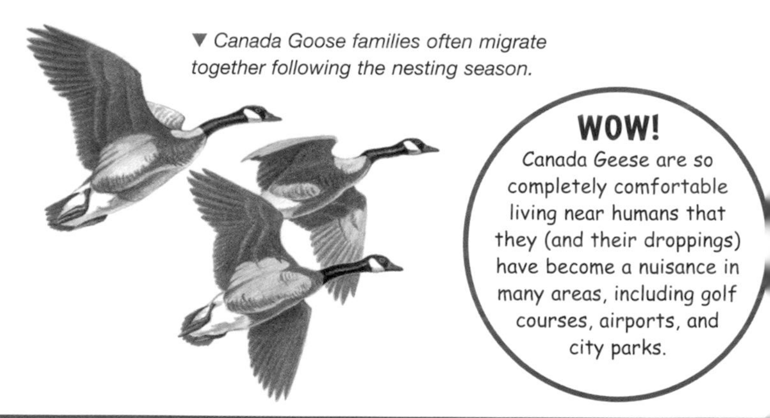

▼ *Canada Goose families often migrate together following the nesting season.*

WOW!

Canada Geese are so completely comfortable living near humans that they (and their droppings) have become a nuisance in many areas, including golf courses, airports, and city parks.

Find it: Canada Geese occur from remote tundra and marshy regions to fields and wetlands. Canadas now maintain a robust year-round population across the middle third of the U.S. Some birds are nonmigratory.

summer

winter

year-round

occasional

/// ocean range

☐ date seen

MUTE SWAN

Cygnus olor **Length: 60"**

Look for: Our only large white swan with an orange bill, the Mute Swan was introduced to North America from Europe in the mid-1800s. The orange bill has a black knob on top, also unique. The Mute Swan swims with its neck in a graceful S-shaped curve, bill pointing down, and sometimes fluffs out its wing feathers, especially when confronting an intruder.

Listen for: Not mute, but does not possess a lovely song either. Grunts and hisses during aggressive encounters. In flight, wings make a low, humming sound.

Remember: You are more likely to see Mute Swans floating on a quiet body of water rather than in flight.

Though we think of a pair of Mute Swans on a park pond as a symbol of love, these birds aggressively drive other waterfowl away from their breeding territory. They are well-dressed bullies who do not play well with others.

◀ *Mute Swans harass native waterfowl, taking entire ponds as their exclusive turf.*

Find it: Found on quiet ponds, lakes, and bays, especially near human habitation. Equally at home on bodies of salt or fresh water. Many birds are nonmigratory residents.

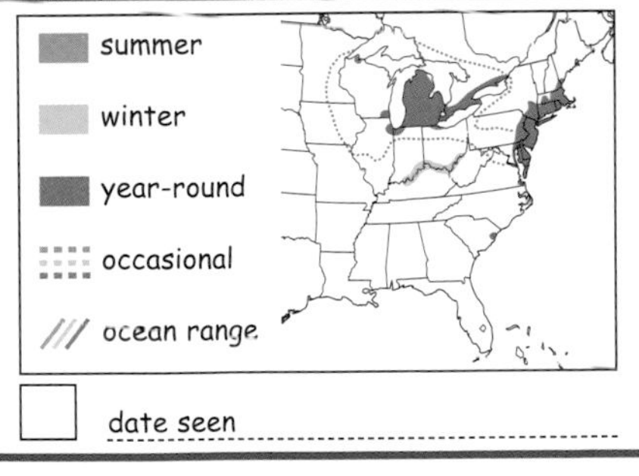

date seen

TUNDRA SWAN

Cygnus columbianus **Length: 53"**

Look for: The Tundra Swan is the smallest of our three white swans. The Tundra's black bill and shorter, straighter neck differentiate it from the orange-billed Mute Swan. The adult Tundra Swan has a small patch of yellow in front of the eye, but this can be difficult to see. Distinguish Tundra Swans from the much rarer Trumpeter Swan by the Trumpeter's much longer, tapered bill.

Listen for: Tundra Swans give high-pitched, nasal *whoo-ooo* calls that almost sound like the trumpeting of a tiny elephant.

Remember: Tundra Swans appear gooselike in flight, but the most similar goose, the Snow Goose, always shows clear black wingtips. Tundra Swans appear all white in flight.

WOW!

If you live anywhere along the Tundra Swans' spring or fall migratory routes, you may hear the flocks of 100 birds or more calling as they migrate overhead.

◄ *Tundra Swans often time their migration just ahead of passing weather fronts.*

Find it: Tundra Swans prefer large bodies of water such as lakes, reservoirs, and bays. Nesting in the northernmost reaches of North America, they follow a variety of migration routes to three primary wintering areas.

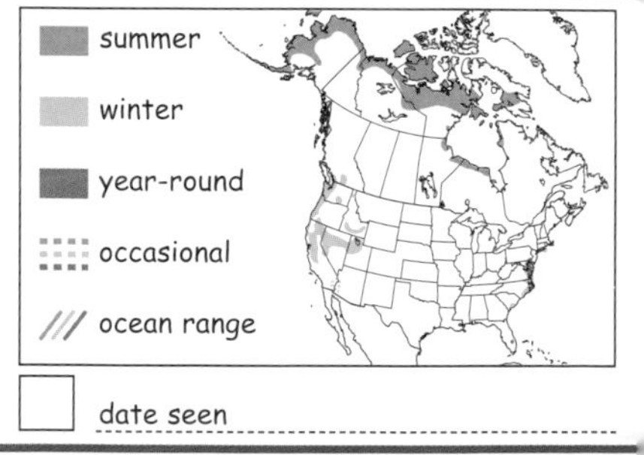

☐ date seen ______________________

TRUMPETER SWAN

Cygnus buccinator **Length: 60"**

Look for: Though rare throughout their range, Trumpeter Swans are hard to miss. It is our largest native waterfowl species (the Mute Swan is as large but is an introduced species in North America). Its all-black bill is long and straight. Adults are all white; young birds show some brownish body feathers, then appear grayish before acquiring the all-white appearance of adults in their second summer.

Listen for: Call is a hornlike tooting, which gives the bird its name. Longer notes sound very much like a trumpet.

Remember: Three things help to separate the Trumpeter Swan from the similar Tundra Swan: the Trumpeter is much larger overall; its bill is long, straight, and all black (Tundras often show some yellow in the bill); and the Trumpeter is much more rare and local than the Tundra.

Hunting and market shooting nearly wiped out the Trumpeter Swan in the late nineteenth and early twentieth centuries.

▲ *Trumpeter Swan pairs stay together for as long as both partners live. Both parents share in the care of the young*

Find it: Trumpeter Swans nest on wooded lakes and ponds in the western mountains and northwestern portions of North America, usually far from human disturbance. Some breeding populations have been re-established in the Midwest through conservation-recovery programs.

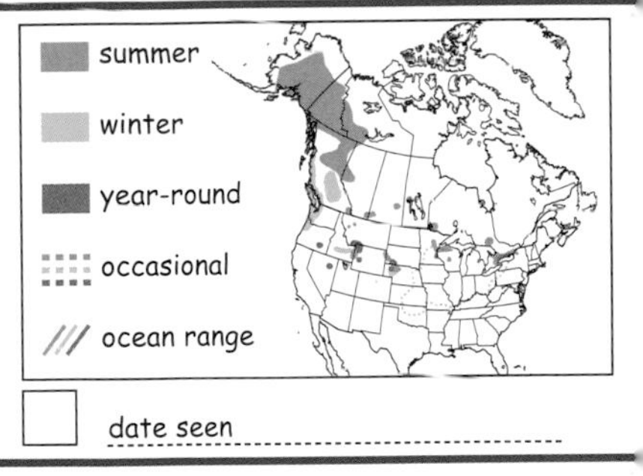

☐ date seen ____________________

WOOD DUCK

Aix sponsa **Length: 18"**

Male (left), female (right)

Look for: A medium-sized duck that may hold the title of Most Beautiful among waterfowl, the male Wood Duck, in breeding plumage, is a rainbow of colors. In late summer, males in eclipse (that is, dull and drab) plumage resemble females. Wood Ducks are fast fliers and often call in flight.

Listen for: Female Wood Ducks give a high-pitched, excited, two-part whistle: *uh-wheek!* They also give a softer series of rapid whistles: *oh-oh-oh-oh.*

Remember: In flight, Woodies appear long-tailed compared to other common ducks. Wood Ducks will use nest boxes, and human-supplied housing has contributed to this species' population comeback during the past 60 years.

WOW!
Just one day after they hatch, Wood Duck babies leap from the nest cavity.

▶ *Wood Duck ducklings leaping from a nest box. They are so lightweight that they float to a soft landing below.*

Find it: Considered one of our perching ducks for their habit of perching, roosting, and nesting in trees. A scan of the trees and snags in wooded swamps, quiet rivers, and in forested marshes will often reveal Wood Ducks dabbling or sitting quietly.

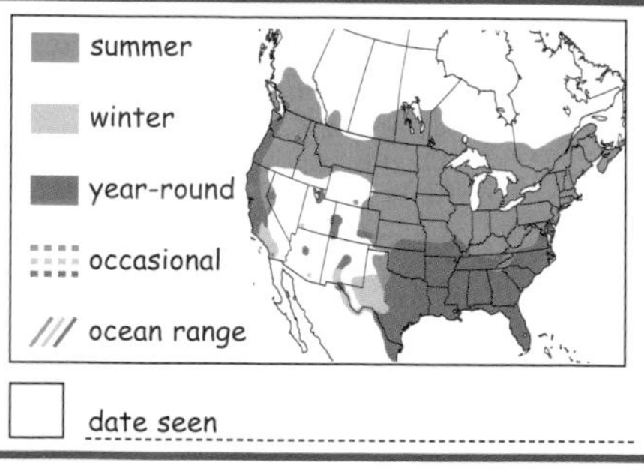

☐ date seen ______

AMERICAN WIGEON

Anas americana **Length: 5"**

Look for: The white crown stripe of the male American Wigeon has earned this species the nickname Baldpate. This field mark contrasts with the dark eye patch and can be seen from great distances. Both male and female appear round-headed and have a pinkish chest and body. The light blue-gray bill has a black tip.

Listen for: Females give a guttural growl. Males make a high whistle that sounds like a squeaky squeeze toy.

Remember: In flight, drake (male) wigeons flash large white patches on the upper surface of their wings. Both sexes have underwings with white centers and both show white bellies.

WOW!
American Wigeons behave like bandits on large lakes. They are known to steal food from other waterfowl.

◀ *An American Wigeon steals a bit of food from a Canada Goose.*

Find it: Look for American Wigeons on large marshes, lakes, and ponds, where they forage near the surface. They also graze for food on land more commonly than other ducks.

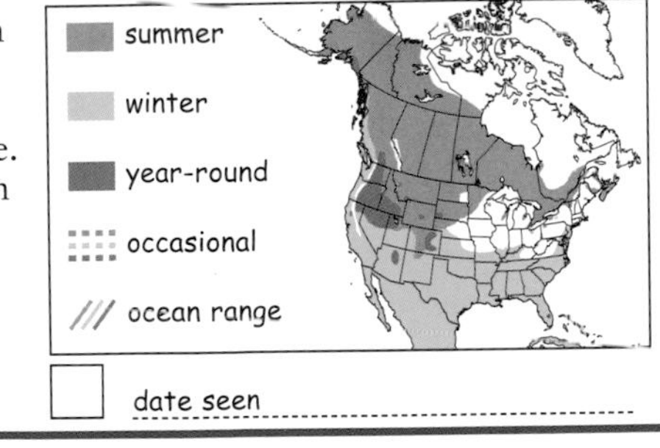

☐ date seen ____________

GADWALL

Anas strepera **Length: 20"**

Look for: At first glance, the Gadwall might look like nothing special. A closer look, especially at the male, or drake, reveals this duck's subtle beauty. The most reliable field mark is the male's black "underpants" (the back end, under the wings), visible from a great distance. The Gadwall's head appears lumpier than other ducks'. Females lack the black underpants but can be told from other plain brown ducks by the light orange spots in the bill.

Listen for: Courting males utter a low, nasal *beep*, often in a series. Females quack like Mallards.

Remember: In flight, both sexes of Gadwall show white on the underside of the wings and a small white square on the upperwing, close to the body.

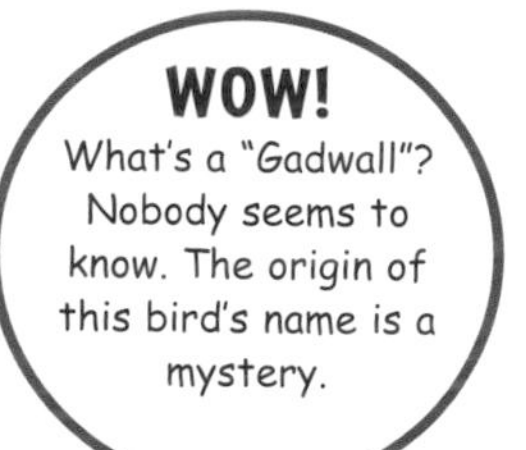

▲ *The Gadwall's white wing patch and the black underpants of the male can be seen from a great distance.*

Find it: Common on marshes and shallow lakes in open (not wooded) settings. Prefers fresh water. Most common from the Great Plains west, but many are found in the East during migration and across the lower third of the U.S. in winter.

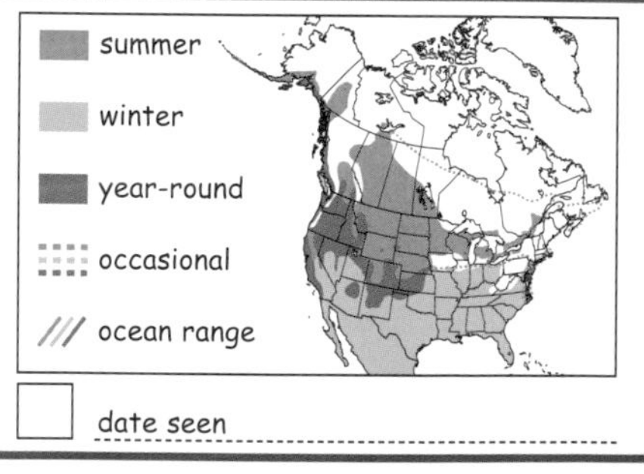

AMERICAN BLACK DUCK

Anas rubripes **Length: 23"**

Look for: The Black Duck is a medium-sized dabbling duck with a dark charcoal body. Males and females are very similar, though the male has a yellowish bill while the female's bill is dull gray. In flight, Black Ducks show mostly white underwings and mostly dark upperwings. This contrast gives a flashing appearance to their flight.

Listen for: A nasal, quacking call: *whap-whap, whap-whap*.

Remember: Black Ducks and Mallards are closely related, and these two species are often found together. Telling a female Mallard from a female Black Duck is difficult: Female Blacks are darker overall and have no orange on the bill.

▼ *The white wing linings showing on this flying American Black Duck are a key field mark for this species.*

WOW!

Mallards and American Black Ducks often breed with each other, creating hybrid offspring. Over time, this interbreeding has caused an increase in Mallards and a decrease in Black Ducks.

Find it: Almost any body of water can host Black Ducks, but they seem to prefer coastal salt marshes and, inland, wooded wetlands. They are found year-round across the Midwest and along the Atlantic Coast from Maine to the Carolinas.

summer
winter
year-round
occasional
/// ocean range

☐ date seen ____________

MALLARD

Anas platyrhynchos **Length: 23"**

Look for: The male Mallard's green head and yellow bill make it one of North America's most recognizable bird species and certainly our most familiar duck. Domestic and semidomestic versions of Mallards exist in many farmyards, parks, and zoos, but few feature the clean-looking plumage of the wild Mallard.

Listen for: Female Mallards give the typical duck call: *quaaack!-quackquackquack!* This sounds very much like Donald Duck of cartoon fame. Males give a high-pitched *queeep!*

Remember: Domestic Mallards can be found almost anywhere. Wild Mallards, with sharp-looking plumage, are wary birds that are quick to flush into the air.

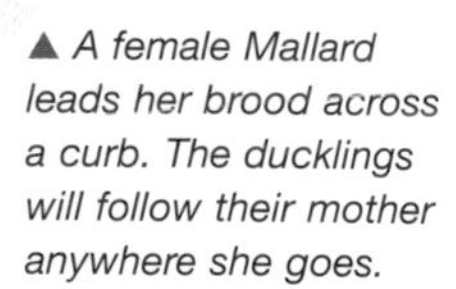

WOW!
Mallards are strong fliers and have been clocked at speeds of up to 60 mph!

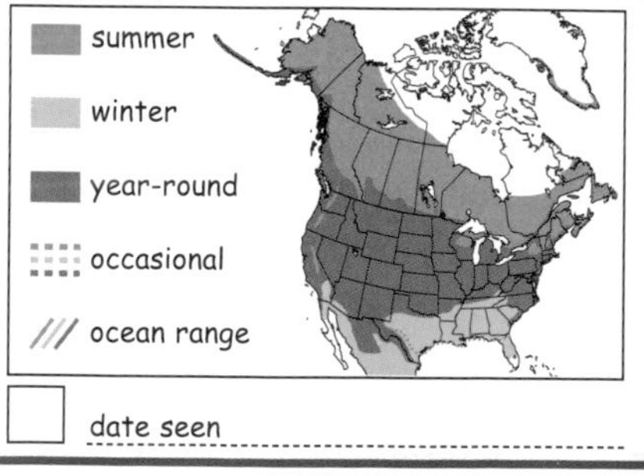

▲ *A female Mallard leads her brood across a curb. The ducklings will follow their mother anywhere she goes.*

Find it: Mallards can be found in almost any freshwater habitat. Some live on suburban ponds and nest in the shrubbery around buildings. Often seen in pairs, which form in fall or winter and last through the breeding season.

summer
winter
year-round
occasional
/// ocean range

☐ date seen

NORTHERN PINTAIL

Anas acuta **Length: 21"**

Look for: The long neck and long tail of the Pintail (especially of the male) give it an elegant shape that is distinct from other ducks. The male's chocolate brown head and white neck and breast are reliable as field marks in all seasons except late summer, when molting males resemble females. Female Pintails are toasted-marshmallow brown overall with dark gray bills.

Listen for: Females give a typical ducklike *quack*. Males give a high-pitched *twoot-twoot* and a very unbirdlike rising and falling buzzy whistle.

Remember: Pintails are skittish—they are often the first birds to flush into flight from a pond or wetland containing lots of ducks.

▼ *A male Northern Pintail stretches its foot and wing past his long tail feathers.*

WOW!

The male Northern Pintail's tail has two long central tail feathers that extend well past the other tail feathers. Other names for this species include Spring-tail, Spike-tail, and Sharp-tail Duck.

Find it: Very adaptable in their choice of habitat, Northern Pintails can be found in marshes and farm fields, and on prairie ponds and mud flats. They migrate throughout the East in spring and fall.

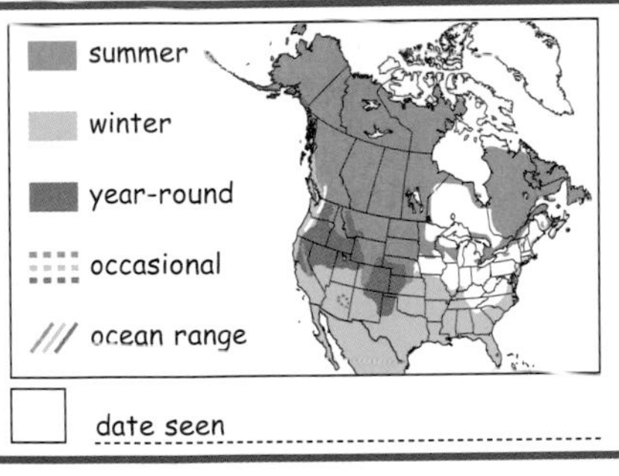

☐ date seen ______________________

BLUE-WINGED TEAL

Anas discors **Length: 15"**

Look for: Named for the light blue wing patch that both sexes show in flight, the Blue-winged Teal is a delicate-looking small duck. The male's crescent-shaped white face patch on a blue-gray head is distinctive among our common ducks. Females are plain overall and are best identified by their shape rather than by any obvious field marks.

Listen for: Blue-winged Teal males utter a high, peeping whistle. Females give a froglike *querk*.

Remember: The blue wing patches are not visible on swimming or resting birds, but if you scan a flock of ducks on a pond, you may be able to pick out the teal by their small size.

▼ *A wintering flock of Blue-winged Teal in a mangrove swamp.*

WOW!

Summer Teal is another name for this species because it seems to time its migration to avoid cold weather, migrating earlier in fall and later in spring than other ducks.

Find it: The Blue-winged Teal prefers freshwater ponds and marshes but can be found on almost any body of water, especially in migration. They migrate late in spring and early in fall. Most winter in Latin America.

summer
winter
year-round
occasional
/// ocean range

☐ date seen ____________

CINNAMON TEAL

Anas cyanoptera **Length: 17"**

Look for: Adult male is rich cinnamon red overall with a demonic-looking red eye. Female is similar to female Blue-winged Teal but plainer, especially in the face. This teal species is also notable for its larger bill, which is similar in shape and size to a Shoveler's schnoz. In flight, Cinnamon Teal shows large white underwing patches; male shows bright blue shoulder patches.

Listen for: Male gives whistly peeps; female quacks nasally, sounding very similar to female Blue-winged Teal.

Remember: A male Cinnamon Teal's bright coloration may be hard to miss, but females can go unnoticed. If you see a very plain female teal among a flock of puddle ducks, check to see if her bill is noticeably larger. If so, you can impress your friends by pointing out a female Cinnamon Teal!

WOW!

It is **not** true that the spice we call cinnamon comes from ground-up male Cinnamon Teal. That would not be cool at all.

◀ *Cinnamon Teal use their large bills to strain food from shallow water.*

Find it: Like other teal, the Cinnamon Teal prefers shallow water in ponds, marshes, sloughs, and flooded fields. Cinnamons are western birds and are rare east of the Great Plains.

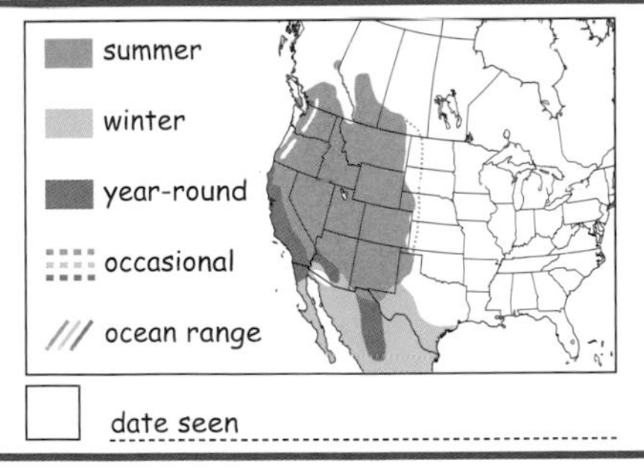

☐ date seen ____________________

GREEN-WINGED TEAL

Anas crecca **Length: 14"**

Look for: The Green-winged is our smallest teal and one of our smallest ducks. The male's handsome breeding plumage holds several key field marks: a dark rust-colored head with a swoop of green through the eye, a vertical white bar on the gray side, and a prominent horizontal patch of custard yellow below the tail, visible at great distances. Females show this custard patch too but are otherwise a rich brown overall.

Listen for: Males give a high *peep!* Females utter high, nasal quacks in a series.

Remember: Flocks of Green-winged Teal are fast fliers, and small flocks maneuver in unison with amazing skill. Distant birds can often be identified by their small size.

WOW!

Other names for the Green-winged Teal include Common Teal, Mud Teal (it often forages on mud flats), Red-headed Teal, and Winter Teal (it winters farther north than other teal species).

◀ *Twisting and turning in flight, Green-winged Teal are among the swiftest of ducks.*

Find it: Common and widespread, and preferring shallow, muddy water to deep lakes, Green-winged Teal often perch out of the water on land or fallen logs, so they are common in habitats with lots of snags and vegetation.

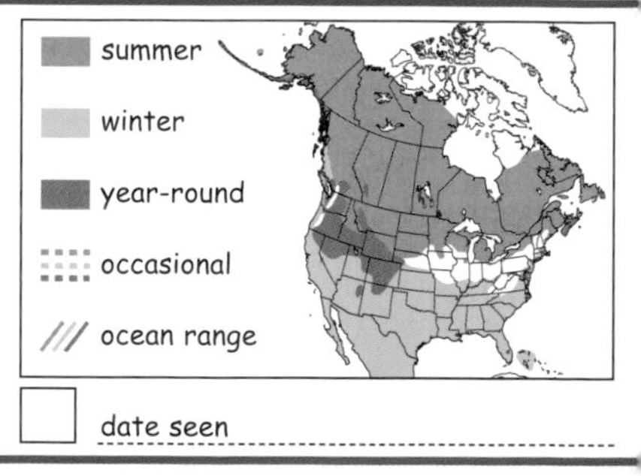

date seen

NORTHERN SHOVELER

Anas clypeata **Length: 19"**

Look for: From a distance the Northern Shoveler looks like a Mallard with a big nose. It's this bird's large, shovel-like bill that earned it its name. The male has a solid dark green head, large black bill, white chest, and rusty sides. The female is plain light brown overall with a large orange and gray bill. In flight, the male shoveler's huge light-blue wing shoulder patch is obvious.

Listen for: Males give a hollow-sounding *took-took*, uttered in pairs.

Remember: At a distance, the best field mark for the shoveler is the huge rusty flank (side) patch surrounded by white.

WOW!

Imagine having to strain all of your food out of muddy water. The shoveler's bill is a huge filter with specially adapted comblike teeth that strain out tiny organisms—delicious!

▼ *A female shoveler straining food items out of a billful of water.*

Find it: Shovelers prefer shallow water, such as ponds and marshes, where they swim with their huge bills submerged as they strain food from the muddy water. During migration, they can be found on almost any body of water.

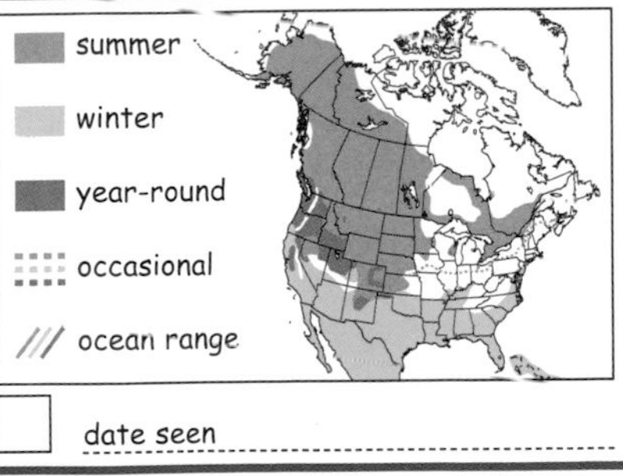

☐ date seen ____________________

SURF SCOTER

Melanita perspicillata **Length: 20"**

Male

Female

Look for: Of our three scoter species, it is the Surf Scoter that is most easily identified. The male is black bodied with large white patches on the forehead and back of the head. In breeding plumage, the large orange and white bill of the male stands out. Adult females are dark brown with splotchy white patches on the head and face.

Listen for: Usually silent. Females give some croaks. Male's wings whistle in flight. Few birders hear these sounds.

Remember: From a distance, Surf Scoters look bulky and dark, with big heads and large bills. The adult male's white head patches are distinctive.

WOW!
Duck hunters refer to the Surf Scoter as the Skunk-head Coot for the male's striking black-and-white head pattern.

◀ *Surf Scoters fly in wavering lines low over the water.*

Find it: Because they nest in the far North, Surf Scoters are most familiar to birders as wintering sea ducks along both coasts and on the Great Lakes. Large, wavering dark lines of scoters can be seen flying low over the water.

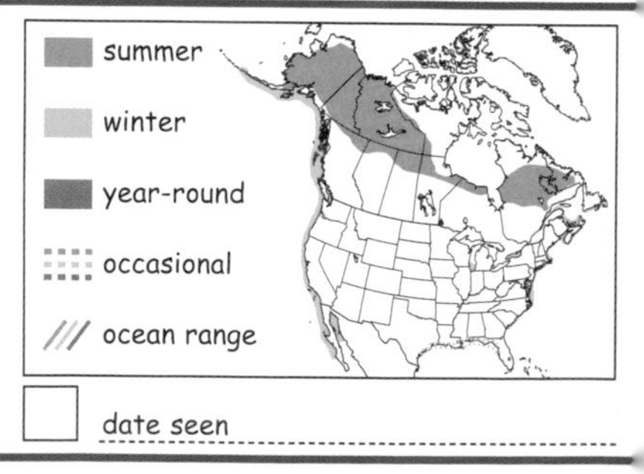

☐ date seen ____________

REDHEAD

Aythya americana **Length: 20"**

Female (left), male (right)

Look for: A medium-sized diving duck named for its deep rust-red head. The male's black breast contrasts with a pale gray back. His gray bill is tipped in black, and the eyes are yellow. Female is uniformly gray-brown overall with a pale eye-ring and a pale area surrounding the base of the bill.

Listen for: Quacks like, well, a duck. Also makes some catlike mewing sounds.

Remember: Very similar to the larger Canvasback, which has a bright white back. In mixed flocks of these two species, Redheads look dull by comparison.

WOW!

Redheads are most active at night, spending days resting on the water. Imagine diving for your food at night!

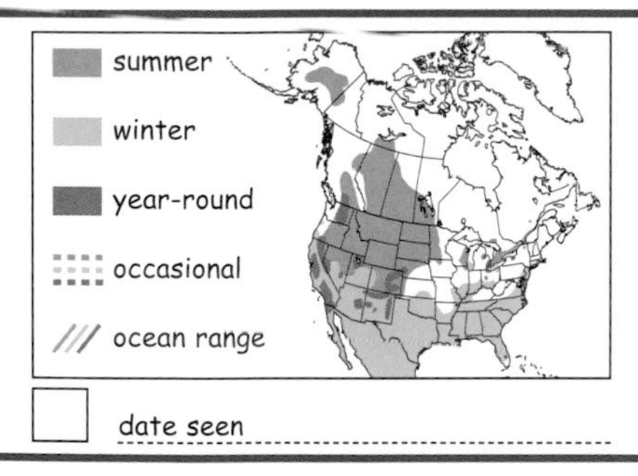

▶ *Redheads are excellent divers. They get much of their food by diving for submerged aquatic vegetation.*

Find it: More widespread in winter, when flocks can be found on open water, especially lakes and freshwater and saltwater bays. Breeds in freshwater marshes throughout the Great Plains to Alaska. Winters across the southern half of the U.S.

summer

winter

year-round

occasional

ocean range

☐ date seen

RING-NECKED DUCK

Aythya collaris **Length: 17"**

Female

Look for: This small peak-headed duck could be named the Ring-billed Duck for the white and gray rings on the male's black-tipped bill. Instead, it is named for the rarely seen subtle ring around the male's neck. A male in breeding plumage has a black head and back, and gray sides separated from the black chest by a bar of white. The female is brown-gray overall but has a white eye-ring and the same peaked head shape as the male.

WOW!
The Ring-necked Duck is named for a brown neck ring visible only on a bird in the hand. This is a holdover from the era of shotgun ornithology, when birds were shot, then examined, then named.

Listen for: Mostly silent. Females give a nasal, burry *errr-errr.* Males in courtship utter a high whistle.

Remember: Ring-necked Ducks appear very similar to both Greater and Lesser Scaup. But scaup have all-white sides while the Ring-neck's flanks are gray. And scaup have plain blue-gray bills.

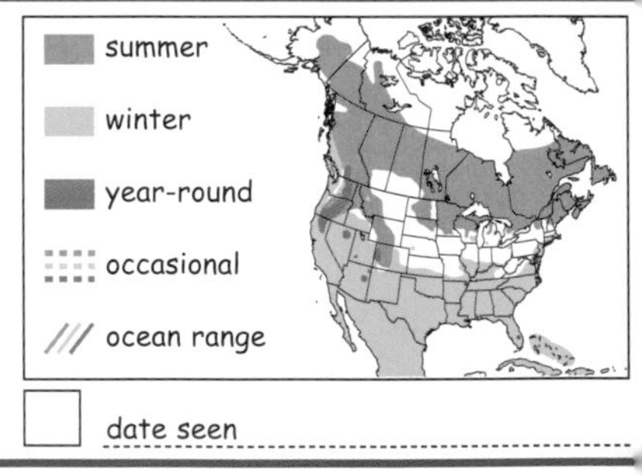

▶ *With its peak-headed profile, the Ring-necked Duck may have been the inspiration for the cartoon character Daffy Duck.*

Find it: Look for Ring-necks on small wooded ponds at all seasons, often mixing with other diving duck species.

summer
winter
year-round
occasional
ocean range

☐ date seen ____________

LESSER SCAUP

Aythya affinis **Length: 16½"**

Look for: Of our two scaup species, the Lesser is the one more commonly seen inland. Male Lesser Scaup are dark-headed and light-backed. The female has a brown head with a white ring around the base of the bill. Lesser Scaup of both sexes can be separated from Greaters by head shape: the Lesser has a less rounded head, with a peak at the back. The Greater has a slight peak at the front.

Listen for: Mostly silent. Female gives a loud, raspy *grr-grr*. Male whistles in courtship (not like *that!*).

Remember: From a distance, the bright white sides and light back of scaup are distinctive. Telling scaup apart by head color is not reliable.

◀ *A female Lesser Scaup surfaces with a mussel. She will swallow it whole, and her stomach muscles will crush the shell.*

WOW!
Both scaup species are known as Bluebills for the blue-gray bill color. The name scaup is derived from scallop, referring to the shellfish these ducks eat.

Find it: Lesser Scaup winter in huge flocks on inland lakes. Unlike other ducks, they are not often found in mixed-species flocks. When feeding, Lesser Scaup submerge with a diving plunge and bob back to the surface a few seconds later.

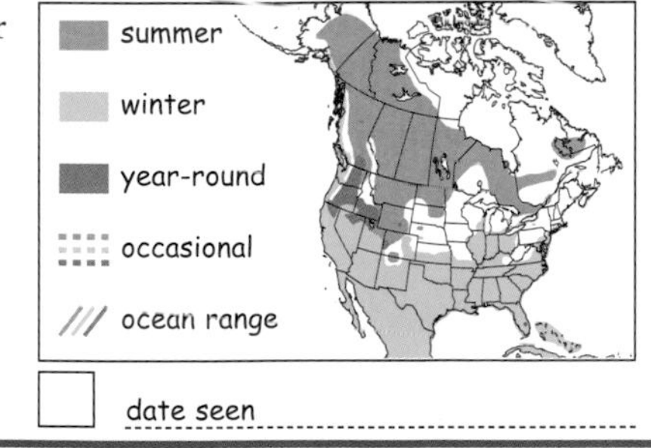

date seen ________

RUDDY DUCK

Oxyura jamaicensis **Length: 15"**

Look for: The Ruddy Duck seems to have a big head and oversize bill for its body size. The rusty-bodied male in breeding plumage has a bright blue bill, dark head, and bright white cheeks, making him hard to mistake for anything else. The best field mark for the less distinctive female is her dirty white cheek divided by a horizontal line.

WOW!
Ruddy Ducks are also known as Stifftails for the male's spiky tail feathers, which are often raised up vertically while he is swimming.

Listen for: Mostly silent except during courtship, when the male utters an otherworldly series of clicks perhaps best described as a water sprinkler that burps. The male Ruddy Duck also produces a variety of other sounds during courtship.

Remember: Ruddy Ducks are not the best fliers. In fact, they will dive underwater to escape danger rather than take to the air.

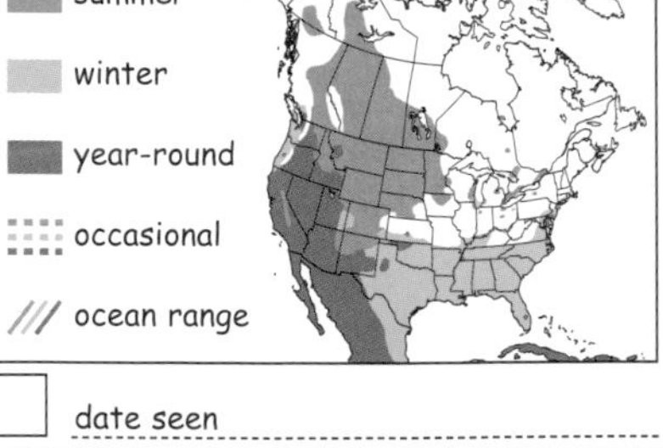

▶ *A courting male Ruddy Duck rapidly beats his bill against his breast, producing a peculiar drumming sound.*

Find it: A common duck found on ponds, marshes, bays, and lakes, the Ruddy Duck rarely mixes with other species in a flock. Huge winter flocks occur on lakes throughout the southeastern and south-central U.S.

summer
winter
year-round
occasional
/// ocean range

☐ date seen

COMMON MERGANSER

Mergus merganser **Length: 25"**

Male (left), female (right)

Look for: Our largest merganser, the male Common Merganser appears mostly snow white from a distance, with a dark green head, long, slender orange bill, and black upper back. The female is gray bodied with a sharply contrasting rusty head, a white throat, and a shaggy crest.

Listen for: The female gives a deep froglike croak, *uhhnk-uhhnk*. The courting male sounds like someone gargling mouthwash. He also calls while swimming in splashy circles around the female.

Remember: The long white body and long slender bill of the male Common Merganser are good field marks separating this species from other male green-headed ducks.

WOW!
A Common Merganser scans for fish with its head underwater. When it sees one, it dives after it, propelling itself with its feet. The sharp, toothlike projections on its bill hold the slippery prey tight.

▲ *The serrated edges of the Common Merganser's bill are not teeth, but they function like teeth for holding slippery fish.*

Find it: The Common Merganser prefers large, open bodies of fresh water, especially lakes and large rivers. Wintering flocks on ice-free water can be large. Even at great distances, the male's low-slung white body is distinctive.

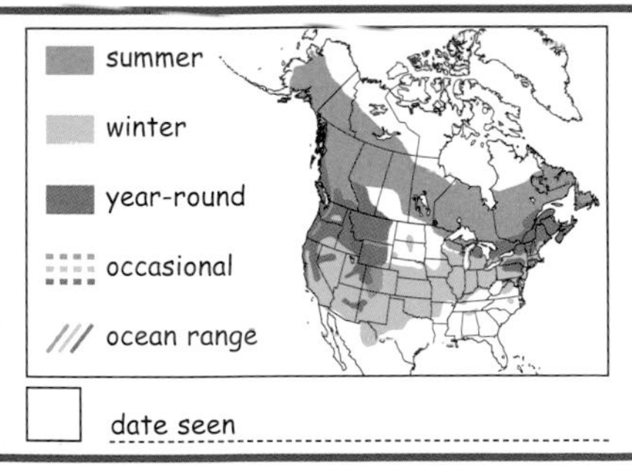

☐ date seen ____________

RED-BREASTED MERGANSER

Mergus serrator **Length: 23"**

Look for: Our medium-sized merganser is the one with the shaggiest crest. Males have a white collar (a good field mark on distant birds) and the reddish breast for which they are named, a bright orange bill, and red eye. Females are similar to female Common Mergansers but are less clearly marked—there is no distinct dividing line between rusty head and gray body.

Listen for: Usually silent. Female utters a harsh *kerr-kerr.*

Remember: From a distance, the Red-breasted Merganser male looks much darker than the male Common Merganser. The Red-breasted's dark head and breast are separated by an obvious bright white collar.

WOW!

Red-breasted Mergansers are fast, low fliers and are surprisingly fast swimmers too. With powerful strokes from their feet, they can pursue and catch the speediest of fish in shallow water.

▶ *Courting male Red-breasted Mergansers go into a frenzy when a female swims up—even in the dead of winter.*

Find it: Red-breasted Mergansers are most commonly found in migration on large bodies of water, but they prefer to winter in shallow saltwater bays along the coasts.

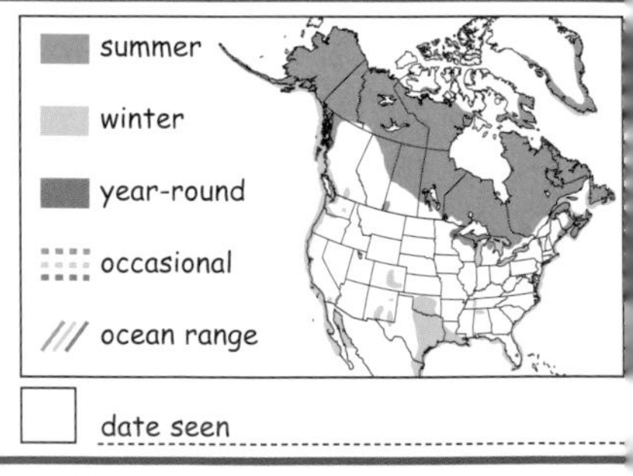

date seen

HOODED MERGANSER

Lophodytes cucullatus **Length: 18"**

Look for: Our smallest merganser species. Even from a distance, the male Hooded Merganser's bright white, fanned head crest outlined in black is easy to spot. Even when his crest is not fanned out, the bold black-and-white pattern remains visible. The female has the same big-headed, thin-billed appearance but with a body in tones of rusty brown.

Listen for: Displaying males give a low, croaky *how loooong!* Females give a short, froglike croak. In flight, the Hooded Merganser's wings produce a high whistle.

Remember: The male Hooded Merganser is similar to the male Bufflehead, but the Hooded has the crest outlined in black, orange flanks, and the two black vertical bars on the sides of the breast.

WOW! The Hooded Merganser can change its head shape depending on whether its crest is raised or lowered.

▶ *A female Hooded Merganser has a duckling ready to leave the nest in a natural cavity in a maple tree.*

Find it: Hoodies are usually seen in pairs or in small flocks. They prefer small wooded ponds, lakes, and swamps. This habitat, and their tendency to dive to avoid danger rather than take flight, makes them easy to overlook.

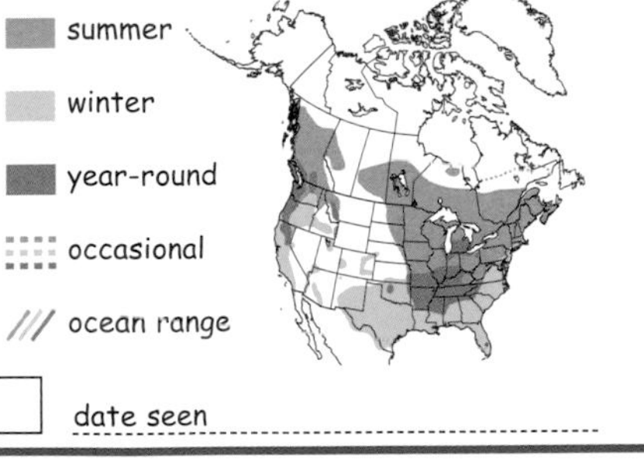

☐ date seen ____________

BUFFLEHEAD

Bucephala albeola **Length: 13½"**

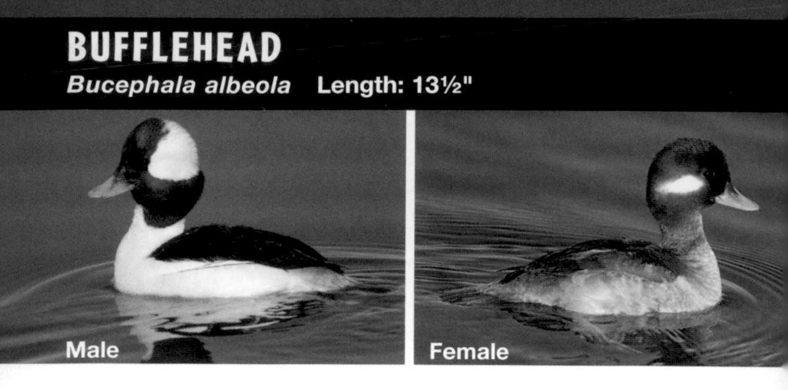

Look for: The Bufflehead is our smallest common duck, and its small size and bold black-and-white markings are distinctive. A male in breeding plumage shows a large patch of bright white on the back of a black head and a white body with a black back. From a distance, the Bufflehead shows a lot of white and appears large headed and small billed. The female resembles the male but has only an oval of white on the side of the head.

Listen for: Mostly silent. Courtship sounds include a series of flat, nasal grunts.

Remember: The Hooded Merganser is similar to the Bufflehead, but its white head patch is outlined in black, and it shows much less white overall.

WOW!
The Bufflehead is a cavity nester that is small enough to nest in holes excavated by Northern Flickers. How cool that a duck can nest in a hole made by a woodpecker!

◀ *The Bufflehead gets its name from the male's large-headed appearance, like a buffalo.*

Find it: Buffleheads nest near ponds in the north woods but winter across a vast portion of the continent in a variety of watery habitats, from large lakes to saltwater bays. They do not often mix with other duck species.

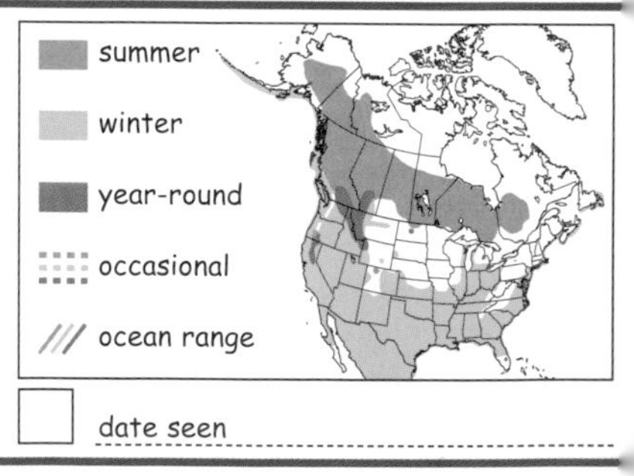

☐ date seen

WILD TURKEY

Meleagris gallopavo **Length: 37" (female) to 46" (male)**

Look for: The Wild Turkey is a surprisingly large bird, with a featherless head and all-dark body perched atop stout pink legs. The dark body feathers can show shades of green, orange, and blue in direct sunlight, and the bare head skin also changes color, especially on males during their tail-fanning courtship display. Females are smaller, with gray, not pink, heads.

WOW!
The Wild Turkey can fly surprisingly fast for such a large bird, but only for short distances. In flights of less than a mile, it may reach speeds of 55 to 60 miles per hour.

Listen for: Displaying males give a loud, distinctive gobble. Females call their young (known as poults) with a sharp *tuk! tuk!*

Remember: The Wild Turkey's huge size, bald head, and dark coloration are distinctive.

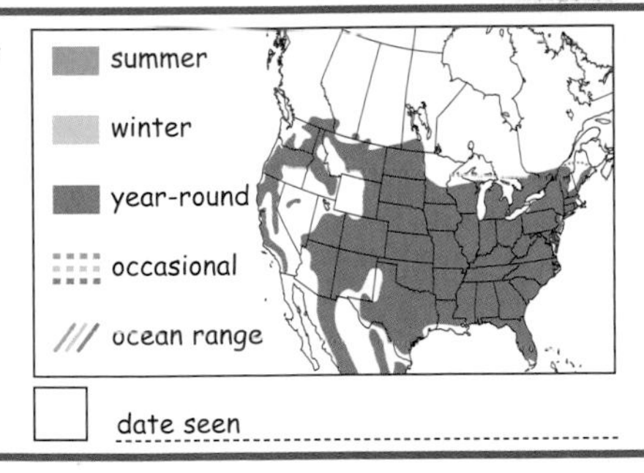

▶ *Young male Wild Turkeys have shorter "beards" hanging from their chests.*

Find it: The Wild Turkey seems to prefer mixed woods that contain acorn-producing oaks. Spring flocks may hold courtship displays in open fields or woodland meadows. Flocks roost in trees at dawn and dusk.

summer
winter
year-round
occasional
ocean range

☐ date seen ____________

RUFFED GROUSE

Bonasa umbellus **Length: 17"**

Look for: This game bird is decked out in browns, grays, and blacks that are perfect camouflage for its woodland habitat. Sexes are similar. The Ruffed Grouse occurs in two plumage varieties: gray and red (rusty orange).

Listen for: The male Ruffed Grouse performs a drumming display, beating his wings against the air while perched on a log, deep in the woods. This sounds like a muffled engine starting up: *whup-whup-whup-whup*, slowly at first, then faster and faster until it dies off.

Remember: Ruffed Grouse tend to stay still for as long as possible as danger approaches, relying on their cryptic coloration to conceal them.

WOW!

In spring, territorial male Ruffed Grouse are surprisingly aggressive. They have been known to challenge (and peck at) hikers, bicyclists, and even cars driving on roads through their territories.

▼ *Ruffed Grouse are named for the male's black neck ruff, which he puffs out during his drumming display.*

Find it: Widespread in woodland settings, this is our most easily encountered grouse species. During spring and summer, locate males by their drumming sounds. In summer, you may see females leading chicks across woodland clearings.

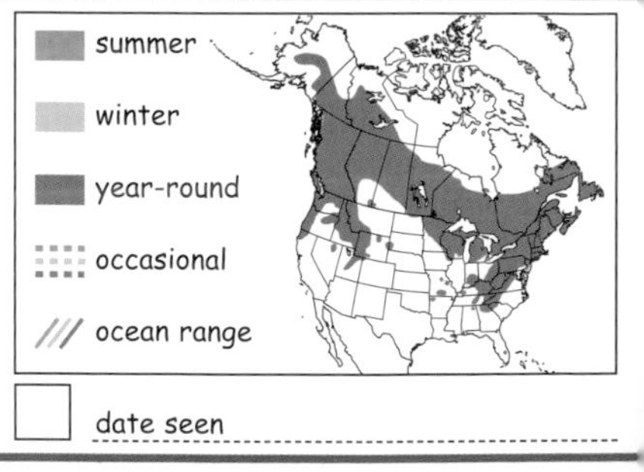

☐ date seen ____________

DUSKY GROUSE and SOOTY GROUSE

Dendragapus obscurus, Dendragapus fuliginosus **Length: 20"**

Look for: Until 2006 these two large dark grouse species were a single species, the Blue Grouse. Males of both species are dark bluish brown with bright yellow eye combs and neck sacs, which can be inflated during courtship displays. Females are dark gray, perfect cryptic coloration for their woodland habitat.

Listen for: Males of both species give a series of low-pitched whoops during courtship displays.

Remember: Don't be fooled by the photos of displaying male grouse. It's fairly rare to see the males with tails fanned and neck sacs inflated. Use range to help tell these two species apart.

▲ *These birds survive in winter on a diet consisting almost exclusively of conifer needles. Yum!*

Find it: Both species are year-round residents of mature woodlands. Dusky prefers more open habitat, including burned areas and meadows. Its range covers more interior montane forests. Sooty inhabits semi-open fir and spruce woods, and its range is closer to the Pacific Ocean.

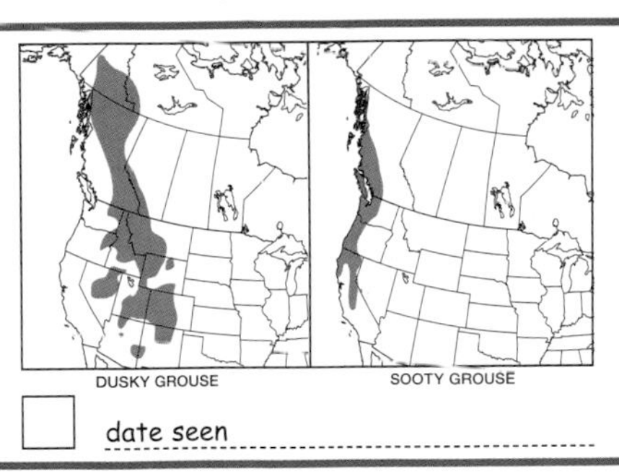

☐ date seen ____________

RING-NECKED PHEASANT

Phasianus colchicus **Length: 21" (female) to 32" (male)**

Look for: A large long-tailed bird most often seen walking on the ground, the Ring-necked Pheasant is not as shy as other game birds. The male is distinctive with his rusty body, white neck ring, red face, and showy tail feathers. Females are the same shape but are tan overall and have shorter tails.

Listen for: Only the males call, uttering a harsh, two-syllable *craa-cahh!* Pheasants' wings make a loud ruffling sound as the birds burst into flight from a concealed spot.

Remember: It's hard to confuse a male Ring-neck with any other bird. To help separate the drab female from prairie grouse species, look for her long tail and elongated body shape.

WOW!
The Ring-necked Pheasant was introduced to North America from Asia as a game bird. In many areas, the Ring-necked Pheasant population is supplemented by the birds raised on game farms.

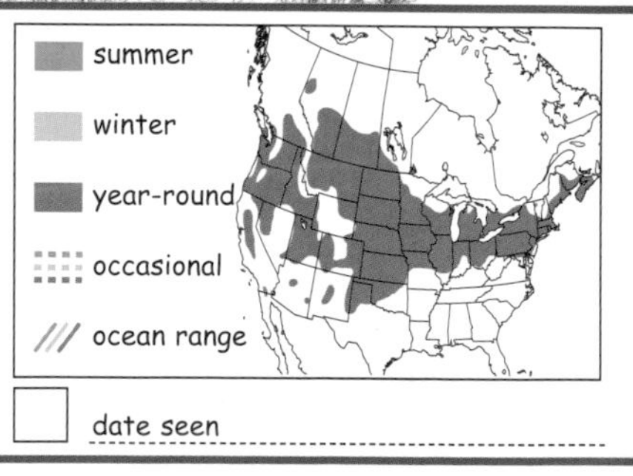

▶ *The rough double crow of the Ring-necked Pheasant is a signature sound of the prairie spring.*

Find it: Widespread in open farmland habitat, Ring-necks prefer old-fields, grain fields, marshy grasslands, and hedgerows. They can often be seen foraging by field edges along rural roads. They prefer to run from danger rather than fly.

summer
winter
year-round
occasional
/// ocean range

☐ date seen ________

CALIFORNIA QUAIL and GAMBEL'S QUAIL

Callipepla californica, Callipepla gambelii **Length: 10", 10½"**

Look for: These two smallish chickenlike birds appear very similar from the wings up. Males and females have distinctive teardrop-shaped black head plumes. Males of both species have black faces outlined in white. The unscaled butter yellow belly is diagnostic of the male Gambel's, while the male California's belly is scaled with dark edges on tan feathers. Females are dull gray-brown overall.

Listen for: California's three-syllable call is *Chi-CAH-go!* Also gives *clucks*, *pits*, and a loud *krrrr!* Gambel's main vocalization is a loud, descending *kaaah!* Also a four-part *kah-KAH-kah-kah*.

Remember: It can be hard to remember which quail species is which, despite their limited range overlap. Two ways to distinguish them: the Gambel's Quail is sometimes called "redhead" by hunters, and they "gamble" with their lives by living in drier desert habitat.

WOW!
One day after hatching, downy quail hatchlings leave the nest and follow their parents. These tiny tots are supercute!

▼ *While foraging, coveys of both of these quail species will post a lookout to watch for approaching danger.*

Find it: Both are nonmigratory residents. California Quail are more coastal in distribution and prefer greener, more vegetated habitat, including chaparral, coastal scrub, woodland edges, and parks. Gambel's prefer drier desert habitat, including canyons, and open areas with scattered brush.

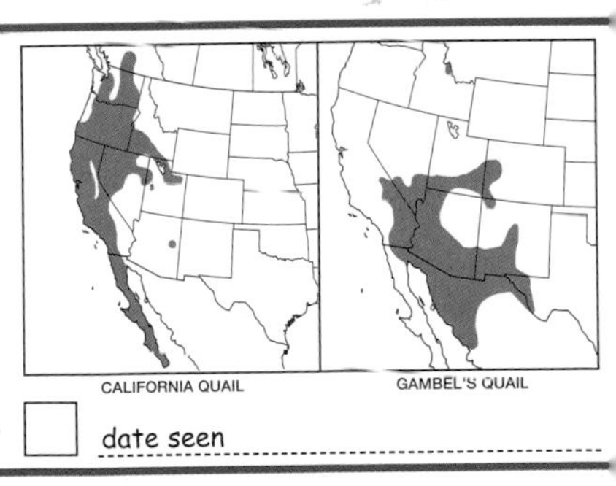

☐ date seen ____________________

NORTHERN BOBWHITE

Colinus virginianus **Length: 10"**

Female (left), male (right)

Look for: The masked appearance of the male Northern Bobwhite is the best visual field mark for this species, but this quail is often heard before it is seen. The female has a pale tan version of the male's mask. Populations of Northern Bobwhites are declining in many areas.

Listen for: The Northern Bobwhite says its own name—at least the *bobwhite!* part. In spring and summer, males give this call repeatedly from an exposed perch. Members of a quail flock, or covey, utter a variety of whistling calls.

Remember: To avoid danger, bobwhites usually run for cover and will fly only when surprised or when running is not a safe option.

WOW!

To guard against predators at night Northern Bobwhite coveys settle down in a circle with all birds facing out. If one bird sees a predator, it explodes into flight, which tells the nearby birds to do the same.

▼ *Bedding down in a tight circle, each bird facing out, a covey of bobwhites prepares for the night.*

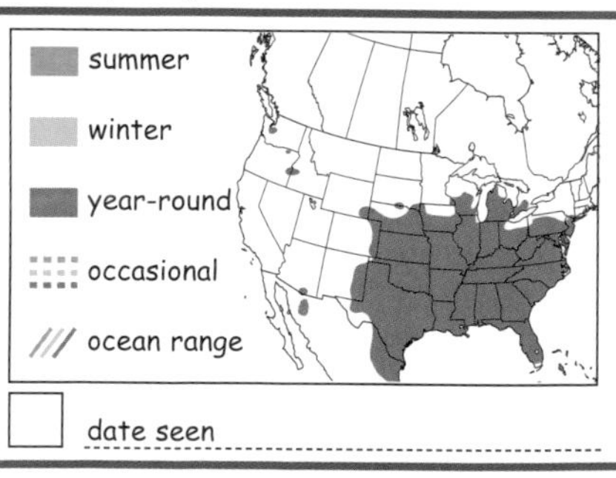

Find it: Listen for the distinctive *bobwhite!* call in old pastures, farm fields with hedgerows, open-understory woods, and grasslands, and scan nearby perches such as low snags and fenceposts for the one bird that acts as the sentinel for the covey.

summer
winter
year-round
occasional
/// ocean range

☐ date seen ____________________

COMMON LOON

Gavia immer **Length: 32"**

Look for: The Common Loon is a large, long diving bird seen most often on big lakes, where it rides low in the water. In summer, the adult's black head, body, and bill contrast with a bright white chest and black-and-white-checkered back. In winter, Common Loons are drab gray above and white below.

Listen for: The haunting high yodel of the Common Loon echoes across wooded lakes in the North. It also gives a laughing *ha-ha-ha-ha* call.

Remember: You may confuse a distant Common Loon on the water with a Double-crested Cormorant. The loon has a heavier overall appearance; a thicker, shorter neck; and a heavier, dark bill.

WOW!
Hollywood movie directors often dub the call of the Common Loon into their movies. Too often, however, they insert it in the wrong setting, such as in the desert or in a junkyard in New York City!

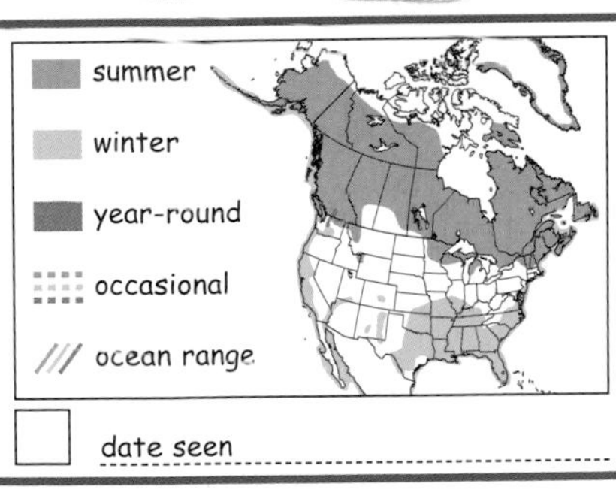

▶ *Rearing up like a cobra, a male Common Loon defends his nest from an intruder.*

Find it: Summer finds Common Loons on large, quiet bodies of water. They winter on waters along both coasts and fly in small flocks low over the water. During migration, loons may be found on large rivers, lakes, and reservoirs.

summer
winter
year-round
occasional
ocean range

☐ date seen ______

PIED-BILLED GREBE

Podilymbus podiceps **Length: 13"**

Look for: Our most widespread and common small grebe, the Pied-billed is usually seen alone (not in flocks) swimming on quiet waters where it dives and pursues small fish underwater, propelled by its feet. Tawny brown overall in all seasons, it gets its name from the black band around its pale bill, a field mark present only in summer.

Listen for: For such a small bird, the Pied-billed Grebe has a big voice, giving a long, rapid-fire series of grunts, toots, and hoarse barks.

Remember: The Pied-billed's brownish overall color, thick bill, and all-dark eyes set it apart from our other small grebes.

▼ *A Pied-billed Grebe can compress its plumage, exhale, and sink without a ripple.*

WOW!

To escape danger, the Pied-billed will submerge and swim a long distance underwater. It may then raise only its head above the water, watching quietly for the danger to pass.

Find it: Almost any North American pond or lake can host Pied-billed Grebes, but they prefer water with concealing vegetation. The Pied-billed Grebe's small size, drab coloration, and solitary nature make it easy to overlook.

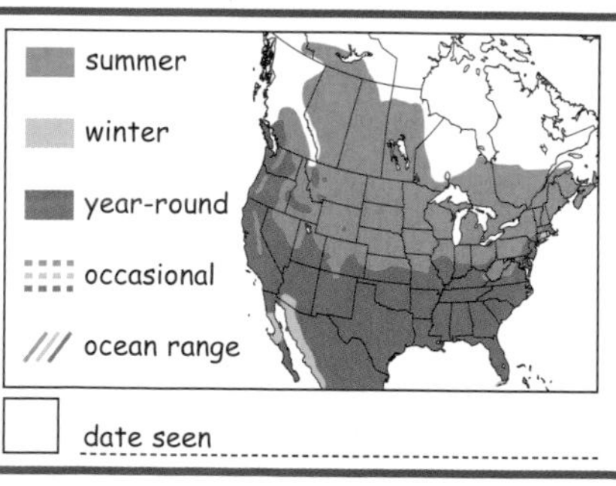

☐ date seen ....

HORNED GREBE

Podiceps auritus **Length: 14"**

Nonbreeding

WOW!

In all seasons the Horned Grebe has bright red eyes. This gives birds in colorful breeding plumage an especially freaky look.

Look for: The Horned Grebe is named for the golden yellow patch of feathers on the sides of its dark head. You won't see this on birds in winter, when they are dark gray and white. The Horned Grebe's white face and neck help to set it apart in winter. Breeding-plumage bird has a black crown, throat, and back with a rusty neck and sides.

Listen for: Mated pairs give a high-pitched twittering duet. In summer their other common call is a sputtering *pitpitPRE-ahhh!*

Remember: Horned and Eared Grebes both have small, thin bills, but the Horned Grebe almost always has a white tip on its dark bill. The Pied-billed Grebe has dark eyes and a stocky bill.

▲ *A lucky birder might see a Horned Grebe in breeding plumage, but winter plumage (right) is more commonly seen.*

Find it: Look for Horned Grebes swimming, diving, and nesting on freshwater lakes. Many spend the winter in coastal (saltwater) bays. Others winter on reservoirs and big lakes. During migration they can be found throughout the East.

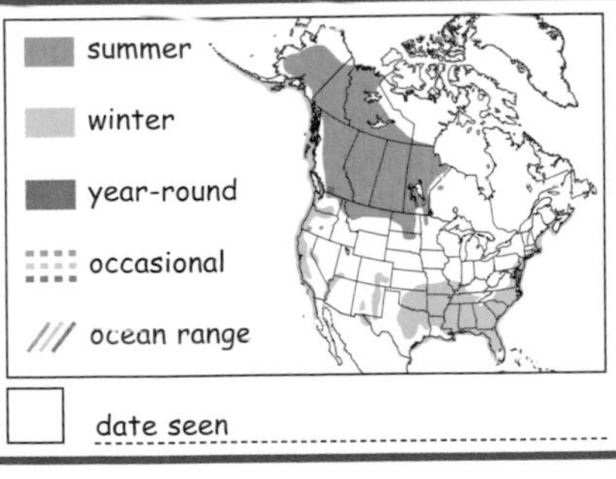

☐ date seen ____________

WESTERN GREBE and CLARK'S GREBE

Aechmophorus occidentalis, Aechmophorus clarkii **Length: 25"**

Look for: By far our largest grebes, these two species are very similar: both are dark bodied with long white necks, black caps, and yellow bills. The black cap of Western Grebe encloses the red eyes. On Clark's Grebe the eye is surrounded by white. Clark's bill is brighter yellow.

Listen for: Western: loud, double *crik-crik*. Clark's: a single-note *creek*.

Remember: The Western Grebe has dark surrounding its eyes, like sunglasses, guarding it from the "western" sun.

▼ Males of both of these two large grebe species perform an exciting courtship dance, running across the surface of a lake with their necks extended. This never fails to impress the ladies.

WOW!

One of these grebe species is among a handful of species discovered and named during the Lewis and Clark expedition in the early 1800s. Guess which one . . . **duh!**

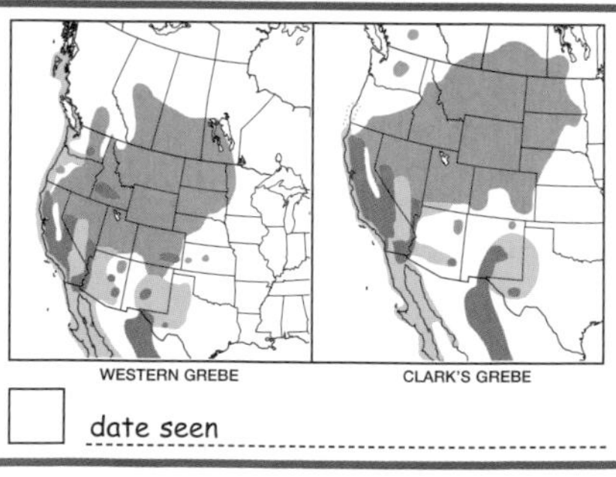

Find it: Ranges and habitat preferences of the two species overlap: in all seasons, both can be found on large lakes and freshwater bays. Western Grebes commonly winter on salt water, while Clark's prefer fresh water all year.

WESTERN GREBE   CLARK'S GREBE

☐ date seen

AMERICAN WHITE PELICAN

Pelecanus erythrorhynchos **Length: 62"**

Look for: This huge white bird (with a nine-foot wingspan!) has black flight feathers and a large yellow-orange bill, making it hard to mistake for any other bird. Often found in flocks foraging in shallow lakes as well as soaring high overhead.

Listen for: Not very vocal, except in breeding colonies, where they utter low, croaking grunts.

Remember: The White Pelican differs in a few important ways from its cousin the Brown Pelican. The White Pelican forages in cooperative flocks, which herd schools of fish to waiting, open bills. The Brown Pelican dives for its food. White Pelicans are inland nesters, while Brown Pelicans are closely associated with coastal saltwater areas.

A White Pelican's pouch can hold more than two and a half gallons of water. Imagine holding that much water in your throat. Now imagine all that water full of squirming fish. Yum, sushi!

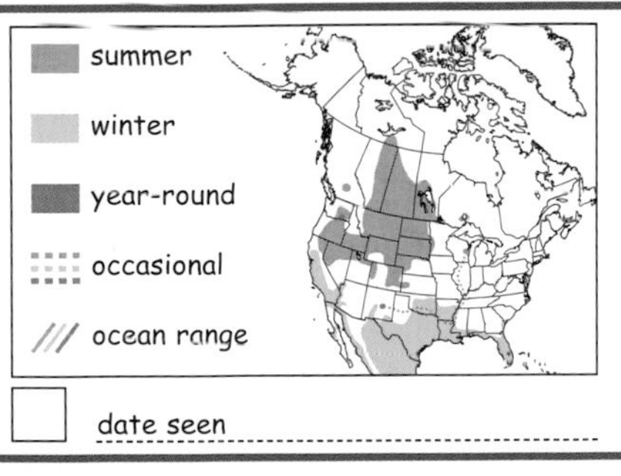

▶ *A White Pelican guards its chick in the nesting colony.*

Find it: During spring and summer, White Pelicans breed in big colonies, usually on islands in large inland lakes. They range far from the colonies to forage on shallow lakes and marshes. They winter along the coasts.

summer

winter

year-round

occasional

/// ocean range

☐ date seen ______________

BROWN PELICAN

Pelecanus occidentalis **Length: 51"**

Look for: Though it's not as large as the White Pelican, the Brown Pelican is still a huge bird. Its mostly brown body, dusty gray wings, and yellow crown patch are minor field marks compared to the oversize pelican bill with its expandable pouch. Breeding-season adults have darker brown necks and more yellow on the heads.

Listen for: Except for young birds in the nesting colonies, these birds are nonvocal.

Remember: Brown Pelicans are large-headed, large-billed, slow fliers. Soaring low over the water on their vast wings, they are hard to confuse with any other bird.

► *A Brown Pelican fishes by folding its wings and plummeting, scooping up the fish in its enormous gular pouch.*

WOW!

Harmful chemicals nearly wiped out the Brown Pelican. In the early 1970s, only a small population remained. Thanks to the banning of DDT and other chemicals, Brown Pelicans have made a strong recovery.

Find it: Common along both coasts, but fairly unusual inland or far from salt water, the Brown Pelican has expanded its range northward. Watch for them perched on pier pilings or in small groups soaring low over the water.

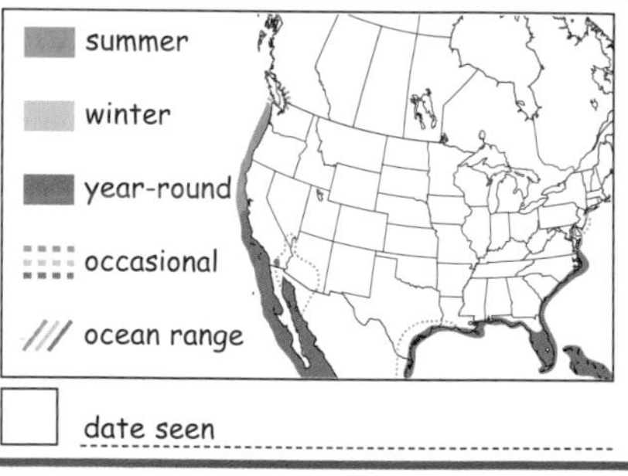

☐ date seen ______________

DOUBLE-CRESTED CORMORANT

Phalacrocorax auritus **Length: 33"**

Adult, breeding

Drying wings

Look for: Our most common cormorant continent-wide, the Double-crested is a long and slender all-dark bird with a pale yellow-orange bill and throat patch. Its crests are apparent only in the breeding season, and even then they are difficult to see. Cormorants float low in the water and dive after fish, pursuing them underwater. Flocks of migrant cormorants fly overhead in wavering Vs. Young birds have pale breasts.

Listen for: Nonvocal away from nesting colonies.

Remember: The Double-crested is the only cormorant commonly found on inland bodies of water. The Anhinga is similar but has many white feathers on its back and swims with only its head above water.

WOW!
In some areas, cormorant populations are controlled because of their perceived impact on populations of game fish.

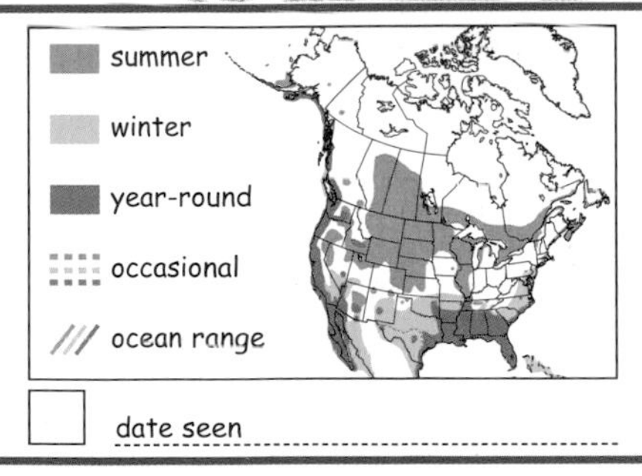

▶ *Although cormorants are often blamed for eating sport fish, the majority of their diet is fish unwanted by people.*

Find it: Large gatherings of cormorants occur on large lakes, along rocky coasts, and on snags along rivers and bays. After diving for fish, Double-crested Cormorants frequently perch near the water and extend their wings to dry them out.

summer
winter
year-round
occasional
ocean range

date seen

ANHINGA

Anhinga anhinga **Length: 35"**

Look for: A long thin neck and small head, thin yellow bill, and white feathers on the wings and back set the Anhinga apart from the closely related cormorants. Anhingas commonly perch with their wings spread and swim with only their heads above water. Adult males are mostly black overall, while adult females and young birds have tan heads and necks atop black bodies.

WOW!
The Anhinga is known by many names, including Snakebird (for its snakelike head), Water Turkey (for its fanned tail), and Black Darter (for its ability to spear fish with its sharp bill).

Listen for: A series of rapid clicks that sound mechanical, like a sewing machine.

Remember: Anhingas are excellent fliers, often soaring high in the sky. When it soars, the long, pointed wings and long tail and neck make the Anhinga look like a flying X.

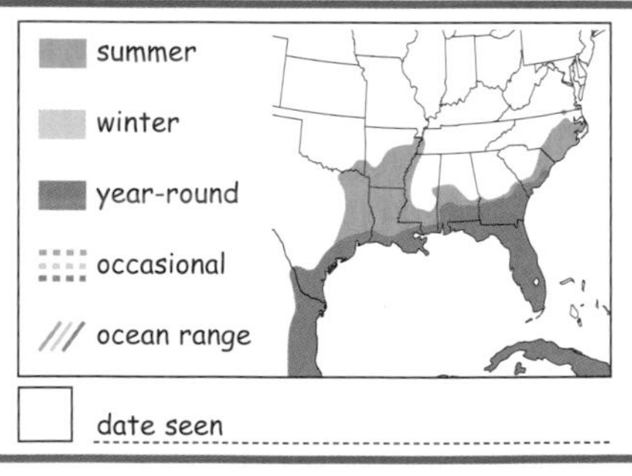

▶ *An Anhinga spears a fish, tosses it into the air, and swallows it whole, headfirst.*

Find it: On slow-moving rivers, lakes, ponds, and swamps in the Southeast, Anhingas often perch on snags near or in the water, hanging their wings out to dry. Anhingas nest in colonies with other wading birds.

summer
winter
year-round
occasional
ocean range

☐ date seen ______

GREAT BLUE HERON

Ardea herodias **Length: 46"**

Look for: A tall, long-legged, grayish bird with a long, yellow, daggerlike bill used to spear fish and other prey. Great Blues are slow and stately in their movement, whether wading or flying. Flying birds show black wingtips, and the legs trailing behind give the bird a superlong appearance.

Listen for: A deep, croaking *graaaak*, usually given in flight, that sounds more like a belch than a bird's call.

Remember: Other dark wading birds resemble the Great Blue but are much smaller, including the Green Heron, Tricolored Heron, Little Blue Heron, and the Glossy Ibis.

▶ *Great Blue Herons nest in large colonies called rookeries. Their large, bulky stick nests are easy to spot in early spring.*

Many Great Blues migrate south in winter, but a few may linger. When water freezes they may hunt in nearby fields for rodents and small birds.

Find it: Widespread and increasingly common. Always found near water, especially clear, calm water where the Great Blues can easily see and hunt for fish, crayfish, frogs, and other prey.

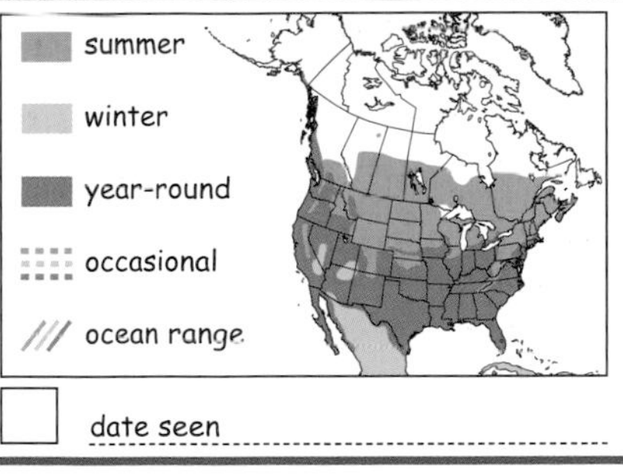

☐ date seen ____________

GREAT EGRET

Ardea alba **Length: 39"**

Look for: The Great Egret is a tall, slim, all-white bird with a long yellow bill and long all-black legs. It can appear extremely tall when the long neck is extended, or smaller when the neck is pulled in close to the body. Lacks the long showy plumes of other herons and egrets.

Listen for: Low, unmusical croaks and harsh squawks.

Remember: Other white egrets are smaller. The Snowy Egret has black legs with yellow feet. The white-morph Reddish Egret has gray legs and a pink and black bill. The Cattle Egret is half the Great Egret's size and has a stubby bill by comparison.

Adult, breeding

WOW!

Great Egret plumes adorned women's hats in the late 1800s, and this nearly resulted in the species' extinction. Conservation efforts saved the egrets.

▲ *If you see a group of wading birds together, the long-legged Great Egret is likely to be in the deepest water.*

Find it: Our most common large white wading bird, the Great Egret is often seen standing completely still, watching for a passing fish, frog, or snake, on marshes, sloughs, and lakes. They may wander far north of their normal range in late summer.

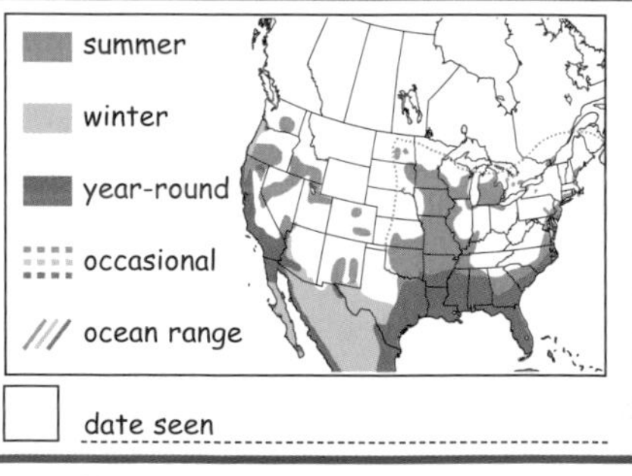

☐ date seen

SNOWY EGRET

Egretta thula **Length: 24"**

Look for: Perhaps the best field mark for this all-white bird are its yellow "slippers"—golden feet at the bottom of black legs. Yellow also appears on the lore, the area between the eyes and the slender black bill. During the spring and summer, adult Snowies grow long, lacy plumes on their heads, necks, and tails. This is their showy breeding plumage.

Listen for: A series of gurgling croaks that sounds like someone is getting sick.

Remember: The Snowy is the most delicate looking of our white egrets, with its fine-pointed bill, feathery plumes (in spring and summer), and overall slender appearance. It often forages with other species of wading birds.

▼ *By rapidly patting the water's surface with its foot, a Snowy Egret attracts fish and stirs up other food from the muck.*

Find it: Snowies prefer wide-open marshes, ponds, and shores—either freshwater or saltwater habitats will do. They wander north in late summer. In winter, northernmost birds move south to coastal areas.

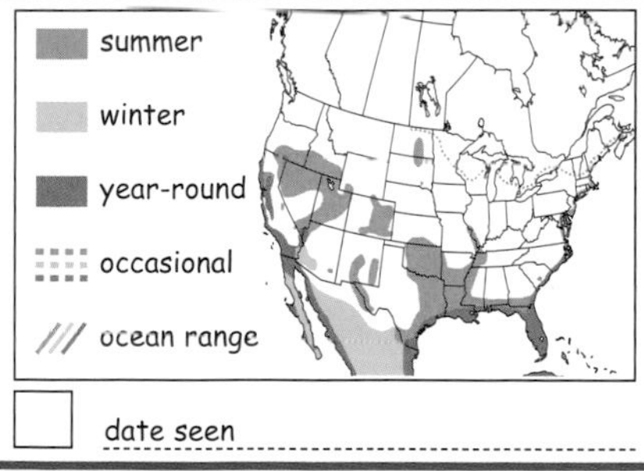

☐ date seen ____________

CATTLE EGRET

Bubulcus ibis **Length: 20"**

Look for: The shortest, stockiest of our white egrets and herons, the Cattle Egret's most reliable field mark is its stout yellow bill. In winter, the Cattle Egret's plumage is all white, but during the breeding season, an adult shows rusty patches on the head, breast, and back.

Listen for: Cattle Egrets are not very vocal birds, except in the nesting colony, where they utter a series of guttural quacks that sound more like a pig than a bird.

Remember: A chunky, white wading bird in the middle of an open field is likely to be a Cattle Egret. A tall, slender, white wading bird standing in water is more likely to be a Great or Snowy Egret.

WOW!

The Cattle Egret is native to Africa but has expanded to areas around the world. The first Cattle Egrets arrived in North America in the early 1950s and have since spread across much of the continent.

▶ *Cattle Egrets follow large grazing animals and even tractors, eating the insects frightened from the grass.*

Find it: The well-named Cattle Egret can be seen foraging in fields and pastures near cattle and other livestock, eating insects disturbed by the animals' movement. Cattle Egrets nest in colonies, often with other herons and egrets.

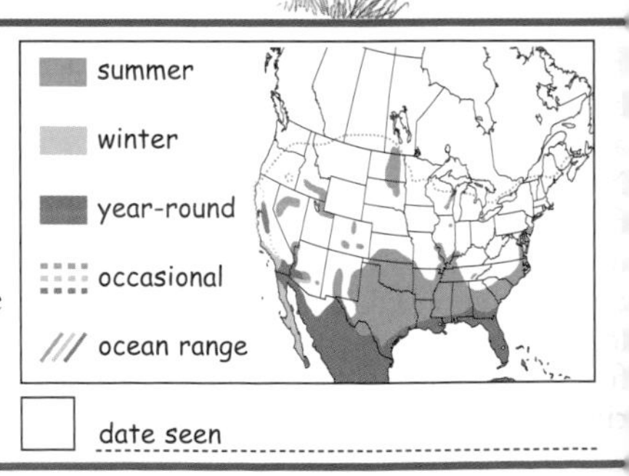

☐ date seen ____________

BLACK-CROWNED NIGHT-HERON

Nycticorax nycticorax **Length: 25"**

Look for: A chunky bird with a gray body, black crown and back, bright orange-red eyes, and sturdy yellow or pink legs. Black-crowned Night-Herons often perch or stand in a hunched-over pose. Crown and back color are black, differing from the closely related Yellow-crowned Night-Heron (adults of which have yellow crowns and gray wings and backs).

Listen for: A loud, low-toned *qwock!* often uttered in flight can identify Black-crowns flying unseen at night.

Remember: Young night-herons of both species are streaky overall and look different from adults. The yellow in the bills of young Black-crowns helps to tell them from young Yellow-crowns.

▲ *The Black-crowned Night-Heron's enormous eyes allow it to find its prey in almost total darkness.*

WOW!

Black-crowned Night-Herons nest on every continent on earth, making them one of the few truly worldwide bird species.

Find it: During the day both Black-crowned and Yellow-crowned Night-Herons can be found roosting in trees in their nesting colonies. At night they leave the colonies to forage in nearby ponds, rivers, and marshes.

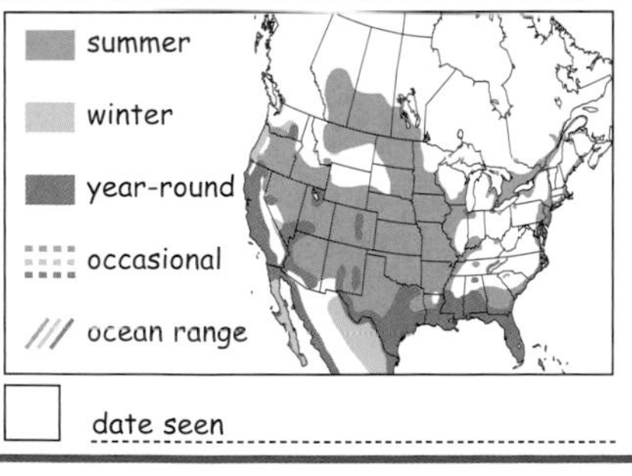

☐ date seen ______

GREEN HERON

Butorides virescens **Length: 18"**

Look for: This very small, dark heron has the perfect coloration to blend in with habitat along the water's edge: dark tones of reddish brown and deep green, with bright yellow-orange legs. Though its body is stocky, its bill is slender and long, perfect for spearing small fish, its preferred food.

Listen for: A loud, hoarse *skowp!* Often given when the bird is spooked into taking flight.

Remember: Most easily confused with the night-herons and bitterns, but the Green Heron is much darker overall and shows all-dark wings in flight.

WOW!
Green Herons may use bait (such as bread or corn that's been fed to ducks, or even a piece of vegetation) to attract curious fish to within striking distance.

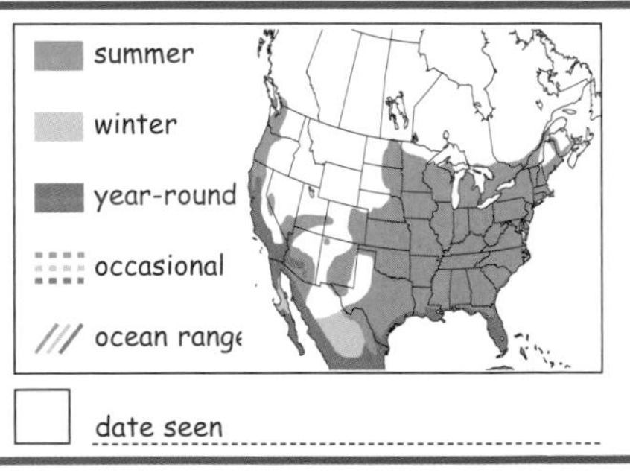

▶ *A Green Heron bait-fishing with a piece of bread.*

Find it: Quiet wooded streams, small ponds, and almost any body of fresh water that is lined with thick vegetation may host a Green Heron. Scan the trees and snags along the water's edge to find a Green Heron sitting quietly.

summer
winter
year-round
occasional
ocean range

☐ date seen ______

GLOSSY IBIS

Plegadis falcinellus **Length: 23"**

Adult

Look for: A large, long-legged, dark-overall bird with a downward-curving bill, the Glossy Ibis gets its name from the glossy appearance of its plumage, with its shades of metallic green, black, purple, pink, and rust. Breeding birds show more rusty coloration on the neck and body.

Listen for: Not a very vocal species, but gives a nasal *uhn-uhn-uhn-uhn* call when flushed.

Remember: The Glossy Ibis is hard to tell apart from the White-faced Ibis, which is common in summer in the Great Plains and western United States. In breeding season, the Glossy has a white outline on a dark face while the White-faced has a white outline on a red face.

WOW!
The Glossy Ibis finds its food by feeling it with the tip of its long bill. As it probes in the mud of a shallow marsh, it senses a food item (such as a beetle or other insect) and snaps it up.

◀ *Glossy Ibises probe deep in the mud. When something moves, they grab it with the tip of their sickle-shaped bill.*

Find it: The Glossy Ibis is commonly found in marshes and wetlands. They are almost always found in flocks, wading in shallow water and probing in the mud for food. This gives them a hunched-over look, like workers picking vegetables in a field.

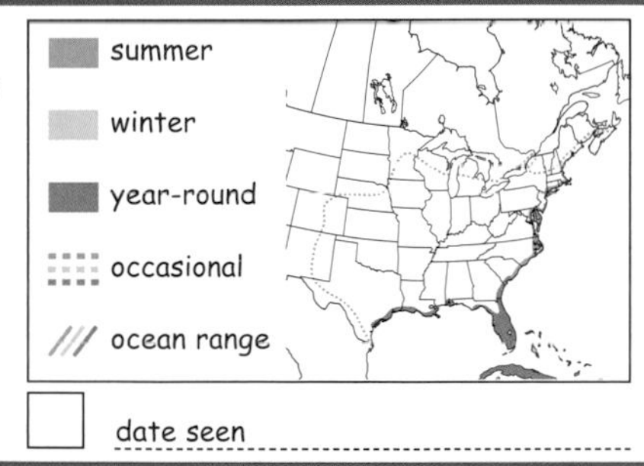

☐ date seen ____________

WHITE IBIS

Eudocimus albus **Length: 25"**

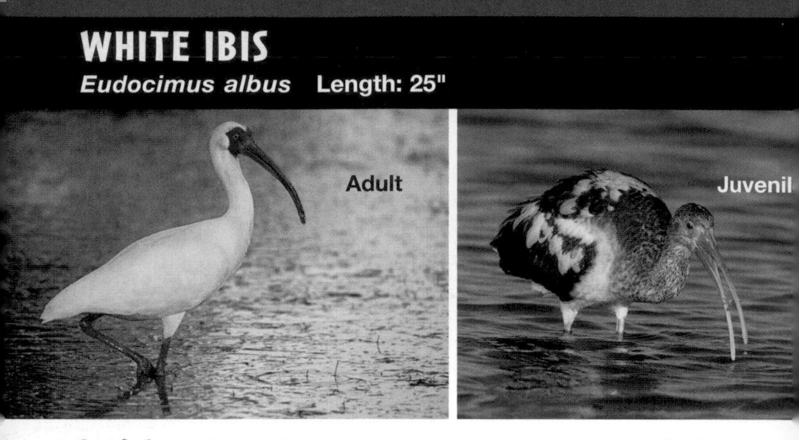

Look for: The adult White Ibis is an impressive bird: all-white body, black wingtips, red legs, and down-curved red bill set off with a blue eye. Young birds are a mixture of brown and white but still show the red bills and legs. In flight, the black wingtips stand out on white wings.

Listen for: A dull-sounding *uhn!* Most commonly heard from the nesting colonies.

Remember: Even viewed at a great distance, the decurved shape of the bill is an excellent field mark to separate this species from other large white wading birds.

▼ *Unlike other ibises, the White Ibis is comfortable sharing its habitat with humans, especially in Florida.*

Find it: Common in a variety of watery habitats, White Ibises prefer to be in groups, and small flocks are commonly seen foraging for insects on beaches, parks, golf courses, and large expanses of suburban lawn.

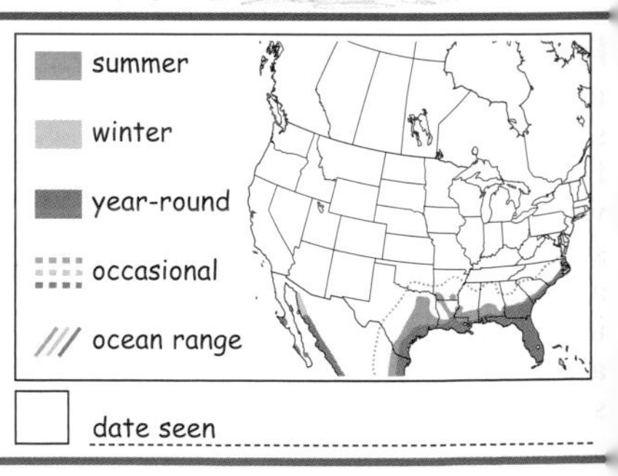

date seen

ROSEATE SPOONBILL

Platalea ajaja Length: 32"

WOW!
One of the most beautiful sights a birder can see is a flock of Roseate Spoonbills at sunset flying back to the nesting colony.

Adult

Look for: The Roseate Spoonbill is hard to confuse with anything else. Its rose pink coloration is stunning, and its spoon-shaped bill is unlike that of any other North American bird. Young birds are light pink, attaining the more intense adult colors by their third year. An up-close look at its massive bill, bald head, and red eye might convince you that this bird looks best from a distance.

▼ *The spoonbill's bill is full of nerve endings, allowing it to feed by "feel."*

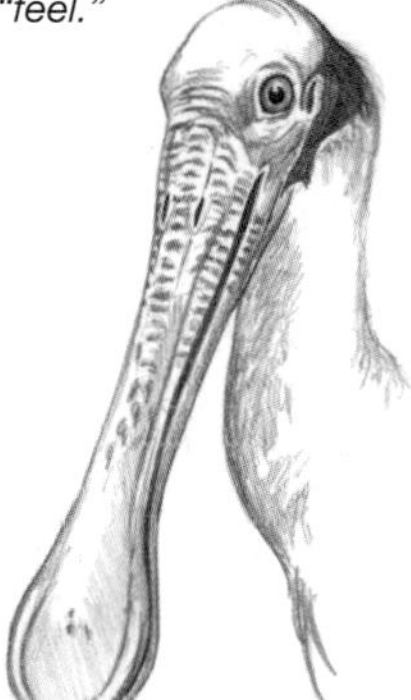

Listen for: Not very vocal but does utter a series of low grunts on a single pitch.

Remember: If you see a flock of large pink birds along the coast, from Florida to Texas, they are almost certain to be spoonbills.

Find it: Common within its coastal range in shallow bodies of both fresh and salt water, where it forages by sweeping its head back and forth and snapping its bill shut when it feels a prey item—usually a small fish.

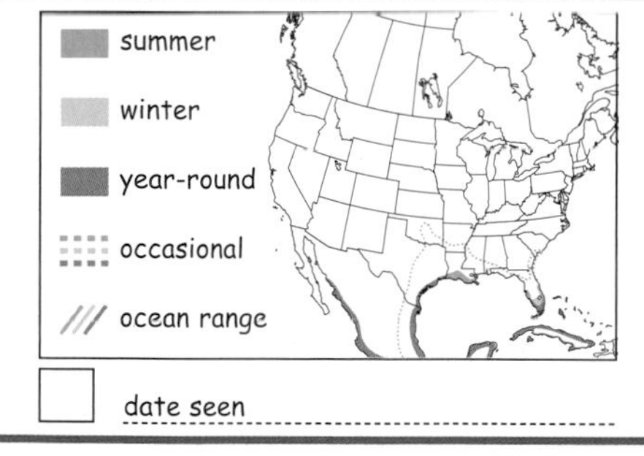

WOOD STORK

Mycteria americana **Length: 40"**

Look for: A very large bird with a white body and wings that are starkly black and white. The large down-curved bill and gnarled, featherless head make the Wood Stork a candidate for "ugliest bird." Wood Storks are strong, if slow, fliers and are frequently seen soaring high overhead.

Listen for: Mostly silent except for some hissing and bill clattering in nesting colonies.

Remember: The Wood Stork is much larger than the White Ibis; has a dark head, unlike the white egrets; and wades instead of swimming, like the much larger White Pelican does.

▼ *A close look at a Wood Stork's bald head hints at this species' close relationship to the Turkey and Black Vultures.*

Find it: Commonly found in southern swamps, lagoons, ponds, and roadside ditches. Colonies nest in large stands of trees, such as cypress. Populations are declining throughout the Southeast because of habitat destruction and alteration.

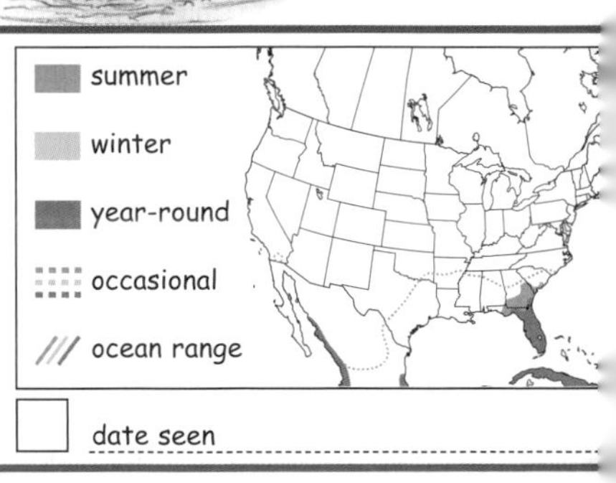

date seen

SANDHILL CRANE

Grus canadensis **Length: 42"–46"**

Adult

Look for: A tall, completely gray bird with a red patch on the head and a white patch on each cheek. In flight, Sandhills show both a long neck and long legs. Young Sandhills lack the red crown, and in summer many adults show a lot of rusty brown body feathers.

Listen for: A low, rattling *graaaahk!* Often compared to the rattle of an American Crow, but louder. This call from flocks high overhead may be your first clue to the presence of Sandhill Cranes.

Remember: The Sandhill Crane differs from the Great Blue Heron in three obvious ways: the Sandhill has a large bustle of feathers on the tail, it has a much stockier body, and it flies with the neck extended, not folded in.

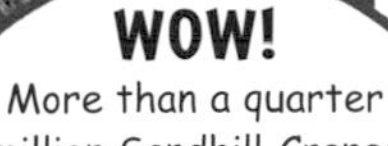

WOW!

More than a quarter million Sandhill Cranes stop over during spring migration along the Platte River in Nebraska. Imagine the noise made by this many large birds as they all take flight at dawn!

▶ *With loud calls and springing leaps, a courting pair of Sandhill Cranes prepares to mate. Cranes mate for life.*

Find it: Widespread and common, especially during spring and fall migration, when large flocks may be seen overhead or foraging in fields, wet prairies, and large marshes.

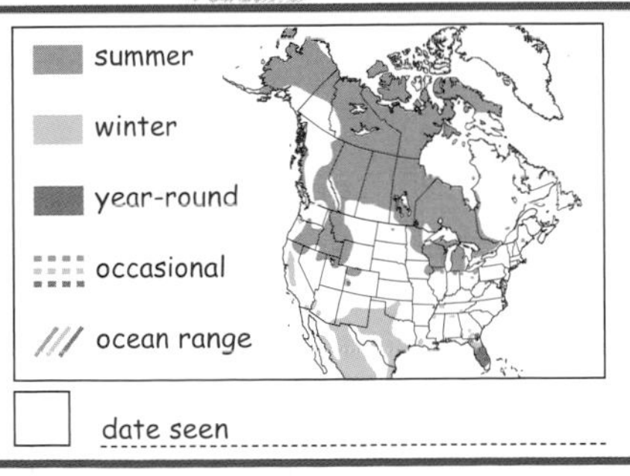

☐ date seen

TURKEY VULTURE

Cathartes aura **Length: 26"**

Adult

Adult

Look for: The Turkey Vulture soars with its wings held slightly above horizontal (called a slight dihedral). From below, the Turkey Vulture (TV) shows two-tone wings—black in front, silver in back. The flight silhouette combines long wings and tail with a tiny head (hawks and eagles appear much larger headed in flight).

Listen for: Normally silent, but will hiss and groan when danger approaches, especially near the nest.

Remember: Turkey Vultures *rock*! When they fly, TVs rock or teeter back and forth, capturing every bit of rising air. So if you see a large raptor rocking in the sky, chances are it's a Turkey Vulture.

WOW!
One of the Turkey Vulture's defenses is to puke on an intruder. Trust me: You do not want to get vulture puke on your clothes—you cannot get the smell out!

▲ *A Turkey Vulture prepares to deliver a load of predigested roadkill to its nestlings in a hollow log.*

Find it: Look up in the afternoon sky in summer and you'll likely see a Turkey Vulture. They are found in open or semi-open habitats but will even venture into heavily wooded areas for their food, which is dead animal flesh, or carrion.

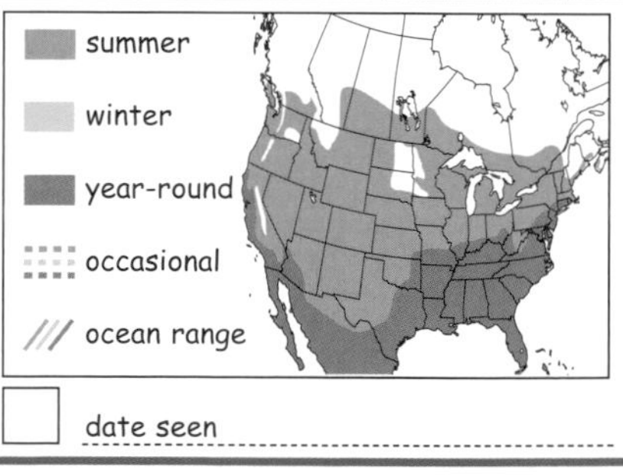

BLACK VULTURE

Coragyps atratus **Length: 25"**

Look for: The Black Vulture is the smaller of our two vultures and in flight looks much more compact and shorter tailed than the Turkey Vulture. The Black Vulture has a featherless gray head (Turkey Vultures have red heads) and when viewed from below shows all-black wings with white tips (primaries).

Listen for: Black Vultures are usually silent except near the nest or when threatened, when they issue guttural hisses and barks.

Remember: Black Vultures fly like they are worried they are going to fall out of the sky: lots of fast flapping in between short glides. They hold their wings flat, not slightly raised as Turkey Vultures do.

WOW!
Where both species occur, Black Vultures, although smaller, often bully the Turkey Vultures and drive them away from a feeding site.

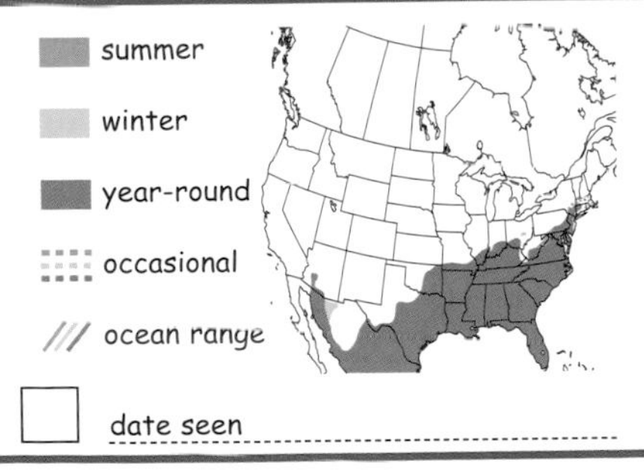

▶ *A Black Vulture tends its two chicks at a nest inside a decaying tree stump.*

Find it: Common in open country and coastal lowlands, the Black Vulture feeds at dumps and even at open dumpsters in cities. Will roost with Turkey Vultures on large power line towers and in dead trees.

summer

winter

year-round

occasional

ocean range

☐ date seen ____________

BALD EAGLE

Haliaeetus leucocephalus **Length: 31"**

Look for: It's hard to mistake a full-adult Bald Eagle for any other bird, with its huge white (not bald) head, large yellow bill, and white tail. Bald Eagles soar with their wings held flat. They do not rock, or teeter, like Turkey Vultures do.

Listen for: For such a majestic bird, the Bald Eagle has a wimpy voice. It utters a variety of high-pitched, chattering whistles that sound more like an excited puppy than a national symbol.

Remember: It takes four years for a Bald Eagle to reach full-adult plumage. Until then, it wears a very splotchy pattern of brown and white. In flight, though, an eagle of any age will show the massive wings and huge head that help this species to stand out.

◀ *Bald Eagles don't dive like Ospreys, preferring instead to pick fish off the water's surface. This is an immature Bald Eagle.*

WOW!

Now that harmful pesticides such as DDT have been banned, Bald Eagle populations are rebounding. Eagles in the wild are living longer, healthier lives, some as long as 30 years.

Find it: Almost always found near water, Bald Eagles are often seen perched in trees near lakes, along rivers (especially near dams in winter), and at other places that offer easy access to large fish or concentrations of waterfowl.

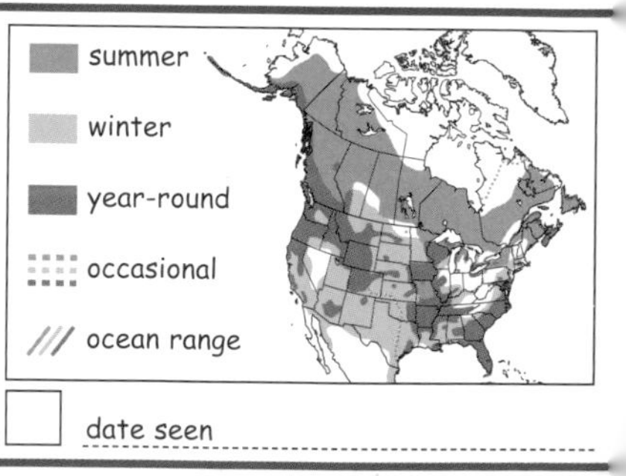

☐ date seen ________

GOLDEN EAGLE

Aquila chrysaetos **Length: 30–40"**

Look for: This magnificent bird soars over the vast open areas of the West on wide dark wings held flat. Adults are almost completely dark brown. A golden hue on the hind-neck feathers gives this bird its name. Juveniles have a bold white base to the tail and white "silver dollar" patches in the middle of their wings.

WOW!
Native Americans greatly admired the Golden Eagle for its powers of flight and its hunting prowess. Eagle feathers are sacred symbols to many native peoples.

Listen for: Rarely vocalizes, though near the nest it will give barking and whistling sounds.

Remember: Young Golden Eagles have well-defined white patches in the wings and tail. Young Bald Eagles look messier—their white is splotched throughout their plumage until about their third year.

▶ *Golden Eagles are often seen perched along highways waiting for an unwary rabbit. They will kill prey as large as Sandhill Cranes and small deer.*

Find it: More common in the West, where it is found from the tundra of the far North, throughout the Great Plains, and south to the deserts and mountains of the Southwest. Always prefers open habitat. In winter they range far and wide in search of food.

summer
winter
year-round
occasional
ocean range

date seen ________________

OSPREY

Pandion haliaetus **Length: 23"**

Look for: Known as the Fish Hawk or Fish Eagle for its preferred food, the Osprey flies with its wings held flat but bent backward at the "wrists." The dark back, white head, and a black raccoon mask make it resemble a Bald Eagle, but the Osprey always has a white breast and belly. The female often shows a dark "bra" band across the chest.

Listen for: Ospreys are very vocal, giving a loud, clear, high-pitched whistle—*tyou-tyou-tyou-tyou*—that sounds like they are giggling.

Remember: A soaring Osprey appears crooked-winged, more like a gull than a raptor from a distance. Eagles and most other raptors rarely show bent-back wings.

WOW!
Bald Eagles often let the Osprey do the hard work of catching a fish; then they'll chase after the Osprey to steal its food—not very nice behavior from our national symbol.

◄ *Young Ospreys remain on the tropical wintering grounds through their second summer, honing their fishing skills.*

Find it: Common near large bodies of water, the Osprey soars over the water looking for fish. When it sees one, the Osprey dives talons-first into the water, then carries the fish to a perch where it can carve up its victim.

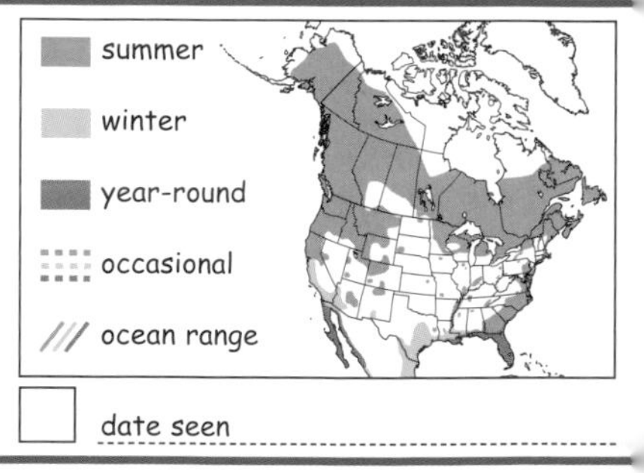

date seen

MISSISSIPPI KITE

Ictinia mississippiensis **Length: 14" Wingspan: 31"**

Look for: The Mississippi Kite is the smallest of our four kite species and also the most widespread. In flight, the adult shows an all-black tail, and the male shows white secondaries on the upperwing. Juvenile birds are streaky dark brown overall.

Listen for: A very high-pitched, whistled *fee-feww!* that descends in tone. It sounds similar to a Broad-winged Hawk's call. A second call is a more percussive (and very shorebirdlike) *fee-titititi!*

Remember: A flying Mississippi Kite can look a lot like a Peregrine Falcon. The kite's wings and body are slimmer than the falcon's. Its flight is light and buoyant compared to the powerful, direct flight of a Peregrine.

WOW!
Mississippi Kites are such good fliers that they can catch large flying insects and eat them in flight! To accomplish this, a kite bends its head down and reaches its feet upward.

◀ *Mississippi Kites catch dragonflies on the wing and devour them in flight.*

Find it: Most common in the Southeast from spring through fall, where it soars high over woodland edges and swamps. Each spring and fall, a few Mississippi Kites are found far outside their normal range. They winter in South America.

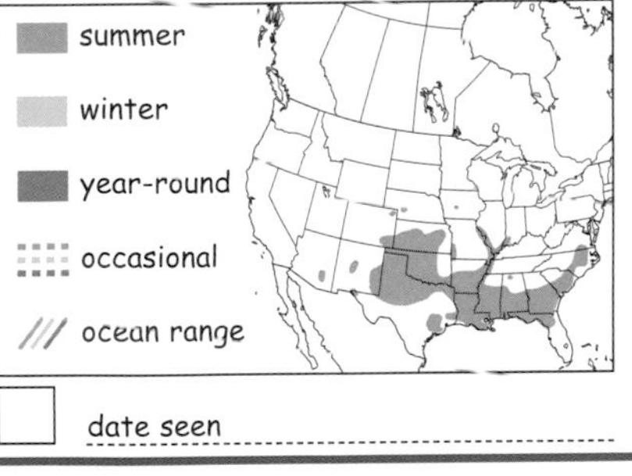

☐ date seen ______________

SHARP-SHINNED HAWK

Accipiter striatus **Length: 13"**

Look for: The Sharp-shinned Hawk is the smaller of our two common accipiters—the bird-chasing hawks. Adults have blue-gray backs and orange-spotted breasts. Young birds are brown overall with large brown blotches on white breasts (young Cooper's Hawks have finer breast streaks). In flight, the Sharpie's head does not extend much past the front of the wings, but the Coop's does.

Listen for: Not a very vocal raptor. The Sharpie gives a high, rapid-fire *tyoo-tyoo-tyoo-tyoo* call, often near the nest.

Remember: In general, Sharpies look smaller headed and thinner legged (when perched) than Cooper's Hawks. They also flap faster when flying.

WOW!
Sharp-shinned and Cooper's Hawks are built with compact wings for short, speedy bursts of flight and for maneuvering through thick woods in pursuit of their main prey: other birds.

▶ *A Sharp-shinned Hawk being mobbed by Blue Jays.*

Find it: You are likely to see a Sharpie rocketing through a woodland clearing scattering panicked songbirds, perched along a wooded edge watching for prey, or soaring in tight circles overhead.

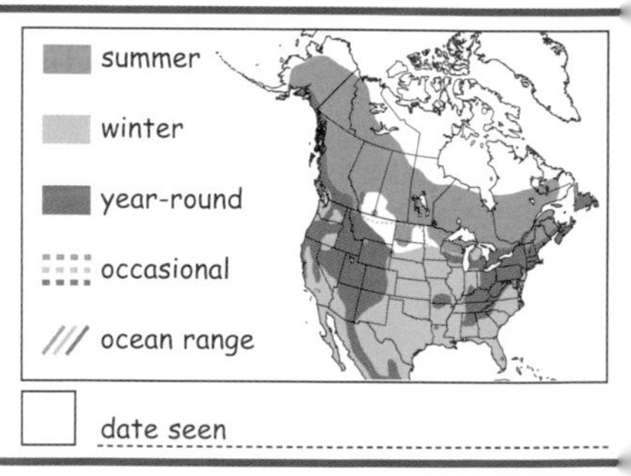

☐ date seen

COOPER'S HAWK

Accipiter cooperii **Length: 19" (female) to 14" (male)**

Look for: The Cooper's Hawk (or Coop) is chunkier than the Sharp-shinned. Bigger headed with thicker legs and a longer tail that often shows a rounded tip (Sharpie tails can appear squared off), the Coop frequently chases and kills larger prey, including Blue Jays, Mourning Doves, and Northern Flickers! Adults have blue-gray backs and reddish orange chests. Juveniles are streaky brown in front with brown backs.

WOW!
To keep the element of surprise on their side, Coops and Sharpies will spiral high to dive on unsuspecting birds. The birds below cannot see the predator coming out of the sun.

Listen for: Not very vocal except when disturbed near the nest; then it utters a flickerlike *ka-ka-ka-ka*.

Remember: In general, Coops fly with slower, stronger flaps than Sharpies, and their heads stick out farther in front of their wings.

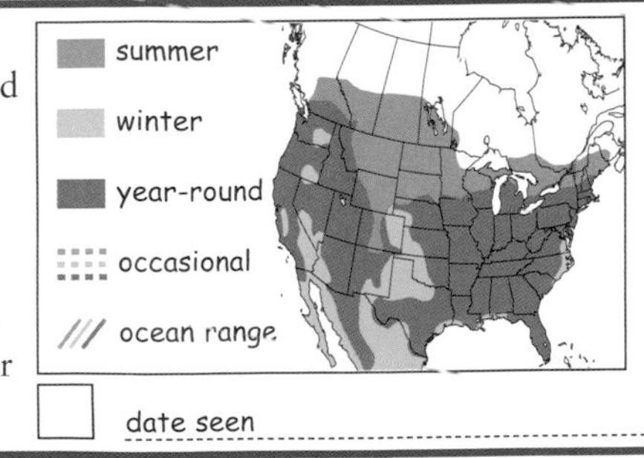

▶ *A Cooper's Hawk hunts at a backyard feeding station, spooking a Tufted Titmouse and a male Purple Finch.*

Find it: Increasingly common in suburban and urban settings, Coops may be more likely to perch out in the open than Sharpies. Look for very thick-looking legs. When you see songbirds scatter suddenly, look for an accipiter.

summer
winter
year-round
occasional
ocean range

date seen

NORTHERN HARRIER

Circus cyaneus **Length: 18"**

Look for: The Northern Harrier is a long-winged, long-tailed, and lanky hawk most often seen gliding low over marshy grasslands, hunting for small mammals and birds. The best field mark is its white rump patch, which harriers have in all plumages. Adult females and juvenile birds are warm brown overall (young birds are almost orange-brown).

WOW!
The feathers that outline its face help to channel sounds to the harrier's ears, much like the feathered faces of several owl species. The harrier is built to hunt by both sight and sound.

Listen for: A chattering, squeaky *chew-chew-chew-chew-chew*, given in a long series. Also gives a thin, high-pitched whistle that slurs downward in pitch, *tssieww!*

Remember: Harriers can change their flight shape. Always look for the long-winged profile and for that distinctive white rump.

▶ *The Northern Harrier's round facial disk of feathers give it an owl-like appearance.*

Find it: Though formerly called Marsh Hawk, Northern Harriers can be found in a variety of open habitats—including along beaches, airport runways, and dry prairies. They often perch on the ground or on low fenceposts.

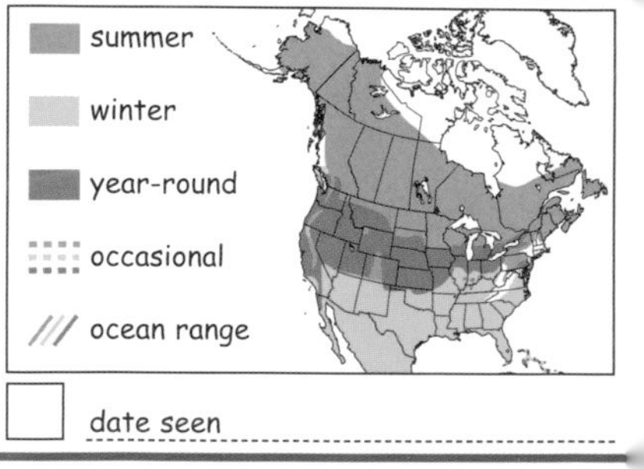

☐ date seen ____________

HARRIS'S HAWK

Parabuteo unicinctus **Length: 20–21"**

Look for: A stunning bird of the desert Southwest, the Harris's Hawk is mostly dark tones of rufous (shoulders and legs) and chocolate (body and head). The dark tail has a broad white base and white tip. Bill has a noticeable yellow base, or cere. Legs are long and yellow.

Listen for: A harsh, loud *raaaack!*

Remember: The all-dark body, rufous legs, and bold white tail base are field marks unique to this species.

WOW!
Two male Harris's Hawks may mate with one female, and all three will work to raise the young.

▶ *Family groups of Harris's Hawks hunt cooperatively. Some birds flush the prey while others pursue and kill it.*

Find it: Uncommon in mesquite woods and thorny desert habitat. Also found along wooded rivers. Often seen perched in small groups of three or more birds.

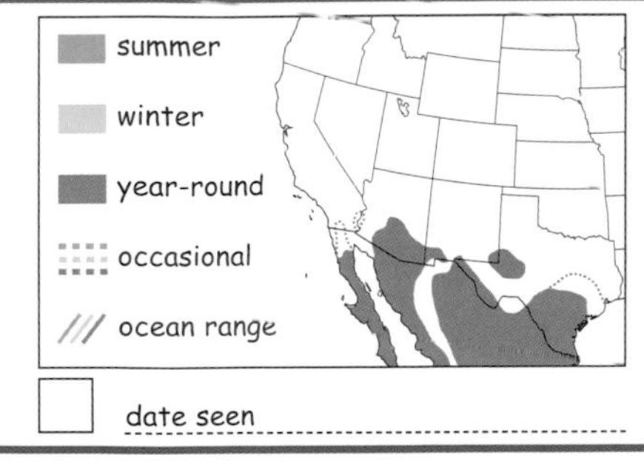

date seen

RED-SHOULDERED HAWK

Buteo lineatus **Length: 17"**

Look for: Smaller than a Red-tailed but larger than a Broad-winged, the Red-shouldered Hawk has a chunky body and long wings. Best field mark for this bird in flight is the pale crescent near each wingtip, called *wing windows* by birders.

Listen for: One of our most vocal raptors, the Red-shouldered Hawk gives a high scream: *keeyah!* This call is repeated often, especially by flying birds during spring and summer. Blue Jays may imitate this call, even using it to scare other birds away from bird feeders.

Remember: Red-shoulders flap their wings quickly two or three times in between glides, much like a Cooper's Hawk does.

◀ *Red-shouldered Hawks perch at the edge of the woods, watching for the movement of prey on the ground.*

Find it: Common in southeastern woodlands, less common in the North, this hawk is often heard before it is seen. Prefers wooded habitat near water in spring and summer, and woodland edges in winter.

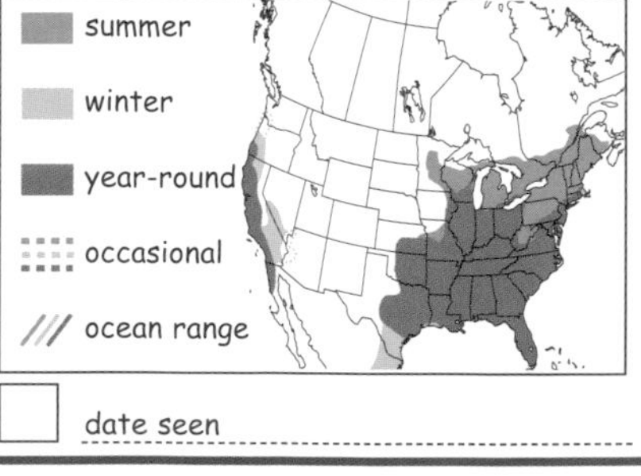

BROAD-WINGED HAWK

Buteo platypterus Length: 15"

Adult, in flight

Look for: The Broad-winged Hawk is our smallest buteo (it's smaller than an American Crow) and has a very compact shape in flight. From below, it appears very pale, with a black outline on wings and contrasting black-and-white bands on the tail. Adults have rusty chests and "wingpits." Young birds often lack the distinctive chest and tail markings of adults.

Listen for: In spring, listen for the Broad-winged's high-pitched, two-note whistle: *tee-teeeee!*

Remember: Soaring hawks can be confusing, but if your bird has broad black-and-white tail bands and a black outline on white wings, you've got a Broad-winged Hawk.

WOW!
More than 400,000 Broad-winged Hawks were counted passing by a fall hawk-watching station in a single day in Veracruz, Mexico!

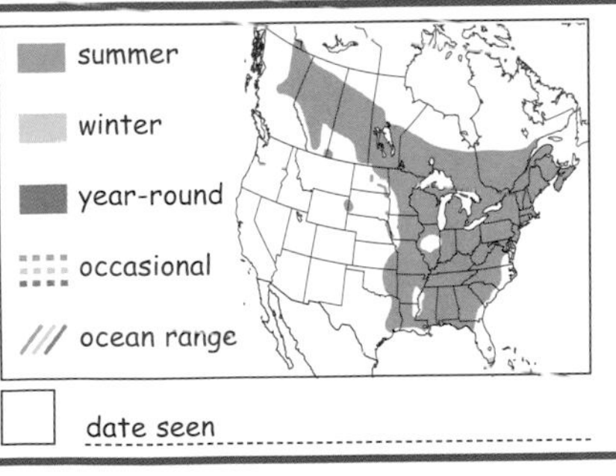
► *Riding rising areas of hot air, known as thermals, a kettle of Broad winged Hawks gains altitude in fall migration.*

Find it: In fall, Broad-winged Hawks gather into large flocks as they head to South America for the winter. Scan the sky on warm, sunny days between mid-August and mid-October for their soaring flocks, called *kettles*.

summer
winter
year-round
occasional
ocean range

☐ date seen ______

RED-TAILED HAWK

Buteo jamaicensis **Length: 19"**

Look for: A large dark-backed raptor. The adult's rust-colored tail (which earned this species its name) is most obvious on soaring birds. On perched birds the dark belly band and light-colored chest often stand out, and they often show a white V across the upper back. Red-tails may soar in large smooth circles or they may hover in place, scanning for prey.

Listen for: A high-pitched scream that sounds like steam escaping from a pipe—*fscheeeew!*—is most often uttered by flying birds.

Remember: Redtails are the most widespread of our large soaring hawks. Their coloration can vary from very light to very dark, so overall size and shape is a good starting point to ID them.

WOW!
The Red-tailed Hawk's wild-sounding call is often used in place of the Bald Eagle's call on TV and in movies.

► *Most perched Red-tailed Hawks show the white V of backpack straps across their backs.*

Find it: Look for Red-tails perched or soaring where they can command a clear view of open habitat where small mammals live. In late winter, look for pairs perched close together. The larger one is the female.

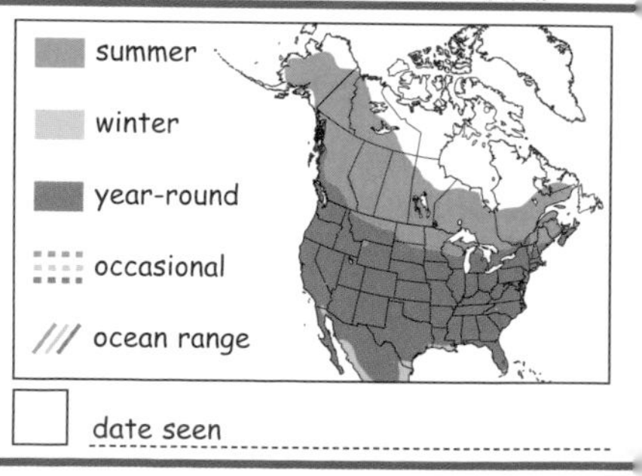

☐ date seen

SWAINSON'S HAWK

Buteo swainsoni **Length: 19"**

Adult

Look for: Compared to other buteos, such as the Red-tailed Hawk, the Swainson's Hawk has a thinner body and longer, more tapered wings in flight. The white inner linings on the wings (in flight) are a key field mark, as is the dark brown bib above a white breast. Soaring hawks hold their wings above horizontal, similar to a Turkey Vulture, but do not rock like a TV.

Listen for: The Swainson's raspy, descending cry—*cree-yaah*—sounds like a cross between a Red-tailed Hawk's scream and a cat's *meow.*

Remember: Swainson's Hawks come in both light and dark color variations (morphs), so base identification on overall size and shape in flight.

WOW!
Swainson's Hawks sometimes migrate in huge flocks. One wintering flock in Argentina contained more than 12,000 birds!

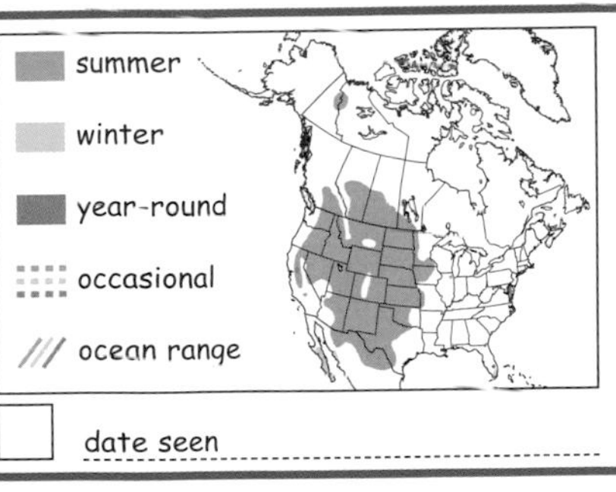

▶ *Swainson's Hawks are primarily insect eaters. Grasshoppers are a favorite food.*

Find it: Swainson's Hawks spend the winter eating insects in South America, and in spring they head back to the Great Plains, where they nest in isolated groves of large trees.

summer
winter
year-round
occasional
ocean range

☐ date seen

FERRUGINOUS HAWK

Buteo regalis **Length: 23–24"**

Look for: A large pale soaring hawk with long narrow-tipped wings. Dark brown leg feathers form a noticeable dark V on birds viewed overhead. Flight feathers (primaries and secondaries) always appear clean and white. Perched birds appear rufous backed and pale breasted.

Listen for: Call is a soft, whistled *keeewww.*

Remember: The wings of the Ferruginous Hawk are longer and more tapered than those of the Red-tailed and Swainson's Hawks. This gives this species a "wingier" look than other buteos have.

WOW!
When bison roamed the plains, some Ferruginous Hawks built their nests from bison bones and lined them with bison "patties."

◀ *Ferruginous Hawks may wait on the ground for an unwary mammal or snake to come out of its burrow.*

Find it: Uncommon across the dry, open plains and prairies of the West, where it hunts for small mammals, birds, and snakes.

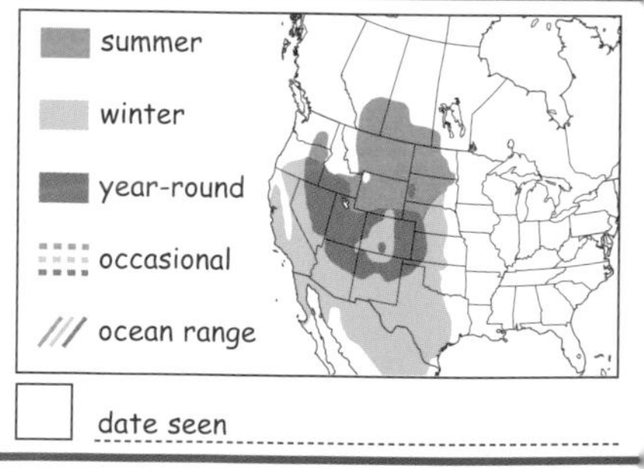

☐ date seen ______________

PEREGRINE FALCON

Falco peregrinus **Length: 17" Wingspan: 41"**

Look for: Large and powerful, the Peregrine looks like a kestrel on steroids. It flies with smooth but shallow wing strokes, easily generating great speed. Overall it appears dark. Young birds are brown overall, and adults have gray backs and white breasts barred with black. All Peregrines have black mustaches, though they can be hard to see on distant birds.

Listen for: Normally silent but near nest will utter a loud, raspy *klee-klee-klee-klee!*

Remember: When a Peregrine flies into sight, waterfowl and shorebirds usually take off in a panic. This is a good clue to the presence of this powerful hunter, so scan the skies for a Peregrine whenever you see fleeing birds.

WOW!

Peregrine Falcons may reach speeds of 200 miles per hour when diving for prey. They use their balled-up talons to knock out their prey, then catch the hapless, falling bird before it hits the ground or water.

▲ *A Peregrine Falcon pair courting high above a city, where they will nest on a building ledge.*

Find it: Found worldwide but not common anywhere, Peregrines like to perch up high (on towers, cliffs, buildings, bridges), where they can watch for prey.

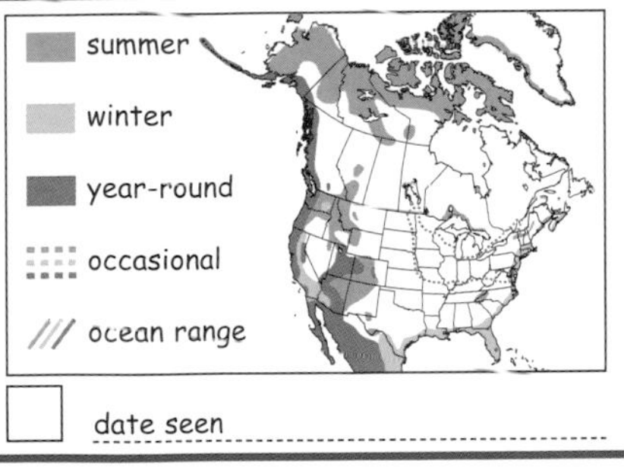

☐ date seen ______________

MERLIN

Falco columbarius **Length: 11–12"**

Look for: This small dark falcon is slightly larger than the more common American Kestrel. A perched Merlin looks darker and stockier than a Kestrel. There are several races of Merlin across North America, but all have a uniformly dark back and a banded tail.

Listen for: Call is a high-pitched *kee-kee-kee-kee.*

Remember: Merlins usually fly purposefully—as if they're late for something. They rarely hover in one place, like Kestrels do.

WOW!
Perch-and-wait hunters, Merlins watch for a passing bird and then pursue and capture their prey in flight.

▶ *Merlins are increasingly nesting in urban settings across the Midwest, living on a diet of House Sparrows.*

Find it: Prefers open habitat throughout its range. Open conifer forests, prairie grassland with scattered trees, coastal marshes. Can be seen anywhere in migration.

☐ date seen ______

AMERICAN KESTREL

Falco sparverius **Length: 10" Wingspan: 22"**

Look for: Our smallest falcon, the American Kestrel is a familiar sight across North America, hovering over meadows or sitting on a power line, flicking its tail backward. Adult females are mostly streaky brown overall, with black stripes on rusty backs. Adult males are more colorful, with bluish wings and headbands and rusty chests, tails, and backs. Both sexes have black mustache stripes.

Listen for: The American Kestrel's primary call is a high-pitched, excited-sounding *killy-killy-killy-killy*.

Remember: In flight, falcons usually show pointed wings. Kestrels appear lighter and "bouncier" than other, larger falcons.

WOW!
The spots on the back and sides of a kestrel's head look like a face. These markings are meant to fool predators into thinking the kestrel is looking at them and is prepared for an attack.

▶ *False eyespots on the kestrel's nape make it appear to be looking at you with eyes in the back of its head.*

Find it: Kestrels love open grassy areas where they can hunt for small mammals and large insects. Watch for them perched along fence lines and power lines and in the tops of trees. A small falcon hovering in place is almost surely a kestrel.

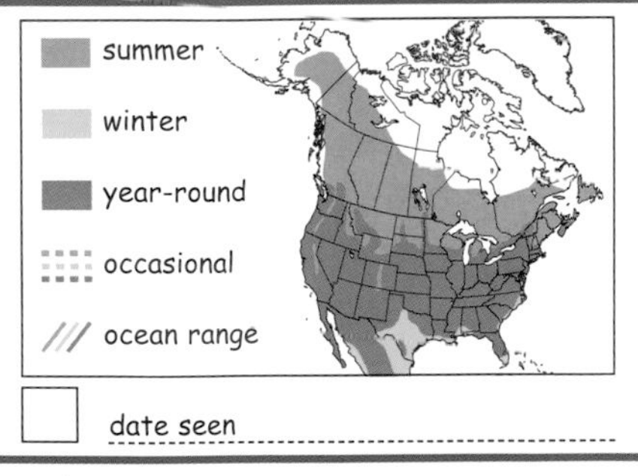

date seen

AMERICAN COOT

Fulica americana **Length: 15"**

Look for: The American Coot is a sooty gray football-shaped chunk of a bird with a stubby white bill and a white-and-rust forehead patch. Although the coot looks and acts like a duck, it's actually a member of the rail family.

Listen for: Coots are noisy birds, uttering a variety of grunts, cackles, chatters, and croaks.

Remember: Coots bob their heads forward and back when swimming. In order to fly, coots have to scamper across the surface of the water to gain enough speed to become airborne.

WOW!

Hatchling coots can swim very well within a few hours of hatching. Coots are aggressive birds that often steal food items from nearby ducks. Bullies!

◀ *American Coot chicks look like bald men in turtleneck sweaters. Their bright heads stimulate the parents to feed them.*

Find it: Widespread on lakes, ponds, and marshes all across North America, coots are as at home walking on land as they are swimming on water. Suburban ponds and golf courses seem to be especially attractive to coots.

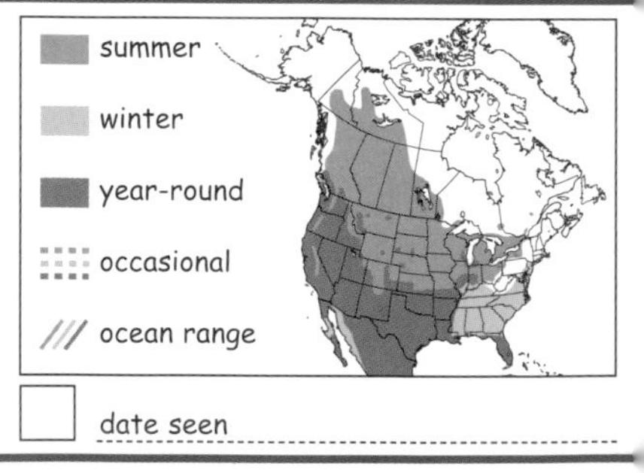

COMMON MOORHEN

Gallinula chloropus Length: 14"

Look for: A charcoal gray bird with a rusty back, white body striping, and a red bill, the Common Moorhen is equally at home swimming in water and walking on land. Both sexes have the bright red bill and forehead shield, though the male's is brighter.

Listen for: Whinnying, squeaky notes in a series, slowing down near the end. Sounds like someone is torturing a frog. Also, a high, sharp *peek!*

Remember: Distant moorhens can be told apart from coots by the white horizontal side stripes and extensive white on the tails.

WOW!
Q: What looks like a duck but is not a duck?
A: A Common Moorhen, which is more closely related to cranes and rails than to ducks (even though they swim like ducks).

◀ *How a Common Moorhen swims so well without webbed feet is a mystery. Perhaps its bobbing head helps it motor along.*

Find it: You may hear a moorhen before you see it, but look for it in or near the dense vegetation along the edge of freshwater marshes and ponds doing a "funky chicken" motion as it swims or walks.

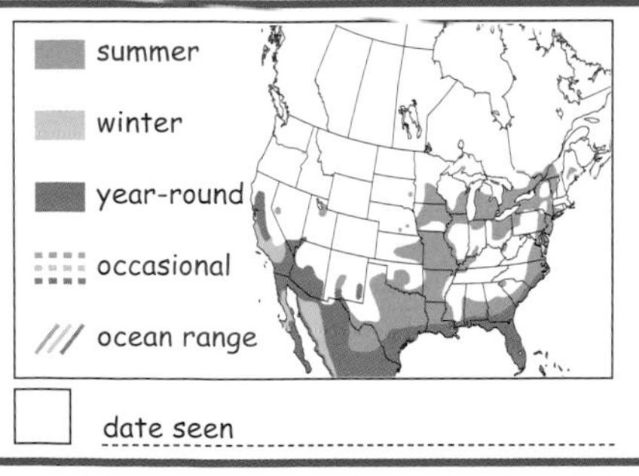

☐ date seen ______

PURPLE GALLINULE

Porphyrio martinica **Length: 14"**

Look for: While not very widely distributed, the Purple Gallinule is likely to make you say, "Wow!" Its football-sized body is decked out in hues of purple, blue, and green, and its legs and bill tip are bright yellow. A red stripe covers the base of the bill, and a light blue shield covers the forehead.

Listen for: A series of high-pitched peeps and nasal clucks that sound like a maniacal chicken.

Remember: The Purple Gallinule is like a Technicolor version of the Common Moorhen. From a distance, the gallinule looks all dark while the moorhen shows white in the wings. Seeing the purple-green body should be fairly easy except in poor light.

▼ The Purple Gallinule's long yellow toes allow it to step lightly over the lily pads without sinking or swimming.

WOW!
Another name for the Purple Gallinule is Swamphen.

Find it: The Purple Gallinule's long narrow toes, when spread out, distribute the bird's weight, enabling it to walk on lily pads and other floating vegetation in wetlands and freshwater marshes.

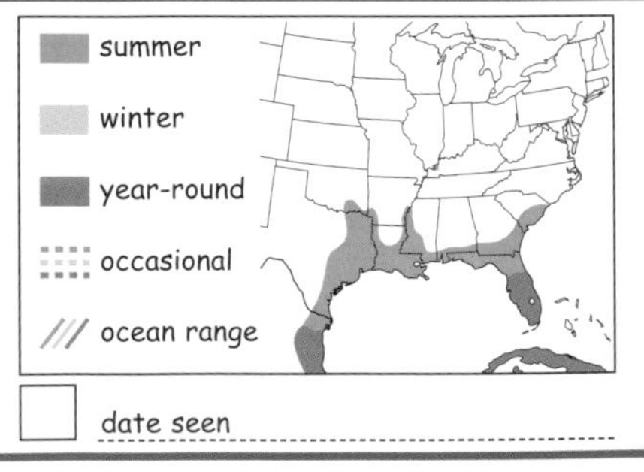

☐ date seen

SEMIPALMATED PLOVER

Charadrius semipalmatus **Length: 7¼"**

Adult, breeding

Look for: Among our small plovers, the Semipalmated Plover is the most frequently encountered, especially in spring and fall migration. It looks like a miniature Killdeer but has only one dark breast-band (the Killdeer has two). The Semi's dark back helps it to blend into the wet mud flats it prefers.

Listen for: The Semipalmated Plover gives three different calls. In flight, it gives a rising, throaty *too-wee*. Threatened birds give a rapid *doi-doi-doi-doi-doi* and a dry, burbly *dwiip*.

Remember: The Killdeer is a full three inches larger than a Semipalmated Plover. Other small plovers are less common inland.

▼ *Semipalmated Plovers are commonly seen on mud flats pulling marine worms, like Robins on a lawn.*

Find it: Though they nest far to the north, Semipalmated Plovers are commonly seen in muddy wetlands, on mud flats, and on beaches in migration, where they scamper along plucking food items from the mud's surface.

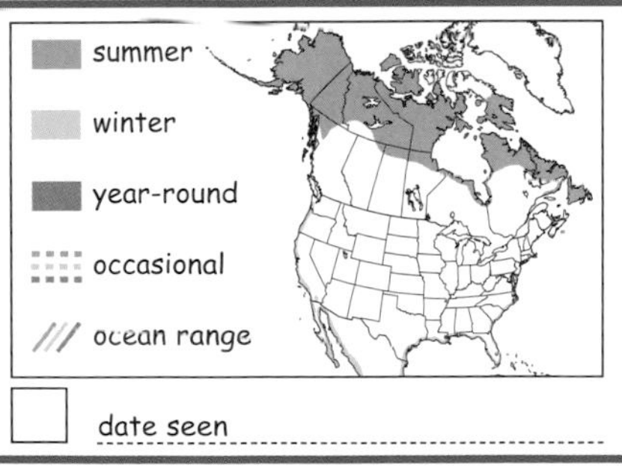

☐ date seen ____________

PIPING PLOVER

Charadrius melodus **Length: 7¼"**

Look for: A small, pale, and stocky plover, the Piping Plover has yellow legs and a plain face that help distinguish it from its more common relatives. In spring and summer, the adult has an orange bill with a black tip and a single narrow breast-band that may or may not connect in front.

Adult, breeding

Listen for: This bird could be called the Peeping Plover for the variety of mellow, whistling *peeps* it utters.

Remember: The closely related Semipalmated Plover is much more widespread and common than the Piping Plover. Semipals have dark brown backs and boldly marked heads and faces. Piping Plovers are pale.

WOW!
Recovery efforts are helping to protect Piping Plover nesting areas on heavily used beaches alon the Atlantic Coast. If you visit there, be sure to obe the signs and avoid disturbing these birds.

◀ *Piping Plover chicks are able to run and pick up food on their first day out of the egg. The parents simply supervise them.*

Find it: Piping Plovers nest on dry sandy beaches along the Atlantic Coast and on dry flats near lakes and rivers in the upper Great Plains. Look for them on dry pale ground that matches their back color.

summer
winter
year-round
occasional
/// ocean range

☐ date seen

KILLDEER

Charadrius vociferus **Length: 10½"**

Look for: A large, noisy shorebird that is often found far from the shore, the Killdeer has two black breast-bands on a white chest. In flight, the Killdeer looks long and slender and shows a bright orange rump and black wingtips and tail tip.

Listen for: The Killdeer is said to call its name as it flies, but really it sounds more like *tee-deee, tee-deee*. High-pitched and loud, the Killdeer's call is hard not to notice (that's why its Latin name means "vociferous").

Remember: No other commonly encountered shorebird has the Killdeer's double black breast-bands.

WOW!

Killdeer perform a distraction display, to lure predators away from their nest, flopping around on the ground and faking a broken wing.

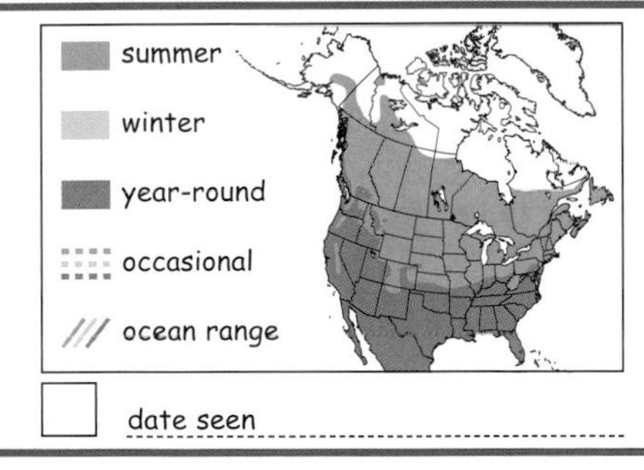

▶ *A Killdeer distracts attention from its well-camouflaged eggs. It flashes its rump, flutters, and calls.*

Find it: The widespread and common Killdeer can be found on open grassy meadows, on ball fields, along gravel roads, at airports, and on mud flats or beaches. Look for them running, foraging, and calling to one another.

summer
winter
year-round
occasional
/// ocean range

☐ date seen ____________

AMERICAN OYSTERCATCHER

Haematopus palliates **Length: 18"**

Look for: The American Oystercatcher is a large, stocky, boldly marked shorebird with a big, straight orange-red bill jutting out from a black head. In flight, the oystercatcher looks white below and black above and seems to flash black and white as it flies. Adult males and females are identical.

Listen for: Long call is a series of loud, high-pitched *wheeeeps* followed by a rapid *dididididididi* dropping in tone and slowing. Short call is a loud *peep* or *weep!*

Remember: You might confuse the oystercatcher in flight with a Willet—both flash a lot of white in the wings. But the oystercatcher's large orange-red bill should stand out, even from a distance.

WOW!
The oystercatcher uses its powerful bill to open the shells of oysters, clams, and mussels. If you've ever tried to do this, you know how hard it is. Imagine doing it without hands!

◀ *An American Oystercatcher pries a limpet away from a rock as its chick looks on.*

Find it: Always found near salt water, the American Oystercatcher is a resident along the Atlantic and Gulf coasts, where it frequents beaches, mud flats, marshes, sandbars, and islands.

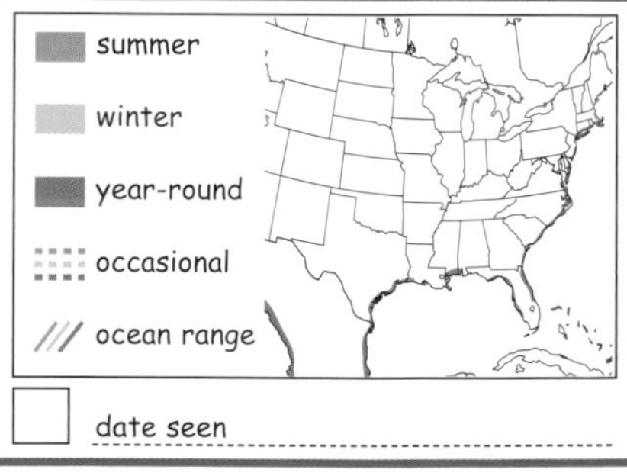

☐ date seen

BLACK-NECKED STILT

Himantopus mexicanus **Length: 14"**

Look for: Named for its long slender legs, the Black-necked Stilt is a delicate-looking bird with a needlelike bill and slender overall appearance. Black above, white below, with bright pinkish red legs.

Listen for: Loud, frantic-sounding *kleep-kleep-kleep!* Often given when disturbed.

Remember: The stilt has a straight bill. The winter-plumaged American Avocet looks similar but is larger and has an upturned bill.

WOW!
Though they look delicate, Black-necked Stilts are tough birds. They spend most of their lives, including nesting, on hot, sun-baked ground, such as salt flats.

◀ *Stilts use their thin, needlelike bills to pick up their prey.*

Find it: Common in shallow ponds, marshes, and mud flats, where it walks delicately, picking up small insects and crustaceans. Often in mixed flocks with American Avocets. Most stilts winter in coastal habitat.

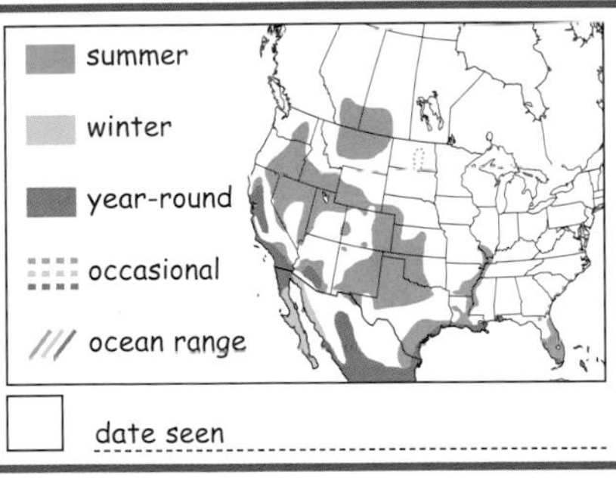

☐ date seen ____________

AMERICAN AVOCET

Recurvirostra americana **Length: 18"**

Look for: The graceful curves of the American Avocet's bill and neck serve it well as it feeds while walking along, sweeping the bill back and forth in shallow water. In spring and summer, adults have rusty heads and necks. In winter, the heads are gray. Black wings are divided by white.

Listen for: A clear, high-pitched *pwee-eep!*

Remember: In winter plumage, the American Avocet's head is gray, making it look more like the Black-necked Stilt than it does in summer, when its head is rusty. Check the bills and legs. Avocets have long blue-gray legs and upturned bills. Stilts have pink legs and straight bills.

WOW!

The bill of the female avocet is more sharply upturned than that of the male. Why? No one knows.

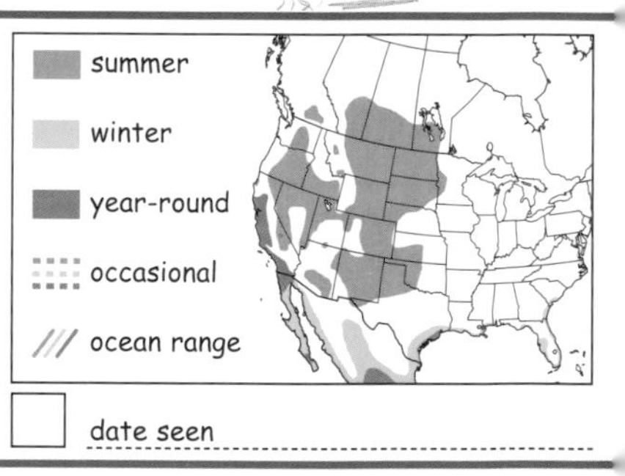

▶ *An American Avocet calls an alarm when danger threatens the nest.*

Find it: Avocets are found in shallow marshes, lakes, and wetlands. They nest in loose colonies. They are often found in flocks, sometimes associated with Black-necked Stilts.

summer
winter
year-round
occasional
/// ocean range

☐ date seen ----------------

WILLET

Catoptrophorus semipalmatus **Length: 15"**

Adult, nonbreeding

Look for: In all plumages, the Willet is nothing special to look at until it flies and flashes its boldly patterned black-and-white wings. A large, chunky shorebird with a straight, stout bill, the Willet changes its mottled brown breeding plumage for conservative gray in winter.

Listen for: A loud ringing call in which the Willet says its name: *pill-will-willet!*

Remember: Compared to other common shorebirds, Willets appear very plain overall but with a solid build—like a shorebird that's been working out with weights.

WOW!
Three weeks after young Willets hatch, the mother abandons them to the care of the dad, who takes care of the young for several weeks until they are able to fend for themselves.

▶ *Willets look like big, generic gray shorebirds until they open their stunning black-and-white wings.*

Find it: Very common along the Atlantic and Gulf coasts, where it prefers saltwater marshes, swampy meadows, and beaches. Western birds nest on inland marshes and grasslands near water.

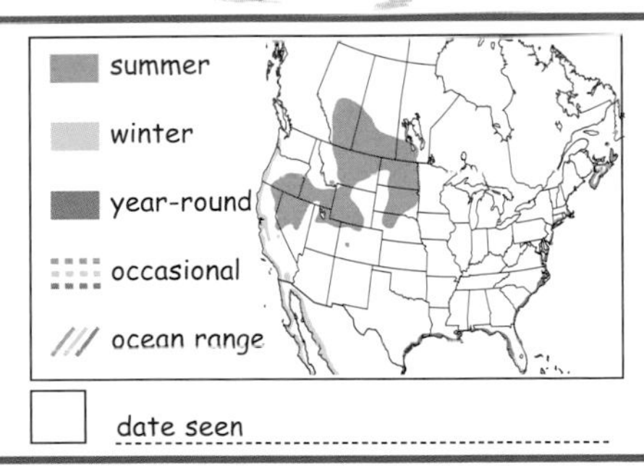

☐ date seen ____________________

GREATER and LESSER YELLOWLEGS

Tringa melanoleuca, Tringa flavipes **Length: 14", 10½"**

Look for: Greaters are larger than Lessers, but size can be hard to judge unless both species are side by side. Greaters also have a longer, thicker bill, especially at the base, that is often two-tone. Lessers appear delicate in every way, including the all-dark needle-thin bill. Both have long, bright yellow legs.

Listen for: Voice is the best way to tell these birds apart. Greaters tend to call in three or four loud, clear notes: *teww-teww-teww-teww!* Lessers call *less,* that is, their calls are shorter, with one or two mellow notes: *ti-teww, ti-teww.*

Remember: Greaters have longer, thicker bills and louder, longer calls, and you have a *greater* chance of seeing a Greater Yellowlegs in cold weather.

▶ *The yellowlegs' names refer to the relative size difference between the two species.*

Find it: During migration both species can be found on marshes (saltwater or freshwater) and mud flats and along streams and rivers. Lessers seem to prefer smaller bodies of water. Greaters remain farther north in winter.

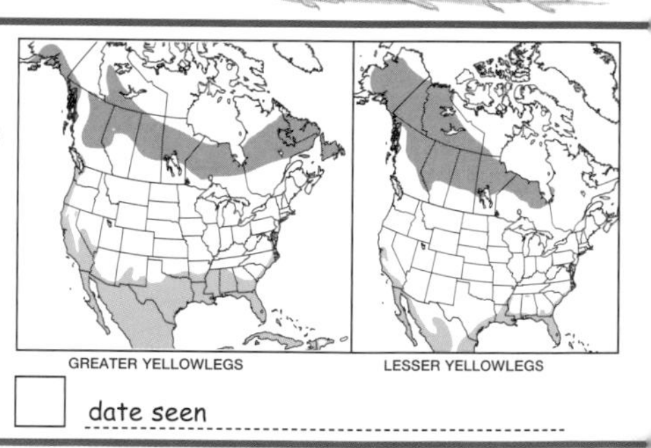

☐ date seen ______________________

SOLITARY SANDPIPER

Tringa solitaria **Length: 8½"**

Look for: A slender shorebird with dark wings and back dotted with tiny white spots, the Solitary Sandpiper is often found alone, as its name suggests. Its white eye-rings give it a spectacled look, as though it were wearing a pair of white glasses.

Listen for: High-pitched, piercing whistle, *peet-deet, peet-deet-deet,* often given in flight or when alarmed.

Remember: Solitaries are darker overall than Spotted Sandpipers and have longer necks and more distinct white eye-rings than Spotties.

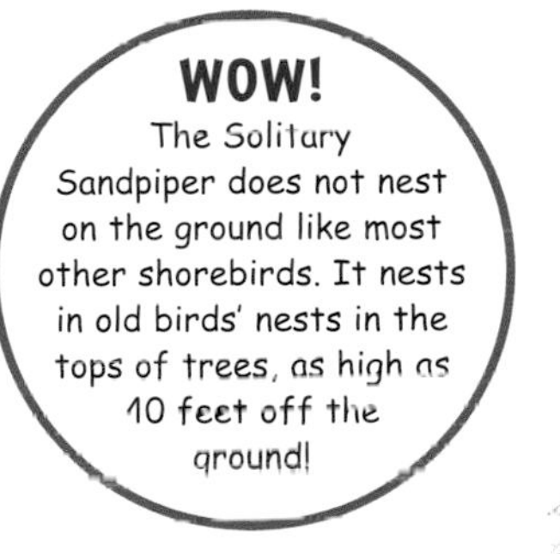

▶ *A Solitary Sandpiper tends its eggs in a twig nest originally built by a Gray Jay.*

Find it: Solitary Sandpipers prefer the muddy edges of ponds, creeks, and wetland marshes. They often bob their tails in a manner similar to that of the Spotted Sandpiper.

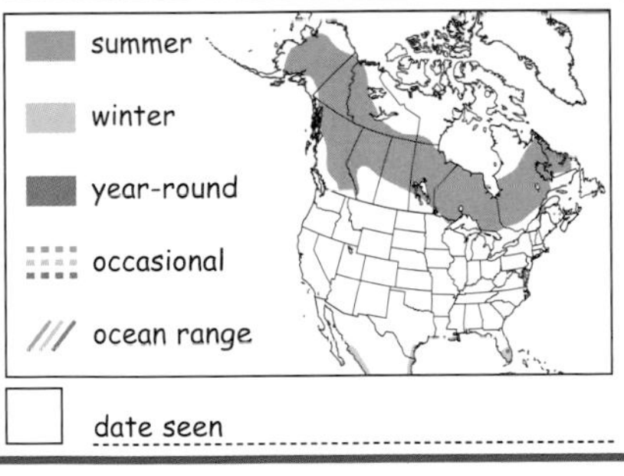

☐ date seen ____________

MARBLED GODWIT

Limosa fedoa **Length: 18"**

Look for: The largest of our three godwit species, the Marbled Godwit shows cinnamon underwings in flight. Its overall warm brown appearance in all plumages lacks obvious contrast. The long upcurved bill is pink at the base and black at the tip.

Listen for: Call is a two-note *god-WIT!* or a harsh single note: *kerr.* Flight display song sounds like *TheRabbitTheRabbit The Rabbit!*

Remember: Similar to Long-billed Curlew but with an upturned bill (curlew's bill is long and downward-curving).

▼ *Marbled Godwits use their long bills to probe for food hiding beneath water or mud.*

WOW!

Probing deep in the ground with their long bills, godwits locate their food by touch. The godwit's diet includes crabs, earthworms, leeches, grasshoppers, and plant tubers.

Find it: Breeds in wet meadows in the prairie potholes region of the upper Great Plains. Winters in flocks on beaches, tidal mud flats, and marshes in coastal regions.

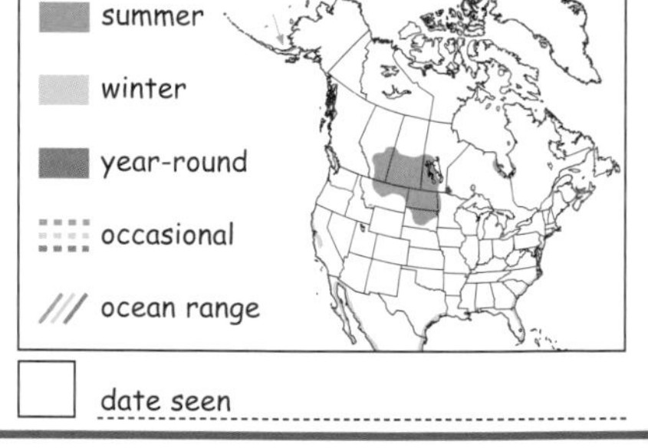

☐ date seen

WHIMBREL

Numenius phaeopus **Length: 17"**

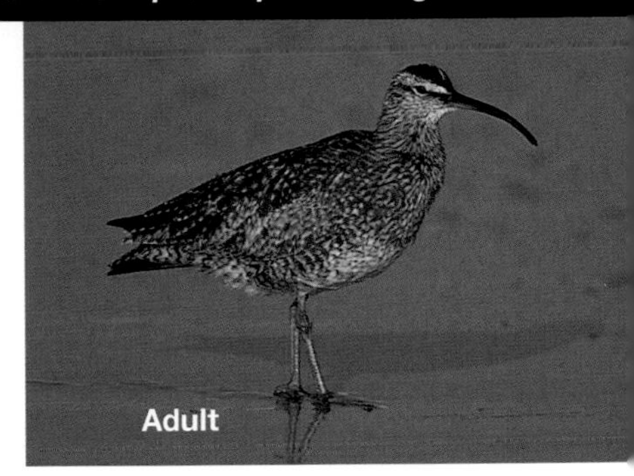

Look for: A large gray-brown bird on sturdy gray legs with a down-curved bill. Unlike other large shorebirds, Whimbrels do not show obvious field marks in flight, appearing rather plain overall. The dark stripes on the Whimbrel's head also set it apart from other large, long-billed shorebirds.

Listen for: A series of six or seven clear, whistled notes on the same pitch: *tu-tu-tu-tu-tu-tu-tu.*

Remember: Flying Whimbrels show no obvious field marks. They are plain Janes.

WOW!

Another name for the Whimbrel is Short-billed Curlew. This name must have been a comparison to the Long-billed Curlew of western North America, which has a really long bill.

◀ *The Whimbrel's long, curved bill is useful for handling prickly prey such as fiddler crabs.*

Find it: Most common on both coasts, where migrant flocks can be found in grassy marshes and on mud flats. Look for Whimbrels as they forage, picking up insects with their long decurved bills.

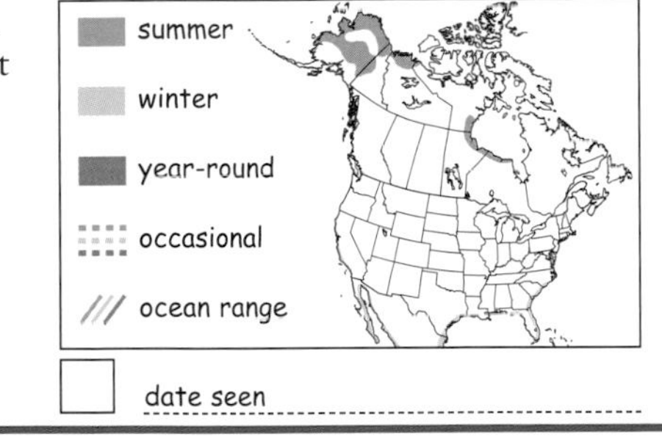

☐ date seen ____________

RUDDY TURNSTONE

Arenaria interpres **Length: 9½"**

Look for: From April through September, the Ruddy Turnstone has a zebra-striped face and breast and rusty (or ruddy) wings. The rest of the year, the bird wears a much-faded version of this plumage. Its short legs remain orange year-round.

Listen for: A very ternlike *klew!* alarm call, plus an unmusical chatter, *kkkkkkkrrrkkkkkrr*.

Remember: Even in its dull winter plumage, the Ruddy Turnstone retains some of the pattern on its head and breast.

WOW!

The turnstone is named for the way it finds food, using its bill to turn over stones. Its Latin name translates to "sandy place interpreter" for its habit of calling out to warn other birds of danger.

◀ *A Ruddy Turnstone flips over a stone.*

Find it: Sandy or rocky beaches, breakwaters, and jetties are the preferred habitat of the Ruddy Turnstone. Most commonly found along coasts in migration and winter, rarer inland.

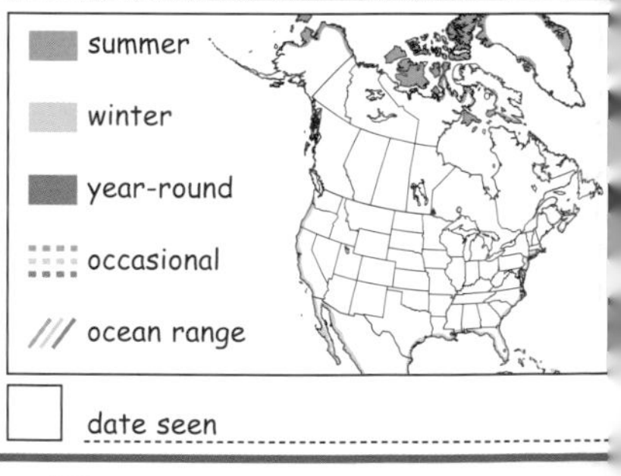

☐ date seen ______________

SANDERLING

Calidris alba **Length: 8"**

Look for: Those small, plump, pale shorebirds running back and forth with each wave on the beach are Sanderlings, the stereotypical birds of the seashore. We almost never get to see the Sanderling's rich rufous breeding plumage, worn only from May through August on its Arctic breeding grounds. The rest of the year the Sanderling is pale gray above and bright white below, with black legs and bill.

Listen for: A short, sharp *queet* or *queet-queet-queet* often uttered in flight.

Remember: Sanderlings are paler and more active and nimble than most other small sandpipers.

WOW!

Sanderlings have an uncanny ability to time their feeding dashes between waves. They pursue a wave as it retreats, probe for exposed sand crabs, then run away as the next wave comes in.

▲ *Sanderlings probe in the mud for food. They always seem to be in a great hurry; perhaps it's because their prey moves so fast.*

Find it: On sandy beaches along oceans and lakes, Sanderlings will be found dashing to and fro like wind-up toys. Often found in large flocks during migration and in winter.

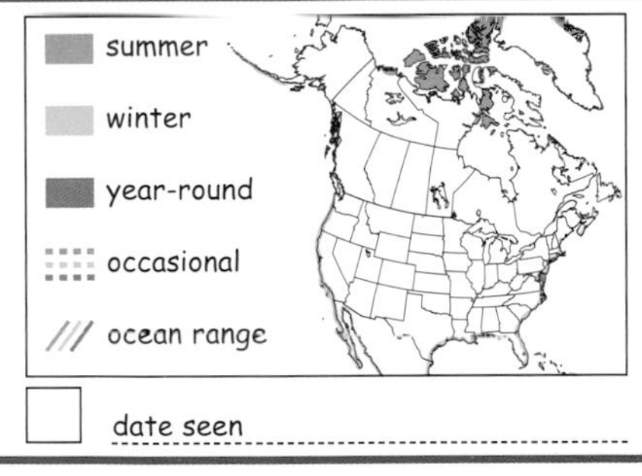

☐ date seen ____________

LEAST SANDPIPER

Calidris minutilla **Length: 6"**

Look for: A tiny (sparrow-sized!) shorebird, the Least Sandpiper can be hard to tell apart from the other small shorebirds known as peeps, which include the Western and Semipalmated Sandpipers. In good light, the Least's yellow legs and uniformly brown back are good field marks.

Listen for: A burry, rising *preep!* The Least's call is higher pitched than other peeps'. Peeps are named for their *peep* calls.

Remember: Mixed flocks of peeps can be hard to sort out by species. Size differences are subtle. With practice you can sort out the Leasts from the other peeps by their slightly drooping fine-tipped bills and yellow legs.

WOW! The Least Sandpiper is well named. It's the world's smallest sandpiper.

◀ *Least Sandpipers must watch out for the ever-hungry Laughing Gulls, which are large enough to swallow them whole.*

Find it: More common inland than other peeps, the Least Sandpiper can be found in small flocks on mud flats around lakes, ponds, and rivers. Leasts are methodical foragers that pick and probe for food in the mud.

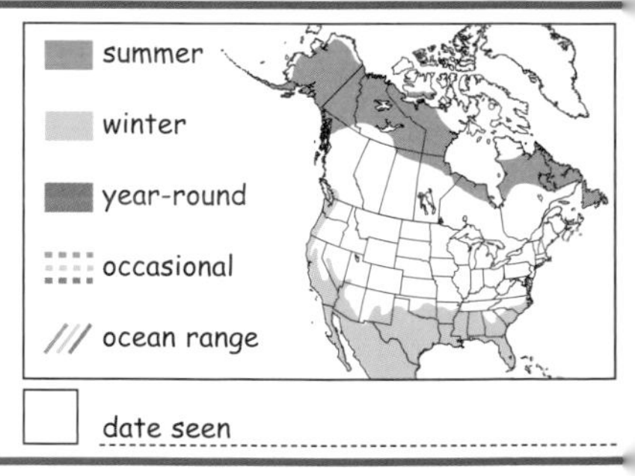

☐ date seen ____________

SEMIPALMATED SANDPIPER

Calidris pusilla Length: 6¼"

Look for: The Semipalmated Sandpiper is another of our common (and hard to identify) peeps. Semis have straight, round-tipped bills and black legs. Winter adults are drab gray overall, while breeding adults and fresh-plumaged first-year birds have a browner overall appearance.

WOW!
Semipalmated Sandpipers get their name from their partially webbed toes. This is not a field mark, however, and it's a feature most birders never see.

Listen for: Flight call is a short *brip, brip,* lower and less whistly than the Least Sandpiper's. Also gives a loud *whe-do-do-do-do-do* call in flocks, which sounds like a mini car alarm.

Remember: The Semipalmated Sandpiper's bill is thicker than the Least Sandpiper's and has a stubbier point. In general, Semis look grayer overall than Leasts (which appear browner overall).

► *Like other birds, Semipalmated Sandpipers can sleep with half the brain while the other half remains alert for danger.*

Find it: Found on mud flats and beaches, while traveling between its Arctic nesting grounds and South American wintering grounds, often in large flocks mixed with other peeps. Will forage in shallow water.

summer
winter
year-round
occasional
ocean range

☐ date seen

SPOTTED SANDPIPER

Actitis macularius **Length: 7½"**

Adult, breeding

Look for: This medium-sized shorebird retains its spots only during spring and summer. Fall and winter birds have dirty but unspotted breasts. The best field mark for the Spotted Sandpiper is its tail-bobbing walking style as it forages along the water's edge. Its white eye line is prominent and helps set the Spotted apart from the similar Solitary Sandpiper, which has a white eye-ring.

Listen for: Loud, ringing *peet-peet!* Often uttered in paired notes but in a longer series *(weet-weet-weet-weet)* when the bird is excited or startled.

Remember: When spooked into flight, Spotties fly with a flap-flap-flap-sail rhythm, much like a Chimney Swift.

WOW!
Another name for the Spotted Sandpiper is Teeter Peep, for the way it walks.

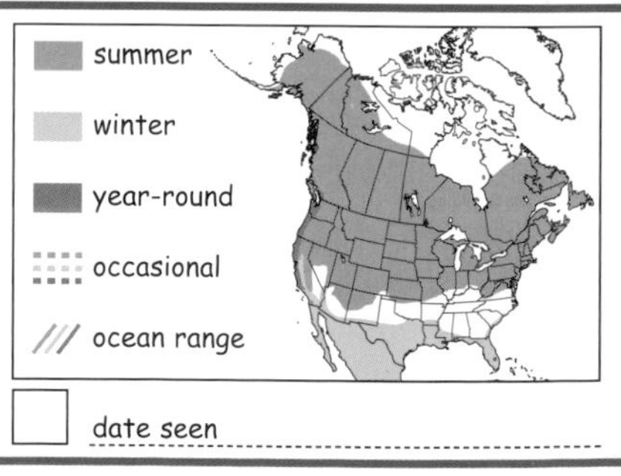

▶ *The Spotted Sandpiper's stiff-winged, stuttering flight is a good giveaway to its identity.*

Find it: Scan along the edge of almost any pond, small stream, or muddy riverbank, and you are likely to see a Spotted Sandpiper. Spotties usually forage alone, sometimes in pairs.

summer
winter
year-round
occasional
/// ocean range

☐ date seen ____________

AMERICAN WOODCOCK

Scolopax minor **Length: 11"**

Look for: A football-sized bird, the American Woodcock is a shorebird that lives its life in and near woods. On the ground, it looks large bodied, big headed, with large black eyes and a long bill. Its eyes are set far back on the head, allowing it to watch for danger from behind and above as it probes for earthworms.

Listen for: The male's elaborate courtship display begins with a series of nasal *peent!* calls. Then he takes flight, wings whistling *too-too-too-too* as he flies upward in a spiral. As he tumbles back to earth, he twitters and burbles.

Remember: Before you see a woodcock, you may hear the whistle of its wings as it flies past, or you may hear one *peent*ing at night.

▲ *An American Woodcock subdues a wriggling earthworm.*

Find it: Woodcocks favor wet woods for nesting and roosting, with nearby fields or meadows for nighttime foraging. The best time to find this species is in spring and summer, when males perform their "sky dance" display.

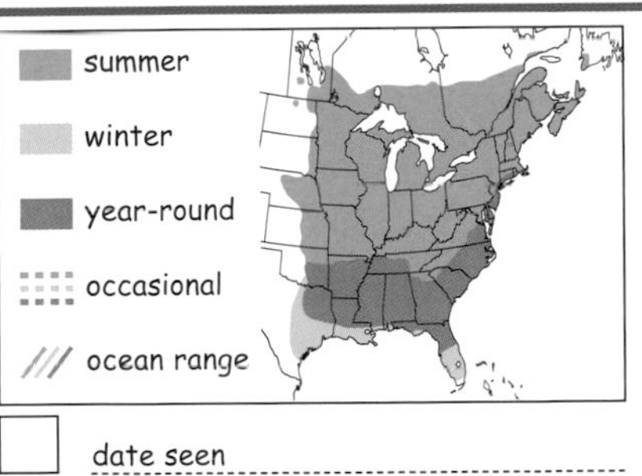

date seen ______________________

WILSON'S SNIPE

Gallinago delicata **Length: 10½"**

Look for: Dressed in the perfect camouflage for its grassy, marshy habitat, the Wilson's Snipe lurks, unseen by most birders. The stripes on the snipe's head and back help camouflage it.

Listen for: Our snipe has three sounds: Its startled-into-flight call is a harsh *skretch!* Its perched display song is a kestrel-like *ki-ki-ki-ki-ki*. And its courtship or winnowing display includes a long series of rising notes: *woo-woo-woo-woo*, a sound given in flight that is actually produced by the wind flowing over specialized outer tail feathers.

Remember: Wilson's Snipe looks superficially like a dowitcher, but it is much less active and rarely found on open mud flats in flocks, as dowitchers are.

WOW!

The snipe hunt is a practical joke named after the Wilson's Snipe. The victim is told to wait for a snipe to appear in a remote place and to use an empty sack and a stick to catch it.

◀ *As a Wilson's Snipe performs its courtship display flight, it tilts from side to side, and the sound made by wind passing through its tail feathers changes tone.*

Find it: Any muddy meadow or boggy wetland can host a snipe in migration. Before you see it, you may spook a snipe into a rapid zigzagging flight. It may be gone before you recover your senses.

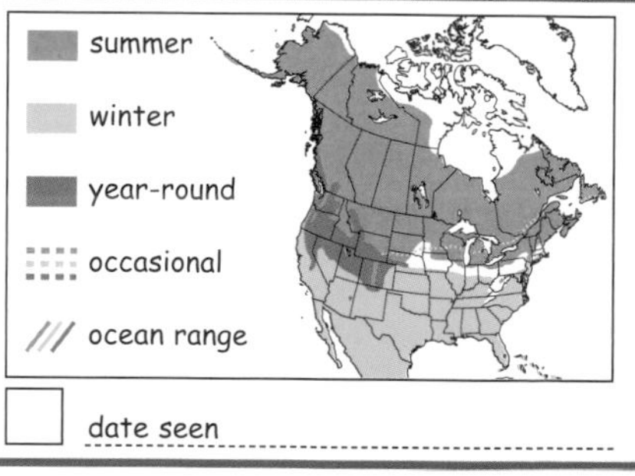

☐ date seen

LONG-BILLED DOWITCHER

Limnodromus scolopaceus **Length: 11½"**

Look for: Our two dowitcher species, Long-billed and Short-billed, are almost impossible to tell apart in the field. Both are chunky-looking birds, rich orange-brown in spring and summer, and gray overall in fall and winter. Both have long, straight bills. It's easier to tell them apart by where you find them (see Remember).

Listen for: Calls are the best way to separate the two dowitchers. Long-billeds give a high-pitched *peep!* or *pee-deep!* Short-billeds say *tew-tew-tew,* usually in threes.

Remember: *S* stands for Short-billed and salt water. Long-billeds are more commonly found in freshwater habitats (think *L* for Long-billed and lake).

WOW!

It wasn't until the 1950s that ornithologists confirmed that Short-billed and Long-billed Dowitchers were two separate species.

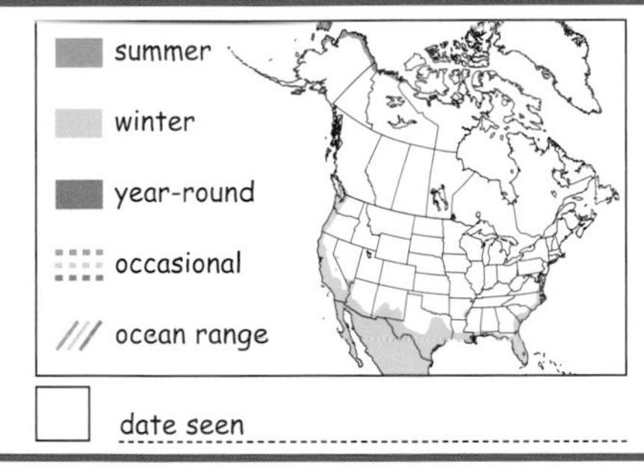

▶ *Dowitchers like these Long-billeds feed by rapidly probing the mud. They look like feathered sewing machines.*

Find it: Look for dowitchers in large flocks on mud flats, shallow lakes, and ocean beaches (Short-billed), using their bills to probe for food. Long-billeds are more common inland during spring and fall migration.

summer
winter
year-round
occasional
ocean range

☐ date seen ____________

LAUGHING GULL

Larus atricilla **Length: 17"**

Adult, breeding

Adult, winter

Look for: A medium-sized and slender gull with a long dark bill and dark legs (both can be orange-red in breeding plumage). In spring and summer, the Laughing Gull's head is black. In winter, the head appears to be dirty white, but the gull retains the slate gray back and dark wingtips.

Listen for: This bird is named for its call, which sounds very much like someone laughing loudly, in either single notes—*Ahh! Ahhh!*—or in a series—*Ah-ah-ah-ah-ah!*

Remember: Identifying gulls in nonbreeding plumage can be difficult. Overall size and shape is a good way to sort out nonbreeding-plumaged birds. Laughing Gulls appear slender overall and have a light, buoyant flying style.

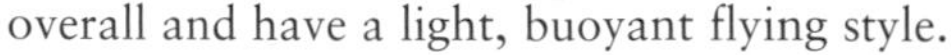

▶ *A Laughing Gull harasses a Forster's Tern, trying to force it to drop its catch.*

Find it: Abundant along the Atlantic and Gulf coasts in the warm months, the Laughing Gull is a creature of saltwater habitats. Flocks can be seen foraging on beaches and loafing on dunes, piers, and breakwaters.

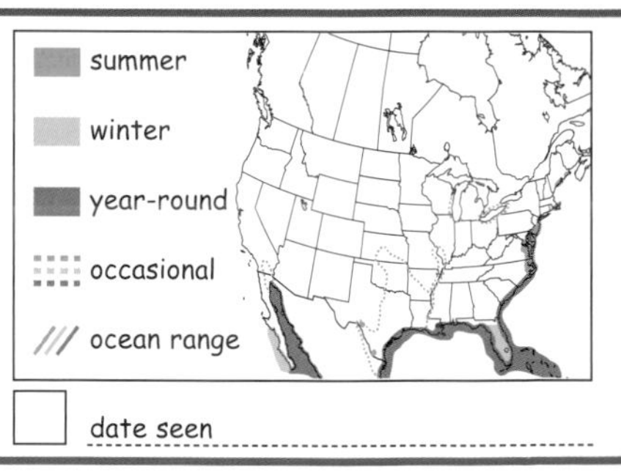

date seen

FRANKLIN'S GULL

Leucophaeus pipixcan **Length: 14½"**

Look for: The wingtips of Franklin's Gulls show a pattern of white-black-white. Breeding adult shows a black hood, a broken white eye-ring, and a red bill, and it may show rosy pink on the breast. Underwing is mostly white with black wingtips.

Listen for: Laughing cries similar to those of Laughing Gull, but higher pitched. A flock of calling Franklin's Gulls sounds like a bunch of teenage girls all crying *Ewwww!*

WOW!

Franklin's Gulls migrate **way** south and spend the winter along the Pacific Coast of South America!

Remember: This is the common black-hooded gull found on the northern Great Plains. Bonaparte's Gulls have white-tipped wings edged in black.

▼ *Franklin's Gull flocks regularly follow tractors plowing fields. They scarf up insects disturbed by the plowing*

Find it: Often found in spring and summer inland in large flocks in plowed or flooded fields, marshes, and lakes. Makes floating nest of vegetation on deep lakes. Migrates in flocks.

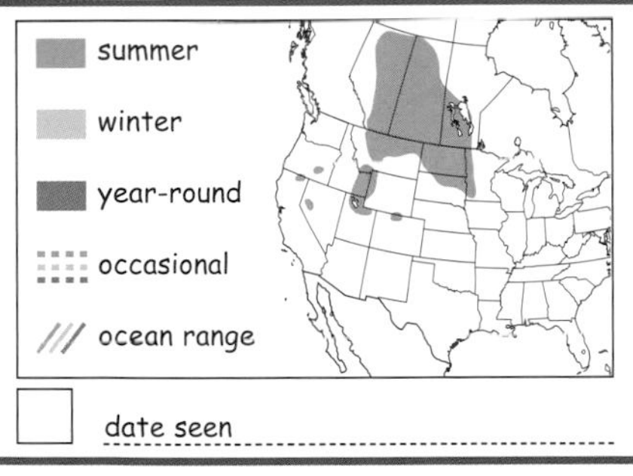

☐ date seen ______________

BONAPARTE'S GULL

Larus philadelphia **Length: 13"**

Look for: The Bonaparte's Gull is our smallest commonly encountered gull. It has a buoyant, ternlike flight, showing large white triangles on the leading edge of its gray wings. Black headed during spring and summer, in winter plumage the Bonie looks like it's wearing headphones.

Listen for: A very high-pitched, squally *nyahhh!* that sounds appropriate for a gull this small.

Remember: The bouncy flight and flashes of white in the wings of the Bonaparte's Gull are field marks that work well, even in poor light or from a great distance.

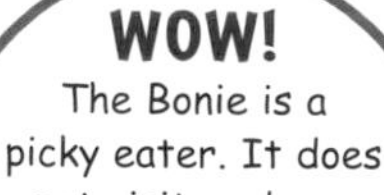

WOW!

The Bonie is a picky eater. It does not visit garbage dumps for food as our other common gulls do.

► *"Pie slices" on the wingtips and "headphones" are good winter field marks for immature (left) and adult (right) Bonaparte's Gulls.*

Find it: Most commonly seen in winter along the coasts, but present inland during migration on rivers, lakes, and reservoirs. Often present in small flocks but rarely associating with other, larger gulls.

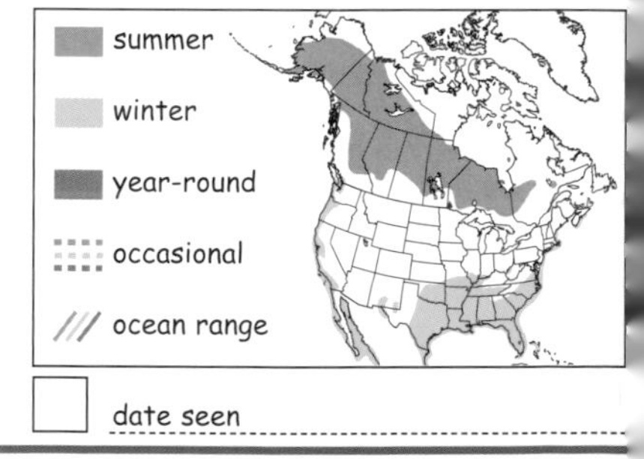

☐ date seen

HEERMANN'S GULL

Larus heermanni **Length: 19"**

Look for: The breeding-plumaged adult's dark gray body, black legs and tail, bright white head, and deep red bill make this gull easy to identify. Nonbreeding Heermann's Gull has a gray head. In flight shows bold white and black tail.

Listen for: Nasal call is *awww!* Also high-pitched squeals.

Remember: No other western gull has the combination of black legs and red bill.

WOW!
Heermann's Gulls are very aggressive birds. They often steal food from pelicans and other birds.

▶ *Heermann's Gulls regularly pirate food from Brown Pelicans.*

Find it: Common within its coastal range, this species is dedicated to the ocean and rarely found far from the coast. Heermann's Gulls breed in early spring in Mexico and move north along the Pacific Coast in summer, following flocks of Brown Pelicans.

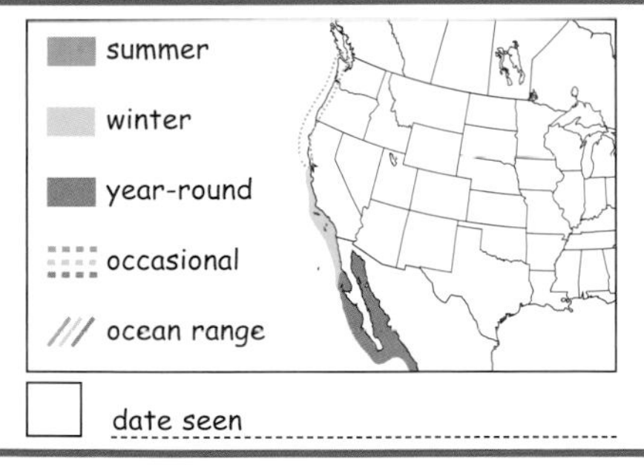

date seen

CALIFORNIA GULL

Larus californicus **Length: 21"**

Look for: Size-wise this gull is in between the larger Herring Gull and the smaller Ring-billed Gull and can be easily confused with either of them. Dark eyes and yellowish legs are key field marks for the adult California Gull. In flight it shows half-circles of white on black wingtips. Head is streaked with brown in nonbreeding plumage.

Listen for: Typical hoarse gull cries, but pitched higher than those of Herring Gull and deeper than those of Ring-billed Gull.

Remember: Say it with me: "Medium-sized with dark eyes." Adults are easier than mottled first-year youngsters to separate from Herring and Ring-billed Gulls.

WOW!
If the Beach Boys had been birders, their hit song might have gone "I wish they all could be California Gulls!"

◀ *In 1848 California Gulls saved the crops of Mormon settlers in Utah when they devoured hordes of crop-eating grasshoppers.*

Find it: Nests on inland lakes and marshes and forages in farm fields, dumps, parks, and lakes, even in urban settings. Winters along the Pacific Coast, often resting on docks, on beaches, and in parks.

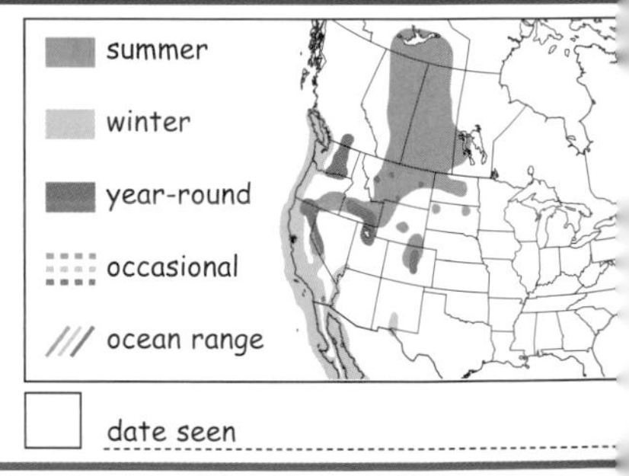

☐ date seen ____________

RING-BILLED GULL

Larus delawarensis **Length: 17½"–19"**

Look for: Part of a group of large white-headed gulls, the Ring-billed Gull retains its namesake field mark (a clean, black ring on the bill) in all seasons, but that alone is not enough for positive identification because the Herring Gull also has a ring on the bill. Overall size, and bill size, is smaller than the Herring Gull, which is often confused with the Ring-billed.

Listen for: Basic call is a high-pitched, nasal, and squeally *klee-ear!*

Remember: Ring-billed Gulls and Herring Gulls both have pale gray mantles (upper back and wings), white heads, and rings on the bills, but Ring-bills are nearly eight inches shorter and look shorter necked than Herrings.

WOW!
The Ring-billed Gull's habit of hanging around fast-food restaurants has earned it the nicknames McGull, Dumpster Gull, and French-fry Gull.

▲ *A young Ring-billed Gull begs for a French fry.*

Find it: The Ring-bill is equally at home on ocean beaches, near freshwater reservoirs and dams, at landfills, near fast-food restaurants, and loafing in a large parking lot. More common inland than other gull species.

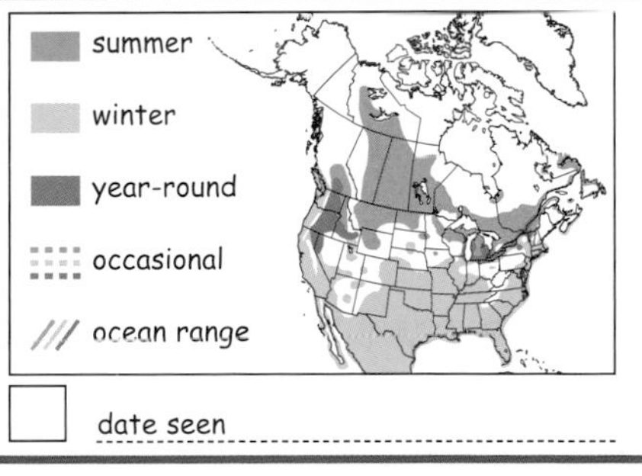

date seen

HERRING GULL

Larus argentatus **Length: 25"**

Look for: Our most common large gull continent-wide, the Herring Gull is heavy bodied and heavy billed. In summer adult plumage, the head is clean white, the bill is yellow with a red spot, and the legs are pinkish. Winter adults have "dirty" heads.

Listen for: Call is similar to other common gulls but a lower-pitched and hoarser two-syllabled *oww-uh! oww-uh!* Also utters a rapid, tuneless *uh-uh-uh*.

Remember: Herring Gulls take four years to reach adult plumage. They start out all brown and get whiter and cleaner looking as they age.

WOW!
Herring Gulls sometimes carry large shellfish such as mussels, clams, and oysters in the air and drop them onto a hard surface in order to break them open and eat them.

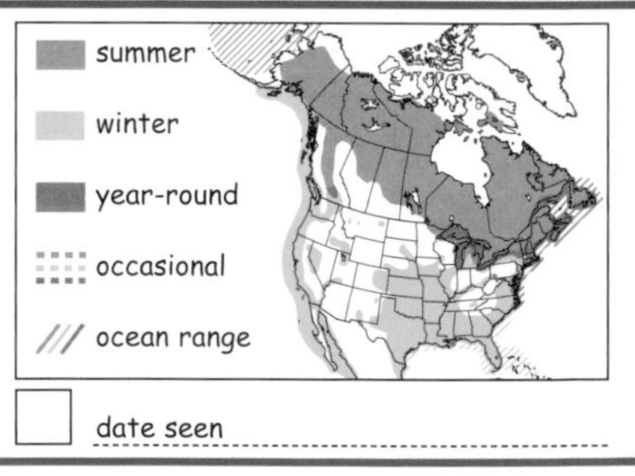

▲ *A Herring Gull prepares to drop a clam on the hard asphalt. When the shell breaks, the gull will descend to eat.*

Find it: Herring Gulls are usually found near large bodies of water, though they also visit large landfills. In winter, they can be found on inland rivers, particularly near dams.

summer
winter
year-round
occasional
ocean range

☐ date seen

GLAUCOUS-WINGED GULL

Larus glaucescens **Length: 26"**

Adult (center), with juveniles

Look for: Named for its pale appearance, this very large gull has pinkish legs, a bold yellow bill, and medium gray back and wings edged in white.

Listen for: This species has two very different calls. One is typical gull-like screams with a few nasal *ka-ka-ka*s thrown in. The other is a piercing cry that sounds like a bull elk trying to imitate a Common Loon.

Remember: Compared with our other large pale gulls, the Glaucous-winged looks fairly plain and unpatterned overall. Its wingtips match its back in color.

The Glaucous-winged Gull is not choosy about its mates. It hybridizes with several other large gulls where their ranges overlap.

◀ *Glaucous-winged Gulls often nest on rooftops in coastal towns along the Pacific.*

Find it: Very common in its preferred coastal habitat, including rocky shores, beaches, bays, and garbage dumps.

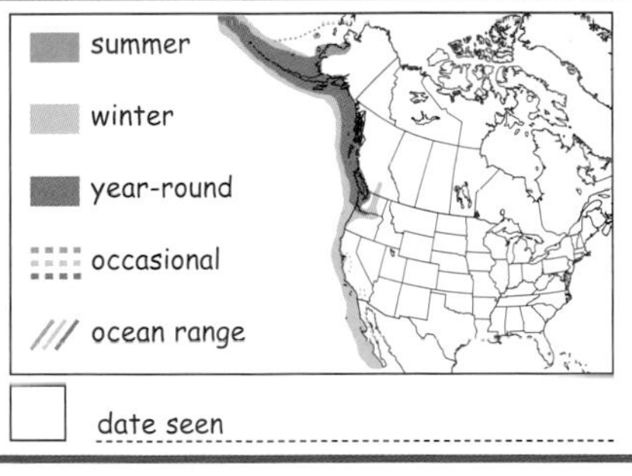

date seen

WESTERN GULL

Larus occidentalis **Length: 25"**

Look for: The Western Gull is the only common large dark-backed gull in its range. The heavy yellow bill stands out. Adult has a dark gray back and pink legs. First-year birds are all dark brown, becoming more cleanly marked as they reach breeding age in the fourth year.

Listen for: Low, hoarse calls that sound like an angry Chihuahua's barks.

Remember: Western Gull is a big bird with a dark back and a huge yellow bill. Heermann's Gull is black backed but much smaller.

WOW!
These birds boldly hang around California sea lion colonies, waiting to scavenge dead pups and other yummy things.

◀ *Western Gulls often nest near other breeding birds so they can steal their eggs and food.*

Find it: Common along the Pacific Coast at all seasons, but rare inland. Found on beaches, docks, breakwaters, parks, parking lots, and dumps.

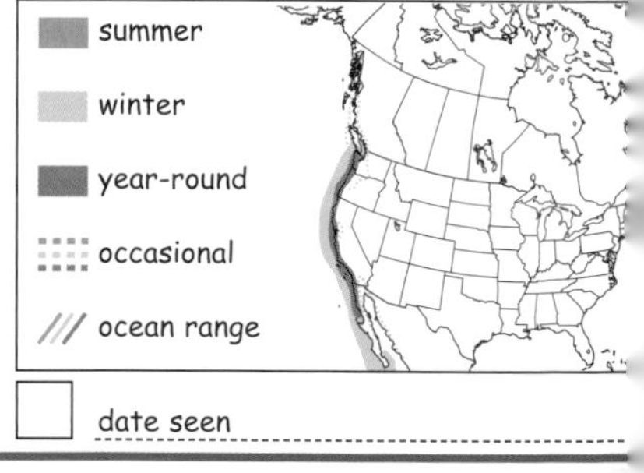

☐ date seen ____________

GREAT BLACK-BACKED GULL

Larus marinus **Length: 30"**

Look for: The Great Black-backed Gull is a huge bird, and its size alone is often enough to pick this species out of a flock of loafing gulls. Big headed with a massive bill, the Great Black-backed dwarfs other common gulls. The dark back of adult birds is an excellent field mark, as are the pink legs. Young birds have checkered backs and wings for their first two years.

Listen for: Common call is a deep, nasal *owwh! owwh!* But also utters a variety of honks and chortles.

Remember: This is the only gull in the East with the combination of black back and pink legs. In the West, there are other gulls with these features.

WOW!
The Great Black-backed Gull can be confused with a Bald Eagle when seen from a distance. But the gull has a white head and breast, while an adult Bald Eagle has just a white head.

▼ *Great Black-backed Gulls often act like hawks, preying on small seabirds such as puffins.*

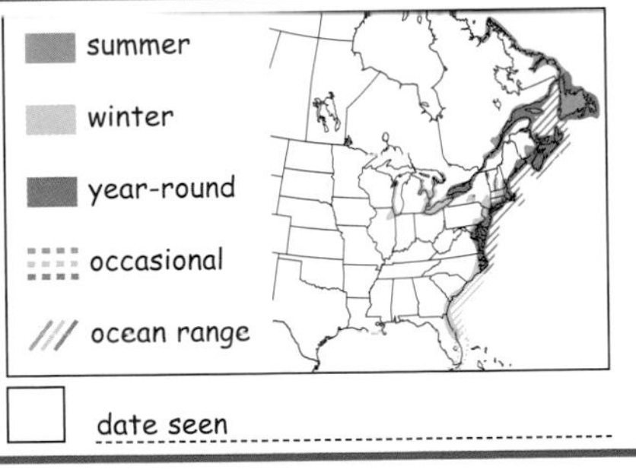

Find it: Very common along the entire East Coast, usually near salt water. In winter Great Black-backs can be found inland on the Great Lakes and near landfills and reservoirs.

summer
winter
year-round
occasional
ocean range

☐ date seen ______________

FORSTER'S TERN

Sterna forsteri **Length: 13–14½"**

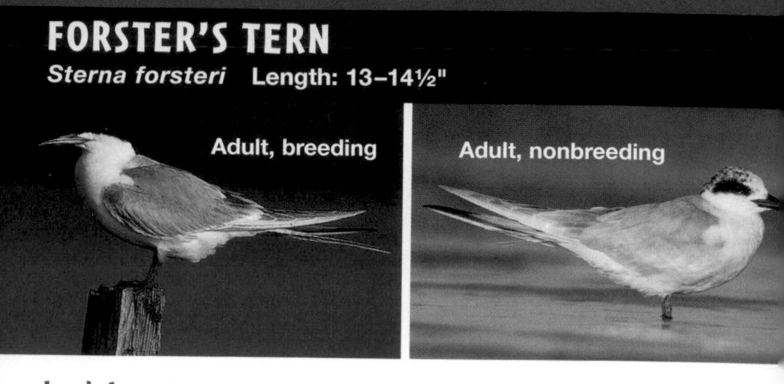

Look for: If you could see a Forster's Tern perched next to a Common Tern, you'd see that the Forster's is slightly larger and bulkier with a longer, orange (not red) bill. In spring and summer, the Forster's underparts are white (not gray), and the tail usually extends beyond the tips of the folded wings on resting birds.

Listen for: A harsh, nasal *kyerrr! kyerrr!* Given both singly and in a series. Calls are buzzier than those of the Common Tern.

Remember: Winter adult and juvenile Forster's Terns lack the dark shoulder bars of the Common Tern. They also have black raccoon-mask eye patches, but this black does not wrap around the back of the head.

Early ornithologists didn't recognize Forster's and Common Terns as two separate species. (They didn't have the benefit of our modern optics and field guides, so we should cut them some slack.)

◀ *In winter, the Forster's Tern* (left) *sports a bandit mask and white wings, while the Common Tern* (right) *has dark wingtips, nape, and shoulders.*

Find it: Forster's Terns love marshes and can be found in both freshwater and saltwater habitats, including bays, lakes, reservoirs, and rivers. Many Forster's Terns winter along the southern coasts of the U.S.

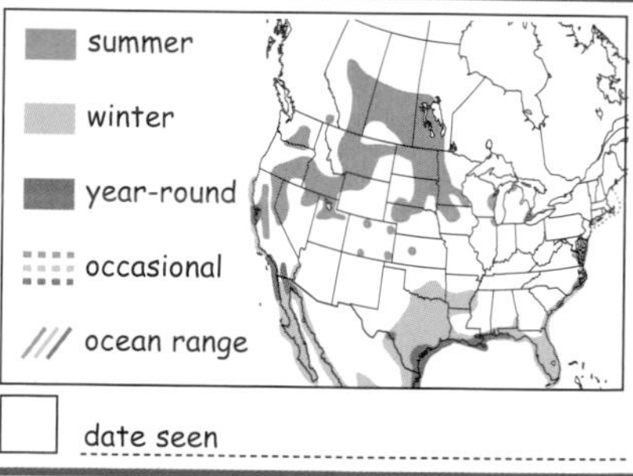

COMMON TERN

Sterna hirundo **Length: 13"**

Adult, breeding

Adult, nonbreeding

Look for: The Common Tern is a medium sized tern with a look-alike cousin in the Forster's Tern. In breeding plumage, the Common has a thicker, redder bill and a gray body (the belly of the Forster's is all white). In winter and juvenal plumages, the Common Tern has a dark shoulder bar and a wraparound black patch behind the eyes.

Listen for: A loud, burry *kee-yarrr, kee-yarr!*

Remember: In winter, adult and juvenile Common and Forster's Terns are very similar, but the Forster's Terns lack shoulder bars and their black eye patches do not connect on the back of the head.

If you get too close to a Common Tern nesting colony, expect to get dive-bombed by angry terns. They might even decorate you with a splat of poop if they really want you to leave.

▼ *Common Terns are specialists at fishing in shallow coastal waters, spotting a fish, then plunging, bill-first, to capture it.*

Find it: In the Northeast, the Common Tern is the most common tern, but elsewhere the Forster's is often more common. Preferred habitat of Common Terns is ocean bays, beaches, lakes, and large rivers.

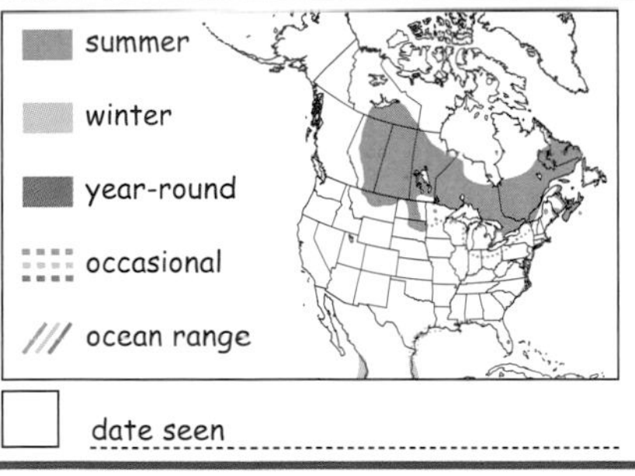

date seen ______________________

ELEGANT TERN

Thalasseus elegans **Length: 17"**

Look for: A medium-sized tern, white bodied with a shaggy black crest and a thin orange-red bill. Black crown covers the forehead on breeding adults. Nonbreeding-plumaged birds show a white forehead.

Listen for: Loud single call is *keek!* Two-note call is a burry *karr-rick!*

Remember: Compared with the similar Royal Tern, the Elegant Tern is smaller, with a thinner, more delicate bill. You could even say it's more elegant-looking!

Adult, breeding

WOW!
Elegant Terns love anchovies. When the anchovy population rises, this species enjoys greater nesting success.

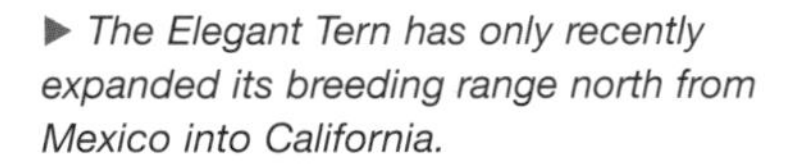

▶ *The Elegant Tern has only recently expanded its breeding range north from Mexico into California.*

Find it: A coastal bird, the Elegant Tern forages in shallow waters in bays and estuaries. After breeding, these birds move north along the coast for the summer.

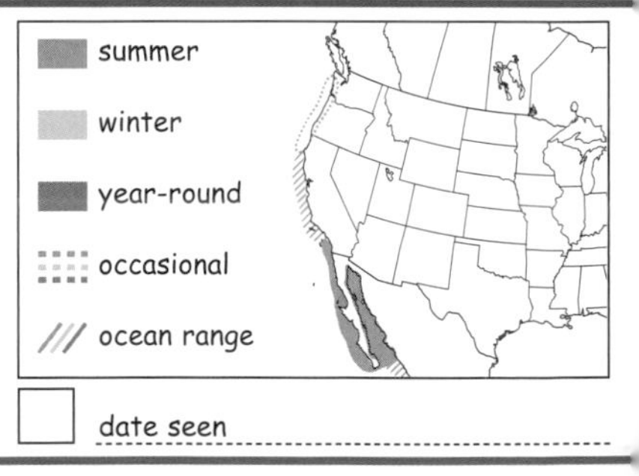

☐ date seen ______________________

CASPIAN TERN

Sterna caspia **Length: 21"**

Look for: The Caspian Tern is a large, gull-sized tern with a heavy red-orange bill. In summer, its striking crown is so dark, it is almost impossible to see its eye. Winter and juvenile Caspians have dirty foreheads.

Listen for: Utters a raspy, ducklike *ahrakk!* Begging call of juvenile is a high-pitched, wavery whistle.

Remember: The Caspian Tern's look-alike relative is the Royal Tern, a bird that is rarely found inland. One field guide calls the Royal's bill "carrot-orange." So if you can remember that the Royals live at the beach (like royalty) counting the "carrots" in their diamonds, you can sort them out from the much less royal Caspians.

WOW!
Young Caspian Terns beg food from their parents months after they are able to fend for themselves. But don't try this with your parents. Trust me, it doesn't work.

◀ *A young Caspian Tern begs noisily from its parent.*

Find it: Our most common large tern found inland on lakes and along large rivers, but also along both coasts. Look for them among flocks of gulls on beaches. Their bright bills will stand out in the crowd.

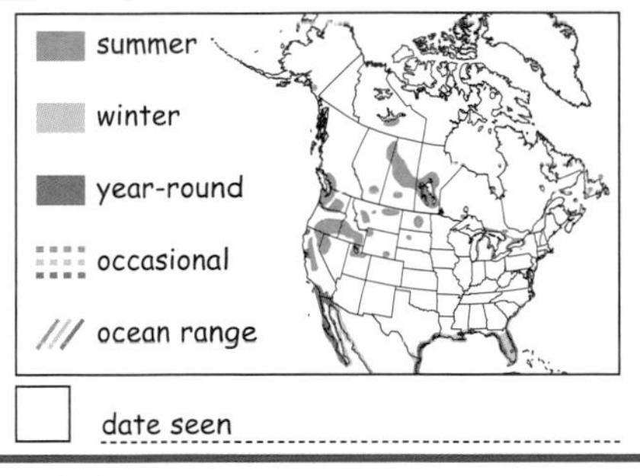

☐ date seen ____________

COMMON MURRE

Uria aalge **Length: 17–17½"**

Look for: A duck-sized, football-shaped bird that is dark above and white below, with a slender, pointed black bill. Flies very fast, with rapid changes of direction.

Listen for: This species is named, apparently, for the low, moaning sound (*mrrrrrr*) it makes at the nest burrow.

Remember: Thick-billed Murre are very similar, but their range is far more northerly. Their bills are also thicker and shorter.

▼ *Murre are very fast fliers but even better divers and swimmers.*

WOW!

Common Murres, like other alcids, are closely related to penguins, except that, unlike penguins, murres can fly.

Find it: Most common near nesting colonies on rocky ocean cliffs and offshore islands. Often seen flying to and from colonies, bringing food to young. Winters at sea.

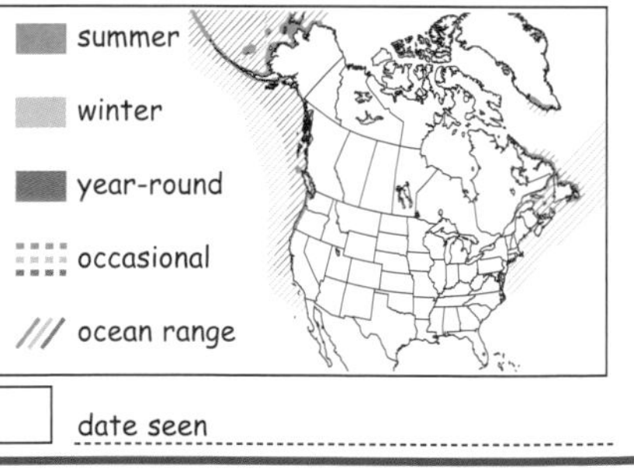

☐ date seen ------------------------------

BLACK SKIMMER

Rynchops niger **Length: 18"**

Look for: One of our most distinctive birds, the Black Skimmer is named for its method of foraging—its elongated lower bill (mandible) plows lightly through the water, feeling for small fish. The skimmer's very long body and bill make its shape easy to pick out in a flock of resting gulls and terns.

Listen for: Skimmers utter a high-pitched, burry bark: *eerrff! eerrff!*

Remember: Skimmers are very graceful fliers, wheeling in tight turns and gliding just above the water's surface. Standing at rest, they appear to lack eyes—their black eyes and black hoods blend together completely.

WOW!
The Black Skimmer's bill is thin and knifelike, perfect for slicing through the water, skimming for tasty fish. The skimmer does not usually see its food. It feels for its food as it skims along.

◀ *When the lower mandible contacts a fish, the skimmer snaps its bill shut.*

Find it: Skimmers prefer to feed on and nest near smooth salt water, such as estuaries, bays, and lagoons, where the calm water makes foraging easier. Flocks resting on beaches will all face into the wind.

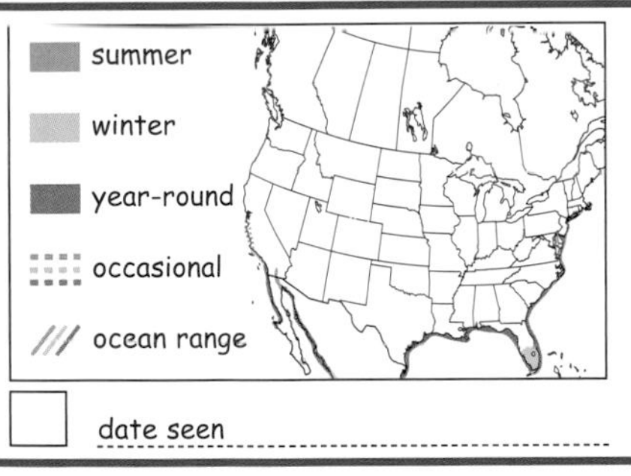

☐ date seen ____________

ATLANTIC PUFFIN

Fratercula arctica **Length: 12½"**

Adult, breeding

Look for: Even people who have never seen a puffin know what a puffin looks like, with its huge rainbow-colored bill, white face, orange feet, and handsome tuxedo plumage. Winter adults have very dull bills and gray faces.

Listen for: At their breeding colonies, Atlantic Puffins give calls that sound *exactly* like chain saws. It's worth the trip to their nesting islands just to hear this.

Remember: The Atlantic Puffin is the only puffin found on the East Coast. Two other puffin species, the Horned Puffin and the Tufted Puffin, are found along the West Coast.

▼ *Conservationists used decoys and sound recordings to attract Atlantic Puffins back to rocky islands where they once nested.*

WOW!

The Atlantic Puffin nearly became extinct because of overhunting of the birds and harvesting of eggs. The population has increased dramatically in the past 30 years through conservation efforts.

Find it: Atlantic Puffins are rarely seen from shore. The best way to see them is to take a charter-boat trip to one of their protected nesting islands off the coast of northern New England.

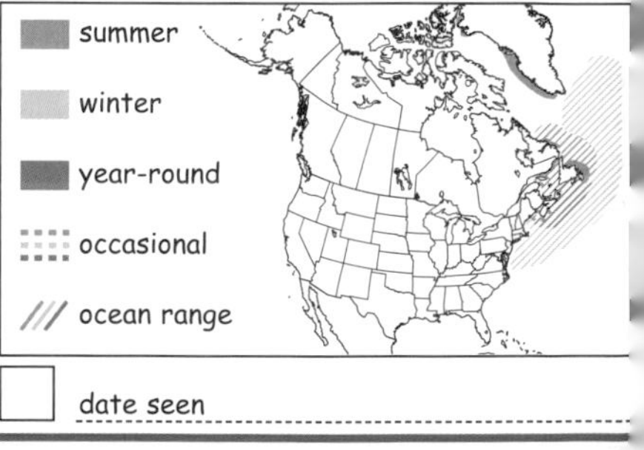

☐ date seen ______________

TUFTED PUFFIN

Fratercula cirrhata **Length: 15–16"**

Look for: A dark-bodied seabird with a massive orange bill. Breeding-plumaged adult has a white face and bright cream-colored tufts on the back of the head.

Listen for: Growls in low tones near nest site, but otherwise silent.

Remember: Adults in summer are unmistakable. Young birds and winter adults are gray bodied with greatly reduced orange on the bill.

◀ *Tufted Puffins dig their nesting burrows in soil on cliffs near the ocean. If no soil is present, they use a rocky crevice.*

Find it: Uncommon in open ocean. Most sightings are at nesting colonies on rocky ocean cliffs and steep grassy slopes on offshore islands.

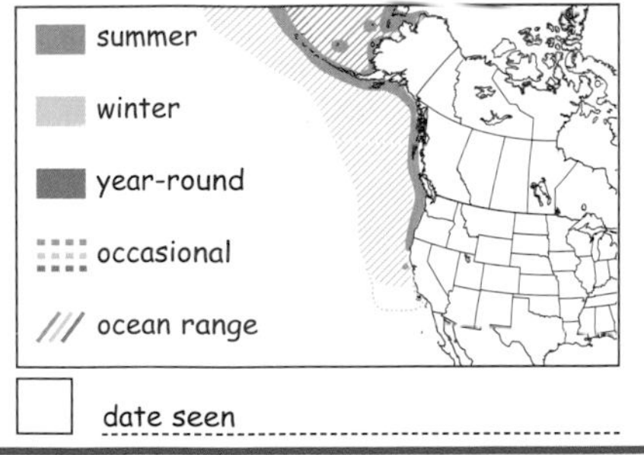

☐ date seen ....

PIGEON GUILLEMOT

Cepphus columba **Length: 13½"**

Look for: Named for its pigeonlike appearance, this small alcid is black in the breeding season with large white wing patches, a pointed black bill, and bright orange legs. In nonbreeding plumage the bird is pale gray below with black wings.

Listen for: Gives shrill peeps and whistles near the breeding colony.

Remember: Look for the large white wing patches, each bisected by a black bar.

► *Most Pigeon Guillemots nest in colonies on rocky cliffs that are inaccessible to predators.*

Find it: Because they nest colonially on sea cliffs, Pigeon Guillemots can be seen flying to and from their nests during the spring and summer months. They prefer nearshore waters, unlike most members of the Auk family.

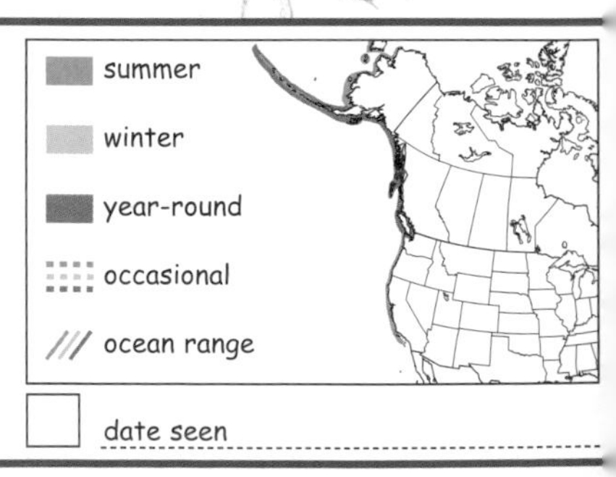

date seen

BAND-TAILED PIGEON

Patagioenas fasciata **Length: 14½–15"**

Look for: Larger and longer than a Rock Pigeon with a white half-collar on nape of neck and a pale gray band on the end of the tail (for which it is named).

Listen for: Call is a very owl-like *whooo-hoo*.

Remember: The Band-tailed Pigeon often lands in trees to forage. Rock Pigeons prefer to land on buildings and other structures.

WOW!
When a Band-tailed Pigeon takes off suddenly, its wings make a loud clapping sound.

▶ *Band-tailed pigeons are adept at clambering around in trees to get at food.*

Find it: Prefers woods containing oaks in mountains, foothills, and canyons, where it forages for fruits, nuts, and berries. Often found in flocks.

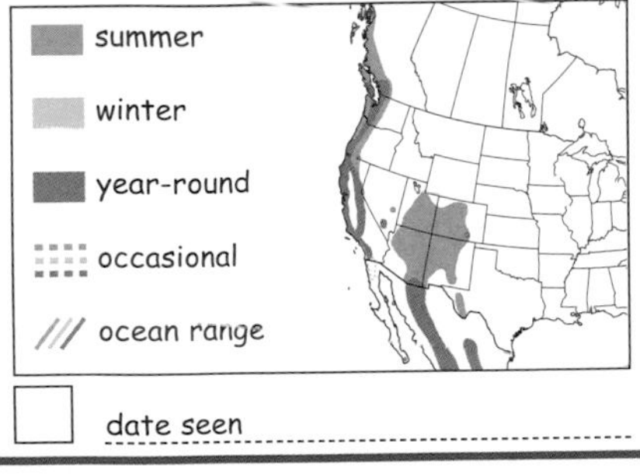

☐ date seen

EURASIAN COLLARED-DOVE

Streptopelia decaocto **Length: 13"**

Look for: Looking like a washed-out Mourning Dove wearing a black bandanna around its neck, the Eurasian Collared-Dove is a recent immigrant to North America. Chunky and pale overall, it lacks the MoDo's long tapered tail.

Listen for: Eurasian Collared-Doves repeat *coo-COO-coo* over and over. The call sounds like a Mourning Dove with a sore throat stuck on Repeat.

Remember: In flight, the Eurasian Collared-Dove looks chunkier than a Mourning Dove and lacks the MoDo's long central tail feathers.

▼ Courtship, mating, and many other aspects of the adaptable Eurasian Collared-Dove's life are carried out on power lines.

WOW!

From an accidental release in the Bahamas in 1974, the Eurasian Collared-Dove has invaded the southeastern U.S. It is so adaptable that it may become one of our most common birds.

Find it: Introduced from Europe, the Eurasian Collared-Dove is colonizing North America at an alarming rate. Common in suburban settings, where it likes to perch on power lines and other exposed places.

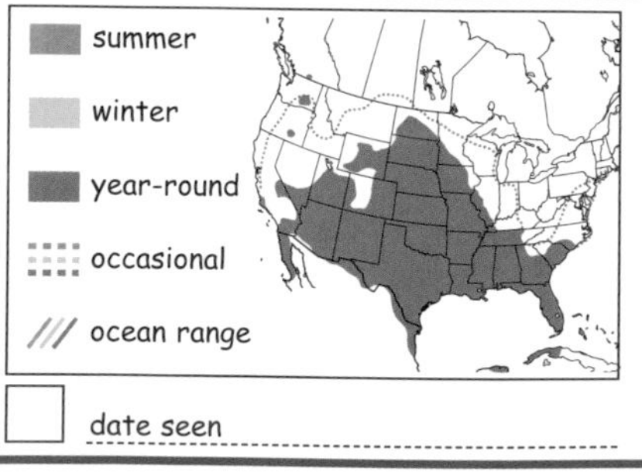

date seen ____________

ROCK PIGEON

Columba livia **Length: 13"**

Look for: The bird formerly known as Rock Dove and Feral Pigeon—and still known by lots of other, less pleasant names—comes in a variety of plumage colors, but the most common is gray bodied and dark headed with orange feet and orange eyes.

Listen for: Rock Pigeons make a low, gurgling *urr-cooooo!*, most often at roosts or nest sites. When Rock Pigeons explode into flight, their wings make a loud slapping sound.

Remember: Rock Pigeons are strong and direct fliers, and they can be confused with falcons in flight. A closer look shows the pigeon's potbellied, broad-winged, small-headed appearance to be different from a falcon's flight shape.

Rock Pigeons have helped in the reestablishment of the Peregrine Falcon. Peregrines that nest in cities eat lots of plump and tasty Rock Pigeons.

▶ A male pigeon struts to court his mate. This pair displays the two wild-type colorations of Rock Pigeon: blue bar and checkered.

Find it: Widespread and common, especially in cities, on bridges, in parks, and near farms with open barns, the Rock Pigeon was introduced to North America from Europe, where its native relatives nest on rocky cliffs.

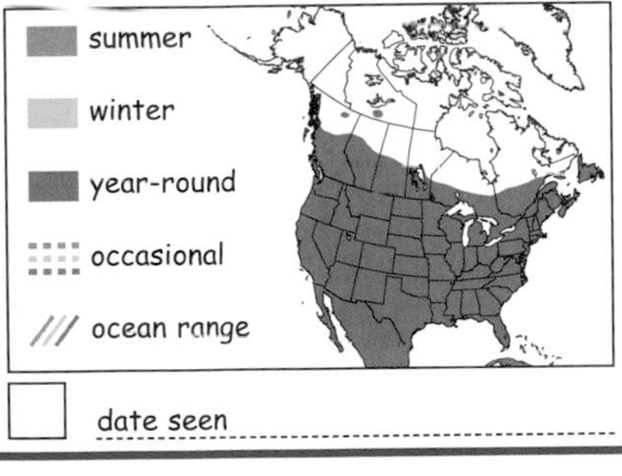

date seen ____________

WHITE-WINGED DOVE

Zenaida asiatica **Length: 11½–12"**

Look for: Named for large white wing patches, which are most obvious in flight, as they contrast with dark flight feathers. Perched birds show a long white wing edge. The tail is rounded with white outer tips.

Listen for: Call sounds very similar to Barred Owl's *who cooks for YOU!* When you hear a day-calling "owl," it may be this species.

Remember: Similar Mourning Dove has a pointed (not rounded) tail and no white in the wings.

▼ *Most White-winged Doves head south in winter.*

WOW!
The White-winged Dove drinks nectar and rainwater from cactus flowers and in doing so helps to spread pollen among cacti.

Find it: A dove of the Southwest that prefers woods along rivers, groves, brushland, and other semi-open habitats. Increasingly found in towns and parks. Range is expanding northward.

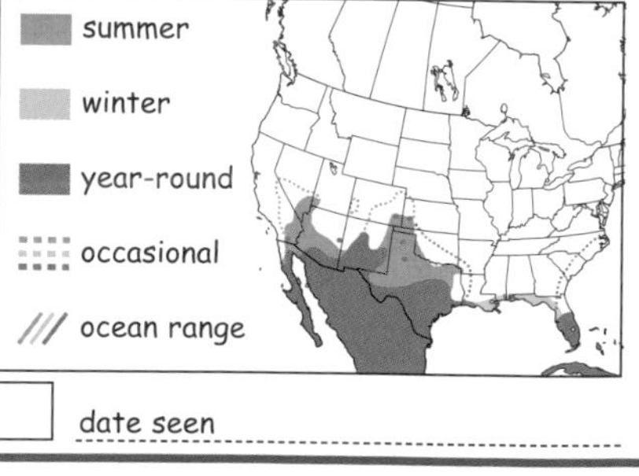

☐ date seen ______

MOURNING DOVE

Zenaida macroura **Length: 12"**

Look for: The slender brown shape of the Mourning Dove, with its long tapered tail, is a familiar sight all across North America. When perched, the dove shows black spots on tan wings, and the adult can show glistening shades of purple, pink, and green on the neck and head. In flight, the Mourning Dove's tail feathers show white tips.

Listen for: This species is named for its sad-sounding cooing: *ah-ooh! whoo-whoo-whoo*. Nonbirders often confuse the Mourning Dove's call with an owl's hooting.

Remember: Their rapid flight makes the MoDo look a little like a hawk, but the wedge-shaped dove tail with white spots (plus the dove's small head) should clinch the ID.

WOW!

The Mourning Dove has a built-in straw! Other birds have to scoop water in their bills and tilt their heads back to swallow. But the MoDo can drink water by sucking it up through its bill.

◀ *Mourning Doves build the flimsiest of nests, often in weird places.*

Find it: You can find Mourning Doves almost anywhere except in thickly wooded habitats. They come to bird feeders and often perch on wires and forage on lawns and gravel roads; they also form huge flocks in agricultural fields during fall and winter.

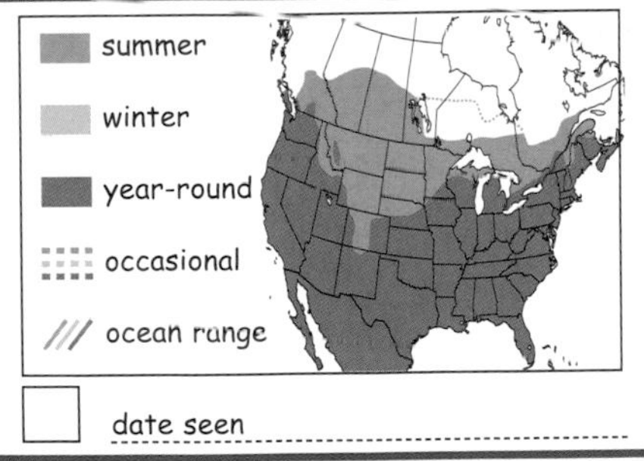

☐ date seen

YELLOW-BILLED and BLACK-BILLED CUCKOO

Coccyzus americanus, Coccyzus erythropthalmus **Length: 12**

Look for: This long and slender skulker can be hard to see despite its size. In flight, the Yellow-billed shows large rusty patches in its wings and appears very long tailed. The similar Black-billed Cuckoo has a smaller black bill and a red eye-ring, and is cream colored below and olive above, with no rust color in the wings.

Listen for: From deep cover, Yellow-billed Cuckoos give a hollow, hoarse, two-syllabled *tee-oo, tee-oo, tee-oo* in a series and a mechanical-sounding *kik-kik-kik-kik-kaKOW, kaKOW*.

WOW!

Because the Yellow-billed Cuckoo commonly sings on humid summer afternoons, the bird's folk name is Rain Crow, as it is said to predict a coming rainstorm.

Remember: A cuckoo's flight shape is long and slender. If you see lots of white below, white spots in the tail, and rust in the wings, you've got a Yellow-billed. If the bird is plainer, it's a Black-billed.

◀ *Yellow-billed Cuckoos have specialized stomachs that allow them to eat hairy caterpillars no other birds will touch.*

Find it: Cuckoos are more often heard than seen. The Yellow-billed is a bit less secretive than the Black-billed, which prefers woodlands near water. Both species like thick foliage where they can find caterpillars.

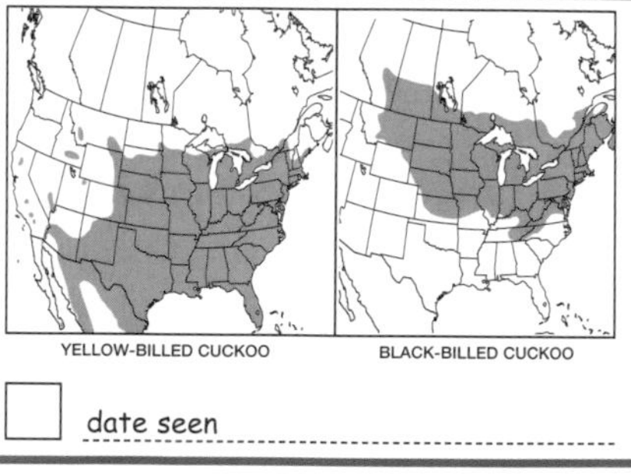

☐ date seen ______________

GREATER ROADRUNNER

Geococcyx californianus **Length: 23"**

Adult

Look for: This large streaky bird with a shaggy head is unmistakable, though it does not look much like the famous cartoon. The Greater Roadrunner has a long tail spotted with white and a head with a crest that it raises and lowers as it stalks through the dry open country while hunting.

Listen for: A very dovelike series of coos, *cu-cu-cu-cu-cuurrrr*, that slows down and drops in tone as it ends. No Greater Roadrunner has ever been recorded as saying *meep-meep!*

Remember: Roadrunners are fast birds when chasing prey —they can run as fast as 15 mph. They prefer to walk or run on the ground rather than flying and will fly only as a last resort.

WOW!

Roadrunners eat almost anything they can catch: lizards and snakes, small rodents, scorpions and tarantulas, and large insects. They'll even leap up to catch hummingbirds at nectar feeders.

▶ *Far from its cartoon image as prey to the coyote, the Greater Roadrunner is a wily and voracious predator itself.*

Find it: Greater Roadrunners prefer very dry habitats, such as brushy desert and chaparral. Watch for them moving methodically through a habitat, hoping to startle a lizard, snake, or mouse into revealing itself.

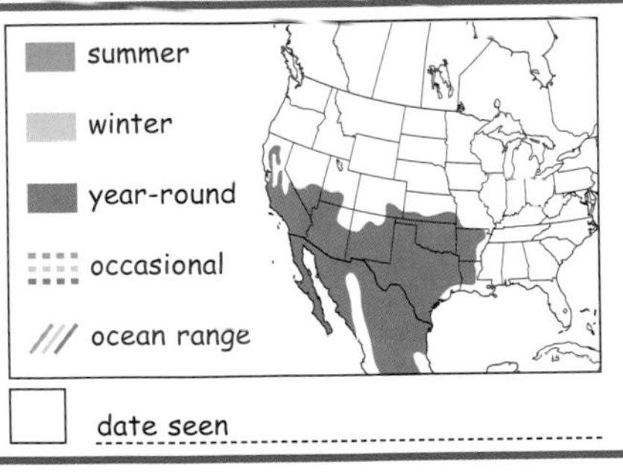

☐ date seen ______________________

SHORT-EARED OWL

Asio flammeus **Length: 15"**

Look for: This bird could be called the No-eared Owl, since its ear tufts are almost impossible to see. Its warm brown color, round face, and flopping, mothlike flight are better field marks. Because it often hunts during daylight hours in open country, the Short-eared Owl is easier to see than our other owls.

Listen for: Gives a hoarse, scraping bark in flight. Also utters a series of short *kek-kek-kek-kek* sounds.

Remember: In flight, the Short-eared Owl might be confused with the Long-eared Owl or Barn Owl. Barn Owls are white below and much paler overall. Short-ears show large buffy patches on their upper wings.

▼ *Hunting by sight and sound, a Short-eared Owl flies low over a meadow on a winter afternoon.*

WOW!

Short-eared Owls can clap their wings! Males use the wing clap as part of their courtship display, and both adults use it to scare predators and intruders away from the nest.

Find it: Short-eared Owls are grassland-loving birds and, where the hunting is good, multiple birds may be found, especially in winter. They use tundra, prairies, old-fields, and recovering habitats such as strip mines.

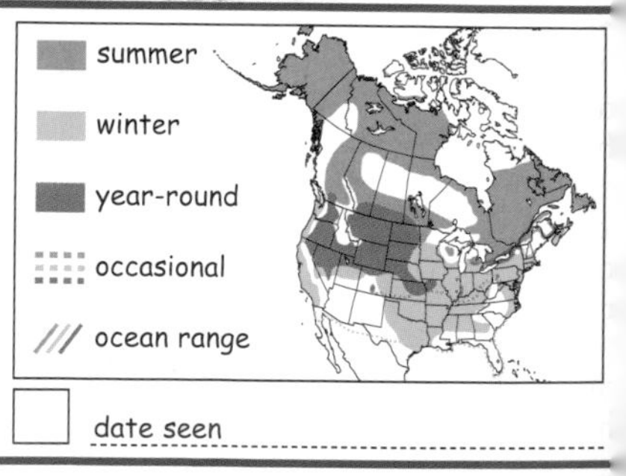

date seen ______________

BARN OWL

Tyto alba **Length: 16"**

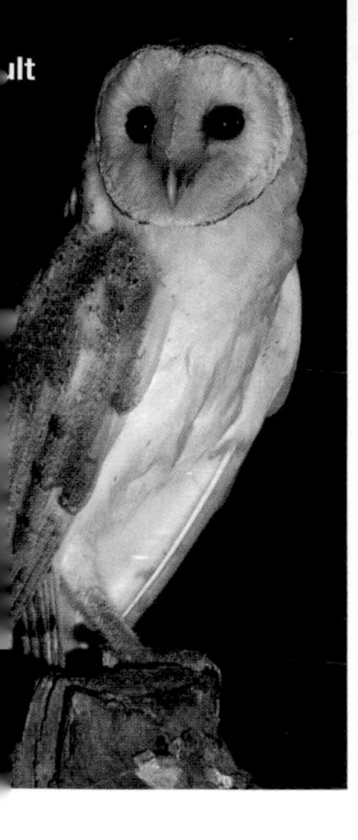

Look for: The heart-shaped white face, brilliant white underparts, and cinnamon brown back make the Barn Owl one of our most distinctive night birds.

Listen for: A shrieking, downward hiss that sounds like a woman screaming in terror. You are more likely to hear a Barn Owl's call on a dark night than to see one, and once you do you'll never forget it.

Remember: The Barn Owl's low, mothlike flight might be confused with the flying styles of the Short-eared Owl or the Northern Harrier, but only the Barn Owl is all white underneath—giving it a ghostly appearance the others lack.

▼ *Barn Owls are great friends to the farmer. They will clear a barn of rats and mice.*

WOW!
The feathers on the face of the Barn Owl form a heart-shaped bowl. This shape channels sound waves to the owl's ears, helping it to locate prey even on the darkest nights.

ind it: Barn Owls are n decline over most of heir range because of oss of habitat and .dequate nesting sites old barns, hollow rees). Look for them at usk and dawn, flying ow to the ground as hey hunt in open areas.

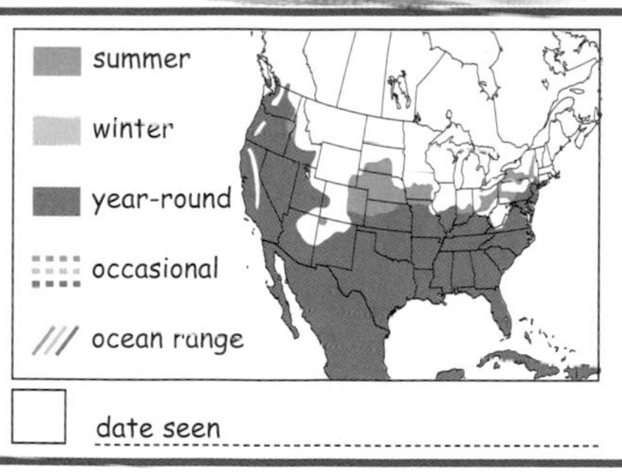

☐ date seen ______

GREAT HORNED OWL

Bubo virginianus **Length: 23"**

Look for: This huge nighttime predator is a common, year-round resident all across North America, nesting in woodlands and hunting woodland edges. Its large size, tufted round head, and horizontally striped breast and belly are excellent field marks.

Listen for: Deep, booming pattern of five hoots, usually *whoo-who-who, whooo-whoo*. Great Horneds begin courtship calling in midwinter, so this is a good time to listen for their far-carrying calls.

Remember: Few other raptors look as large and bulky as the Great Horned Owl. Their big-headed, horned shape is distinctive.

▲ *Great Horned Owls are about the only predators that kill and eat skunks.*

WOW!
Although the female is larger than the male, he has a deeper voice and more complex call. If you hear two Great Horneds calling to each other, the voice with the deeper, lower notes belongs to the male.

Find it: During the day, Great Horneds roost in deep cover but may be harassed loudly by groups of crows. They are best located by sound, but at dawn and dusk they often perch up high in trees, on towers, and on buildings.

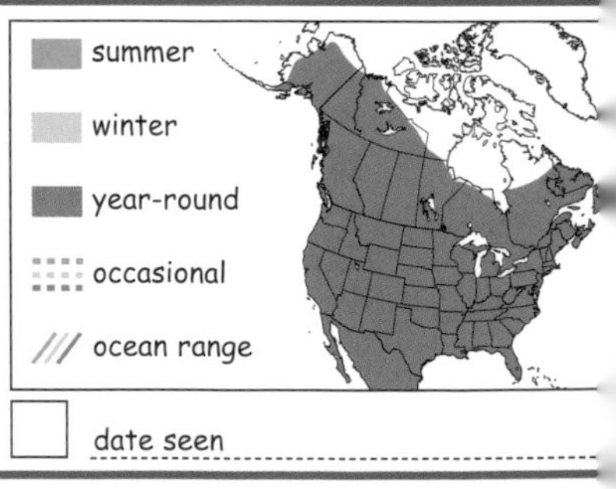

date seen

Strix occidentalis **Length: 17½–18"**

Look for: The Spotted Owl is a medium-sized, round, dark brown owl with dark eyes. It gets its name from the large white spots on its belly.

Listen for: Loud, high-pitched hoots in groups: *whoo-whowho-WHOO.*

Remember: The Spotted Owl has spots on the breast. The very similar Barred Owl has bold dark vertical streaks on its breast.

WOW!
Spotted Owls hunt mostly at night, capturing small mammals on the ground and even catching bats in the air.

▲ *A Spotted Owl in its daytime roost in a tree.*

Find it: Rare in old-growth forests of the mountain West and pine-oak forested canyons of the Southwest. Most often seen while roosting during daylight. Most birds are resident but some at high elevations move lower in winter.

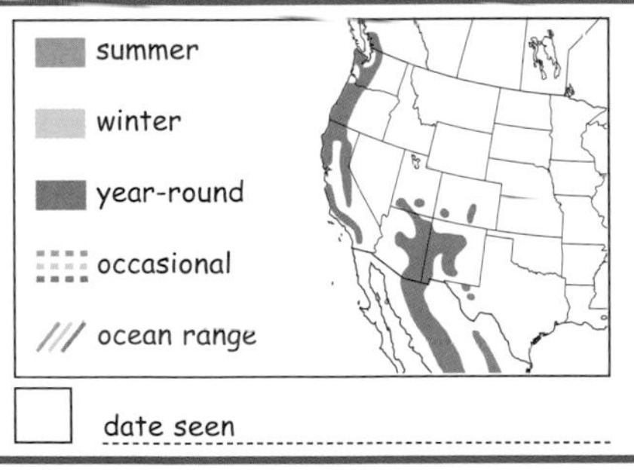

date seen ________

BARRED OWL

Strix varia **Length: 21"**

Look for: Mostly nocturnal and heard far more often than it is seen, the Barred Owl may be active on cloudy days, particularly when feeding hungry nestlings in late winter and spring.

Listen for: *Who cooks for you! Who cooks for you-all!* The Barred Owl's low-pitched hooting call is one of the better known among birders. Barred Owls will call at any time of year, and they will even call during the day in response to a loud noise.

Remember: Among the three common owls of eastern woodlands, the screech-owl has the highest voice and the Great Horned the lowest. The Barred Owl's call is pitched between these two and always follows its distinctive eight-note pattern.

▼ *Barred Owls will wade in creeks to catch crayfish to eat.*

WOW!
In western North America, the Barred Owl is expanding its range, and in some areas this is forcing out the closely related but much smaller and rarer Spotted Owl.

Find it: Most common in the Southeast but present across eastern North America in mixed woodlands, especially woods near water.

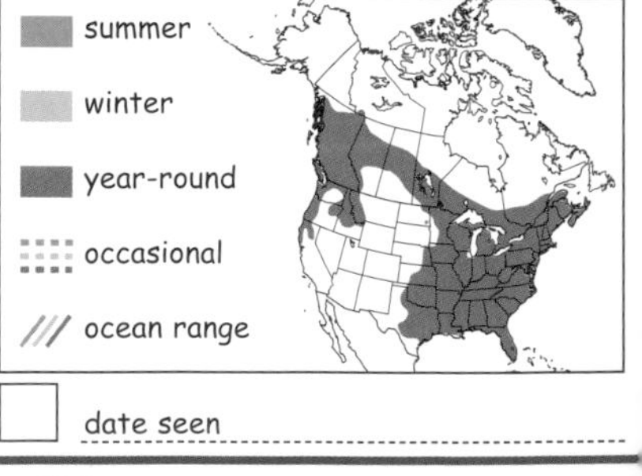

date seen

GREAT GRAY OWL

Strix nebulosa **Length: 26–28"**

Look for: A huge dark owl with a round face and piercing yellow eyes. Streaking on breast is smudgy. Black-and-white pattern on chin is distinctive. Pale feathers between eyes and bill form a large gray X.

Listen for: A series of ten or fewer deep-toned hoots, which get softer near the end.

Remember: The Great Gray has a large, rounded head, unlike the more common Great Horned Owl's, which shows ear tufts.

Great Gray Owls appear huge, but if you stripped all their feathers off, you'd see that their bodies are actually much smaller than you'd think.

▶ *Using their facial disks to focus sounds to their ears, Great Gray Owls can home in on their prey by sound. They may try to capture prey they hear deep below the surface of the snow.*

Find it: Uncommon in dense boreal forest and nearby bogs and meadows in the North. Found in mountain clearings and burned-over woods in West. Active hunter in both day and night.

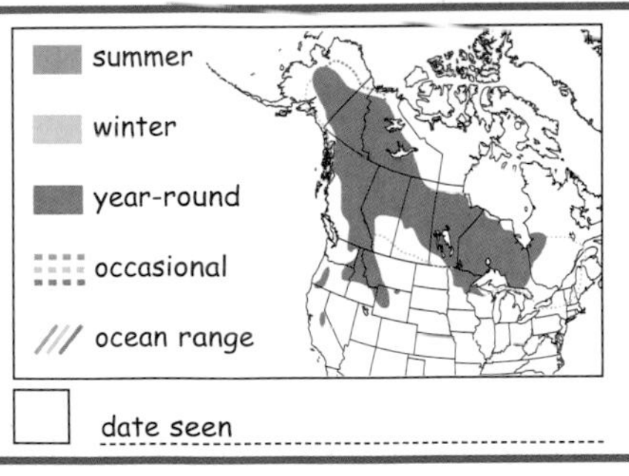

SNOWY OWL

Bubo scandiacus **Length: 24"**

Look for: This nearly all-white bird of the far North is difficult to confuse with anything else. Young birds and adult females are streaked with black. Older adult males are pure white. Some winters when small mammals are scarce in the far North, Snowy Owls head south in search of food.

Listen for: Snowies are usually silent while wintering south of the Arctic nesting grounds, but if you spend the summer on the Arctic tundra, you'll probably hear a Snowy Owl make some noise.

Remember: Snowy Owls that come south in winter may be stressed and hungry, so it's best to enjoy watching them from a distance, where you won't disturb them.

WOW!
In the Arctic in summer, the sun never goes down, so Snowies are more adept than other owls at daytime hunting. They may catch and consume as many as 1,600 lemmings in a single year.

◀ *A male Snowy Owl offers his mate a prey item, hoping to impress her.*

Find it: Every winter some Snowy Owls journey south from the Arctic tundra in search of food, and birders see them perched on fenceposts along meadows and coastal dunes and at other large, open expanses.

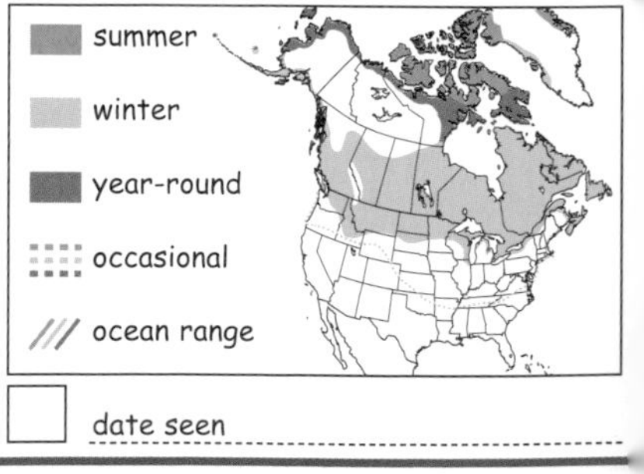

date seen

EASTERN SCREECH-OWL

Megascops asio **Length: 8½"**

"Red" morph

"Gray" morph

Look for: The most common of our small owls, the Eastern Screech-Owl comes in two color versions, or morphs: red and gray, with gray being the more common. Its small size, cryptic coloration, and inactivity during the day make this bird easy to overlook.

Listen for: Eastern Screech-Owls utter a series of high, wavery whinnies that descend in tone. This is often followed by a long trill on a single, lower tone.

Remember: Your first clue to a screech-owl's presence might be hearing its haunting call at dusk. If it sounds close by, try slowly moving a flashlight across nearby trees, watching for the reflection of the bird's yellow eyes.

WOW!

If your backyard is wooded, consider putting up an owl nesting box. Place it high on a tree that has a trunk wider than the box. The box opening (three inches in diameter) should face south.

◀ *An Eastern Screech-Owl peeks out of a nest box to soak up the warmth of the afternoon sun.*

Find it: This species can be difficult to find because it is nocturnal. A cavity nester, it prefers habitats with old trees that have holes to nest in. Found in woodland settings, including suburbs and city parks.

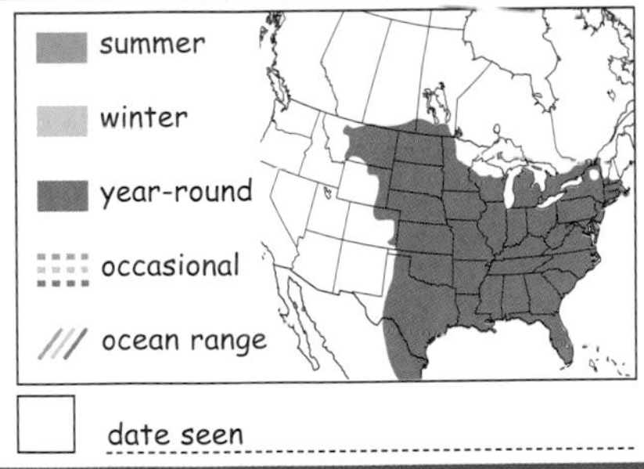

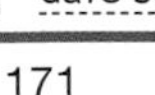

date seen

BURROWING OWL

Athene cunicularia **Length: 9½"**

Look for: A medium-sized long-legged owl that is brown overall with a white throat and white eyebrows. It has white spots on its back and a brown-striped breast. When open, its large yellow eyes stand out. Stands near burrow or on a low perch, often bobbing its head.

Listen for: A mellow, dovelike *coo-HOOO* that does not sound like an owl. Also rattles like a rattlesnake from within its nest burrow.

Remember: The Burrowing Owl is the owl most likely to be seen during daylight in open grassland.

◄ *Burrowing Owls will readily use artificial nest burrows in appropriate habitat.*

Find it: Found in prairies, farmland, and open grassy areas such as airfields, where it nests in underground burrows. Declining because of loss of habitat throughout its range.

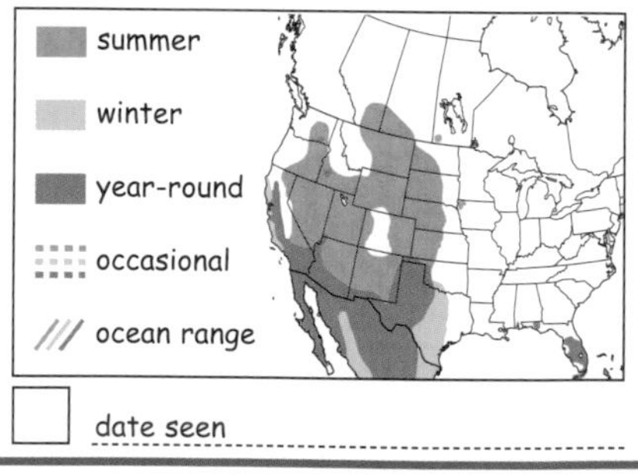

☐ date seen ____________________

NORTHERN PYGMY-OWL

Glaucidium gnoma **Length: 6¾–7"**

Look for: This sparrow-sized owl is round headed and long tailed, with a spotted crown and boldly streaked belly. It bobs its head and flicks its tail while perched. Pursues and catches small birds in flight.

Listen for: A series of toots on one tone, repeated over and over. Bird watchers in the West often imitate the Northern Pygmy-Owl's tooting call to lure chickadees and other birds into view.

Remember: The Northern Pygmy-Owl is longer tailed than our other small owls.

WOW!
The Northern Pygmy-Owl has fake "eyes" on the back of its head.

◀ *Songbirds, often led by chickadees, will eagerly mob a Northern Pygmy-Owl because this tiny owl eats lots of songbirds.*

Find it: Found in a variety of open woods and wooded canyons. Avoids deep woods. Active during daylight, when it perches on interior tree branches, watching for unwary small birds. Responds to imitations of its call.

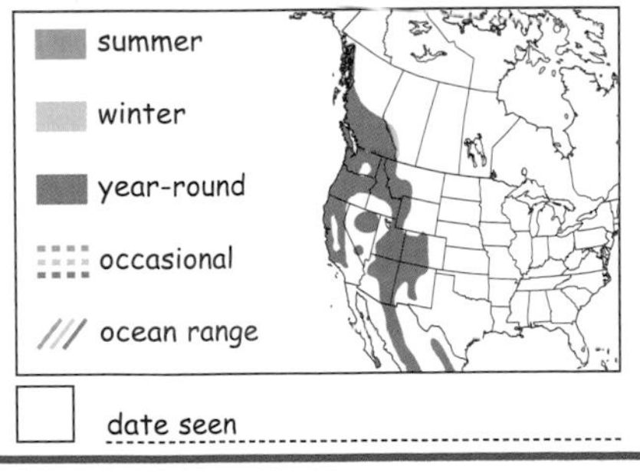

☐ date seen ____________

COMMON NIGHTHAWK

Chordeiles minor **Length: 9½" Wingspan: 24"**

Look for: Overhead in the summer sky, the Common Nighthawk flies with its choppy wingbeats, gliding and swooping after flying insects. Up close, the nighthawk is perfectly camouflaged for daytime perching on the ground or lengthwise along a tree branch. In flight, the bird appears dark overall with a white slash across each wing.

Listen for: Flying nighthawks utter a sharp, nasal call: *beeertt!* Males performing their courtship flight dive from a great height, making a deep, booming noise with their wings.

Remember: Nighthawks are not really hawks at all, though they look hawklike in flight. Scan the sky at dusk for their distinctive shape and flight style.

◄ *Common Nighthawks in a flock during fall migration*

WOW!

Nighthawks are members of the goatsucker family, which got its name from the mistaken impression that these birds sucked milk or blood from goats and other livestock.

Find it: Common in summer, flying night or day over cities, fields, and parks, the Common Nighthawk is often heard before it is seen. In August and September, look for large flocks of migrating nighthawks at dusk.

summer
winter
year-round
occasional
ocean range

☐ date seen ____________

COMMON POORWILL

Phalaenoptilus nuttallii **Length: 7½–7¾"**

Look for: Its cryptic gray-brown coloration lets the Common Poorwill blend in perfectly when roosting on the ground during the day. This smallest of our nightjars has a white throat, large head, and short tail.

Listen for: Call is a loud, ringing, repetitive *puh-REE-wah* or, if you use your imagination, *poor-WILL*.

Remember: Like other nightjars, the Common Poorwill is heard more often than seen. If you *do* see one, it will appear much smaller than a Whip-Poor-Will, with rounder wings and a shorter tail.

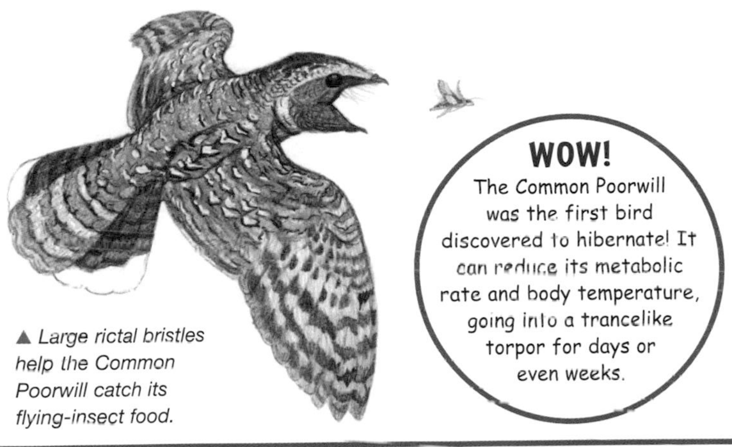

▲ *Large rictal bristles help the Common Poorwill catch its flying-insect food.*

WOW!
The Common Poorwill was the first bird discovered to hibernate! It can reduce its metabolic rate and body temperature, going into a trancelike torpor for days or even weeks.

Find it: Locally common in areas with dry, rocky ground and scattered brush. Roosts by day, hunts from ground at night, flying up to catch insects. Often hunts along roadways where headlights or a flashlight will cause its eyes to reflect bright orange.

summer
winter
year-round
occasional
/// ocean range

☐ date seen ____________

WHIP-POOR-WILL

Caprimulgus vociferus **Length: 10"**

Look for: The Whip-poor-will is heard far more often than it is seen, but you might catch a glimpse of one as it makes short, wheeling flights after a flying insect. In flight, it is silent and compact looking with rounded wings. The male Whip shows a flash of white on the outer edges of the tail.

Listen for: Calls its name over and over again: *whip-poor-WILL, whip-poor-WILL, whip-poor-WILL*. Calls only at night, and only from spring through early fall.

Remember: In the Southeast, the Whip-poor-will is replaced by the larger and browner Chuck-will's-widow. The Chuck also says its name. Voice, not visual clues, is the best way to separate these two closely related species in the area where their ranges overlap.

WOW!
Whip-poor-wills have been timed giving nearly one call per second for more than 15 minutes! That's 1,000 calls in a row without stopping. Don't try this at home!

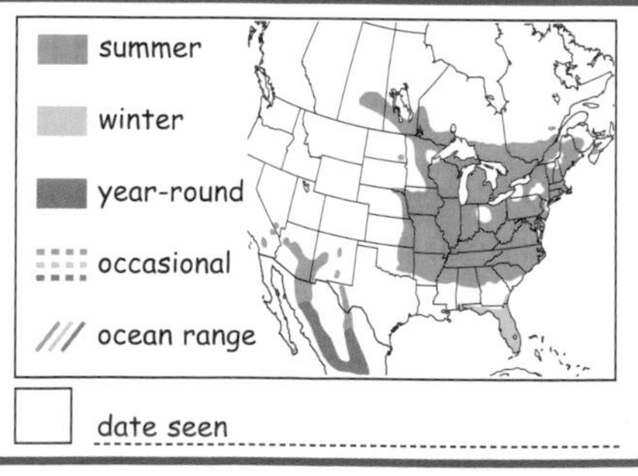

▶ *The Whip-poor-will's large, dark eyes gather light, allowing it to see during its nighttime foraging flights.*

Find it: Whip-poor-wills fill eastern woodland nights with their loud, ringing call. Listen for them to start calling at dusk. During the day, Whips roost on low branches or on the ground, relying on their cryptic coloration for camouflage.

summer
winter
year-round
occasional
/// ocean range

☐ date seen ____________

RUBY-THROATED HUMMINGBIRD

Archilochus colubris **Length: 3¾"**

Look for: The male's red throat (or gorget) shines brightest when in direct sunlight. At other times it can appear black. The female has a white throat, and both males and females have metallic green backs and wings.

Listen for: Ruby-throats utter a series of high-pitched twitters almost constantly. They are named for the humming sound made by their wings.

Remember: The tiny Ruby-throated Hummingbird zips past so fast it can be mistaken for an insect, especially a sphinx moth.

WOW!

Ruby-throats are tiny (weighing less than a penny!) but powerful birds. Their wings flap as fast as 75 beats per second! They are strong fliers, able to migrate 500 miles in one night.

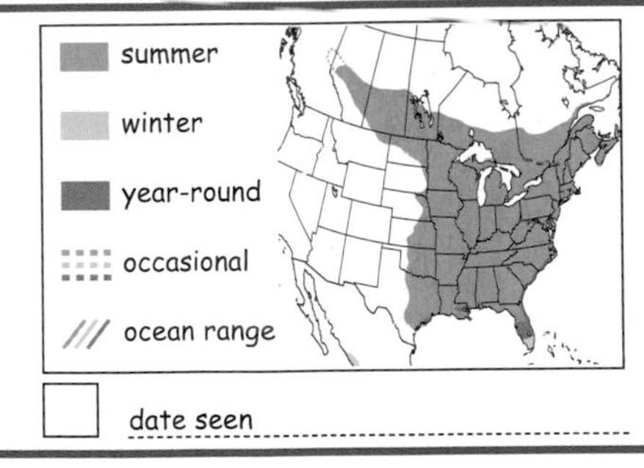

▶ *A male Ruby-throated Hummingbird flashes his red throat in a courtship flight for a perched female.*

Find it: The only common hummingbird in the eastern U.S., the Ruby-throat prefers mixed habitat where nectar-producing plants can be found, including parks, gardens, meadows, and woodland edges.

summer

winter

year-round

occasional

/// ocean range

☐ date seen ______________________

COSTA'S HUMMINGBIRD

Calypte costae **Length: 3½"**

Look for: The male Costa's Hummingbird has a striking, deep purple crown and a gorget that extends downward like a Fu Manchu mustache—unique among our hummingbirds. The female is green above and white below.

Listen for: Call is diagnostic: a high-pitched, tinny *tink-tink-tink-tink*. Male's display call is a rising, then falling *ziiiing!*

Remember: This species is similar to the Black-chinned Hummingbird, which is slightly larger. The male Black-chinned lacks the extended gorget and purple crown of the male Costa's.

This species nests in late winter/early spring in the desert, then migrates west to the Pacific Coast to avoid the desert's summer heat.

▶ *The Costa's Hummingbird is a desert-dwelling bird adept at feeding on desert flowers.*

Find it: Common in low-desert habitats: desert scrub, chaparral, dry washes with flowering plants, backyards, and gardens.

summer
winter
year-round
occasional
/// ocean range

☐ date seen

BLACK-CHINNED HUMMINGBIRD

Archilochus alexandri Length: 3¾"

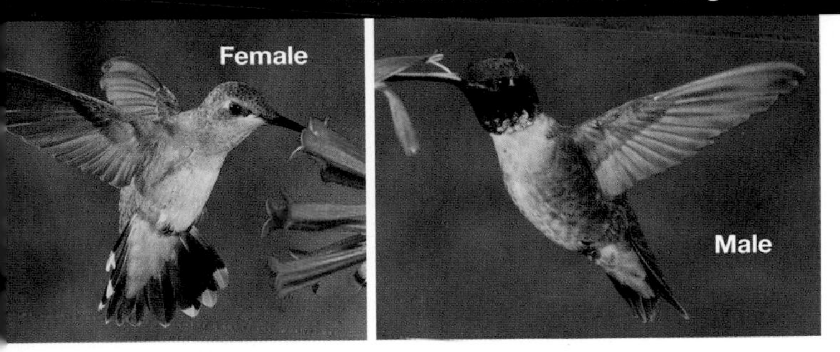

Look for: The male's gorget flashes deep purple in the right light; in poor light it can appear black. Both sexes are bright green on the head, back, and tail. Male appears dark headed and shows a clean white bib below the dark purple gorget. Female looks similar to female Ruby-throated Hummingbird.

Listen for: The wings make a humming sound in flight. Chasing birds give a series of twitters and buzzes. Call is a soft *chew!*

Remember: Black-chinned Hummingbirds pump their tails consistently while hovering.

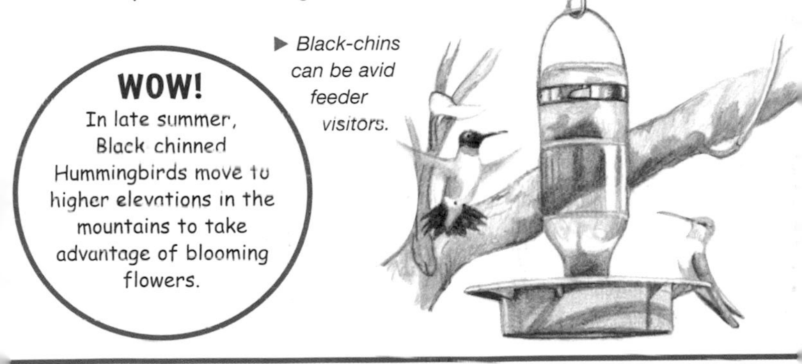

▶ *Black-chins can be avid feeder visitors.*

WOW!
In late summer, Black chinned Hummingbirds move to higher elevations in the mountains to take advantage of blooming flowers.

Find it: This is the most widespread hummingbird in the West during summer. Common in wooded foothills and canyons and along the ocean coast. Largely absent from the West in winter.

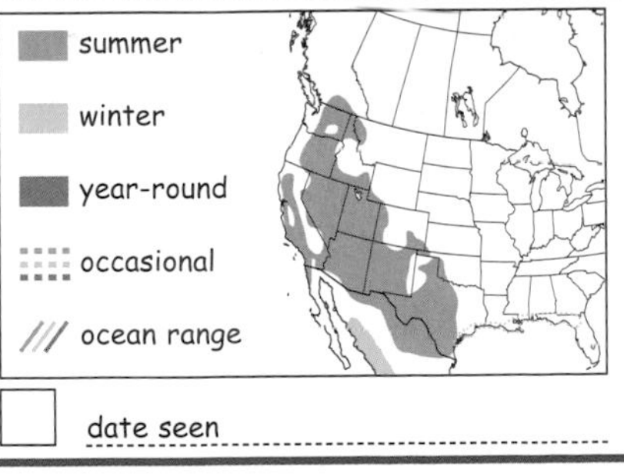

☐ date seen

ANNA'S HUMMINGBIRD

Calypte anna **Length: 4"**

Look for: The male Anna's Hummingbird is our only hummer with a bright red crown. His crown and gorget are brilliant ruby in good light but can appear black in bright sunshine. A greenish back and smudgy green belly make the Anna's appear "messier" than other hummers. The female may have some red spots on the otherwise white throat.

WOW!
Male Anna's can erect gorget and head feathers and move back and forth to flash bright colors at intruders, rivals, and potential mates.

Listen for: One of our most vocal hummingbirds. Gives a series of raspy notes and chattery buzzes in what passes for its song, which the male usually delivers while perched.

Remember: At 4 inches long, the Anna's is larger than most other widespread western hummingbirds.

▶ *Male Anna's hummers are avid, if not gifted, singers, usually singing from a perch in their territory.*

Find it: The Anna's is common along the Pacific Coast all year long (often the only hummingbird present in midwinter) and is found in brushy habitat and open woods with nectar-producing flowers, including parks, gardens, and backyards.

summer
winter
year-round
occasional
/// ocean range

☐ date seen

BROAD-TAILED HUMMINGBIRD

Selasphorus platycercus **Length: 4"**

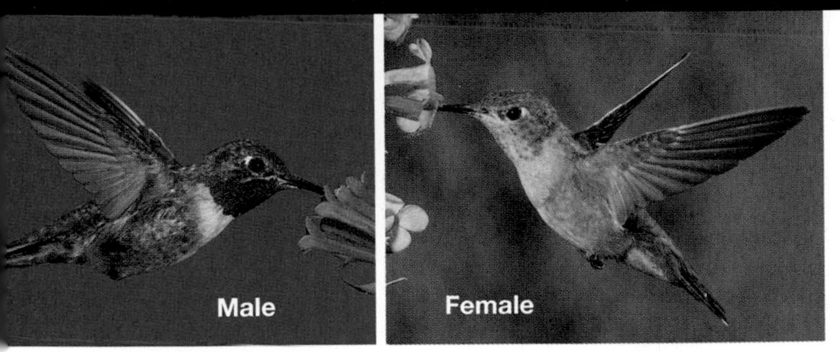

Look for: Males and females are metallic green above with a white breast and belly. Male has a magenta gorget. Female has a speckled throat and buffy sides. This hummingbird is often heard before it is seen because the wings of the adult male produce a loud, whistled trill as he flies.

Listen for: In addition to the wing trill of the male in flight, Broad-tails produce a high-pitched chattering song, usually from a perch. Call is a loud *chit!*

Remember: This species is the most commonly encountered hummer in the Rocky Mountains.

WOW!

Broad-tails arriving on territory in early spring before flowers are blooming survive on sap at sapsucker wells.

▶ *When the western mountain meadows are full of blooming flowers in summer, Broad-tailed Hummingbirds are easy to find.*

Find it: The species is common throughout the inland mountain West, preferring mountain forests, meadows, and wooded canyons.

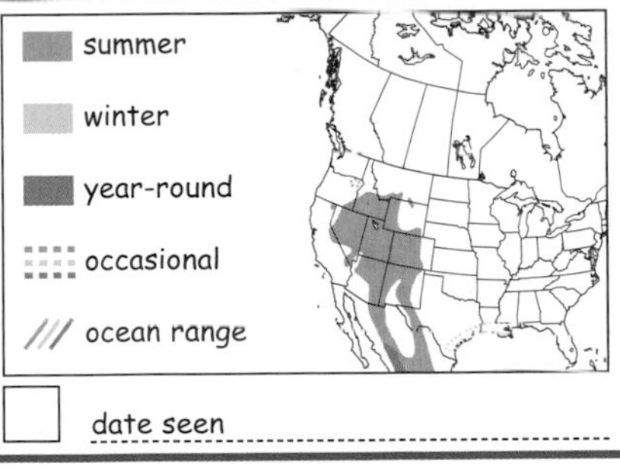

☐ date seen ____________

RUFOUS and ALLEN'S HUMMINGBIRD

Selasphorus rufus, Selasphorus sasin **Length: 3¾"**

Look for: These two small hummingbirds are very similar, and females are nearly impossible to tell apart. The male Rufous lives up to its name in color: bright rufous-orange overall, including the back. The male Allen's has a green back and orange tail.

Listen for: Both vocal and nonvocal sounds of the two species are similar. Aggressive flight call: *zee chippity chippity*. Call note: *chuk*. Wings of males in flight make a high-pitched trill.

Remember: Rufous: none of our other hummers has a rufous back. Allen's: looks like a Rufous Hummingbird but with a *green* back.

WOW!
Individuals of both species can be found wintering in the eastern and southeastern U.S. but the Rufous is the more likely one to see.

▶ *Rufous Hummingbirds breed as far north as Alaska, where late-spring snows can make finding flower nectar more difficult.*

Find it: During the spring and summer: wooded, brushy areas, canyons, open forest, parks, gardens, backyards. In migration both species seek out mountain meadows with flowering plants. Breeding range of Allen's is mostly within California. Rufous is more widespread.

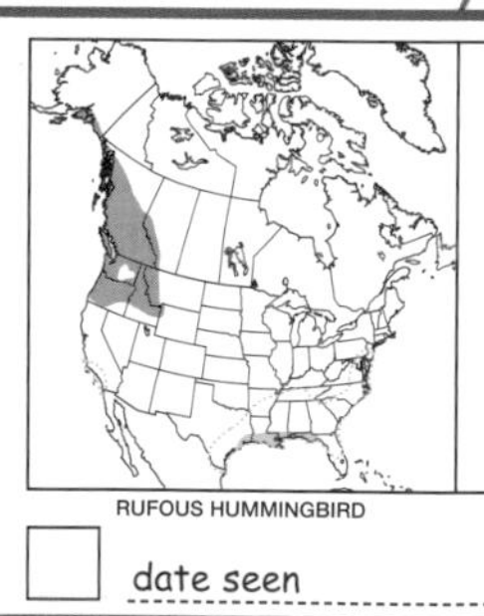

RUFOUS HUMMINGBIRD

ALLEN'S HUMMINGBIRD

☐ date seen

CALLIOPE HUMMINGBIRD

Stellula calliope **Length: 3¼"**

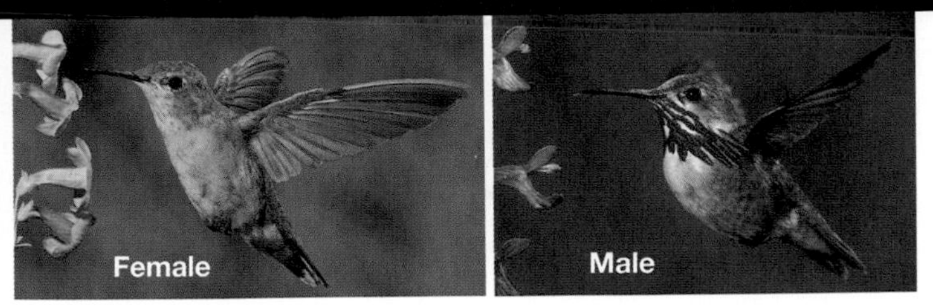

Look for: A very small short-billed hummingbird that's shiny green above and white below. Adult male has a gorget that appears striped with magenta feathers. Female is plainer overall with a white spot at the base of the bill.

Listen for: Song is a high whistle. Calls consist of high-pitched chips and buzzes in a series.

Remember: Besides its small size, another good field mark is this: when perched, the Calliope Hummingbird has wingtips that extend past the tip of the tail.

The Calliope is our smallest North American bird species, but it can survive the very cold nights in the Rocky Mountains where it nests in the summer.

► *Male Calliopes perform a striking U-shaped courtship flight to impress prospective mates.*

Find it: Uncommon in summer in mountain meadows and wooded canyons, often near streams. In migration can be found in lowlands. Will visit backyard feeders.

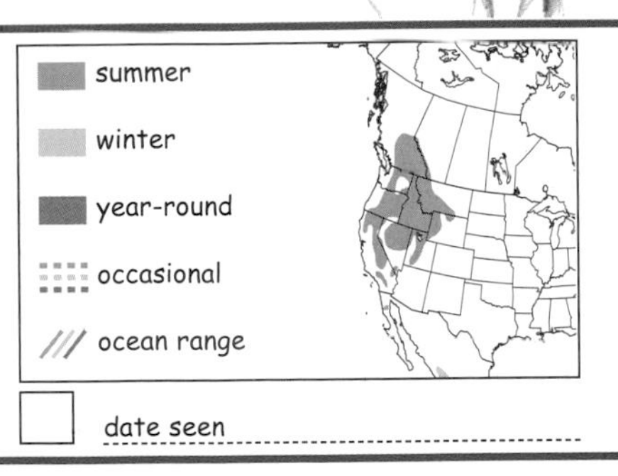

date seen

CHIMNEY SWIFT

Chaetura pelagica **Length: 5¼"** **Wingspan: 14"**

Roosting

In flight

Look for: Look! Up in the sky! It's a flying cigar! That's what the flight shape of the Chimney Swift looks like: a cigar-shaped body with long slender wings. Flying style is rapid flapping, with short glides in between. Adults have pale throats, but overall the Chimney Swift looks sooty gray. In early fall, huge flocks of swifts may gather at a large roost site, swirling above it before dropping in for the night.

Listen for: As they fly, Chimney Swifts emit a high, sputtering chatter that wavers in pitch.

Remember: Swifts are faster, more jerky fliers than swallows, and compared to swallows, the Chimney Swift appears stub tailed.

WOW!
Based on a ratio of body size to wing size, Chimney Swifts have the longest wings of any bird.

◀ *A Chimney Swift pair performing a courtship flight, gliding with wings held in a V.*

Find it: Most common in cities, where they use their namesake chimneys and other hollow structures for nesting and roosting. The Chimney Swift is the only eastern swift species.

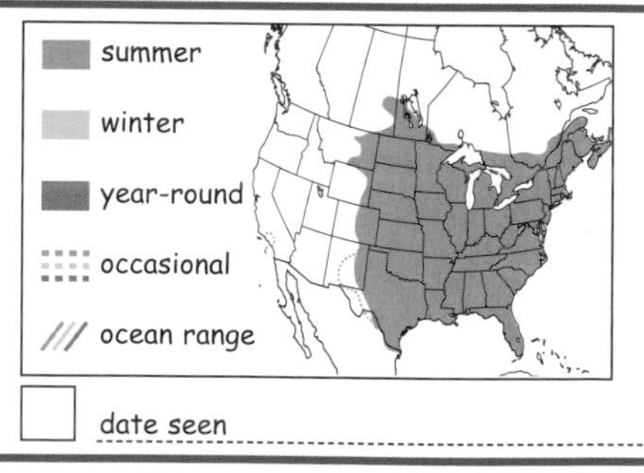

☐ date seen

WHITE-THROATED SWIFT

Aeronautes saxatalis **Length: 6½"**

Look for: Our only North American swift species with contrasting black-and-white body, the White-throated Swift usually forages in flocks, flying rapidly and chattering loudly. Throat, central belly, and sides of rump show obvious white.

Listen for: Loud, chittering *jejejejeje* dropping in pitch.

Remember: Our other swifts may show pale areas on the throat, but only one of our swifts is truly white throated.

WOW!

A fling on the wing. White-throated Swifts sometimes mate in flight: a male and female meet in the sky, then tumble downward before parting and regaining flight.

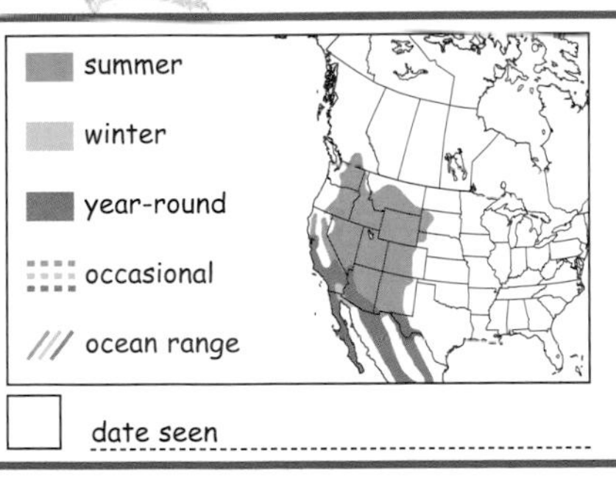

▶ *We know very little about the nesting habits of this species because they nest in such inaccessible places.*

Find it: Look up anywhere near a rocky cliff in the West, and these birds are likely to be present. Most common near nesting sites: cliff faces, arid mountains, canyons. Nearly always seen in flight, calling constantly.

summer

winter

year-round

occasional

/// ocean range

☐ date seen

BELTED KINGFISHER

Ceryle alcyon **Length: 13"**

Look for: This big-headed, shaggy-crested bird is hard to confuse with anything else. The long, daggerlike bill is used to spear fish when the kingfisher plunges from the air into the water. Both sexes have a gray breast-band, but females also have a rusty band (like a bra) across the upper belly.

Listen for: Most common call is a loud, dry rattle, often given in flight: *ptptptrrrrrrr!* This call has been compared to that of a Hairy Woodpecker.

Remember: In flight, the Belted Kingfisher has jerky, irregular wingbeats. Its white underwings and belly flash as it flies. These clues make it fairly easy to identify a Belted Kingfisher at great distances.

WOW!
Kingfishers dig nesting holes in vertical soil or sandbanks, often along streams. These nesting tunnels can be up to six feet long!

◀ *With beaks and feet, Belted Kingfishers excavate a deep burrow in a sandy bank, where they raise their young.*

Find it: A conspicuous bird that perches in the open along streams, rivers, and lakes, the Belted Kingfisher gives a loud, rattling call as it flies. It hovers over the water, making a spectacular plunging dive when it spots a fish.

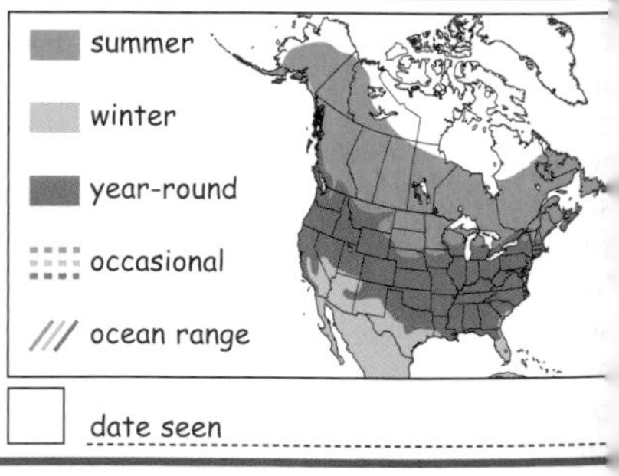

date seen

LEWIS'S WOODPECKER

Melanerpes lewis **Length: 10¾–11"**

Look for: A large dark woodpecker. Adults have a red face, pinkish belly, dark green back and wings, and silver-gray collar. Juveniles have an all-brown face and lack the collar.

Listen for: Not very vocal. Occasionally utters a high-pitched, sputtery chatter. Drumming is very rapid.

Remember: Unlike the undulating flight of other woodpeckers, the Lewis's flight is even, like a crow's.

▼ *Lewis's Woodpeckers are experts at flycatching: flying out from a perch to nab a passing insect.*

WOW!
The Lewis's Woodpecker is one of the species discovered and named during the Lewis and Clark expedition to the West in the early 1800s.

Find it: Uncommon in summer in open woods, orchards, riparian groves, and in logged or burned woodlands. In summer, open habitat is better for flycatching. Winters in loose flocks near a reliable source of food—such as acorns—in woods of oaks or other nut-bearing trees.

summer
winter
year-round
occasional
/// ocean range

☐ date seen ---------------------------------

ACORN WOODPECKER

Melanerpes formicivorus **Length: 9"**

Look for: Loud, active, and boldly marked, the Acorn Woodpecker is hard to miss. Both males and females are black backed and have a facial pattern resembling clown makeup (black and yellow face, white eye, black bill, red head patch), but male has a more extensive red crown. White rump and wing patches stand out in flight.

Listen for: A very vocal bird. Call is a raucous *wake-up! wake-up! wake-up!* Also sputtering *churr*s.

Remember: Other black-backed woodpeckers of the West lack the Acorn's clown makeup.

Male

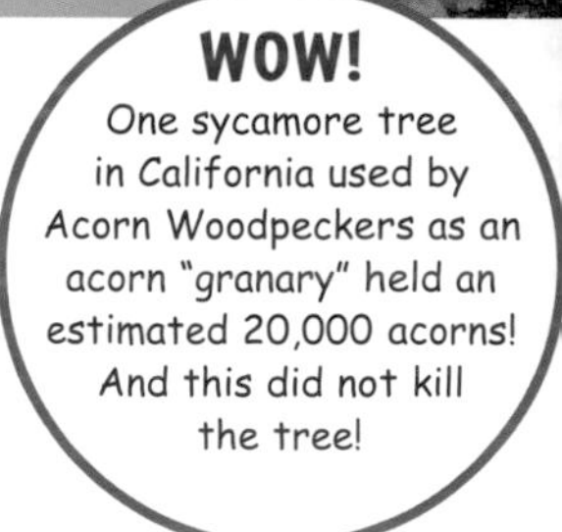

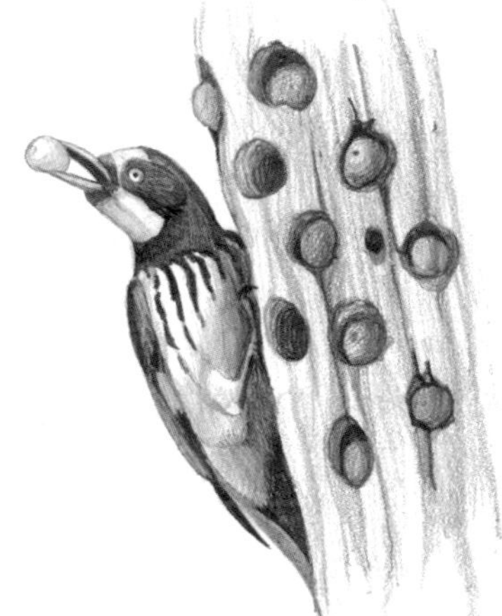

◀ *Acorn Woodpeckers drill holes in tree bark into which they insert individual acorns as a "pantry" of food for later in the year.*

Find it: As the name suggests, the Acorn Woodpecker usually associates with acorn-bearing oaks of several species. Lives in small, very social, and conspicuous colonies in mixed oak-pine woods, canyons, foothills, and in wooded residential areas.

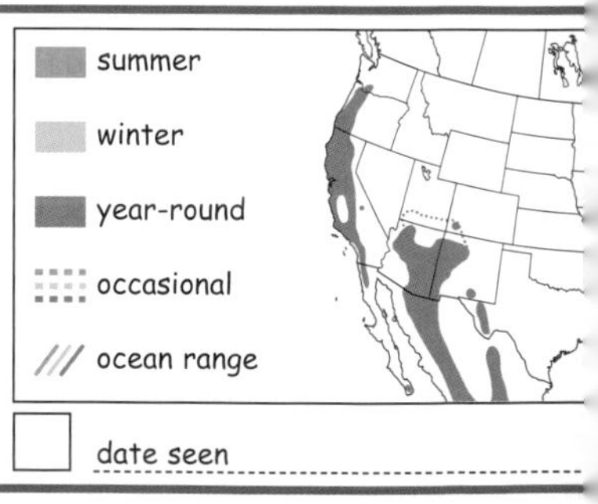

☐ date seen ______

RED-HEADED WOODPECKER

Melanerpes erythrocephalus **Length: 9¼"**

Look for: The adult Red-headed Woodpecker has an all-red head and a body that goes black-white-black from back to tail. In flight, the white wing patches flash and are an excellent field mark on distant birds. Young Red-heads have brown heads until their first full spring.

Listen for: Call is an excited-sounding *queerp!* Also gives a *chrrr* call in flight. Vocalizations are similar to those of the Red-bellied Woodpecker, but the Red-headed sounds more excited.

Remember: Only the Red-head has an *all-red head*. The male Red-bellied Woodpecker has a mohawk of red across the top of the head from front to back.

WOW!
Red-headed Woodpeckers are excellent fliers, and they use their aerial skills to catch flying insects.

◄ *Most battles between Red-headed Woodpeckers and European Starlings occur over nest cavities. The woodpecker usually wins.*

Find it: Locally common in stands of trees or open woods, the Red-head is easy to see when it flies. When perched, though, it often sits quietly and may be overlooked. It is loosely colonial, so where you find one bird, there may be others.

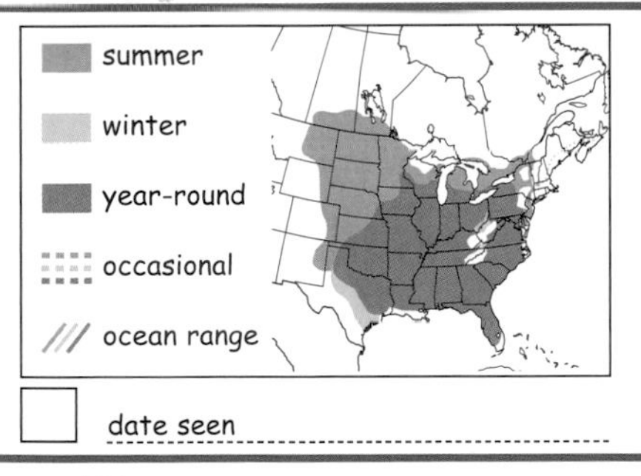

☐ date seen ____________

RED-BELLIED WOODPECKER

Melanerpes carolinus **Length: 9¼"**

Look for: All Red-bellied Woodpeckers have zebra-striped backs; the male has a band of red on the head from the bill to the nape. Females have red napes only—their crowns are brown. Flight is strong and swooping and shows flashes of white in the wings and tail.

Listen for: Red-bellies make a variety of sounds, including a short call that rises in tone: *quiirrr*! And a longer call: *ch-ch-ch-chirrrrrrrr*. Avid drummers, they sometimes use metal chimneys and drainpipes for maximum noise.

Remember: The Red-bellied Woodpecker is named for a field mark that is very hard to see. Any red on the belly is hidden from view as the bird is perched against a tree trunk or branch.

WOW!
This species will eat almost anything, including insects, small fish, tree frogs, and even other birds and bird eggs!

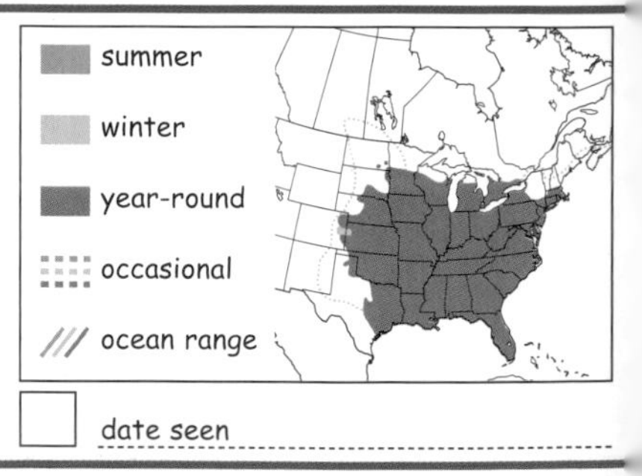

▶ *Red-bellied Woodpeckers often rule the feeder. Here, one threatens a White-breasted Nuthatch.*

Find it: Vocal and active, these birds are not hard to find when they are present in a habitat. They prefer mixed woodlands, but are common in all habitats throughout the Southeast and are expanding northward.

summer
winter
year-round
occasional
/// ocean range

☐ date seen

GILA WOODPECKER

Melanerpes uropygialis **Length: 9¼"**

Look for: A tan-bodied medium-sized woodpecker with a zebra-striped back and wings and a plain face. Male has a round red cap (female lacks this). Dark eye and bill stand out on tan head.

Listen for: Call is a burry, rising *churr!* Also gives a series of squeaky calls: *earp! earp! earp!* Drumming is loud but spaced out and slows near the end.

Remember: Farther to the east, in Texas and Oklahoma, lives the very similar Golden-fronted Woodpecker, the male of which has a golden forehead and nape. Similar Ladder-backed Woodpecker has a striped face.

▶ *Many cavity-loving species, including martins, owls, and flycatchers, use the nest holes that Gila Woodpeckers excavate in saguaro cacti.*

Find it: A woodpecker of the low southwestern deserts, where it prefers desert washes, riparian woodlands, and towns, especially in areas with saguaro cacti.

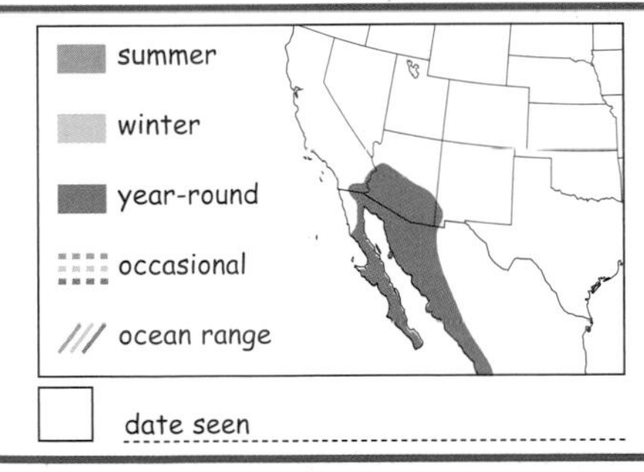

☐ date seen ____________

NORTHERN FLICKER

Colaptes auratus **Length: 13"**

Look for: The Northern Flicker occurs in two distinct forms: Yellow-shafted (a mostly eastern bird) and Red-shafted (mostly western), named for the color on the undersides of the wings. Other field marks include a black breast-band, black belly spot, red nape spot, and a black mustache (on Yellow-shafted males).

Listen for: The flicker's main call is a high-pitched, explosive *kleer!* Or a long, even series of maniacal-sounding notes: *bir-bir-bir-bir-bir*. A nasal *wik-a-wik-a-wik-a-wik* call accompanies courtship and territorial displays. Frequently drums loudly.

Remember: The flash of color in its wings and the brilliant white rump can ID a flicker even at a great distance.

▶ *A Northern Flicker eats plant seeds in winter when insects are not available.*

Find it: Prefers open areas of scattered trees, such as parks, cemeteries, and backyards, over deep woods. Of all of our woodpeckers, this is the one most likely to be seen on the ground, where the flicker loves to eat ants.

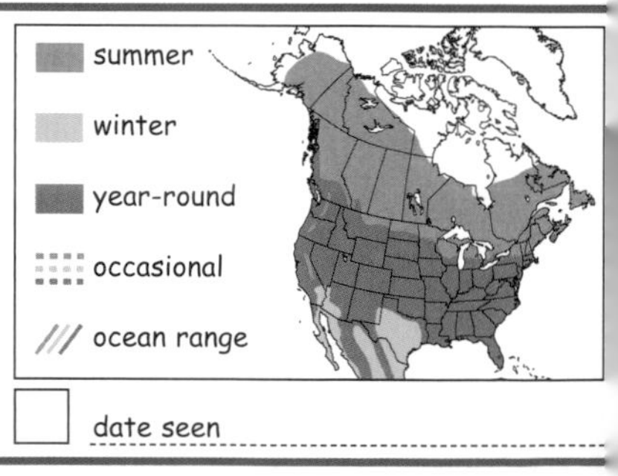

☐ date seen ________

RED-BREASTED SAPSUCKER

Sphyrapicus ruber **Length: 8½"**

Look for: Of our four sapsucker species, only the Red-breasted has an all-red head and breast. The back and tail are loosely zebra striped, and the black wings have a bold white stripe that is vertical on perched birds. Adults are similar. Juvenile birds are dusky headed.

Listen for: Loud, descending *cherrr!* Nasal in tone. Also a wild-sounding series of squeals. Drumming is irregular clusters of beats: *ratatat-tatat-rata-tatatat.*

Remember: This is one of the few woodpecker species in which the adult males and females look alike.

WOW!
Once considered part of the same species as the Yellow-bellied Sapsucker, the Red-breasted will interbreed with the Yellow-bellied and the Red-naped in the limited areas where their ranges overlap.

▶ *Sapsuckers will chase away other birds and animals that try to mooch a meal at their sap wells. This moocher is a Rufous Hummingbird.*

Find it: Common in mixed deciduous-coniferous woods and groves along the western edge of North America. This sapsucker species is less migratory than its closely related species.

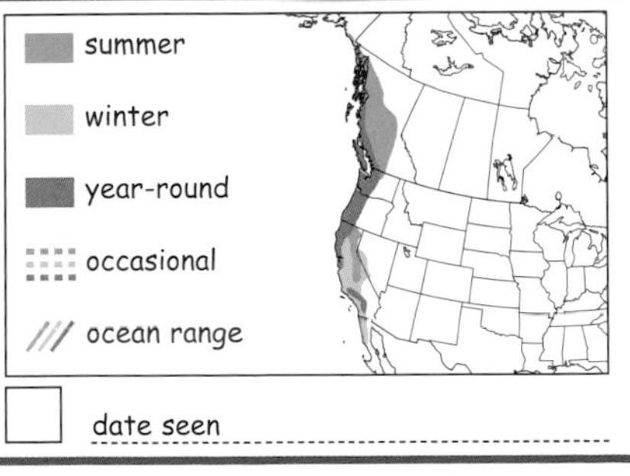

☐ date seen ______________

YELLOW-BELLIED SAPSUCKER

Sphyrapicus varius **Length: 8½"**

Male

Female

Look for: The yellow belly for which this bird is named is visible only on flying adult birds. Adults have red crowns, and males add a red throat to the mix. The best field mark is the long vertical white slash along the wing. First-year birds have mottled brown heads and breasts.

Listen for: Usually silent in winter but vocal on the breeding grounds and in spring migration, when it utters a wheezy, catlike *meeyaah!* Light tapping as it drills sap holes can be a good way to locate this species.

Remember: This is our only eastern sapsucker, but there are three closely related sapsucker species in the West: the Red-breasted, Red-naped, and Williamson's Sapsuckers.

WOW!

It's no joke! The Yellow-bellied Sapsucker has a silly name. The truth is they do eat sap, and their sap holes are also used by many other birds, mammals, and insects as sources of food.

▼ *An adult male Yellow-bellied Sapsucker defends its sap wells from a curious Ruby-throated Hummingbird.*

Find it: Yellow-bellied Sapsuckers are especially fond of fruit trees and other trees that readily ooze protective sap. They can be creatures of habit, returning to visit old sap holes year after year. Look for their rings of sap holes on tree trunks.

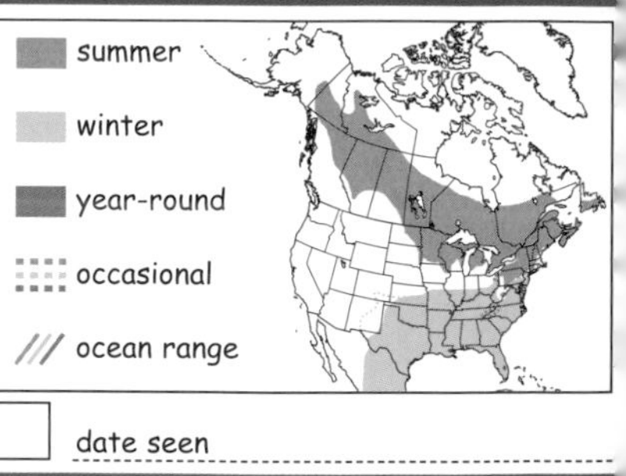

☐ date seen ____________

LADDER-BACKED WOODPECKER

Picoides scalaris **Length: 7¼"**

Look for: It's hard to miss the horizontal black-and-white "ladder" stripes across the back of the Ladder-backed Woodpecker. The boldly marked face shows a white cheek. Adult male has a red crown.

Listen for: Common call is a high-pitched *peek!* Also a laughing rattle call, dropping off in pitch at the end. Drumming is a fast drumroll.

Remember: A woodpecker with a zebra-striped back in arid southwestern habitat is likely to be this species.

WOW!

Two of the folk names for the Ladder-backed Woodpecker are Cactus Woodpecker (for its preferred nest site) and Mexican Woodpecker (for its range). Neither one of these is as descriptive as Ladder-backed.

▲ *Mated pairs of Ladder-backs forage together but concentrate on different parts of the tree: males on larger limbs and trunks, females on smaller ones.*

Find it: A permanent resident in the dry mesquite and wooded scrub habitats of the Southwest, the Ladder-backed Woodpecker fills the same ecological niche as the Downy Woodpecker, which does not occur in this specific habitat.

summer
winter
year-round
occasional
ocean range

☐ date seen ____________

HAIRY WOODPECKER

Picoides villosus **Length: 9"**

Look for: The Hairy Woodpecker looks like a supersized version of the Downy Woodpecker. The best field mark for separating the two is bill length. The Hairy has a large bill that's as long as its head is wide.

Listen for: Hairy Woodpeckers even *sound* bigger than Downies. Their call note, *peek!*, is sharper and louder than the Downy's, and the Hairy's rattle is lower and more emphatic and *does not drop in pitch*. They also drum on hollow branches.

Remember: Despite their larger size, Hairy Woodpeckers can be shier than Downies. When a Hairy sees you approaching, it may scoot around to the back of the tree trunk it's on before flying off.

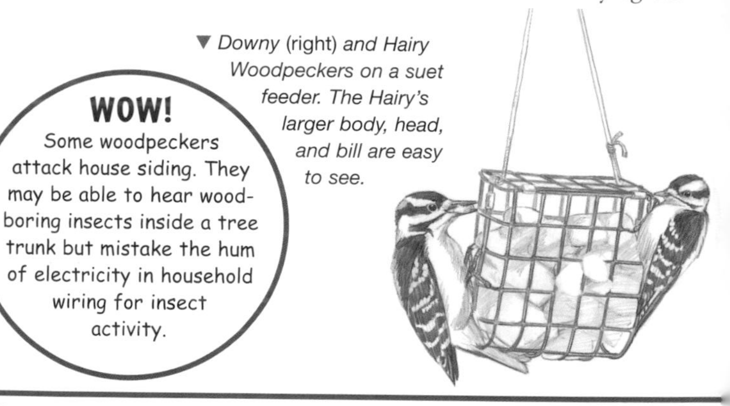

▼ *Downy* (right) *and Hairy Woodpeckers on a suet feeder. The Hairy's larger body, head, and bill are easy to see.*

WOW!

Some woodpeckers attack house siding. They may be able to hear wood-boring insects inside a tree trunk but mistake the hum of electricity in household wiring for insect activity.

Find it: Hairies can be found year-round across the continent, but they need larger trees than Downies do for foraging and nesting, so they are not as common in open areas. At bird feeders, Hairies eat suet and peanuts.

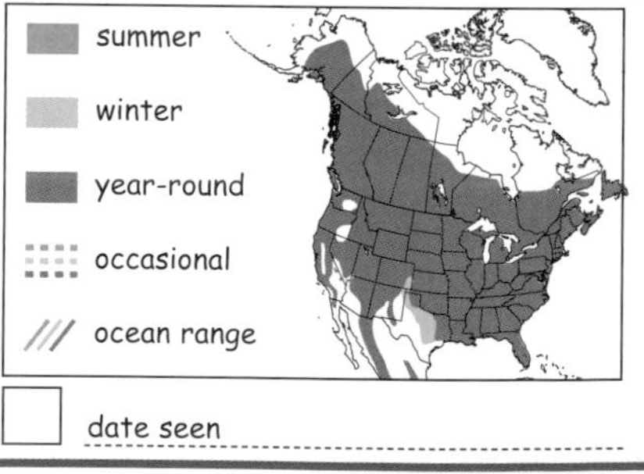

☐ date seen

DOWNY WOODPECKER

Picoides pubescens **Length: 6½"**

Look for: The Downy is the smaller of our two common black-and-white woodpeckers (and North America's smallest woodpecker). Its bill is short, equal to about half of the width of its head. The male Downy has a small red spot on the back of its head (females have no red).

Listen for: Downies give a high-pitched *pik!* call as well as a long, ringing rattle, *trrrrrrrrrrrr!*, that descends in pitch (the Downy's rattle goes *down*). Also drums on resonating hollow branches.

Remember: Downy is "dinky," Hairy is "huge." Telling these two species apart is easy once you remember this size difference. The Hairy's bill is as long as its head is wide.

Every fall, male and female Downies each excavate their own roost holes. These are the cozy places where they spend the chilly winter nights.

► *The Downy Woodpecker's winter diet includes larvae it extracts from galls on goldenrod stems.*

Find it: Widespread and common wherever there are trees for foraging and nesting. Downies often perch on weed stalks and do not require the large trees needed by the Hairy Woodpecker. Frequents backyard bird feeders.

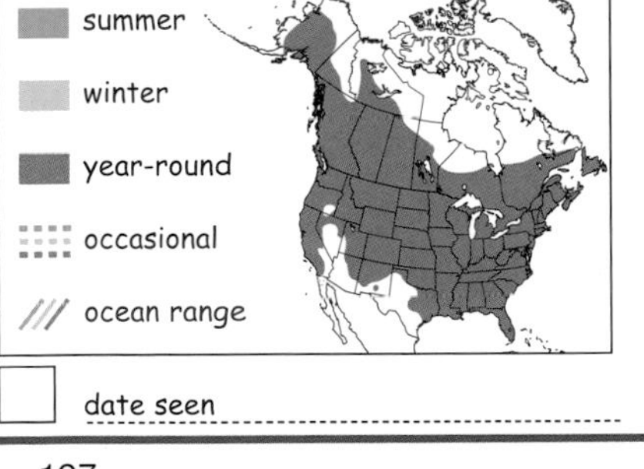

☐ date seen ______________________

PILEATED WOODPECKER

Dryocopus pileatus **Length: 17"**

Male

Female

Look for: The Pileated is difficult to confuse with any other bird because of its large size and obvious red crest. Perched, the Pileated appears mostly black except for the head and neck. The bright white underwings are an excellent field mark on distant flying Pileateds.

Listen for: Pileateds are very vocal; the series of notes in its long, loud call sounds like someone laughing. It is similar to the call of the Northern Flicker, but louder and more excited. In flight, they may utter single or double notes of the same quality. Also drums loudly, with a pattern that slows down as it ends.

Remember: The much rarer (and possibly extinct) Ivory-billed Woodpecker has a white bill and black throat, and shows huge white wing patches when perched.

▲ *In fall and winter, Pileated Woodpeckers rely on fruits and berries for much of their diet.*

Find it: As the woodlands of North America have returned, so has the Pileated Woodpecker. It is most common in mature woods with large trees but also found in city parks and suburban neighborhoods.

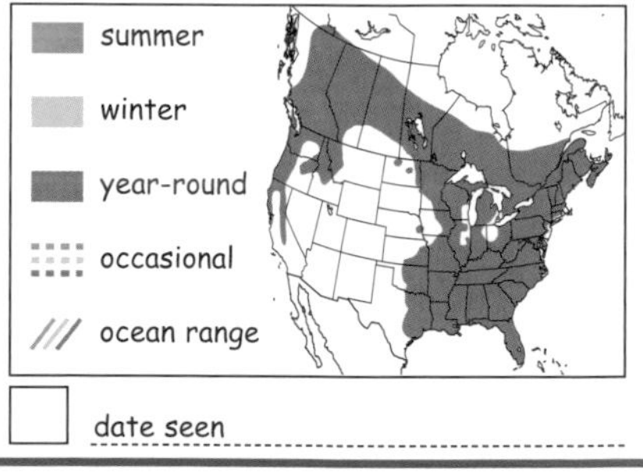

☐ date seen

EASTERN and WESTERN WOOD-PEWEE

Contopus virens, Contopus sordidulus **Length: 6½", 6¼"**

Look for: The Eastern Wood-Pewee is similar to the Eastern Phoebe but smaller. Two obvious pale wing bars, lack of a distinct eye-ring, and a lack of tail wagging are good ID clues for the pewee. Western: dark "vest" on chest is stronger than that of Eastern, wing bars are less distinct.

Listen for: The song of the Eastern Wood-Pewee is *pee-ah-WEE!* or *PEE-yerr!* Western: nasal *pee-yerrr*. Pewees often sing throughout the day and continue even after dusk, when many other birds are silent.

Remember: Eastern Wood-Pewees do not wag their tails while perched, but many birders have noted that wood-pewees often flick their wings right after landing on a perch. Tail wagging = Eastern Phoebe. Wing flicking and no tail wagging = Eastern Wood-Pewee.

▼ *Even though it's exposed, the Eastern Wood-Pewee's nest is inconspicuous, resembling a lichen-covered knot on a branch.*

WOW!

The Eastern Wood-Pewee's western cousin is the Western Wood-Pewee. Fortunately, these two look-alike species have separate ranges and different voices, otherwise we'd never be able to tell them apart.

Find it: You are likely to hear a wood-pewee before you see it. Look for Easterns in summer along woodland edges, perched high in trees on internal bare branches, from which they make regular foraging flights. Western: prefers pine-oak forests, mature woods, woodland edges.

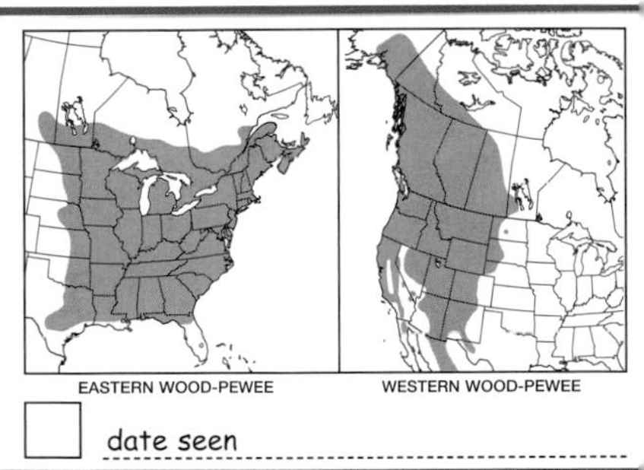

☐ date seen ____________

ACADIAN FLYCATCHER

Empidonax virescens **Length: 5¾"**

Look for: Among the small, drab, and confusingly similar *Empidonax* flycatchers, the Acadian Flycatcher is perhaps the most commonly encountered. It has the two pale buff wing bars and eye-ring of its fellow Empids, but its body shows more contrast. Its head often shows a peak at the back. Acadians sing frequently, so learning their call is the best way to identify them.

Listen for: A high, sharp call that sounds like a squeeze toy: *peet-ZUP!* At dawn and dusk it also gives a longer, more elaborate song that is very sputtery and explosive.

Remember: The Acadian Flycatcher is the "hungry" flycatcher. It orders a "pizza" when it calls: *peet-ZUP!*

▼ *The Acadian Flycatcher's nest looks like debris caught in a limb. It is a well-woven basket hung between two twigs.*

WOW!

Acadian Flycatchers spend summers with Wood Thrushes and Ovenbirds, then migrate across the Gulf of Mexico to spend the winter in tropical cloud forest with quetzals and bellbirds.

Find it: In woodlands, especially near water, the Acadian usually perches on a branch halfway up a tree that is inside the forest (rather than on the edge). From there it calls and makes short flights out to capture insects.

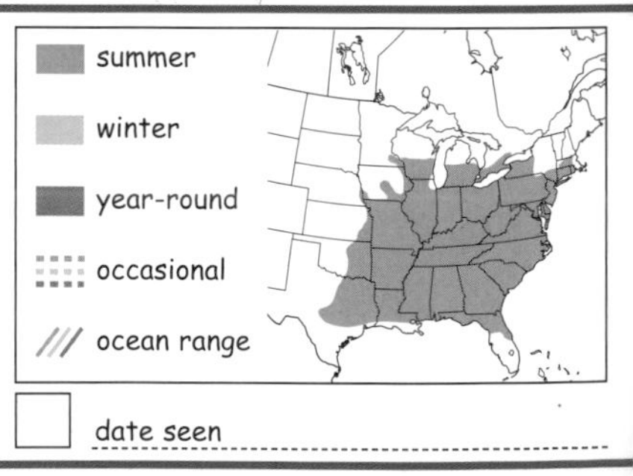

☐ date seen ______________

LEAST FLYCATCHER

Empidonax minimus **Length: 5¼"**

Look for: As its name implies, this is a small flycatcher, olive green above, pale greenish below, with a bold white eye-ring and two obvious white wing bars.

Listen for: A loud, emphatic *che-bek!* repeated in rapid succession. Call note: *whit!*

Remember: It's really hard to tell the Empidonax flycatchers apart. But most ID experts agree that the Least Flycatcher is the smallest and grayest of the bunch.

WOW!

Least Flycatchers don't like American Redstarts as neighbors. They regularly chase Redstarts (which compete for food) out of their nesting territories.

◀ *The Least Flycatcher's voice may lack melody, but it's the best ID clue for this species.*

Find it: Common in summer in mature deciduous and mixed woods, where it often perches on a bare branch in the middle of a tree or lower. Often heard before it is seen.

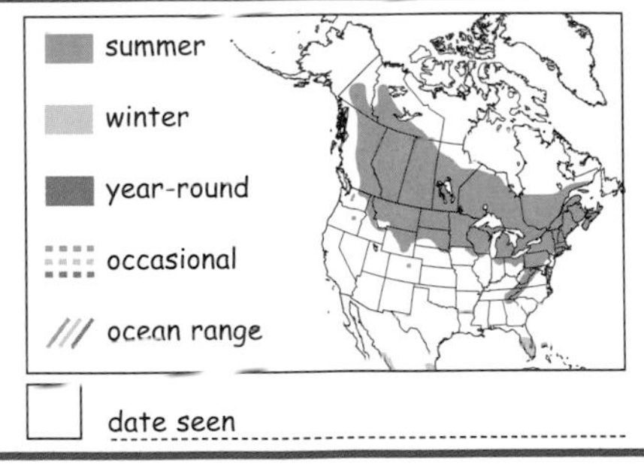

☐ date seen ____________

WILLOW FLYCATCHER

Empidonax traillii **Length: 5¾"**

Look for: Olive green above, dusky pale yellow below with a barely noticeable eye-ring and weak, drab wing bars. Bill is longish and dark above, yellow below. Like other Empids, often heard before it is seen.

Listen for: The Willow's *FITZ-bew!* call is explosive and sounds like the bird is sneezing.

Remember: Voice is the best way to separate this species from its close relatives.

◀ *Don't be fooled! Willow Flycatchers are not* always *found in willows. This bird sings from a streamside cottonwood.*

WOW!
The Willow Flycatcher and its close relative the Alder Flycatcher were considered a single species (Traill's Flycatcher) until the 1970s.

Find it: The Willow Flycatcher spends the summer in wet willow thickets, brushy old-fields, and woodland edges. Its distribution is mostly south of the Canada border, while the similar Alder Flycatcher is more common north of the border.

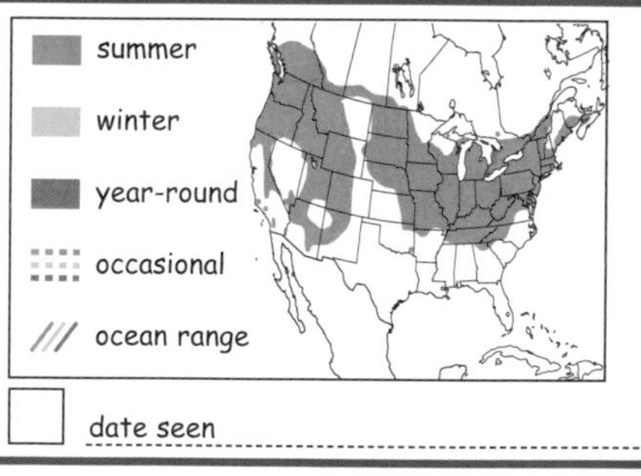

date seen

BLACK PHOEBE

Sayornis nigricans **Length: 6¾–7"**

Look for: Dark above, white below, the Black Phoebe is our only flycatcher with a black breast. Except for its flycatching behavior and wagging tail, it looks much like a Dark-eyed Junco.

Listen for: A thin, whistled *fee-bee, fee-bew.* Call note is a *chep!*

Remember: Like other phoebes, the Black Phoebe bobs its tail constantly while perched.

WOW!
Some Black Phoebes have been observed catching and eating small fish!

▶ *Black Phoebes build their nests out of mud and grass and line them with animal hair. The mud helps the nest stick in place.*

Find it: Always found near water and a year-round resident in most of its range. Along streams, ponds, and lakes, in towns, and even near farmyard troughs, the Black Phoebe can be found perched low and catching insects just above the water's surface.

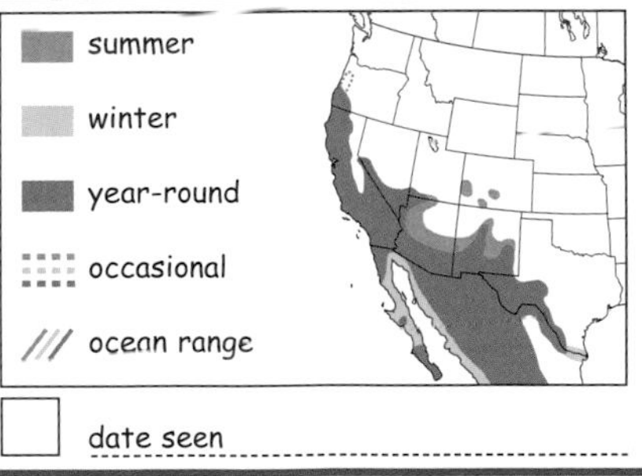

☐ date seen ____________

EASTERN PHOEBE

Sayornis phoebe **Length: 7"**

Look for: The drab-looking Eastern Phoebe is considered by many birders to be a sign of spring's arrival. Calling out its name and flicking its tail up and down, this medium-sized flycatcher returns in spring as soon as insects are active. Phoebes are dark headed and pale bellied. Juvenile birds can show yellow bellies.

Listen for: Eastern Phoebes call out *FEE-bee!* or *FEE-bree!* They also utter a soft, sweet-sounding chip note.

Remember: Phoebes can be confused with the smaller Eastern Wood-Pewee, but phoebes wag their tails and pewees do not. Pewees always show obvious white wing bars. You have to look really hard to see the phoebe's wing bars.

WOW!
The first North American ornithologist to band birds, John James Audubon tied a small silver thread around the legs of nestling Eastern Phoebes.

◀ *Phoebes will build their mud and moss nests on any ledge that is protected from the weather.*

Find it: Eastern Phoebes prefer wooded habitat near water. They build mud nests under overhanging rocks, under bridges, and in barns and can be reliably found in these settings during warm months.

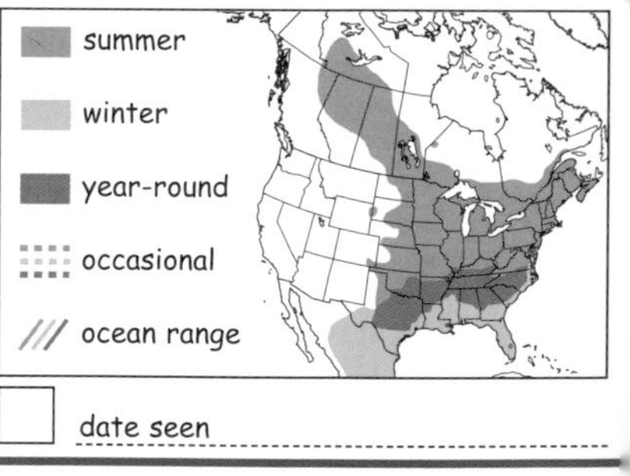

date seen

SAY'S PHOEBE

Sayornis saya **Length: 7½"**

Look for: A medium-sized gray-brown flycatcher with rufous pink sides and a square black tail. Dark eye line stands out on gray head. Upper wings, best seen in flight, are plain gray. Like other phoebes, wags tail while perched.

Listen for: A sad-sounding, down-slurred call: *pee-yerr* or *pyeer*. Sometimes alters the two phrases.

Remember: While our other two phoebes prefer to be near water, Say's Phoebe can survive in much drier habitats.

WOW!
Say's Phoebe has a huge breeding range, from the desert Southwest to the tundra habitats near the Arctic Circle.

▼ *On the windblown prairie of North Dakota, a Say's Phoebe perches near its nest in an abandoned farm building.*

Find it: Prefers open habitat, including prairie, scrublands, canyons, ranches, and parks.

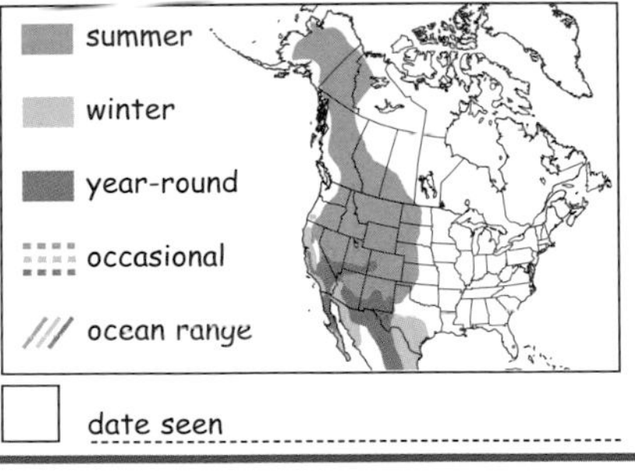

VERMILION FLYCATCHER

Pyrocephalus rubinus **Length: 6"**

Look for: It's hard to misidentify a male Vermilion Flycatcher, with his flaming red crown, throat, and underparts offset by the dark brown mask and upperparts. Adult female is pale gray above and streaky chested with a dark mask and black tail. Some adult females show a pink wash on the belly.

Listen for: Song is a series of rising, staccato notes, ending in a trill: *pit-pitpit-zree!*

Remember: Females can have pinkish or yellowish wash on lower belly. Brown breast streaks, dark mask, and short dark tail are best field marks.

WOW!
The male Vermilion performs an amazing courtship flight above his territory, fluffing out his red feathers while fluttering and singing, then swooping back to a prominent perch.

▶ *When perched, Vermilion Flycatchers wag and flare their tails just like phoebes do.*

Find it: Found near open water in dry habitats with scattered trees in the Southwest. Also found along streams, near ponds. Most noticeable as it sorties into the air after insects and returns to a perch.

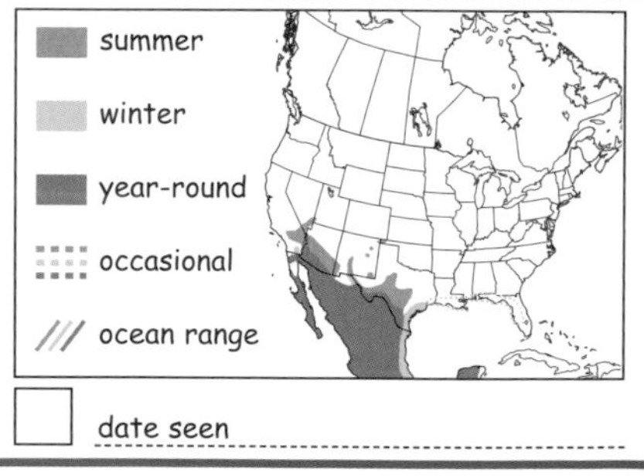

☐ date seen ____________

GREAT CRESTED FLYCATCHER

Myiarchus crinitus Length: 8½"

Look for: This tall flycatcher's most obvious field marks are its crested head, gray face and throat, bright lemon-yellow belly, and rufous tail, but Great Cresteds are most often located by their voices. Three other close *Myiarchus* relatives occur in the West and Southwest, but the Great Crested is the only one in the East.

Listen for: A loud, upslurred *whee-eep!*, given singly or in a sputtering series.

Remember: You might confuse this flycatcher with a kingbird, but the Great Crested is far more colorful on its belly and tail.

▼ *A Great Crested Flycatcher brings food to a hungry nestling.*

WOW!
This flycatcher often includes shed snakeskin in its nest — perhaps to deter predators, but they also simply like the crinkly feel of the skin.

Find it: Great Cresteds prefer older woods but may also be found in backyards and parks with tall, leafy trees. They nest in tree hollows and old woodpecker holes and may occasionally use nest boxes.

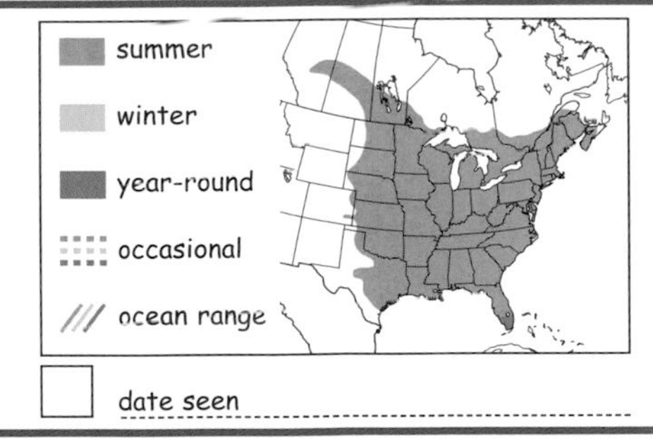

date seen ______

WESTERN KINGBIRD

Tyrannus verticalis **Length: 8¾"**

Adult

Look for: Among our common flycatchers, the Western Kingbird, with its pale gray head, white throat, and lemon yellow belly, stands out. In flight, the black tail shows white edges.

Listen for: Short squeaky call notes. Also a series of explosive, sputtery squeaks like a squeeze toy getting chewed on by several puppies at once: *pick! peepick! pick! peekaboo!*

Remember: Where their ranges overlap, the Western Kingbird prefers more open habitat than the Eastern Kingbird.

WOW!
Western Kingbirds will boldly attack a crow, raven, or hawk that passes near their nesting territory.

◀ *Roadside fences provide an excellent launching pad for catching flying insects.*

Find it: Common in summer throughout the West in open country with scattered trees and along roadside fences in farmyards and towns. Perches in the open, flying out to catch insects in the air.

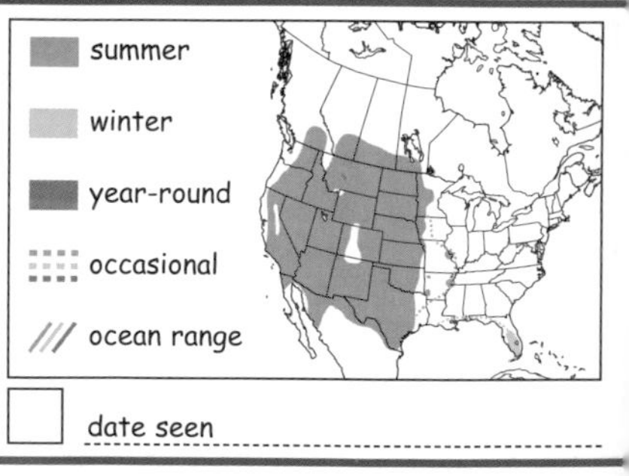

date seen

EASTERN KINGBIRD

Tyrannus tyrannus **Length: 8½"**

Look for: Black above, white below, with a heavy black bill, the Eastern Kingbird looks a bit mean. As its Latin name suggests, it is a tyrant of tyrants, often attacking much larger birds that invade its territory. The obvious white tip to the black tail is a field mark unique to this flycatcher species.

Listen for: Eastern Kingbirds are loud birds (often heard before seen) with an unmusical, zapping call: *ptzeent!*, often given in a rapid series.

Remember: From the Great Plains westward, you can see both Eastern and Western Kingbirds in the same habitats. They are easy to tell apart—Westerns are light gray above and yellow below.

WOW!

If you get a really good look, you'll see a narrow stripe of red feathers on the crown of an Eastern Kingbird. This rarely seen plumage is displayed by the male during spring courtship.

◀ *Eastern Kingbirds overwintering in the Neotropics travel in flocks and feast on the fruit of Cecropia trees.*

Find it: Common in summer in a variety of open and semi-open habitats across the continent, Eastern Kingbirds perch in obvious places—on roadside wires, fences, and treetops—and make aerial forays to catch insects.

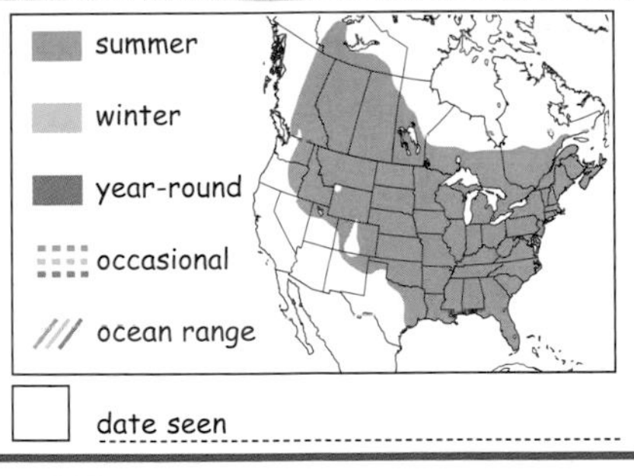

date seen

SCISSOR-TAILED FLYCATCHER

Tyrannus forficatus **Length: 13"**

Look for: Though it is not common in the East, this spectacular pale flycatcher with the elegant, long black-and-white tail is a must-see for every birder. As if the tail were not impressive enough, the Scissor-tailed Flycatcher has a wash of bright pinkish orange on its sides and in its wingpits.

Listen for: Common call is sputtery and rising in tone: *pik-pik-pik-pik-piDEEK!*

Remember: Not all Scissor-tails have superlong tails. Females and juveniles have shorter tails on average.

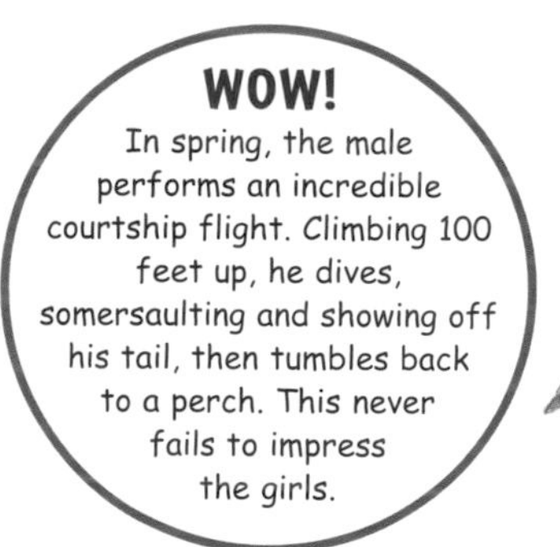

◀ The male Scissor-tailed Flycatcher looks like a bird of paradise as it makes a swooping courtship flight.

Find it: In summer it perches along fence lines and in treetops, making foraging flights to catch insects. In fall, large migratory flocks are common in Texas. Individuals may be found far beyond the normal range in spring and fall.

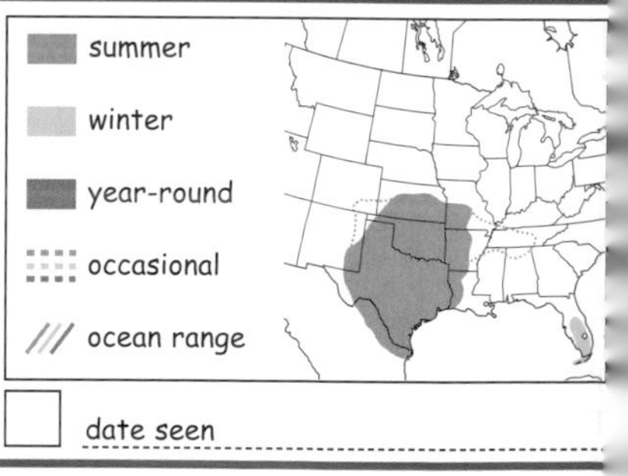

date seen ________

LOGGERHEAD SHRIKE

Lanius ludovicianus **Length: 9"**

Look for: The Loggerhead Shrike is a songbird with a killer instinct, catching a variety of insects, small mammals, reptiles, and other birds with its hooked, hawklike bill. The stout black bill and black mask give this bird a dark-headed look. In flight, the Loggerhead Shrike shows white patches on black wings and tail, similar to a Northern Mockingbird.

Listen for: The Loggerhead Shrike has a surprisingly musical voice. Short, burry whistles are interspersed with harsh, nasal notes and buzzes.

Remember: Look at the black facemasks to tell the Loggerhead Shrike from the (less common) Northern Shrike and Northern Mockingbird. Mockingbirds lack a mask. Northern Shrikes have barely any mask. The Loggerhead's mask seems to connect to the black bill.

WOW!
The Loggerhead Shrike and its relative the Northern Shrike have the folk name of Butcherbird for their habit of impaling prey on thorns or fence wire as a butcher hangs out slabs of meat.

▲ *The last thing many voles see is a Loggerhead Shrike's masked face.*

Find it: Loggerhead Shrikes prefer exposed perches in open country from which they watch for prey. The Loggerhead is disappearing from most of the Northeast. The cause of this population decline is poorly understood.

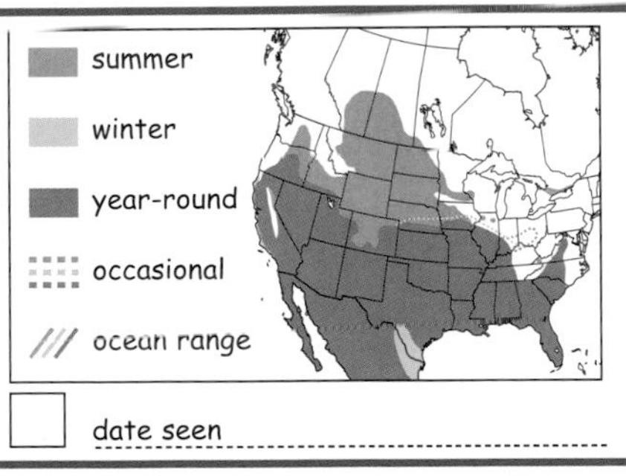

☐ date seen ______________

WHITE-EYED VIREO

Vireo griseus **Length: 5"**

Look for: This bold, vocal, and stocky little vireo is our only common songbird with a white iris (hence the name White-eyed). More obvious, however, are the vireo's yellow spectacles around the eyes, its white wing bars, and its pale yellow sides.

Listen for: The remember-it phrase for the White-eyed Vireo's song is *quick get the beer check!* Or *chik-chik-a-chee-wow!*, given in a harsh, scolding tone, as if the bird is mad about something.

Remember: You are most likely to see a White-eyed Vireo low to the ground in thick cover. Other vireos and warblers are more likely to be found higher up, foraging and singing in trees.

WOW!
Other descriptions of the White-eyed Vireo's song are **Gingerbeer-quick!**, **Take me to the railroad quick!** and **Chick of the village!**

◀ *Listen closely to hear the White-eyed Vireo imitate the songs and calls of other birds in its whisper song.*

Find it: White-eyed Vireos love old overgrown meadows and viney tangles, but they are not skulkers. You may hear a White-eyed before you see it, and the bird can be coaxed into view with a few simple pishes or squeaks.

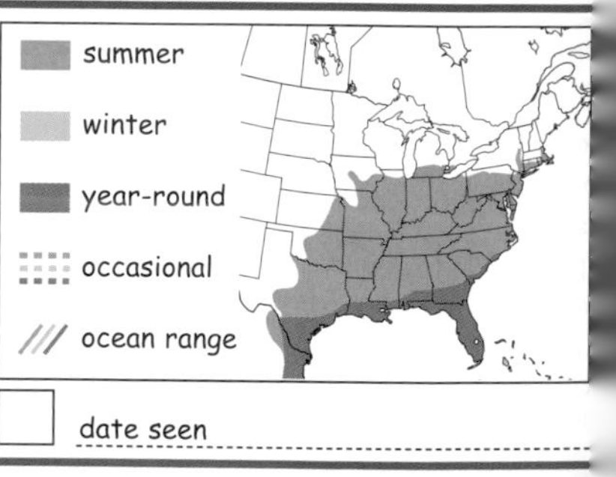

☐ date seen

BLUE-HEADED VIREO

Vireo solitarius **Length: 5½"**

Look for: It might be a stretch to say this bird has a blue head. But the dark gray head and bright white spectacles (connected eye-rings) and throat are distinctive among our common eastern vireos. It looks chunky-headed compared to other vireos.

Listen for: The Blue-headed Vireo's song is a series of high-pitched, sweet phrases that seem first to ask a question (upward-slurred *see me?*), then answer it (downward-slurred *I'm up here!*). Call is a whiny, inquisitive *chew-wee?*

Remember: Formerly known as the Solitary Vireo, the species was split into three separate species: Blue-headed (mostly eastern), Cassin's (from the Rockies to the Pacific Coast), and Plumbeous (in the inland Southwest).

WOW!
Many Blue-headed Vireos spend the winter in the Southeast, where they survive on berries if insects are not available. Most other vireos spend the winter in the insect-rich tropics.

▲ *Male Blue-headed Vireos do a greater percentage of the work of raising the young than do males of other vireo species.*

ind it: The Blue-headed ireo prefers mixed orest (hardwoods and ines) within its breeding ange. Because this pecies is not as active a orager as some other ireos, it is often first ocated by sound.

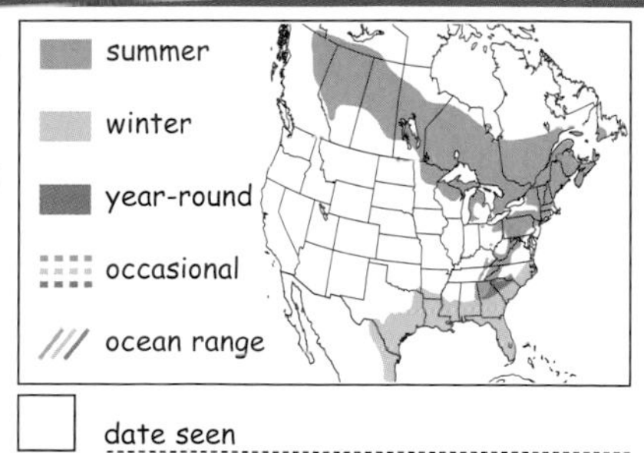

☐ date seen ______________

YELLOW-THROATED VIREO

Vireo flavifrons **Length: 5½"**

Look for: The Yellow-throated Vireo adds a pair of yellow spectacles to its attractive yellow-throated attire. Look closely at this species, and it appears to have been dipped headfirst in yellow paint, leaving the belly white and the back and tail gray.

Listen for: The song is *three-eight, three-eight, cheerio!*, given in a burry, hoarse voice. Call is a harsh, scolding *chur-chur-chur-chur.*

Remember: The Pine Warbler is similar to the Yellow-throated Vireo, but it has a thinner bill and a lightly streaked belly. The Yellow-throated Vireo's bill is thicker than any warbler's, and its big-headed appearance and slow, methodical movements are very unwarblerlike.

WOW!

Pesticides used to control Dutch elm disease may have caused the disappearance of Yellow-throated Vireos from the tall trees in many towns and city parks. Since this practice stopped, the vireo has experienced population growth.

▲ *The Yellow-throated Vireo often looks as if it is taking an imaginary bath as it hops around, singing its hoarse song.*

Find it: Treetop singers and foragers that are more often heard than seen, Yellow-throated Vireos are common in summer in tall hardwood forests, especially in oaks and maples. They winter in the tropics.

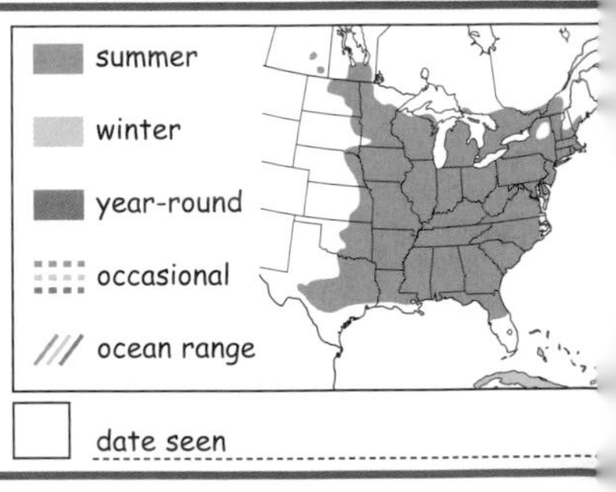

☐ date seen

RED-EYED VIREO

Vireo olivaceous **Length: 6"**

Look for: The Red-eyed Vireo's best field marks are all on its head. The red eye is not the most reliable one, though it can be seen on most adult birds in good light. Better field marks are the dark gray cap, bold white and gray eye lines, and a longish bill.

Listen for: Singsong, two- or three-note phrases given once every two seconds or so make up the Red-eyed Vireo's song: *here I am, up here, in this tree, over here.* Call is a harsh, down-slurred *meww!*

Remember: Many of our vireos can appear to be drab look-alikes. The Red-eye's gray cap, contrasting white eye line, plain wings, and long bill help set it apart.

The Red-eyed Vireo is the only North American bird with a declarative statement as a Latin name. It means "I am green" in Latin.

▲ *One enthusiastic Red-eyed Vireo was recorded singing more than 27,000 times in one day!*

Find it: The Red-eyed Vireo's greenish coloration helps it blend into the foliage, so in summer it is best located by song. In fall, they chase other songbirds and zip around from tree to tree for no apparent reason.

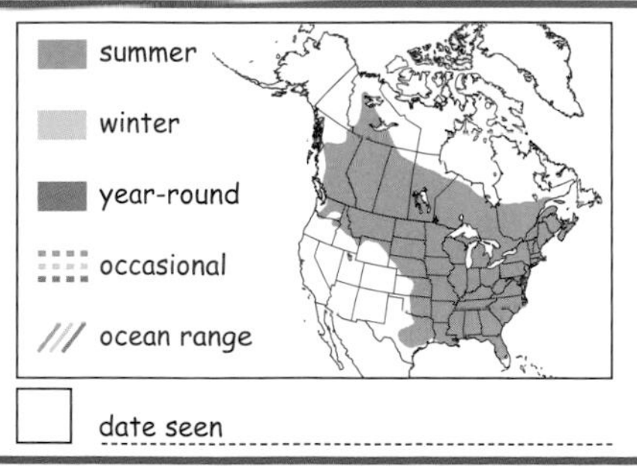

☐ date seen ____________

WARBLING VIREO

Vireo gilvus **Length: 5½"**

Look for: One of our plainest-looking birds, and that's a great clue to its identity. Dull olive-gray above, white below, with a pale eyebrow and no wing bars. Smaller and paler than the Red-eyed Vireo, with a less boldly marked head.

Listen for: Song is a long unbroken series of rich, husky warbled notes, similar in quality to a Purple Finch's song. Often ends on a higher note. Call is a nasal *quah!*

Remember: This species is best identified by its rich, musical song, which it gives in an unbroken series, unlike the short-phrase songs of our other vireos.

WOW!
The Warbling Vireo's song, as taught to me: **If I see you I will seize you and I will squeeze you till you squirt**. At least it's memorable!

◀ *Warbling Vireo nests are often parasitized by Brown-headed Cowbirds.*

Find it: Common in open deciduous woods, mixed coniferous woods, and along wooded waterways in the East. In the West, found in aspen and cottonwood groves, along rivers, and in wooded canyons. Usually heard before it is seen.

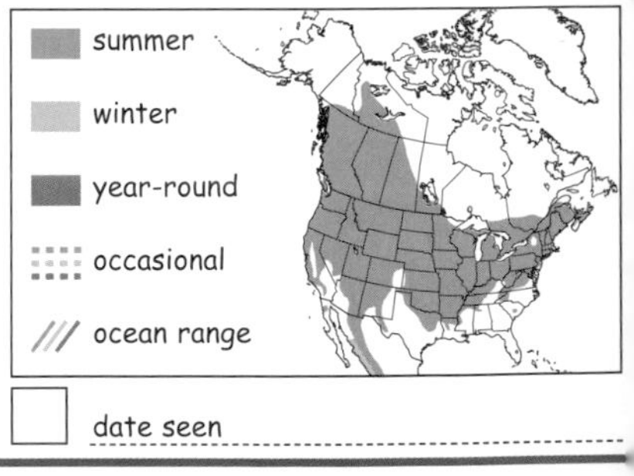

☐ date seen

WESTERN SCRUB-JAY

Aphelocoma californica **Length: 11–11¼"**

WOW!
Western Scrub-Jays will sometimes land on deer to remove and eat ticks.

Look for: A bright blue crestless jay of the West with long all-blue wings and tail, a white throat, dark cheek, blue necklace, and gray-white underparts. Bill is stout and black. Upper back is gray.

Listen for: Call is a harsh-sounding *kressh-kressh* and a rising *jreee!* Also: *shrek-shrek-shrek-shrek*.

Remember: Western Scrub-Jays living in the interior West are less boldly colored on the face and throat than coastal birds.

◀ *This species readily visits bird feeders for peanuts, suet, fruit, and mealworms.*

Find it: Often seen in pairs or small flocks in oak-dominated woods and pinyon-oak woods in a variety of habitats, including canyons, foothills, wooded rivers, and residential neighborhoods.

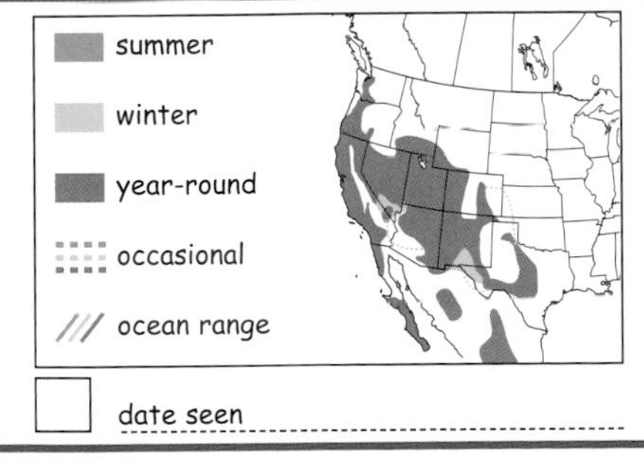

☐ date seen ______________________

BLUE JAY

Cyanocitta cristata **Length: 11"**

Look for: Handsome members of the Corvid family (crows and their relatives), Blue Jays are bold and obvious birds much of the year. In flight, the Blue Jay shows a mostly blue back with white inner wingtips (secondaries) and white outer tail tips.

Listen for: Most common vocalization is a harsh, scolding *jaay, jaay!* Jays also make a variety of other calls, including bell-like whistles and a rusty-gate sound, and they can imitate the calls of raptors, particularly the descending scream of the Red-shouldered Hawk.

Remember: This is the largest blue songbird found in the East. There are several other jays in the West, but only the Steller's Jay is both crested and mostly blue.

◀ *Scolding Blue Jays often alert other birds to the presence of hawks or owls.*

WOW!

While a jay may cache thousands of acorns in its lifetime, many are never dug up and consumed but left to grow into oak trees instead.

Find it: Widespread and common all year round in mixed woodlands, city parks, and backyards, where they will come to bird feeders. Northern birds migrate southward in fall, flying in large loose flocks during the day.

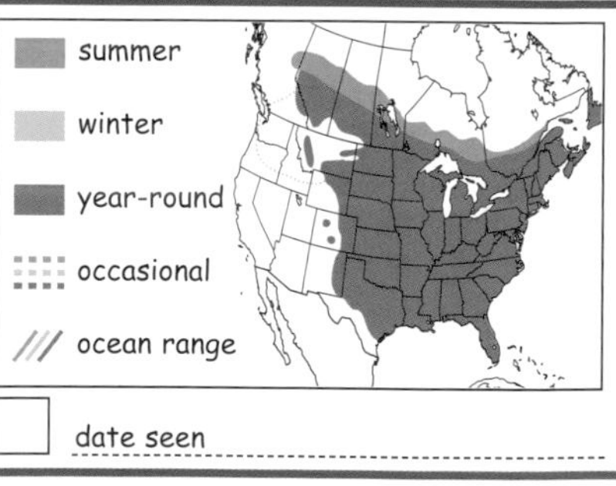

☐ date seen ____________

STELLER'S JAY

Cyanocitta stelleri **Length: 11½"**

Look for: A large, dark, crested bird with a large bill and long tail. The front half of the Steller's Jay is black, the back half is dark blue. Flight is direct with long swooping glides.

Listen for: This very vocal species makes a variety of sounds: a harsh *sharrrr;* a loud, low-pitched *shook-shook-shook;* and a scream that sounds just like that of a Red-shouldered Hawk.

Remember: The all-dark Steller's Jay is unique among its peers. Other jay species of the West lack a crest. The Blue Jay has a pale belly.

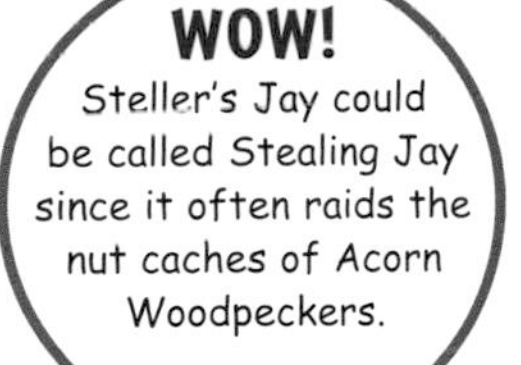

▼ *Steller's Jay flocks often move around by flying in a loose single-file line.*

Find it: Common in coniferous forests and mixed pine-oak woodlands of the West, especially in the mountains.

summer

winter

year-round

occasional

ocean range

☐ date seen ____________________

CLARK'S NUTCRACKER

Nucifraga columbiana **Length: 12"**

Look for: Named for its ability to pry pine nuts out of pinecones with its long black bill, the Clark's Nutcracker is mostly tan-gray overall. Its black wings and tail have prominent white patches, most obvious in flight.

Listen for: A loud, nasal, and harsh *krawww!* and other raspy sounds.

Remember: The similar Gray Jay is more compact-looking with a smaller bill and no white in the wings and tail.

▼ *It's hard to miss the Clark's Nutcracker in flight. Large white patches stand out on the black wings and tail.*

A Clark's Nutcracker may bury as many as 30,000 pine seeds in a single fall season. It will retrieve and eat many of them later in the winter.

Find it: Prefers conifer forests high in the mountains where it is a year-round resident. Often in small flocks. Perches on treetops and in other prominent places. Like Gray Jay, will often approach humans for food handouts.

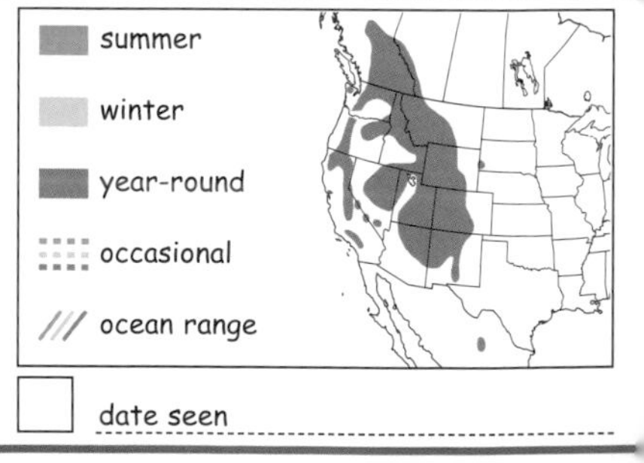

☐ date seen

GRAY JAY

Perisoreus canadensis **Length: 11¼–11½"**

Look for: The Gray Jay is indeed mostly gray with a white forehead and face. But its dark crown and nape and small black bill make it look like a super-sized chickadee. Overall appearance is rounded and "puffed out." Flying birds are plain gray above.

Listen for: Makes many different sounds, from soft whistles to harsh chuckles and almost gull-like cries. Also a nasal *nyah-nyah-nyah*.

Remember: In the western mountains, the similarly gray Clark's Nutcracker has a large pointed black bill and huge white patches on the tail and wings. Gray Jays are small billed.

▼ *Gray Jays will boldly follow hikers in the woods, looking for a handout or an opportunity to swipe some food.*

WOW!

Among the folk names for the Gray Jay: Whiskey Jack, Camp Robber, Carrion Bird, Grease Bird, and Meat Hawk.

Find it: If you are in the woods of the North, it's more likely that a pair of Gray Jays will find *you*. They are year-round residents of spruce and fir forests. Often found near trails, campgrounds, and picnic sites.

summer

winter

year-round

occasional

ocean range

date seen

GREEN JAY

Cyanocorax yncas **Length: 10½"**

Look for: It's hard to mistake the colorful Green Jay for any other bird. With its lime green back, yellow belly, black throat, and bright violet crown and face, it blends in with its tropical habitat.

Listen for: A series of harsh, nasal notes: *shack-shack-shack-shack!* Also other weird sounds, like those made by other jay species.

Remember: This is our only green-colored jay.

◀ *Many birders get their first look at a Green Jay when one swoops in to eat fruit offered at a feeding station.*

WOW!
Green Jays help to spread oak trees by burying acorns and never returning for them.

Find it: Common in dense scrub woods and mesquite thickets in southernmost Texas, where it is a year-round resident. Usually found in flocks. Visits feeders, backyards, and parks for seeds, fruit, and water.

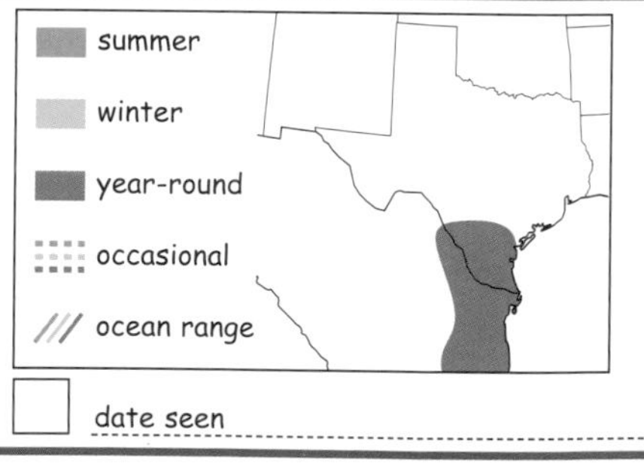

date seen

COMMON RAVEN

Corvus corax **Length: 24"** **Wingspan: 53"**

Adult

Adult, in flight

Look for: This large all-black relative of the American Crow can soar like an eagle, swoop and dive like a falcon, and is adaptable enough to thrive anywhere. In flight it is the wedge-shaped tail that best separates this species from the American Crow (which has a normal fan-shaped tail).

Listen for: Hoarse, croaking calls (*kraaak, kraak!*), unique to ravens, carry a great distance and are often your first clue to the bird's presence. Also makes a variety of other calls, including rattles, gurgles, and toots.

Remember: If you see a Common Raven next to an American Crow, either flying or perched, the raven is the larger of the two. Ravens soar more than crows and look heavier and more powerful in flight.

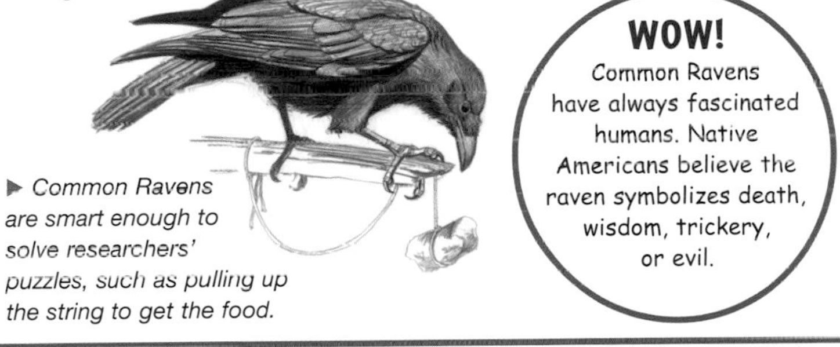
▶ *Common Ravens are smart enough to solve researchers' puzzles, such as pulling up the string to get the food.*

WOW!
Common Ravens have always fascinated humans. Native Americans believe the raven symbolizes death, wisdom, trickery, or evil.

Find it: More common in the North and throughout the West, the Common Raven in the East is found primarily in heavily wooded mountainous areas, above 3,000 feet in elevation.

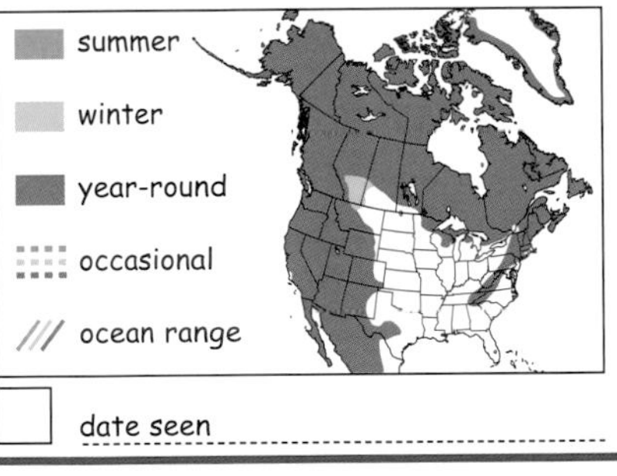

date seen

BLACK-BILLED MAGPIE

Pica hudsonia **Length: 18½–19½"**

Look for: A symbolic resident of the open range in the West, the Black-billed Magpie is named for its least obvious feature. More obvious field marks include its long dark tapered tail, its white belly and shoulders, and the blue-green iridescence on its wings. Flight is smooth and level and clearly shows long tail and white primaries.

Listen for: A series of harsh calls: *rak-rak-rak-rak* and a rising *maag?*

Remember: Our "other" magpie is the aptly named Yellow-billed Magpie, a resident of the central valleys of California.

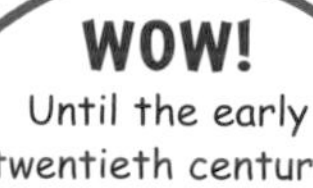

WOW!
Until the early twentieth century, Black-billed Magpies were mercilessly hunted down or poisoned as pests and predators.

► *Being scavengers, Black-billed Magpies rarely pass up an easy meal of roadkill.*

Find it: Common in small flocks in open country with scattered trees or brush, including farmland, rangeland, prairies, and parks. Forages on or near the ground. Often spotted along roadways, where it feeds on carrion (road-killed animals).

summer
winter
year-round
occasional
/// ocean range

date seen ______

AMERICAN CROW

Corvus brachyrhynchos **Length: 17½" Wingspan: 39"**

Look for: Far more common and widespread than the larger Common Raven, the American Crow lacks the raven's large head and bill. In flight, the crow flaps its wings in a smooth, rowing motion and glides, but unlike the raven, it does not soar. A fan-shaped tail further differentiates it (the raven shows a wedge-shaped tail in flight).

Listen for: *Caa-caa-caa!* is the common call of the American Crow, but this species makes many other sounds, including loud rattles, harsh nasal scolds, and high-pitched, rapid gurgles.

Remember: Another Corvid in the East, the Fish Crow, is often found near water (as the name suggests). Only slightly smaller than the American Crow, Fish Crows are best identified by their call, a nasal, two-noted *ah-ahhh!*

WOW!

American Crows have calls to assemble and disperse the flock, to signal that a predator has been sighted, and to indicate distress, such as when a crow is being attacked by a predator.

◀ *American Crows may travel great distances each day to and from their nighttime winter roosts.*

Find it: Incredibly adaptable, the American Crow lives successfully in nearly every habitat, including parks in urban areas. Crows forage on the ground. Outside of the nesting season, crows often gather in huge communal roosts.

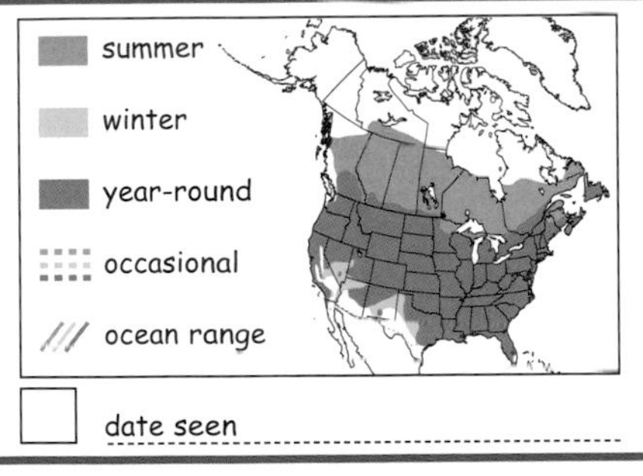

☐ date seen ______________________

TREE SWALLOW

Tachycineta bicolor **Length: 5¾"**

Male (left), female (right)

Look for: In bright sunlight, the back of the Tree Swallow glows an iridescent blue-green, contrasting sharply with the pure white throat and belly. The all-white underparts are the best field mark for separating the Tree from other swallows. In late summer young have gray-brown backs and pale gray breast-bands.

Listen for: The song of the Tree Swallow is surprisingly musical, a series of warbling gurgles: *tia-weet, tia-weet, churweet, weet, weet*. Also utters less musical twitters near nest.

Remember: Tree Swallows are among the earliest returning birds in spring. Many spend the winter in the southern U.S., where they survive the cold by switching their diet from insects to berries.

◀ *A male Tree Swallow uses a white feather to court a female.*

WOW!

Tree Swallows line their nests with white feathers for insulation. If you have Tree Swallows nesting near you, toss out a few white craft-store feathers. The swallows will take them to their nests.

Find it: In summer, Tree Swallows fly over fields and waterways and nest in tree cavities and birdhouses. Their nesting range is expanding south, helped in part by bluebird nest boxes. In fall, Trees join in huge, mixed-species roosting flocks with other swallows.

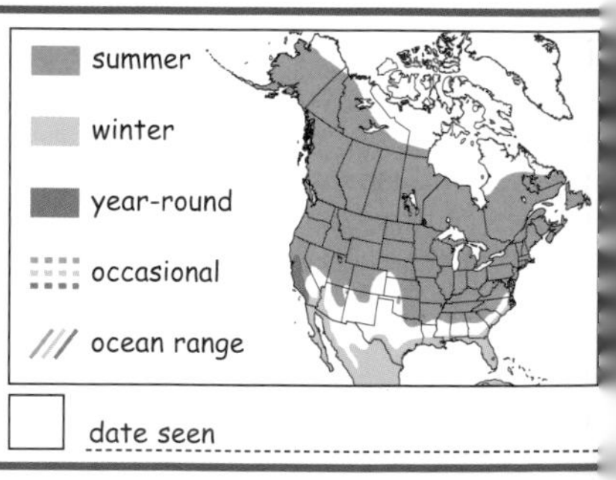

☐ date seen ------------------------

VIOLET-GREEN SWALLOW

Tachycineta thalassina **Length: 5¼"**

Look for: Adult male is bright green and iridescent above and bright white below. The bright white cheeks nearly encircle the eye. In flight the white rump sides show. Female is duller backed, white below. Smaller overall than the Tree Swallow.

Listen for: Song is a rapid twittering: *chit-chit-chit weet*. Call is a thin *chlip*.

Remember: The Tree Swallow lacks the white face of the Violet-green Swallow.

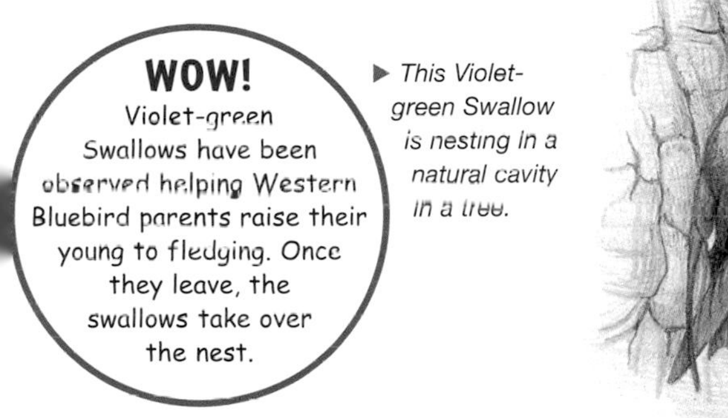

WOW!
Violet-green Swallows have been observed helping Western Bluebird parents raise their young to fledging. Once they leave, the swallows take over the nest.

▶ *This Violet-green Swallow is nesting in a natural cavity in a tree.*

Find it: Common in the West in summer in open woods, mountain forest edges, and in open prairies and towns if nesting sites (boxes, natural tree cavities, rock crevices) are available. Widespread when foraging, often near water. Foraging flocks can be seen high in the air during breeding season.

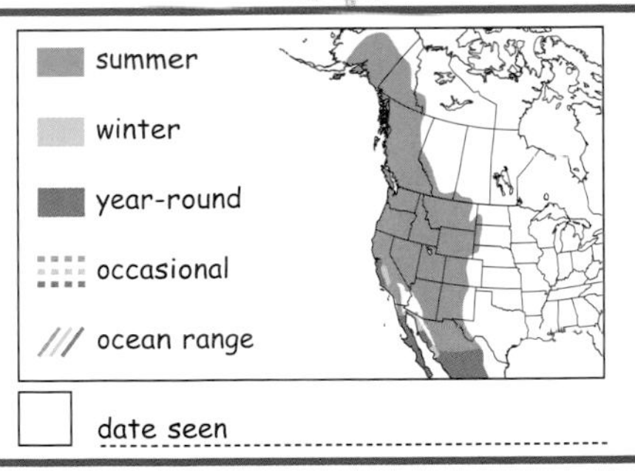

☐ date seen ______

PURPLE MARTIN

Progne subis **Length: 8"**

Look for: The Purple Martin is our largest swallow. Adult males are actually a glossy blue, but in direct sunlight their feathers have a purple sheen. Wings and tail are black.

Listen for: Both male and female Purple Martins sing. Main vocalization is a liquid, warbling chortle given in a series of down-slurred notes: *teer, teeer, teer, jeer-deert, teer!* Males perform a predawn song display in spring, flying high over their nesting colony, hoping to attract females and other martins.

Remember: In flight, Purple Martins resemble European Starlings, but martins fly more smoothly and glide more often. They also show noticeably forked tails.

WOW!

Hundreds of years ago Native Americans placed hollowed-out gourds on poles for the martins to use. The martins returned the favor by controlling flies and wasps around the village.

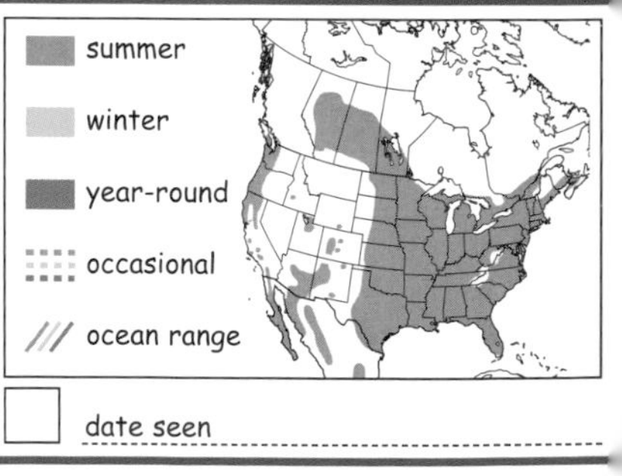

▶ *Circling high above a chosen nest site, a male Purple Martin attracts potential neighbors with his dawn song.*

Find it: Purple Martins are colonial nesters, and in eastern North America, these colonies are made in human-supplied martin houses. In fall, before they migrate to South America, martins form huge roosting flocks at dusk.

summer

winter

year-round

occasional

/// ocean range

☐ date seen ____________________

CLIFF SWALLOW

Petrochelidon pyrrhonota **Length: 5½"**

Look for: The Cliff Swallow looks superficially like a Barn Swallow, but it lacks the Barn's deeply forked tail. The Cliff Swallow has a white forehead, pale collar, and a pale rump that separate it from other swallows.

Listen for: Cliff Swallows make a series of squeaky noises that sound more like R2D2 from *Star Wars* than a bird. They are most often heard at nesting colonies, where dozens of birds congregate.

Remember: Cliff Swallows have a "headlight" (white forehead patch) and a "taillight" (pale buffy rump), which make it easy to pick the Cliff Swallows from a mixed flock of flying swallows.

WOW!
A pair of Cliff Swallows makes their nest one tiny mouthful of mud at a time. They may make as many as 2,100 total trips from the mud source to the nest site during the nest's construction.

◀ *Cliff Swallows nest in colonies.*

Find it: Named for its habit of building its jug-shaped mud nest on cliff faces, today most Cliff Swallows nest under bridges, on dam walls, and inside buildings. This adaptability has expanded the Cliff's range and population.

summer
winter
year-round
occasional
ocean range

☐ date seen ____________

BARN SWALLOW

Hirundo rustica **Length: 7"**

Look for: The graceful Barn Swallow is a common and familiar summer resident across North America. Its deeply forked swallowtail separates this species from our other common swallows. Its rusty face and throat and orange belly help to make this our most colorful swallow.

Listen for: More vocal than most other swallows, Barn Swallows seem to chatter almost constantly. Main sound is a rapid, chortling series of squeaky notes: *pit-pitpit-pit-pit-pitpit.*

Remember: The Barn Swallow's forked tail is unmistakable. Note that it lacks the white forehead and pale rump of the Cliff Swallow.

◀ *Barn Swallows build their mud nests one billful of mud at a time.*

WOW!
The Barn Swallow is also found in Europe and Asia, where it is known simply as Swallow. Early European settlers in North America were pleased to find this familiar bird in their new homeland.

Find it: Barn Swallows can be found wherever there are suitable nest sites, such as old barns and bridge girders. They swoop low over water and follow farm equipment in fields to catch insects startled into flight.

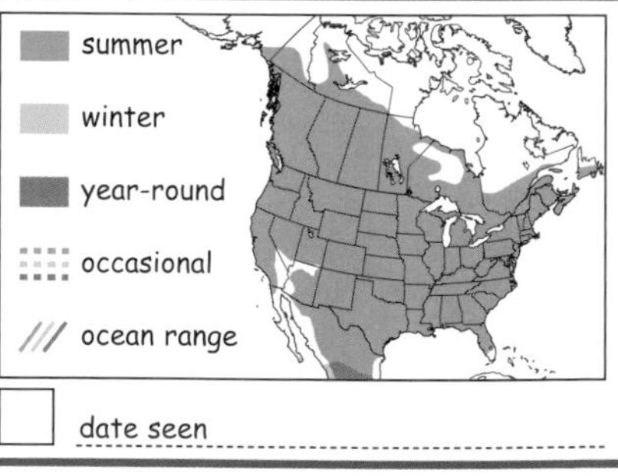

☐ date seen ----------

HORNED LARK

Eremophila alpestris **Length: 7¼"**

Look for: Horned Larks blend into the ground so well that you may not see a nearby flock until it takes flight, bounding and swooping lightly, showing black outer tail edges. The horns on a Horned Lark may not stick up off the crown. A better field mark is the black facemask and breast-band, set off by the yellow throat and pale belly.

Listen for: Song and calls are high, tinkling notes, like tiny pieces of broken glass blowing in the wind. Their weak songs are often drowned out by the wind.

Remember: Winter Horned Lark flocks may include longspurs, Snow Buntings, and pipits. All these birds have plumages that vary by age and season, so scan these flocks carefully.

WOW!
In spring, male Horned Larks hover high above their territories singing their soft, tinkling songs, trying to impress any nearby females.

▼ *When danger threatens, Horned Lark chicks huddle low in their nests, and a lawn mower might pass right over them without hurting them.*

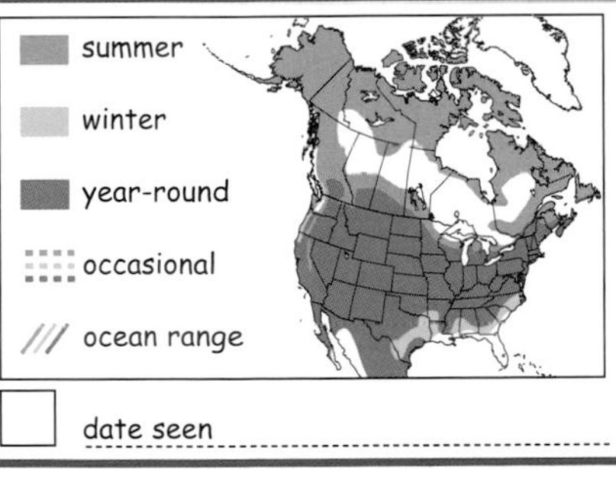

Find it: Summers are spent in the far North on the tundra. In winter, flocks move south, where they prefer open areas such as beaches and farm fields (especially ones with freshly spread manure).

summer
winter
year-round
occasional
ocean range

☐ date seen

BUSHTIT

Psaltriparus minimus **Length: 4½"**

Look for: A small plain gray bird with a long gray tail, a rounded head and body, and a very short black bill. Female has pale yellow eyes; male has dark eyes.

Listen for: Calls consistently in a series of high, thin, and scratchy notes: *zrrr*, *zeeet*, and *tink*. Though it vocalizes constantly, the Bushtit's voice is not loud.

Remember: The Bushtit does have some regional differences in appearance. Birds along the Pacific Coast are browner overall with a brown cap. Some Bushtits in Texas and New Mexico have black "bandito" masks.

WOW!
Bushtits like to party with their friends! Flocks may contain sixty or more Bushtits plus individuals of other species, such as kinglets, warblers, wrens, and chickadees.

◀ *On cold nights, Bushtits may huddle together in a clump to share body heat.*

Find it: Found in active flocks (except when breeding) in a variety of wooded habitats, including mixed oak, pinyon, juniper, chaparral, and riparian woods, and parks and wooded residential areas.

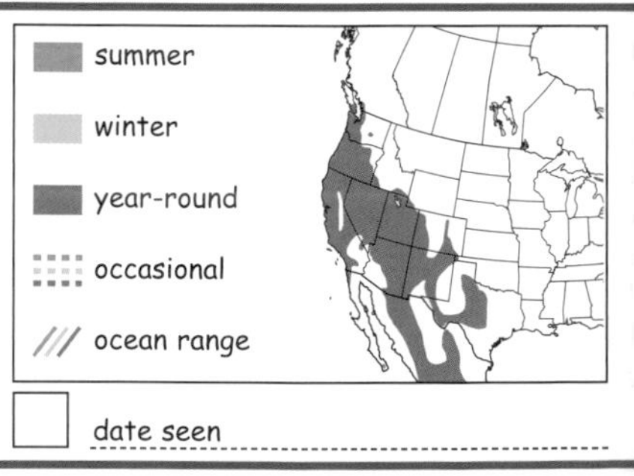

☐ date seen ____________

VERDIN

Auriparus flaviceps **Length: 4½"**

Look for: A chickadee-sized gray bird with a yellow head and a rusty shoulder patch (which may or may not be visible), the Verdin is an active, vocal bird. Juvenile birds are gray overall.

Listen for: Calls incessantly: *tsoot, tsoot* or *tweet-tweettweet-tweet!* Calls are variable.

Remember: Of all the small drab gray birds of the southwestern deserts, only the Verdin has a yellow head.

WOW!
Verdins know how to survive the desert heat. On hot afternoons they find a shady roost and take a siesta, then resume their activity later in the day when the temps are cooler.

▶ *Male Verdins may build several nests each year. One is used for nesting while others may be used for roosting.*

Find it: Though often solitary or in a mated pair, the Verdin's active foraging behavior and constant calling make it easy to notice for a small bird. Found in brushy desert habitats, mesquite thickets, and arid lowlands.

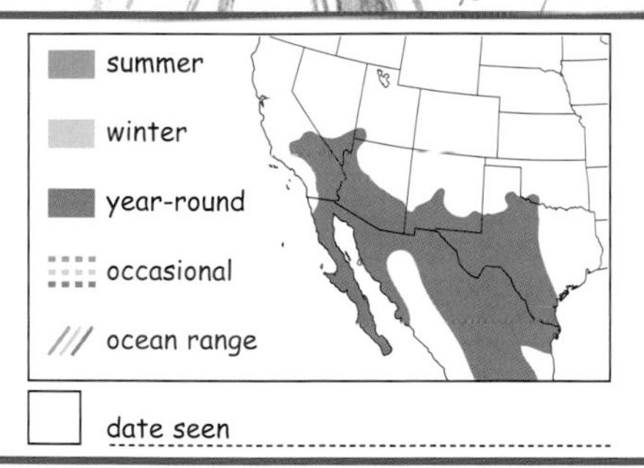

date seen ______________

CAROLINA CHICKADEE

Poecile carolinensis **Length: 4¾"**

Look for: The southern cousin of the Black-capped Chickadee is a near look-alike. Carolinas are smaller headed and less buffy on the sides. They also lack the Black-capped's white hockey stick pattern in the wing.

Listen for: The Carolina Chickadee's song is a high-pitched, sweet *fee-bee, fee-bay* in four or more notes. The call is the familiar *chickadee-dee-dee.*

Remember: Our two common chickadees can be very hard to separate in the field. Range is the best way: in the North it is Black-capped, in the Southeast it's Carolina. Where the ranges meet, it's best to rely on field marks and not voice, since these species learn each other's songs.

WOW!

A Carolina Chickadee nesting in a nest box may hiss at you like a snake and strike at your finger as you check the nest. This behavior is meant to drive off nest predators.

▶ *Carolina Chickadees may alert homeowners when feeders go empty by pecking on windows.*

Find it: Though they can be secretive during the breeding season, chickadees are active and vocal, often forming mixed-species feeding flocks with kinglets, nuthatches, and others. They prefer mixed woods, often with large trees.

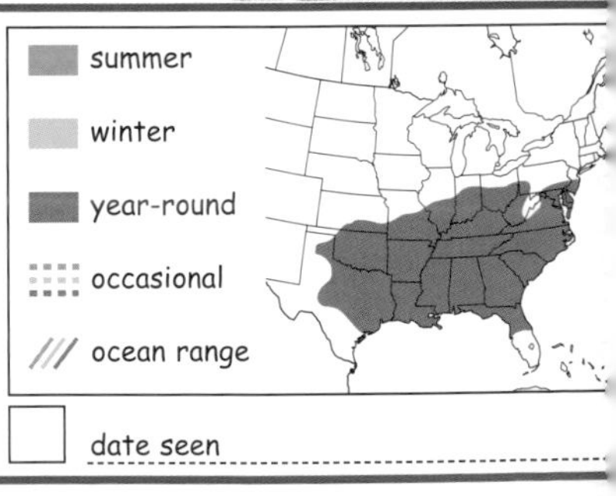

date seen

BLACK-CAPPED CHICKADEE

Poecile atricapillus **Length: 5¼"**

Look for: All of our common chickadees have black caps, but only this one is called Black-capped. Black-caps tend to look bigger headed and chunkier than Carolinas. The white edges of the secondary wing feathers form a hockey stick of white when the wings are folded.

Listen for: Black-caps have a slower, harsher, and lower-pitched *chickadee-dee-dee* call than the Carolina. Song is usually a two-part *fee-beee!*

Remember: The three Bs of the Black-capped Chickadee: they are Bigger headed, Buffier colored, and have Bolder white edges on their wings.

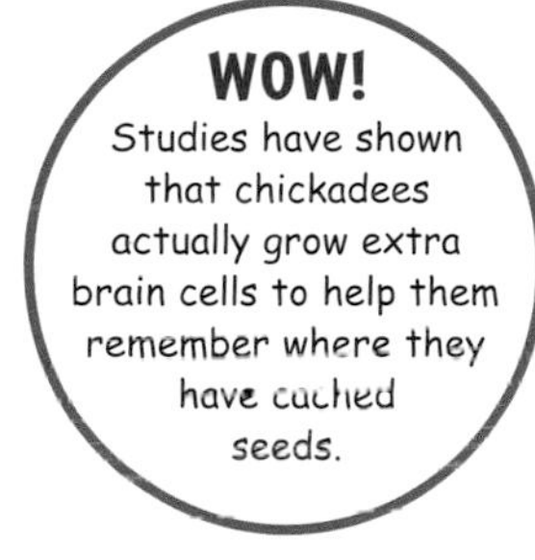

◀ *Black-capped Chickadees attending the nest, one feeding the young, one removing a fecal sac.*

Find it: In areas of overlap with the Carolina Chickadee, Black-caps are found at higher elevations. In some winters, Black-caps are found south of their normal range. They prefer mixed woods but are regulars at feeders.

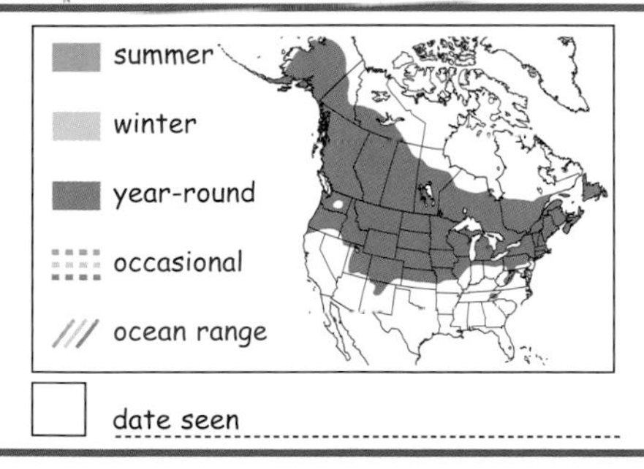

☐ date seen ____________

CHESTNUT-BACKED CHICKADEE

Poecile rufescens **Length: 4¾"**

Look for: With its dark reddish back and flanks, this handsome bird looks just like a Black-capped Chickadee that is starting to rust. White cheek patch is narrower in this species than in other chickadees. Chestnut-backed Chickadees along the central California coast have gray (not rusty) sides.

Listen for: Song is a series of fast, thin chips. Call is a down-slurring *psee-cheer!* Does not have a whistled song like other chickadees.

Remember: In much of the West, if you hear a chickadee's voice, it could be any one of the three widespread species: Black-capped, Mountain, or Chestnut-backed.

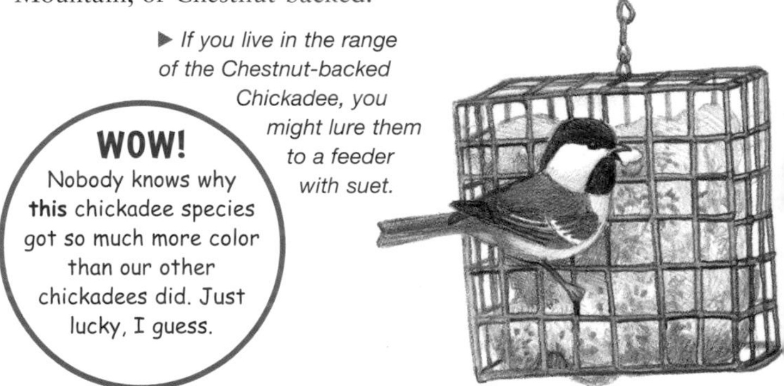

▶ *If you live in the range of the Chestnut-backed Chickadee, you might lure them to a feeder with suet.*

WOW!

Nobody knows why **this** chickadee species got so much more color than our other chickadees did. Just lucky, I guess.

Find it: Prefers moist mixed woods, including pines, firs, oaks, and willows. A year-round resident throughout its range.

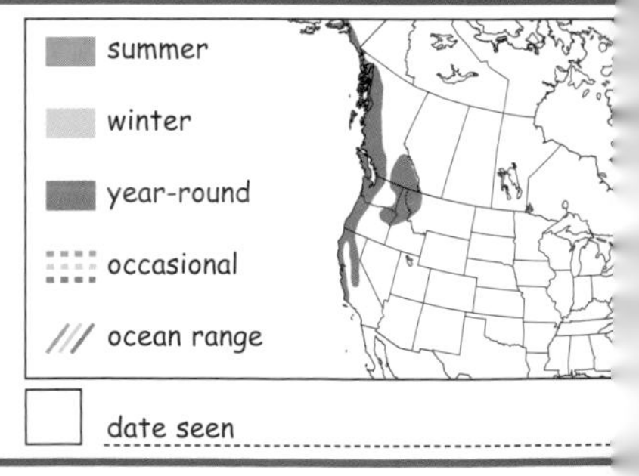

date seen

MOUNTAIN CHICKADEE

Poecile gambeli **Length: 5¼"**

Look for: Very similar to the more widespread Black-capped Chickadee, but an obvious white line over the eye of the Mountain Chickadee is its key field mark. More subtle clues include the smaller bill and the lack of white feather edges on the wings.

Listen for: Song is *I see fee-bee!* Call is a hoarse *chickachickadee-dee-dee.*

Remember: This is the chickadee you are most likely to encounter at higher altitudes in the mountains. Check for the distinctive white eye line.

WOW!

A female Mountain Chickadee will sit tight on her nest. But if really disturbed, she will lunge forward and hiss like a snake to scare away predators.

◀ *When they are not busy breeding or raising young, Mountain Chickadees form mixed-species foraging flocks.*

Find it: Common in summer in western mountain forests of conifers and aspen. Often forages in treetops so can be difficult to spot. May move to lower-elevation habitats in winter.

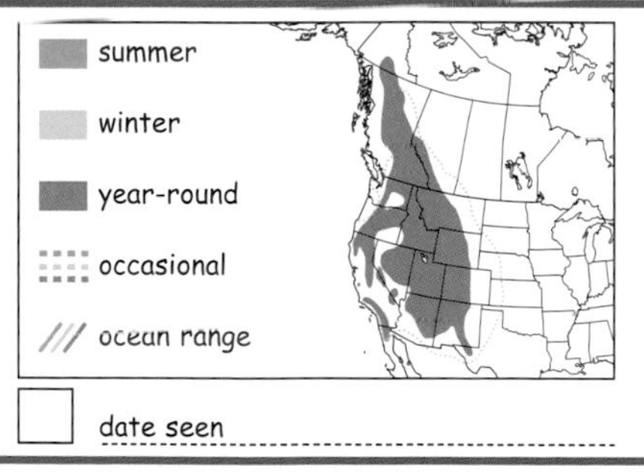

TUFTED TITMOUSE

Baeolophus bicolor **Length: 6½"**

Look for: The Tufted Titmouse's gray-crested head and large black eyes on a pale face give this familiar bird a friendly look. Larger than the chickadees with which it often associates, the Tufted Titmouse shows obvious rusty peach patches on its flanks (sides).

Listen for: *Peter, peter, peter, peter!* is the clear, whistled song of the Tufted Titmouse. Also utters a harsh chickadee-like scold and a variety of short, sweet, whistled calls.

Remember: There are five titmouse species in North America, but only the Tufted Titmouse is found commonly in the East. A close relative of the Tufted Titmouse called the Black-crested Titmouse is found in central Texas.

▶ *Tufted Titmice feed hungry nestlings in an old Downy Woodpecker hole.*

WOW!

Tufted Titmice love to line their nests with soft hair. More than one sleeping family pet (and even some humans) have felt a sudden tug as a titmouse boldly steals a bit of hair.

Find it: Regulars at backyard bird feeders, where they prefer sunflower seeds, peanuts, and suet, Tufted Titmice are both active and vocal. In winter, they may join mixed feeding flocks of chickadees, nuthatches, and others.

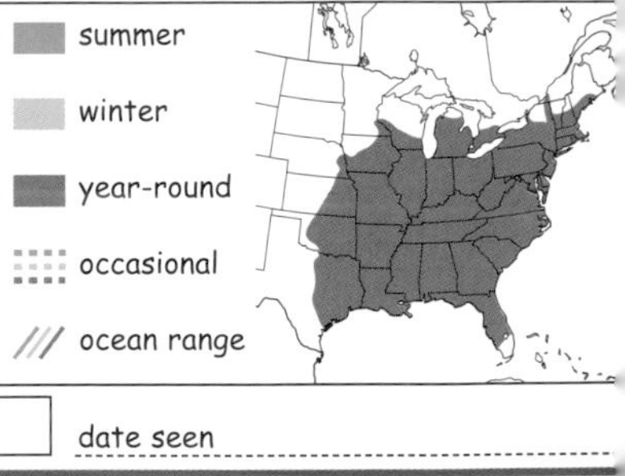

date seen

OAK TITMOUSE and JUNIPER TITMOUSE

Baeolophus inornatus, Baeolophus ridgwayi **Length: 5¾"**

Look for: Both species are short crested and plain gray overall, though Oak Titmouse may be warmer brown. The black eye stands out on the plain face of both species.

Listen for:

Oak: Song is a loud, whistled *wheetywheetywheety*. Call is *sissy-sissy-dee*.

Juniper: Song is low-pitched and rapid: *jee-jee-jee-jee-jee*. Call: *see-deedeedeedee*.

WOW!
These two species were formerly lumped into a single species called Plain Titmouse. That's still a good name, considering how they look!

Remember: The best way to tell these two birds apart is to look at their range maps. It helps to know where you are when you do this, however.

▶ *Like other titmice, the Oak Titmouse holds a nut with its feet while using its bill to crack the shell.*

Find it: Oak Titmice are common in oak and oak-pine woodlands. This is the only titmouse species west of the Sierra Nevada. Juniper Titmice are found in oak-juniper and pinyon-juniper woodlands in the Southwest. Both species are year-round residents and usually occur in small flocks.

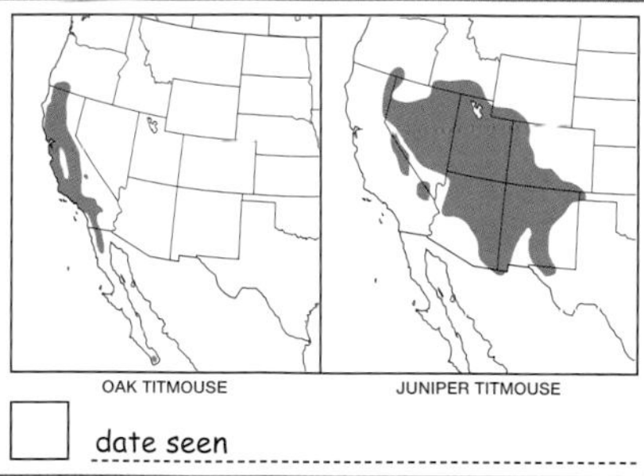

☐ date seen ____________________

WHITE-BREASTED NUTHATCH

Sitta carolinensis **Length: 5¾"**

Look for: The White-breasted Nuthatch has a black crown stripe and a plain white face against which its black eyes stand out. A long, chisel-like bill juts out from its face, and a rusty patch below its tail is obvious. Larger than our other nuthatches, it is also the most commonly seen member of this family.

Listen for: Song is a nasal series of notes on the same tone: *uhh-uhh-uhh-uhh*. Birders often locate this species by the pounding sound of its bill hacking open a seed against a tree trunk or branch.

Remember: The White-breasted Nuthatch has an undulating flight pattern, much like that of a woodpecker. In flight, it shows contrasting black and white in the tail.

▼ *Squint your eyes and you can see how this nuthatch's threat display resembles an owl's face.*

Find it: The White-breasted Nuthatch prefers mature woods of all kinds. It's a common visitor to bird-feeding stations, where sunflower seeds, peanuts, and suet are its favorite foods.

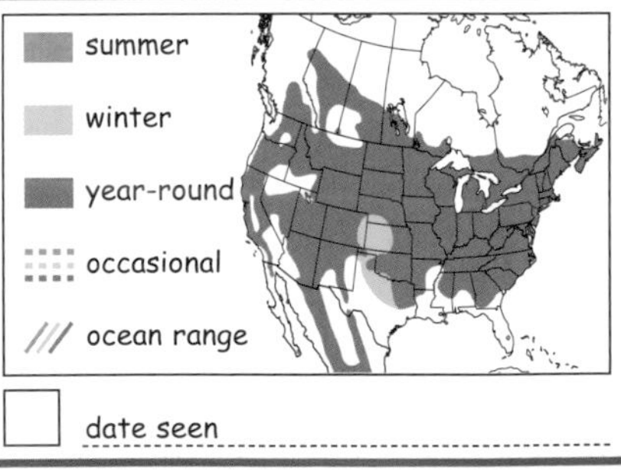

☐ date seen ____________

RED-BREASTED NUTHATCH

Sitta canadensis **Length: 4½"**

Look for: The orange (not red) breast and black line through the eye set apart the Red-breasted Nuthatch from the larger White-breasted. The male has a black cap and bright orange belly; the female has a blue-gray cap and pale orange belly.

Listen for: The song of the Red-breasted Nuthatch has been compared to the tooting of a tiny tin horn. The notes are nasal, each one slurring upward in tone: *yenk, yenk, yenk, yenk*.

Adult male

Remember: Nuthatches have powerful legs and feet and can climb up and down tree trunks and along branches. Red-breasteds look like tiny wind-up toys as they walk along tree trunks and branches. Brown Creepers can only crawl up—never down—because they must use their tails as props to support themselves.

◀ *Red-breasted Nuthatches prefer conifers in all seasons. This one has pried a seed out of a spruce cone.*

WOW!
The name nuthatch describes this family's habit of wedging a seed in a crevice, then hacking the seed open with the chisel-like bill.

Find it: Northern coniferous forests and mountains are the summer home of the Red-breasted Nuthatch. In winter, Red-breasteds can be found in a variety of wooded habitats. Like their White-breasted cousins, they will visit backyard bird feeders.

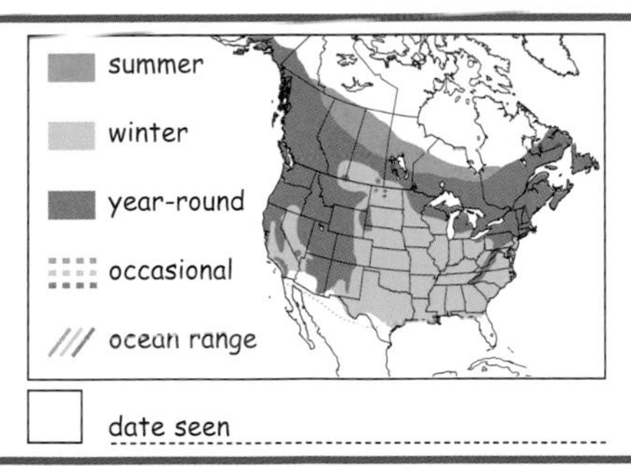

☐ date seen ____________________

PYGMY NUTHATCH

Sitta pygmaea **Length: 4¼"**

Look for: Our smallest nuthatch is aptly named. Its head and bill look big on its small body. Gray back, brown cap, with a black line through the eye. Buffy and gray below. White spot on dark nape.

Listen for: Often heard before it is seen. Calls are high-pitched, clear squeaks: *pee-deedee* or *pee-pee!* Calls constantly.

Remember: This tiny treetop-loving nuthatch can be hard to spot. Learning its call will alert you to its presence and help you tell it apart from White- and Red-breasted Nuthatches.

WOW!
In winter flocks of up to fifteen Pygmy Nuthatches will roost together in a single tree cavity.

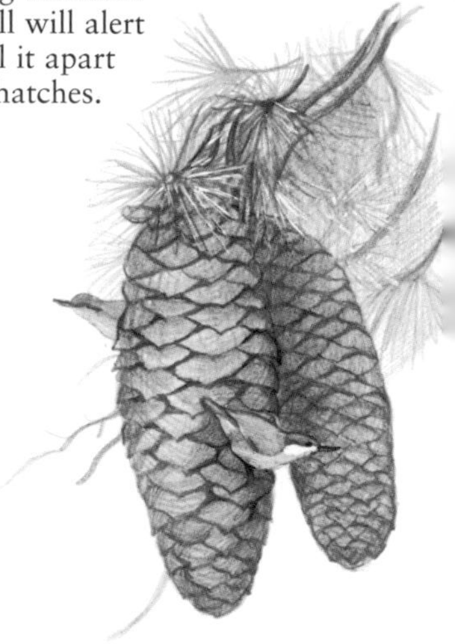

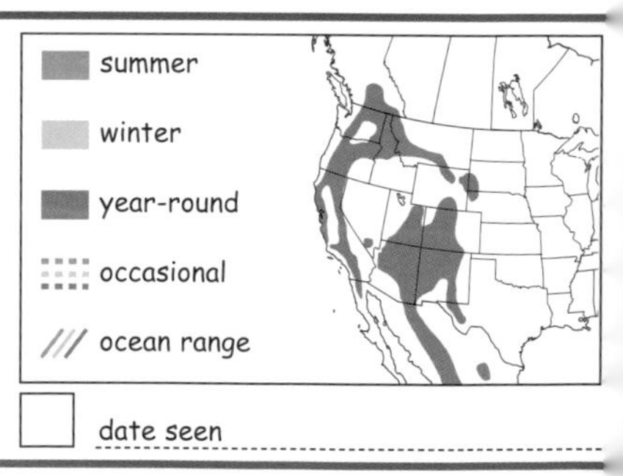

▶ *Pygmy Nuthatches are experts at clambering over pinecones to get at pine seeds.*

Find it: Nearly always found in pine forests, where it forages for seeds and insects, often in the highest branches. Often in noisy mixed flocks of chickadees, titmice, and warblers.

summer
winter
year-round
occasional
/// ocean range

☐ date seen

BROWN-HEADED NUTHATCH

Sitta pusilla **Length: 4½"**

Look for: This tiny nuthatch has a brown cap with a white spot on the back of its neck and a bill that looks too large for the bird's size.

Listen for: If any bird's call ever sounded like a squeeze toy, this one is it. The Brown-headed Nuthatch normally utters a series of high, two-syllable squeaks: *pyee-deet! pyee-deet!* Also gives a single high *queet!* and a burbling series of squeaks.

Remember: A close relative of the Brown-headed is the Pygmy Nuthatch of western pine forests. Fortunately, these look-alike birds are geographically separated, or we'd have a hard time telling them apart.

WOW!
This species has been observed using a tool to get its food! A bird will hold a small piece of bark in its bill and pry up other pieces of bark to get at insects or insect eggs hidden underneath.

◀ *Using a wood chip, a Brown headed Nuthatch pries up a piece of bark, revealing a juicy spider.*

Find it: A year-round resident of southeastern pine forests, the Brown-headed Nuthatch is often found in small family groups and in pairs during the breeding season. In winter they may join mixed feeding flocks.

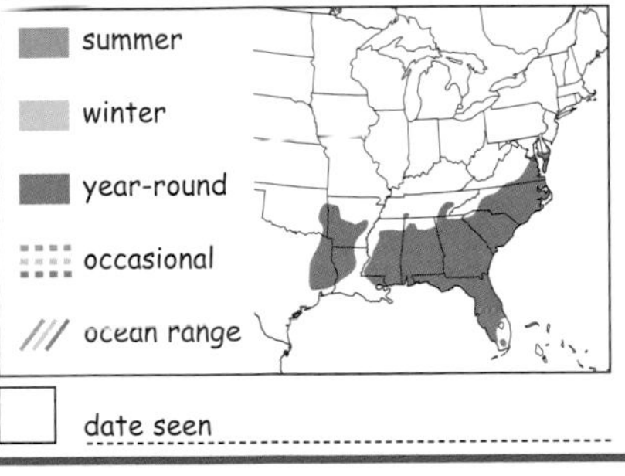

☐ date seen ______________

BROWN CREEPER

Certhia americana **Length: 5¼"**

Look for: Creeping quietly up the trunk of a tree, the Brown Creeper can be extremely hard to spot. Its mottled brown back is the perfect camouflage. The long, stiff tail feathers and thin, curved bill help to separate this species from the nuthatches that forage in a similar fashion.

Listen for: The creeper's call is a high, thin, single note: *seet!* It is similar to the Golden-crowned Kinglet's call, but the kinglet nearly always gives three *seet* notes. The song is a high-pitched musical one that begins with two *seet* notes and ends with a downward-slurring jumble.

Remember: If you don't hear a creeper's call, you may hear its bill or claws scraping against tree bark before you see the bird.

◀ *Brown Creeper nests are hidden beneath loose tree bark.*

Find it: Found in mature woodlands, Brown Creepers creep along the trunks of trees, always climbing upward, very quietly and subtly. When they reach the top of a tree, they fly to the base of another and start over.

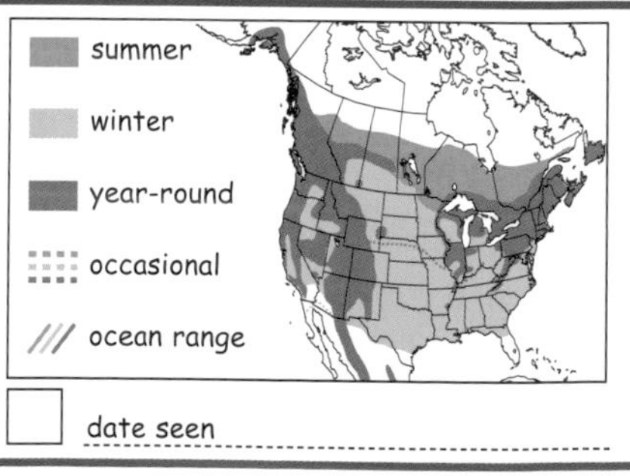

HOUSE WREN

Troglodytes aedon **Length: 4¾"**

Look for: This small chocolate brown bird is plain looking compared to the larger Carolina Wren, but it makes up for this by being very active and very vocal. The House Wren has a faint eye-ring on its otherwise unmarked face.

Listen for: House Wrens have a song that really warbles. It is a long, high-pitched, liquid burble of notes in a jumble with a few harsh notes at the beginning. Also gives a harsh scold call: *chit-chit-rrr-rrr-rrr!*

Remember: The House Wren vocalizes almost constantly and often cocks its tail upward, much like the Carolina Wren. But the House Wren is smaller, darker brown overall, and lacks the Carolina's white eye line.

WOW!
House Wrens can be very feisty when it comes to nest sites. They will pierce and discard the eggs of bluebirds and others who dare to build a nest in a site the House Wren feels it "owns."

▼ *Anything will do as a nest site for the House Wren—even a discarded boot in a toolshed.*

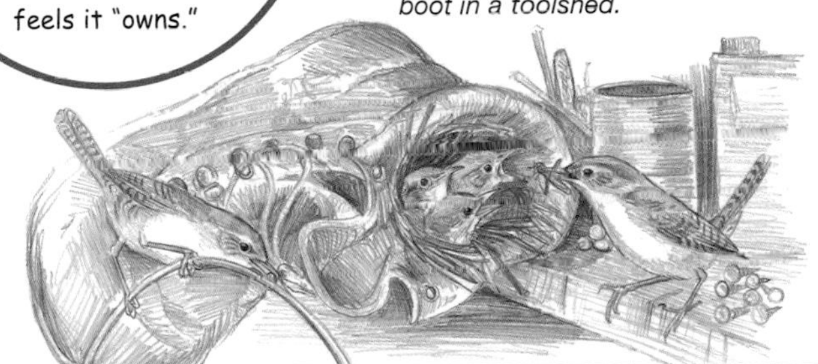

Find it: Most common and vocal in summer, found skulking through thick underbrush and along woodland edges, the House Wren moves (and calls) almost constantly.

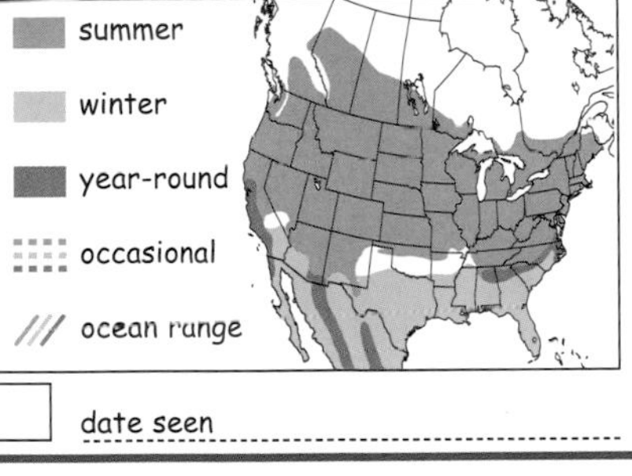

☐ date seen ____________

BEWICK'S WREN

Thryomanes bewickii **Length: 5¼"**

Look for: A medium-sized wren, plain brown above, pale gray below with a bold white eyebrow line. The long gray-brown tail is rounded at the tip with white corners. While perched, it often fans and switches its tail.

Listen for: Very vocal. Song is similar to Song Sparrow's: a few high notes, lower trills, rising buzzes, and ending on a trill. Calls include harsh scolding notes, *pit-pit*, and a dry buzz: *dzzzrrrt*.

Remember: The white corners on the Bewick's Wren's tail are diagnostic. The similar Carolina Wren is warmer brown overall with an all-brown tail.

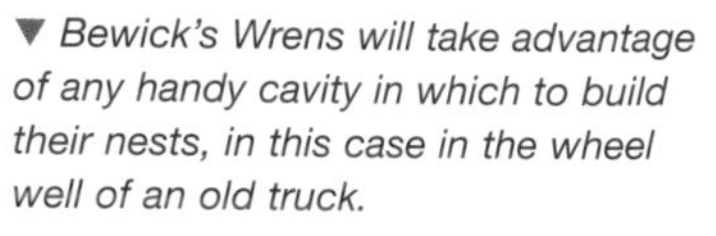

▼ *Bewick's Wrens will take advantage of any handy cavity in which to build their nests, in this case in the wheel well of an old truck.*

WOW!

The Bewick's Wren is not named for the Buick (though it's pronounced the same). It was named by John James Audubon to honor his friend and fellow artist Thomas Bewick (who lived way before there were any Buicks).

Find it: Common in the West in scrublands, woodland and riparian thickets, chaparral, and in towns, parks, and backyards with thick underbrush. Declining steeply in eastern parts of range.

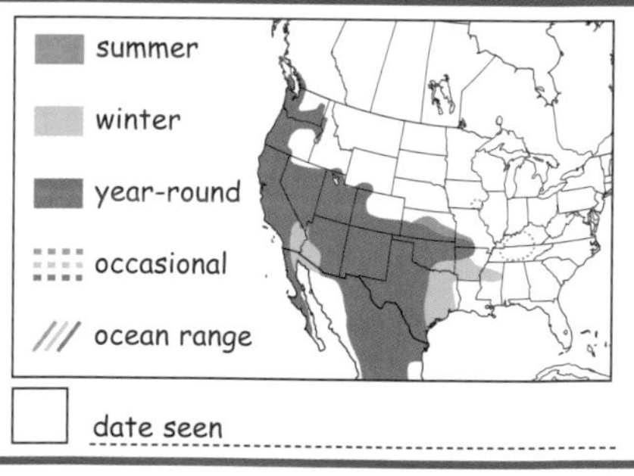

☐ date seen

CAROLINA WREN

Thryothorus ludovicianus **Length: 5½"**

Look for: Our largest eastern wren, the Carolina Wren has a bold white eye line that helps set it apart visually from our other wrens. When perched or foraging, it often holds its tail cocked upward.

Listen for: Often heard before it is seen, the Carolina Wren's most common call is *teakettle, teakettle, teakettle*. Pairs stay together all year and keep in touch throughout the day by vocalizing back and forth. Carolinas also give a variety of *chink* calls as well as a harsh, rolling metallic scold—*cheerrrrrr-rrr-rrr!*

Remember: The House Wren is an inch smaller than the Carolina Wren, darker brown overall, and lacks the buffy chest and white eye line.

WOW!
Carolina Wrens seem to love nesting near humans. Nests have been found inside garages, in old shoes and empty cans, and even in clothespin bags hanging on well-used clotheslines.

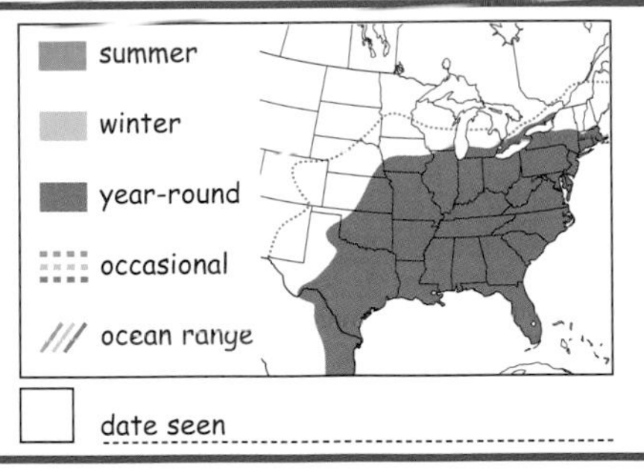

◄ *A Carolina Wren visits its nest inside a toolbox in an old shed.*

Find it: Common in brushy thickets and ravines, woodland edges, and backyards, Carolina Wrens readily come to bird feeders, where they eat sunflower seeds and peanut bits, suet, and fruit.

summer
winter
year-round
occasional
ocean range

☐ date seen

CANYON WREN

Catherpes mexicanus **Length: 5¾–6"**

Look for: Rusty brown body and grayish head contrast sharply with white throat and breast. Bill is very long, slender, and slightly down-curved. Small white dots on head, back, and wings are visible at close range. Usually heard before it is seen.

Listen for: The Canyon Wren is a strong singer, and its song is a series of slurred, clear notes that cascade downward in tone: *tee-tee-tee-tee-tyew-tyew-tyew*. Call is a harsh, burry *jeertt!*

Remember: Our other western wrens lack the contrasting dark belly and white throat and breast of the Canyon Wren.

◀ *The Canyon Wren likes to sing from an exposed perch. The song cascades downward, like it's falling from the mountain.*

Find it: A year-round resident of steep rocky slopes, cliff faces, canyons, and stone buildings. Often near water. Its loud, ringing song is the best clue to this bird's presence in appropriate habitat.

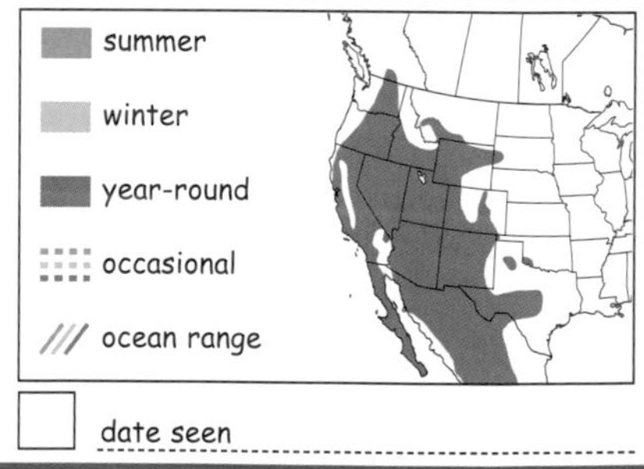

date seen

CACTUS WREN

Campylorhynchus brunneicapillus **Length: 8½"**

Look for: Our largest wren, the Cactus Wren is sometimes mistaken for a thrasher. The bold white eyebrow stands out below the dark brown cap. Dark spots below (especially on breast), white streaks above, buffy flanks. Long tail is edged in black and white dashes.

Listen for: Song is a series of harsh, unmusical notes: *churr-churr-churr-churr,* often speeding up. It sounds like someone trying to start a car. Scold note is *clack!*

Remember: The somewhat similar thrashers, especially the smallish Sage Thrasher, all have longer tails and plain, unstreaked backs.

▲ *The bulky stick nest of a Cactus Wren is an obvious clue to the presence of this species.*

WOW!

Cactus Wrens are big, bold, and curious birds. They have been observed picking insects off the front grilles of parked cars.

Find it: A common year-round resident of arid cactus-mesquite brushlands in the Southwest. Also found in scrubby sagebrush habitat and coastal scrub. Found in pairs or family groups foraging on or near the ground.

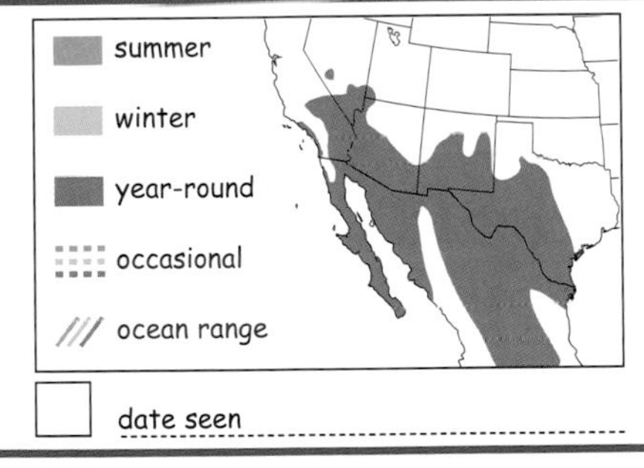

date seen ____________________

AMERICAN DIPPER

Cinclus mexicanus **Length: 7½"**

Look for: A medium-sized, plump, dark gray bird with a brownish face, short upturned tail, and long pale legs. It will perch momentarily on a midstream rock, bobbing (or "dipping") up and down, before slipping into the water to forage.

Listen for: Song is surprisingly beautiful, almost mockingbird-like, consisting of a series of rich, clear whistled notes paired with some buzzy tones with phrases repeated. Call is a raspy *zeet!*

Remember: A good way to spot a Dipper is to look for a midstream rock with lots of poop splats on it, and then wait. Dippers have favorite rocks from which to forage and on which to do other things.

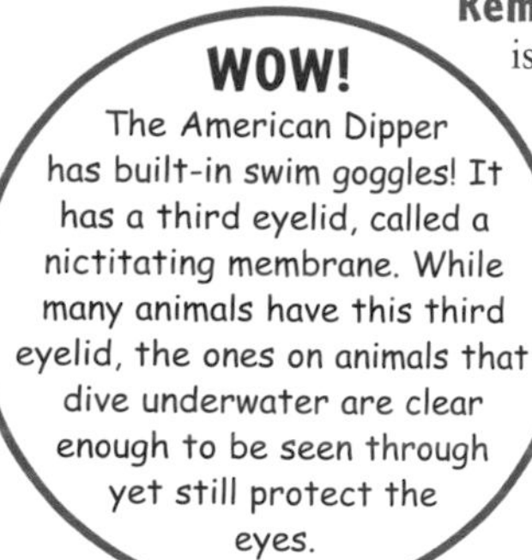

► *Dippers use their strong legs and short rounded wings to propel themselves underwater in fast-moving mountain streams as they forage for aquatic insects.*

Find it: Always found in or near fast-flowing (even roaring) mountain streams in the West. Flight is fast and low over the water. Despite size and constant motion, can be difficult to spot. Resident year-round, but may move to lower elevations in winter.

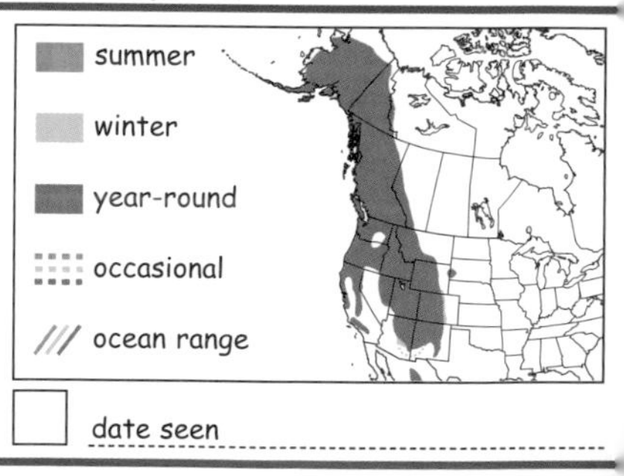

☐ date seen ______

RUBY-CROWNED KINGLET

Regulus calendula **Length: 4"**

Look for: The Ruby-crowned Kinglet is slightly larger than its Golden-crowned cousin and lacks the bold markings on the head and face. The Ruby-crowned has an obvious white eye-ring. The male has the namesake ruby crown patch, which can be hard to see because it is raised only when the bird is excited.

Listen for: The call, *juh-dit-dit!*, is often the first clue to this tiny bird's presence. Song is a long, rich warble.

Remember: No other bird is as tiny, plain, and hyperactive as the Ruby-crowned Kinglet. The eye-ring, plain face and crown, and wing-flitting action set this species apart from the Golden-crowned Kinglet and our warblers and vireos.

▼ *Male Ruby-crowned Kinglets in aggressive displays "blow their tops," revealing a burst of red feathers atop their heads.*

Find it: The Ruby-crowned Kinglet breeds in the woods of the North, where it sticks to the treetops. In fall and winter, it moves south to wooded habitats, where it forages high and low in the vegetation.

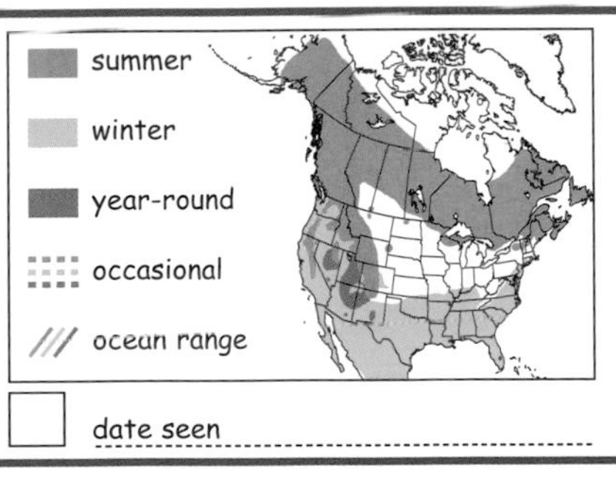

date seen

GOLDEN-CROWNED KINGLET

Regulus satrapa **Length: 3¾"**

Look for: The tiny Golden-crowned Kinglet flits through evergreen trees, flicking its wings open and shut, hoping to startle an insect into moving. The male has an orange patch on the crown (not always visible), and the female has a yellow one. Its face seems to have white stripes coming out from its tiny black bill.

Listen for: The call of the Golden-crowned Kinglet is more commonly heard than its song: a high, thin *seet-seet-seet!* The song is a series of thin *see* notes rising in tone to a thin jumble of notes that falls back down the scale.

Remember: The tiny size, greenish color, and flitting wings can tell you you've got a kinglet. No eye-ring, bold wing bars, and dark crown stripes indicate a Golden-crowned Kinglet.

WOW!

Kinglets of both species are able to hover underneath a branch to glean insects off the undersides of the foliage.

▶ *Golden-crowned Kinglets can be identified by their habit of hover-gleaning—picking small insects off branches while in flight.*

Find it: Nesting in the North or at higher elevations, Golden-crowned Kinglets are most often encountered in winter, when they head south to forage in pines with other kinglets, chickadees, and nuthatches.

summer
winter
year-round
occasional
ocean range

date seen

BLUE-GRAY GNATCATCHER

Polioptila caerulea Length: 4½"

Look for: This species is well named for both its color and its behavior. Its long tail and white eye-ring on a plain face stand out visually. Its active, treetop foraging for gnats and other small insects and its almost constant calling make it easier to spot than many of our small songbirds.

Listen for: The Blue-gray Gnatcatcher's most common vocalization is its call: a mewing *chee-chee-chee*. Song is a lively and messy jumble of high, nasal notes and short warbles and whistles. Song may occasionally incorporate phrases from other birds.

Remember: The Blue-gray Gnatcatcher is our only gnatcatcher that is common outside of the Southwest.

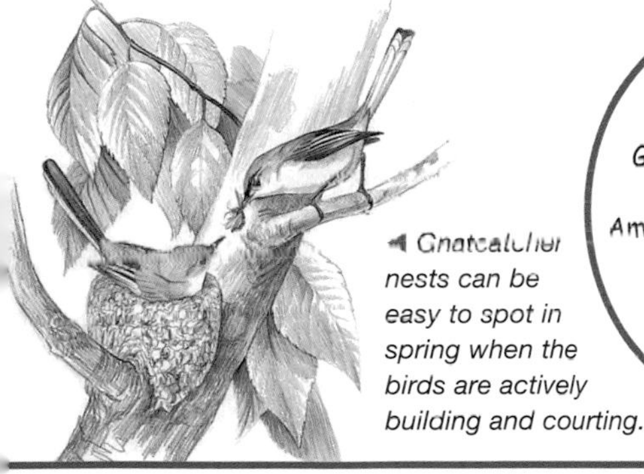

◀ *Gnatcatcher nests can be easy to spot in spring when the birds are actively building and courting.*

WOW!

Many consider the arrival of the Blue-gray Gnatcatcher a better sign of spring than the first American Robin. Gnatcatchers appear in many areas as soon as small insects become active.

Find it: Common in spring and summer in open mixed woodlands and along woodland edges, the Blue-gray Gnatcatcher announces its presence almost constantly with calls, short fly-catching flights, and active movements.

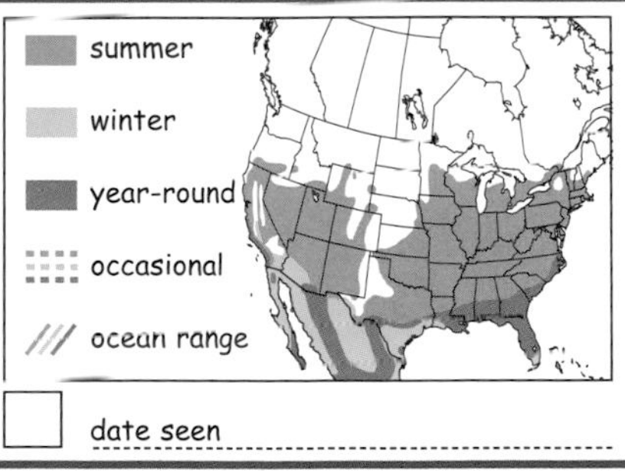

EASTERN BLUEBIRD

Sialia sialis **Length: 7"**

Male

Look for: This beautiful member of the thrush family and its relatives in the West (the Western and Mountain Bluebirds) are the only thrushes that nest in cavities. They appear round bodied and round headed when perched. Males are more boldly colored than females. Juvenile birds are blue-gray, spotted with white.

WOW!
Bluebird populations were decimated in the mid-1900s. Bluebird lovers provided nest boxes, helping North America's three bluebird species recover.

Listen for: The Eastern Bluebird's song is a soft, rich warble given in short phrases: *tur, tur, turley, turley!* Flight call is a two-noted *ju-lee!* When they spot a predator, they utter a down-slurring *tyew!*

Remember: Both Eastern and Western Bluebirds have an orange breast, but on the Western the orange continues onto the back. The Western has a blue throat, while the Eastern's is orange.

◀ *Eastern Bluebirds prefer to nest in open habitat and will readily use nest boxes, especially where natural cavities are scarce.*

Find it: Common in open, grassy settings such as pastures, roadsides, parks, and large backyards, the Eastern Bluebird perches on wires and exposed branches and watches for insect movement below.

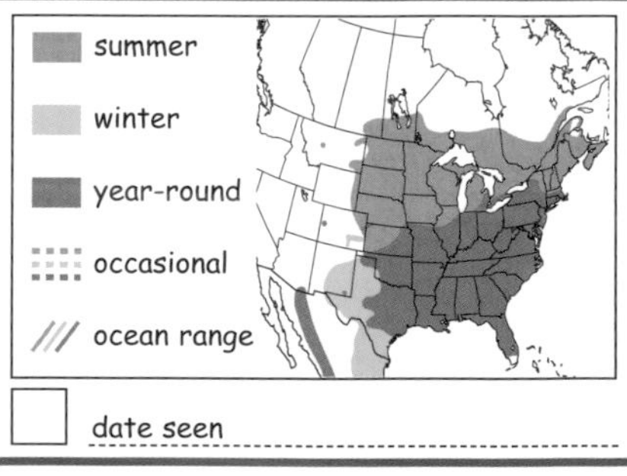

date seen ______________

WESTERN BLUEBIRD

Sialia mexicana Length: 7"

Look for: The male Western Bluebird is a bold combination of blue and rusty orange. Blue on the throat and belly and rust on the shoulders and back are diagnostic. Female is paler overall with a grayish throat and belly. Juveniles are grayish spotted with white.

Listen for: Song is simple, down-slurred whistled notes: *tyew. Tyew-tyew!* Call note is harsher chatter: *chack-chack*.

Remember: Similar to the less colorful Eastern Bluebird, but their ranges barely overlap.

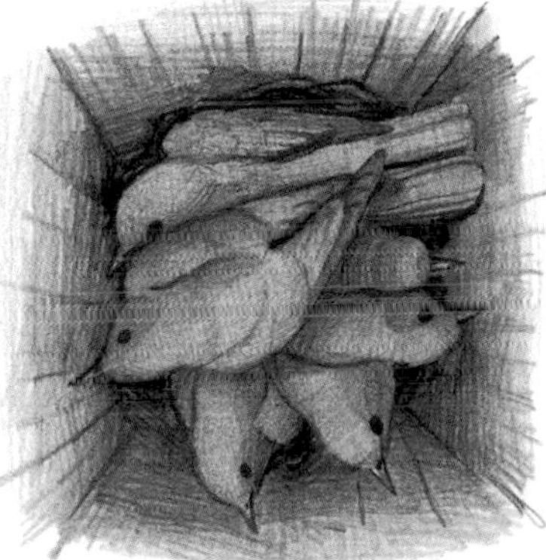

▶ *Western Bluebirds will roost communally on cold nights.*

Find it: Common in the West in open habitat with scattered trees, including open forests, savanna, mountain meadows, farmland, parks, and golf courses. Often seen in small flocks. Regular user of roadside nest boxes. Winters in habitat with berry-producing trees, such as junipers.

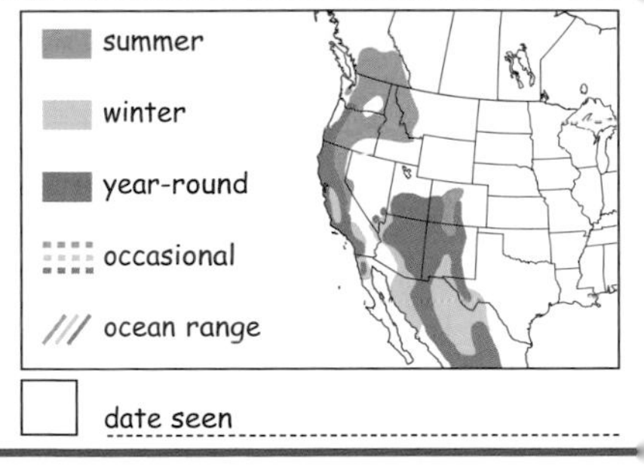

date seen

MOUNTAIN BLUEBIRD

Sialia currucoides **Length: 7¼–7½"**

Look for: Adult male is pale turquoise blue overall; female is gray below and pale blue on wings and tail. Mountain Bluebirds are more slender and have longer wings and tails than our other two bluebirds.

Listen for: Song is a series of soft warbles: *too-toodle, too-too-toodle.* Call is a soft *tchack* or *tchack-it.*

Remember: Unlike our other bluebirds, the Mountain Bluebird has no rust or red in its plumage.

WOW!

Because it prefers wide-open spaces, often far from any signs of human activity, the Mountain Bluebird does not get as much competition for nest sites from European Starlings and House Sparrows as its fellow bluebirds get.

◀ *In winter, Mountain Bluebirds sometimes form huge foraging flocks.*

Find it: Prefers wide-open country with scattered trees in all seasons. Often hovers just above the ground looking for insects. Found in mountain meadows and above the tree line, but also in lower habitats, including prairies, sagebrush, and farmland.

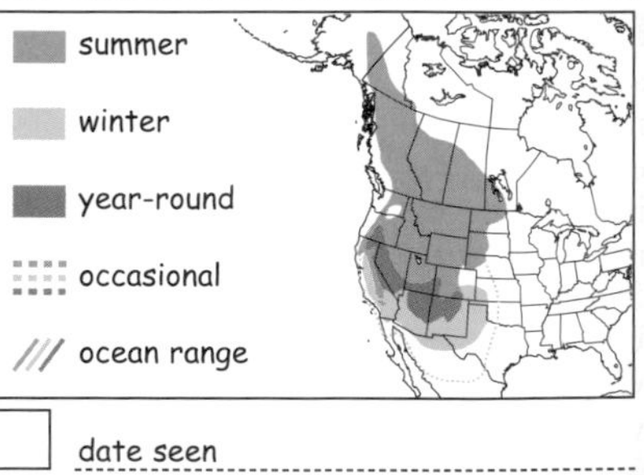

☐ date seen

TOWNSEND'S SOLITAIRE

Myadestes townsendi **Length: 8½"**

Look for: A slender gray member of the thrush family, with a prominent white eye-ring and white edges on its long gray tail. Adult Townsend's Solitaires show varying amounts of buff in the wings.

Listen for: Song is a rich, finchlike warble, slightly hoarse. Call is a high, ringing, single whistled note: *too*.

Remember: The solitaire's short bill, eye-ring, and buff wing patches distinguish it from the Northern Mockingbird and our two shrike species.

Townsend's Solitaires defend winter territories to protect their food source (berries) from other birds. No wonder they're always alone.

◀ *Townsend's Solitaires perch in a very upright posture—a good clue to their identity.*

Find it: An uncommon and solitary bird of western montane forests, woodland edges, and wooded canyons in summer. In winter, it moves to juniper and pinyon forests, often along streams. Usually seen perched at the very top of a tree or snag, watching for passing insects.

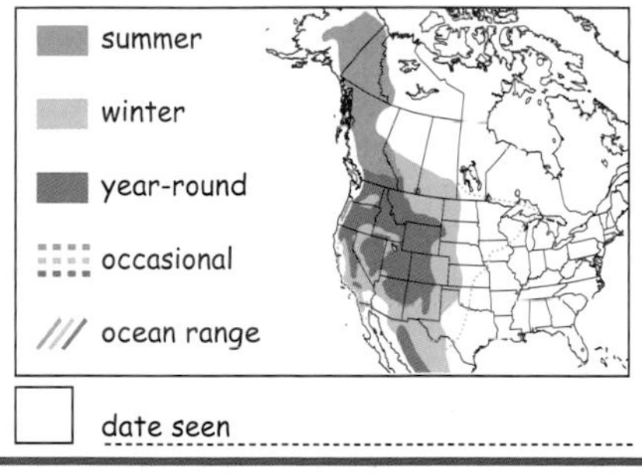

HERMIT THRUSH

Catharus guttatus **Length: 6¾"**

Look for: Several field marks set the Hermit Thrush apart from our other brown woodland thrushes: a lightly spotted breast, a strong buffy eye-ring, a rusty tail, and a habit of raising and lowering its tail when perched.

Listen for: Many birders consider the Hermit Thrush's ethereal, flutelike song to be among the most beautiful of all bird songs. It starts with a clear, low note and spirals upward in a sweet jumble. Two very different call notes are commonly given: a soft *tchup!* and a nasal, rising *vreee!*

Remember: Each of our six brown woodland thrushes has at least one unique field mark. The Hermit's are its eye-ring, rusty tail, and tail motion.

WOW!
The Hermit Thrush's beautiful song has earned it the folk nicknames American Nightingale and Swamp Angel.

◀ *The Hermit Thrush has a unique habit. It raises its tail suddenly, then lets it fall slowly back to a normal position.*

Find it: This is the thrush you are most likely to see here in the winter. They spend the breeding season in northern coniferous forests. In winter and during migration, they can be found in any wooded setting.

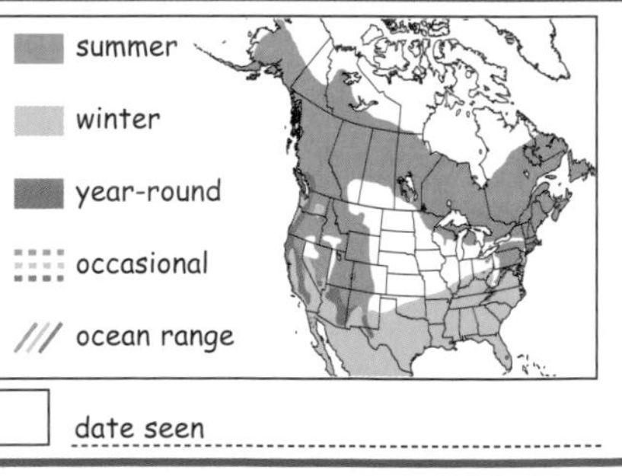

date seen

AMERICAN ROBIN

Turdus migratorius **Length: 10"**

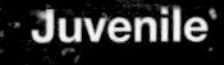

Look for: The American Robin is so familiar in North America that even nonbirders know its name. Robins spend hours foraging for earthworms in our yards, gardens, and parks. Male robins have darker heads and backs; females are paler overall. Juveniles are spot breasted throughout their first summer.

Listen for: Song is a rich, slightly hoarse warble: *cheery-o, churlee, cheery-up!* Calls include a loud *see-seet-tut-tut-tut!* and a thin, soft, down-slurred *tseeeet!* as an alarm call.

Remember: In all seasons and at all ages, robins have orange on their breasts, with dark heads and backs.

WOW!

Robins are often considered the first sign of spring, but not all robins leave their home range in winter, so their appearance is not really a sign of spring.

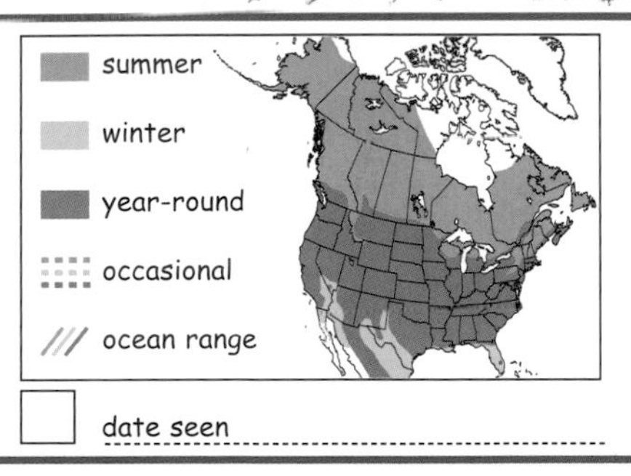

▶ *An adult American Robin jams a juicy earthworm into the gaping mouth of a nestling.*

Find it: Robins are found in a variety of habitats, from suburban parks and backyards to mountain meadows, and are most commonly seen on the ground. In winter, they concentrate in the woods where there are berries and other fruits to eat.

summer
winter
year-round
occasional
ocean range

☐ date seen ______________________

VARIED THRUSH

Ixoreus naevius **Length: 9½"**

Look for: Looking like an American Robin that put on a bunch of makeup and a black necklace, the adult male Varied Thrush also has orange wing bars and an orange eye stripe. Female is paler overall. In flight it shows an orange central wing stripe.

Listen for: Song is a long buzzy whistle given in varying tones at short intervals: *tzzzeeee!* Call is a soft *chup*.

Remember: A Varied Thrush may look a bit like an American Robin, but the two species rarely mingle.

WOW!
Every year a few Varied Thrushes get mixed up in fall migration and end up far to the east. This causes eastern bird watchers to go cuckoo.

◄ *Winter flocks of Varied Thrushes eat mainly berries and fruits, but they also forage on lawns for insects.*

Find it: Common (but often difficult to see) in dense, wet coniferous forests (spruce, hemlock, fir) of the Northwest where it breeds. In winter, flocks of Varied Thrushes occupy moist thickets, brushy ravines, and coniferous forests. Often heard before it is seen.

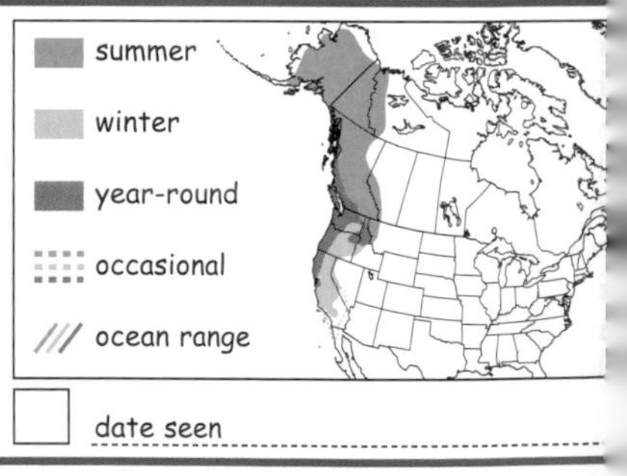

☐ date seen ____________

WOOD THRUSH

Hylocichla mustelina **Length: 7¾"**

Look for: The Wood Thrush is the largest and most boldly marked of our brown woodland thrushes. The large spots on the white breast stand out even in deep woodlands. A bright rusty brown back and white eye-ring also stand out.

Listen for: A beautiful song starting on a clear note or two, followed by a flutey warble: *tu tu ee-oo-laay-yi-yi.* Two common calls are *tu-tu-tu* notes and a staccato *pit-pit-pit-pit* often used as a scold or warning.

Remember: No other brown thrush has the striking breast pattern of the Wood Thrush. It might be more easily confused with the Brown Thrasher, but the thrasher has a long decurved bill, a much longer tail, and two wing bars.

▼ *The Wood Thrush gives its flutelike song in the evening, usually from a well-hidden spot in the mid-canopy.*

Find it: The Wood Thrush breeds in mature deciduous woods, often near water. Easily heard, but can be hard to see in its deep woodland habitat. Look for it foraging on the ground, much like an American Robin.

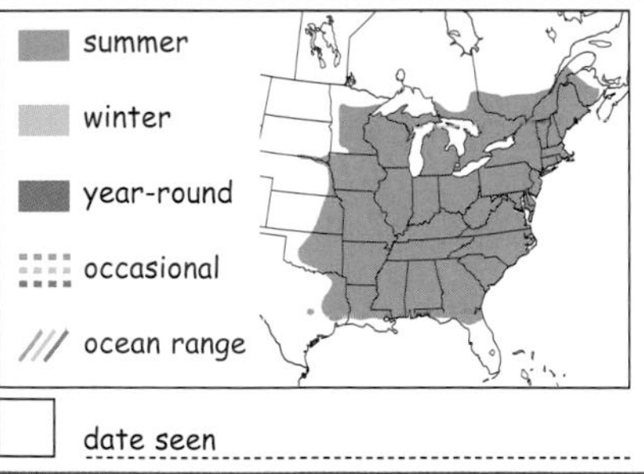

BROWN THRASHER

Toxostoma rufum **Length: 11½"**

Look for: This large, rusty brown bird is often a skulker in thick underbrush. The long decurved bill, twin wing bars, and bright yellow eye help to separate the Brown Thrasher from the Wood Thrush and our other brown woodland thrushes.

Listen for: Song is a series of warbled and whistled phrases in pairs. The old farmer's rendering of the thrasher's song is *see it see it, pick it up pick it up, dig it dig it, plant it plant it*. Call is a metallic *chaak!* that sounds like two giant marbles cracking together.

Remember: The Brown Thrasher is named for its foraging technique. It whips its head back and forth as it walks, using its long decurved bill to thrash leaves and twigs out of the way, hoping to uncover an insect.

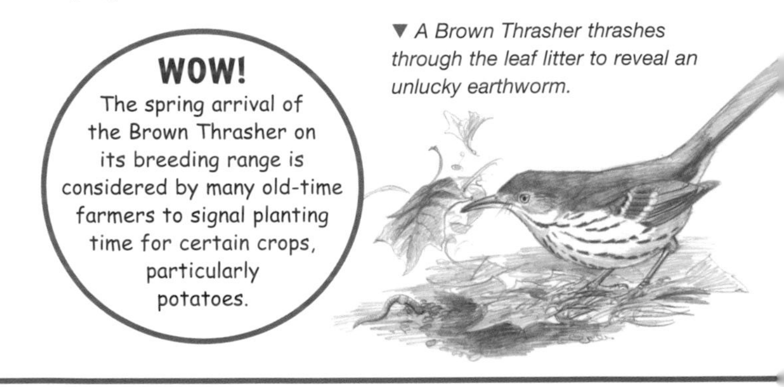

WOW!

The spring arrival of the Brown Thrasher on its breeding range is considered by many old-time farmers to signal planting time for certain crops, particularly potatoes.

▼ *A Brown Thrasher thrashes through the leaf litter to reveal an unlucky earthworm.*

Find it: Shrubby woodland edges, brushy old-fields, and hedgerows, even in suburban settings, are the favored haunts of the Brown Thrasher.

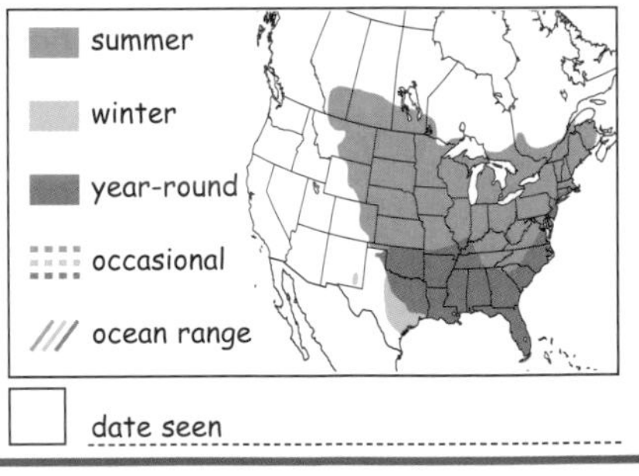

☐ date seen ....

GRAY CATBIRD

Dumetella carolinensis **Length: 8½"**

Look for: This elegant and slender-looking bird is gray overall with a black cap and a rusty crissum (undertail) patch.

Listen for: The Gray Catbird has a variable song comprising chortles, squeaks, and short warbles. But it is named for its call note, a mewing (*meeyah!*) that sounds like a cat with a sinus problem. Also utters a loud, staccato *kak-kak-kak-kak!* as an alarm call.

Remember: Except when singing from an exposed perch on the breeding territory, the Gray Catbird and most of our other mimics prefer to remain near the ground and in or near thick cover.

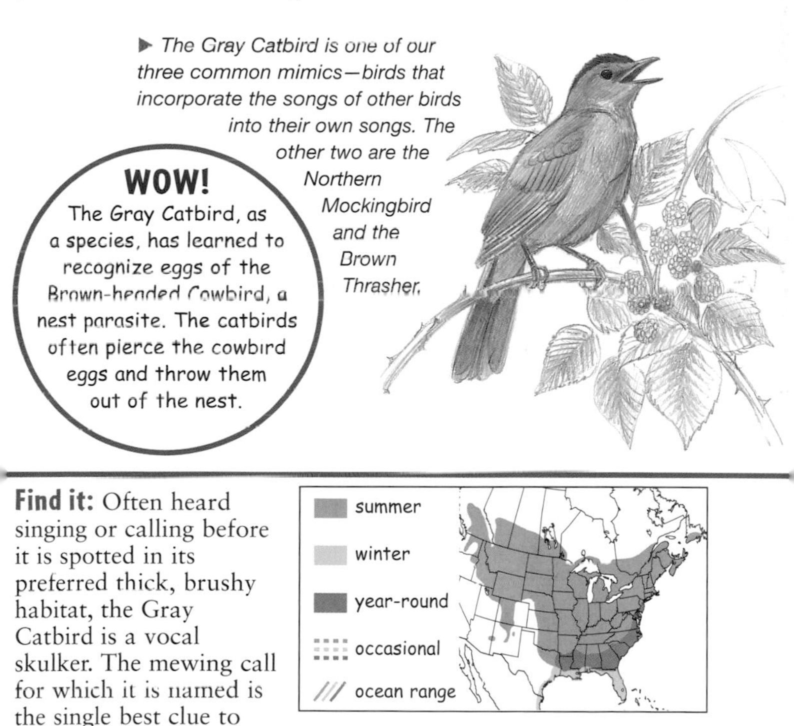

▶ *The Gray Catbird is one of our three common mimics—birds that incorporate the songs of other birds into their own songs. The other two are the Northern Mockingbird and the Brown Thrasher.*

WOW!

The Gray Catbird, as a species, has learned to recognize eggs of the Brown-headed Cowbird, a nest parasite. The catbirds often pierce the cowbird eggs and throw them out of the nest.

Find it: Often heard singing or calling before it is spotted in its preferred thick, brushy habitat, the Gray Catbird is a vocal skulker. The mewing call for which it is named is the single best clue to this bird's presence.

☐ date seen ______________________

NORTHERN MOCKINGBIRD

Mimus polyglottos **Length: 10½"**

Look for: This long, slender, gray, black, and white bird is the king when it comes to variety in its song. It earned its name from its ability to imitate other birds and sounds.

Listen for: Individual Northern Mockingbirds may use more than 200 different songs and sounds, including those of birds and other animals and mechanical sounds. Most common call is a harsh *tchapp!*

Remember: Few other gray birds flash as much white in flight as the Northern Mockingbird. The Loggerhead Shrike is smaller bodied but stockier in appearance, with a short, thick bill and a black facemask.

WOW!

Unmated young male Northern Mockingbirds will sing all night during the spring and summer, especially when the moon is bright, possibly trying to impress the female mockingbirds.

▼ *Mockingbirds will dive-bomb intruders in their territory, including housecats.*

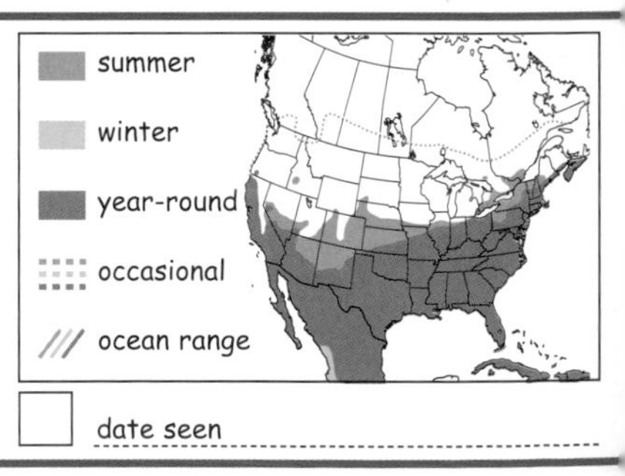

Find it: The mockingbird loves large, open areas, such as lawns and parks surrounded by shrubby undergrowth. As they forage on the ground, mockers fan their wings, flashing the white patches to startle insects into moving.

summer
winter
year-round
occasional
/// ocean range

☐ date seen

CURVE-BILLED THRASHER

Toxostoma curvirostre **Length: 11"**

Look for: The long, curved, all-dark bill and the breast with large, dark, round spots help tell the Curve-billed Thrasher from its desert-dwelling thrasher relatives. Has large orange eyes and a long tail. Some birds in Texas have faint white wing bars.

Listen for: Song is a rich mix of whistles, squeals, squawks, and warbles. Call is a loud, whistled *whit-wheet!* Calls frequently.

Remember: Can be confused with Bendire's Thrasher, which has a shorter bill and triangular (not round) spots on the breast.

◀ *Curve-billed Thrasher pairs remain together all year.*

Find it: The most commonly encountered of our resident desert thrasher species. Found in deserts, arid brushlands, desert canyons, around ranches, and in suburban yards. Usually associated with cholla cactus or prickly pear cactus.

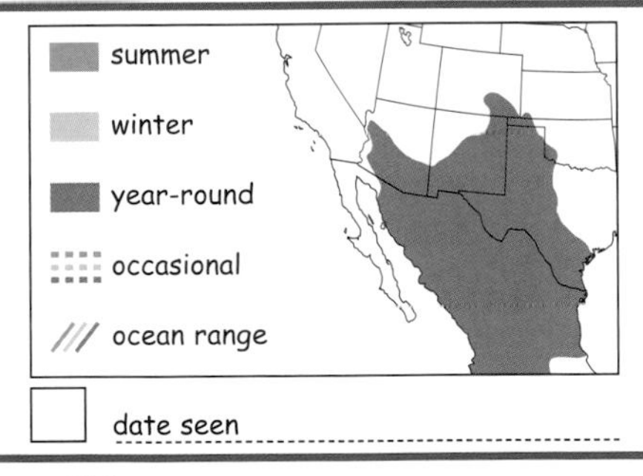

CEDAR WAXWING

Bombycilla cedrorum **Length: 7¼"**

Look for: A warm brown bird, the Cedar Waxwing is named for the small, red, waxy tips on some of its wing feathers. Some waxwings have orange rather than yellow tail tips, a color shift caused by eating certain fruits. Young birds appear streaky and grayer overall.

Listen for: Waxwings vocalize almost constantly, uttering a high, thin *sreee!* It's not much of a song for such a lovely bird, but it is often your first clue that waxwings are around.

Remember: In the North and in winter, the larger, grayer Bohemian Waxwing may be found with Cedar Waxwings. Bohemians have more white in the wing and an obvious rusty undertail. In flight, Cedar Waxwings may be confused with starlings. The waxwing is sleeker and faster and flocks call constantly.

WOW!

Waxwings are sometimes found on the ground, appearing drunk and unable to fly after eating fermented (or too many) berries. They recover their senses after a short period.

▲ *Two Cedar Waxwings pass a wild cherry in a bonding ritual used between mates as well as flock members.*

Find it: Cedar Waxwings are nomads, going wherever the natural berry and fruit crops are plentiful. They occur in flocks at all times except during the late-spring breeding season.

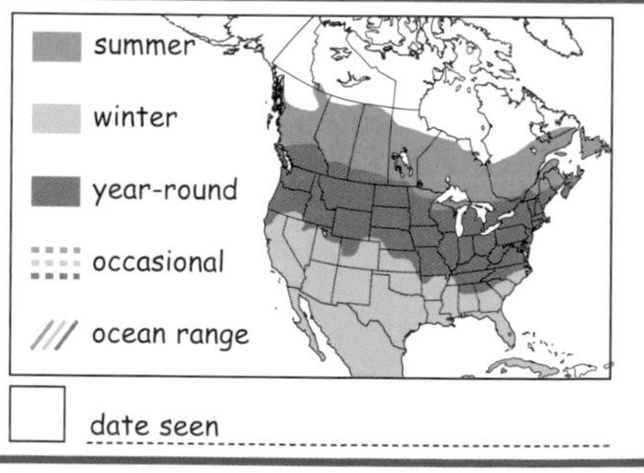

date seen

PHAINOPEPLA

Phainopepla nitens **Length: 7¾"**

Look for: The adult male Phainopepla is a handsome bird: long and sleek and glossy black with a crest and a ruby red eye. In flight, male shows obvious large white wing patches. Adult female is slate gray, also crested and with red eyes. Flight style is flopping and butterfly-like.

Listen for: Song is a series of odd whistles and harsh phrases sung in a random, disjointed way. Call is an up-slurred, froglike *hoit*.

Remember: No other crested birds are so dark or have red eyes!

◀ *Mistletoe is the dietary staple of the Phainopepla.*

WOW!
The name Phainopepla comes from the Greek word meaning "shining robe," in reference to the male's glossy black plumage.

Find it: Common in desert scrub, oak foothills, and mesquite, especially where mistletoe is present. Males especially perch in the tops of trees to sing and flycatch.

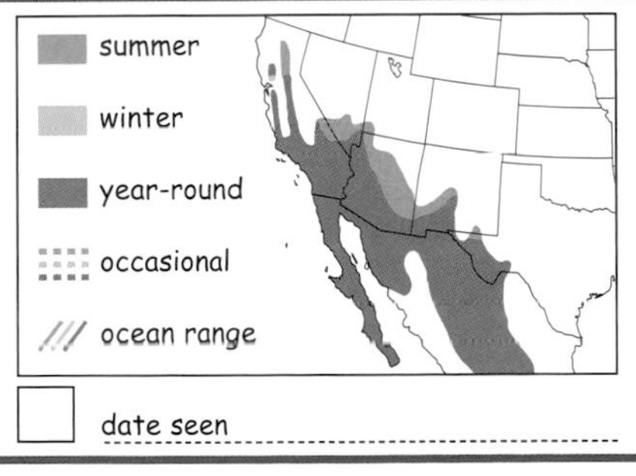

date seen

EUROPEAN STARLING

Sturnus vulgaris **Length: 8½"**

Look for: Despite its reputation as a stealer of nest cavities and a hog at bird feeders, the European Starling, which looks like a blackbird but is not, is quite lovely in both its glossy breeding plumage and its spangled winter plumage.

WOW!
Starlings love to adorn their nest cavities with shiny or colorful things such as coins, bits of plastic, and other birds' feathers.

Listen for: Song is a high-pitched jumble of whirs, whistles, and chatter. Excellent mimics, starlings may incorporate other bird songs and even human sounds into its songs. Calls include a metallic *wrrrsh* and *pink!* (often used when a hawk is sighted).

Remember: In flight, the starling can be confused with the Purple Martin: both birds' wings look triangular, but the starling flaps more often and does not glide with the grace of the martin.

▶ *Starlings prefer to live near humans.*

Find it: Introduced from Europe, the European Starling now occurs in almost every North American habitat but has adapted especially well to living around humans. In fall and winter, starlings form large feeding and roosting flocks.

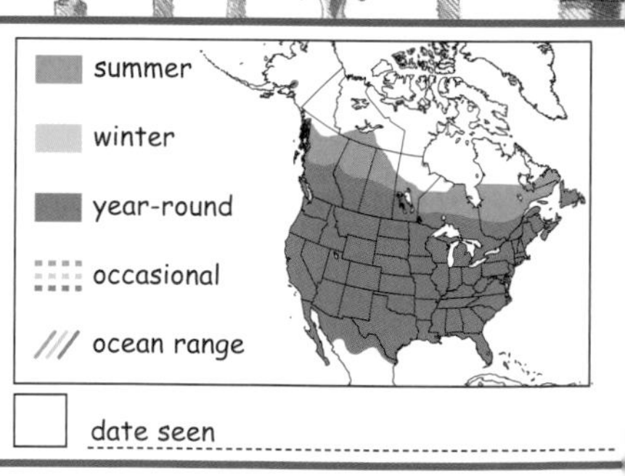

☐ date seen ____________

BLUE-WINGED WARBLER

Vermivora pinus **Length: 4¾"**

Look for: Singing from an exposed perch atop a sapling in an old-field, the male Blue-winged Warbler is a stunning sight. The male's bright yellow head and body contrast with its blue-gray wings, which have large white wing bars. A slim black line runs through the eye. Females are duller overall. In flight, both sexes flash white outer edges in the tail.

Listen for: The male Blue-winged Warbler sings *beee-buzzzz!* A frequently given alternative song is a short trill followed by an up-slurred buzz: *ch-ch-ch-ch-tzeee!* Call note is a sweet, sharp *chick!*

Remember: Several warblers have songs similar to the Blue-winged Warbler's. Time spent listening to and learning warbler songs before the birds return in the spring can really pay off.

◀ *A male Blue-winged Warbler brings his mate a caterpillar while she incubates the pair's eggs.*

WOW!

Blue-winged Warblers sometimes interbreed with the closely related Golden-winged Warbler, producing two general hybrids: the Lawrence's Warbler and the Brewster's Warbler.

Find it: Blue-winged Warblers prefer open habitat such as brushy woodlands, woodland edges, and overgrown pastures for nesting and foraging. Territorial males often sing all day long in spring and summer.

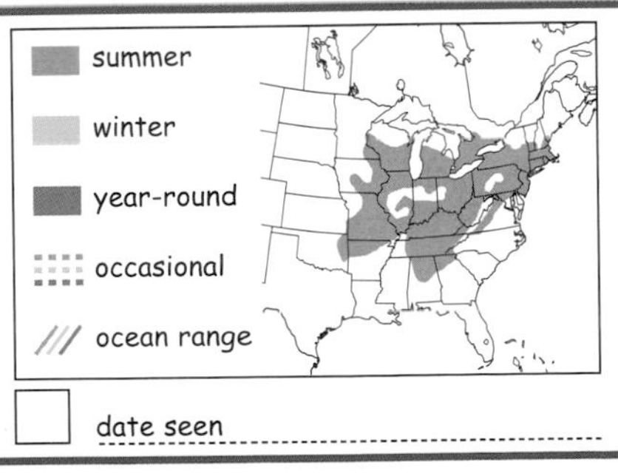

☐ date seen ____________

ORANGE-CROWNED WARBLER

Vermivora celata **Length: 5"**

Look for: A small and very plain warbler with a small finely pointed bill and a light, broken eye-ring. Perhaps the best field mark for this greenish-above, yellowish-below warbler is the yellow undertail.

Listen for: Song is a fast trill that slows in tempo and drops in tone. Call is a faint *tik!*

Remember: Despite its name, the Orange-crowned Warbler's orange crown is not a regularly seen, reliable field mark.

WOW!
The Orange-crowned is one of our warblers that will visit bird feeders for mealworms, suet, and peanut bits.

► *Orange-crowned Warblers are savvy survivors and will drink sap from sapsucker holes in trees.*

Find it: Far more common in the West and North than in the East. In summer, prefers aspen groves, woods with a brushy understory, spruce forest, and streamside thickets. In winter, common in the South in similar habitats as well as in parks and gardens.

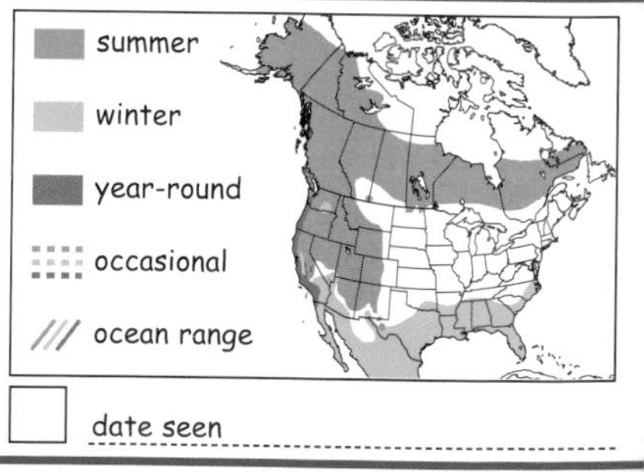

☐ date seen ____________

NORTHERN PARULA

Parula americana **Length: 4½"**

Look for: A tiny warbler of the treetops, the Northern Parula's (pronounced *PAR-you-lah*) yellow throat and white belly are often all you see as the bird forages high above. But look closely and you may see a yellow lower bill and the reddish black neck band.

Listen for: The Northern Parula's song is a musical buzz that rises up the scale, ending with a sharp down-slurred note: *zeeeeeee-zup!* One way to remember the pattern of this call is that it climbs up the ladder and drops (or hooks) over the top.

Remember: No other widespread warbler has the color combination of the Northern Parula: blue-gray head and back, white eye crescents (like white mascara), and the reddish neck band.

WOW!

Northern Parulas love the stringy strands of Spanish moss and other lichens for nest building. The pouchlike nest inside a clump of hanging moss is very hard to spot.

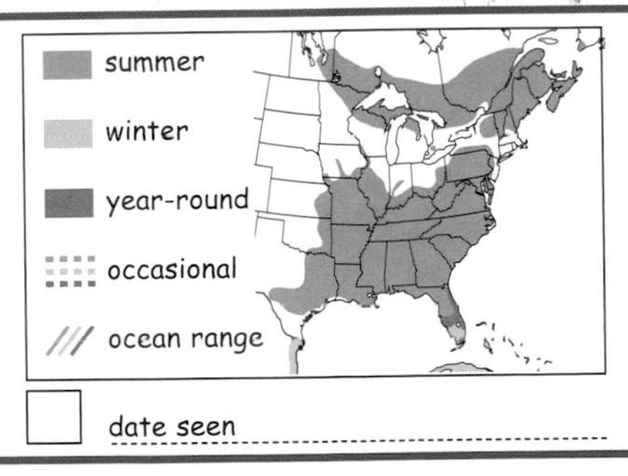

▶ *A male Northern Parula forages in the top of a birch tree.*

Find it: Breeds in large trees often near streams or other bodies of water. A treetop forager, the Northern Parula is best located by its voice, but because it is a small bird, it can be difficult to see.

summer
winter
year-round
occasional
/// ocean range

☐ date seen ----------------

YELLOW WARBLER

Dendroica petechia **Length: 5"**

Look for: This bird is well named because nearly every part of it is yellow except for its eyes, bill, and the male's reddish breast streaks. Females lack the reddish breast streaks, and first-year females can appear drab brown.

Listen for: Song is a variable series of whistled notes: *sweet, sweet, sweet, I'm so very sweet!* Also utters a loud and sharp call: *chip-chip-chip!*

Remember: The Yellow Warbler's dark eyes on a plain yellow face give this species a big-eyed look. The yellow undertail is a good field mark to use when you get only a glimpse of this species. (Most other warblers have white undertails.)

WOW!
Some Yellow Warblers can spot a cowbird egg in their nest. Their response is to start over, often building a new nest on top of the old one.

▶ *A cutaway view of an actual Yellow Warbler nest, including eggs laid by a Brown-headed Cowbird.*

Find it: Yellow Warblers love willow trees, especially near water. But they can be found foraging in a variety of trees, usually at midlevel, making them one of our easiest-to-see warblers. Most birds winter in the tropics.

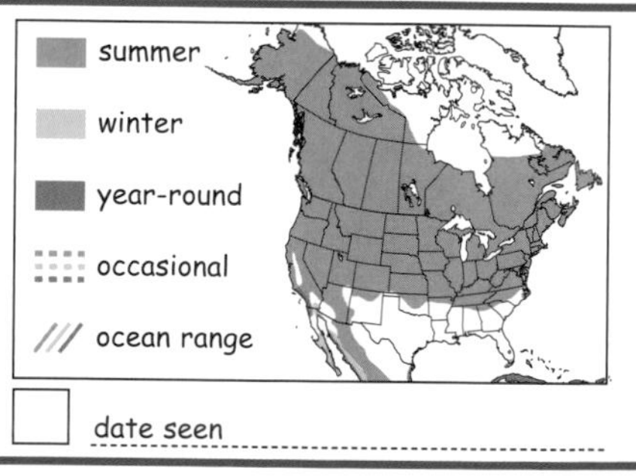

☐ date seen ______________

MAGNOLIA WARBLER

Dendroica magnolia **Length: 5"**

Look for: The male Magnolia Warbler has one of almost every field mark common to our songbirds: eye line, mask, necklace, streaky sides, wing panels, rump patch, tail spots, and black band across tail tip, making it one of our easiest warblers to identify.

WOW!
The Magnolia Warbler was named by early American ornithologist Alexander Wilson, who first saw this species flitting through a magnolia tree during spring migration in 1810.

Listen for: Magnolia Warblers sing a sweet but weak-sounding *pretty-pretty Maggie!* Call note is a very sweet *chet!*

Remember: With the Magnolia Warbler, it's all about the white tail spots—and they seem to love to show them off. Females and fall males lack the breeding male's bold facial markings, but all Maggies have the lemon breast, bold black streaks on flanks, and clean white "underpants" on undertail coverts.

▶ *A territorial male Magnolia Warbler flits high in the spruces, flaring its white-paneled tail in a signal to rivals and potential mates.*

Find it: Maggies breed in young coniferous forests. They are commonly seen as migrants in almost any habitat in the East (but only rarely in magnolias). They often forage in low, shrubby vegetation during migration and are active and easy to spot.

summer
winter
year-round
occasional
ocean range

☐ date seen ______

YELLOW-RUMPED WARBLER

Dendroica coronata **Length: 5½"**

Breeding, Myrtle form

Ma
Audubon's for

Look for: The Yellow-rumped Warbler is well named for its most obvious field mark. Older field guides (and many birders) refer to this species as the Myrtle Warbler (eastern form) and the Audubon's Warbler (western form). Both forms have a yellow rump and white tail patches, but Audubon's has a yellow throat and Myrtle has a white throat.

WOW!
Yellow-rumped Warblers are hardy birds, adapted to live on berries (especially wax myrtle fruits) when cold weather makes insects scarce.

Listen for: Song is a weak trill, usually trailing off toward the end and dropping in tone: *tee-tee-tee-brr-brrbrr!* The call note, a loud *tchep!*, is often a better year-round clue to the presence of this species.

Remember: Cape May and Magnolia Warblers also have yellowish rumps, but they're not as obvious as the butter butt on the Yellow-rumped Warbler.

▶ *A winter Yellow-rumped Warbler eating wax myrtle fruit.*

Find it: Yellow-rumped Warblers spend summers in the northern coniferous forests and at higher elevations. They are very active birds, flitting from tree to tree, flashing their yellow rump patches as they move.

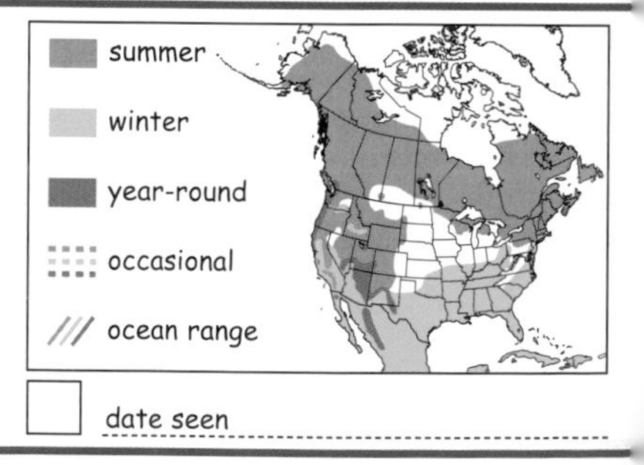

date seen

BLACK-THROATED GRAY WARBLER

Dendroica nigrescens **Length: 5"**

Look for: Bold black-and-white pattern of the adult male's head and the gray back suggest a chickadee. Has heavily streaked chest and flanks and white wing bars. Between the eye and bill is a small spot of yellow. Adult female has a white throat.

Listen for: Song is buzzy and rising: *zeedle-zeedle-zeedle-zeezee* or *Don't ya think I look so pretty?* Call is a strong *tchup*.

Remember: This is the only western warbler that appears to be all black and white.

WOW!
This western warbler occasionally shows up in the East, heating up the rare-bird alerts.

▶ *Black-throated Gray Warblers may visit bird feeders for suet bits and mealworms.*

Find it: Common in summer in oak, oak-juniper, and pine woods in the foothills of western mountains. Forages low in vegetation. May join mixed-species flocks in migration.

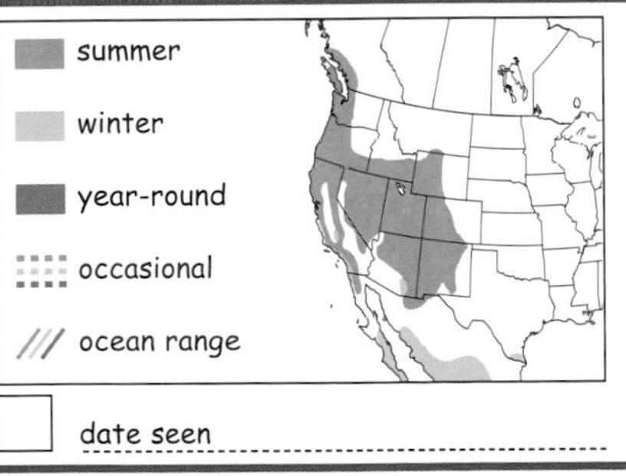

☐ date seen

TOWNSEND'S WARBLER

Dendroica townsendi **Length: 5"**

Look for: Adult male has a boldly marked black and yellow head, with a black cheek and throat. Female has a yellow throat and gray, not black, cheeks. Both sexes have a yellow chest with heavy dark streaking and a white belly.

Listen for: Song is variable, but usually contains a series of high, thin whistles, ending in buzzier notes: *sleazy-sleazy-cheezy!*

Remember: A similar western warbler, the Hermit Warbler, has a plain yellow face and crown.

◀ *Overwintering Townsend's Warblers often join mixed-species feeding flocks, which may include chickadees, nuthatches, and others.*

WOW!

Birds have evolved foraging strategies that allow them to coexist with their neighbors. The Townsend's Warbler's foraging strategy focuses on the food at the very tops of coniferous trees.

Find it: Prefers conifers or mixed conifer-deciduous woods in summer; especially fond of tall trees. In migration and winter (some birds do not migrate) may be found in other habitats.

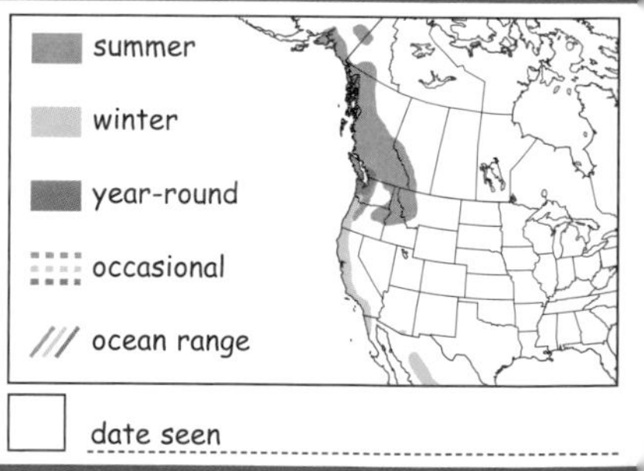

☐ date seen

BLACK-THROATED GREEN WARBLER

Dendroica virens **Length: 5"**

Look for: It's the golden face patch and black throat that are the most noticeable features on the male Black-throated Green Warbler. Only after seeing these field marks do you notice the green upper back, white wing bars, and olive lines around the eyes. Females lack the black throat. In flight, white outer tail feathers are obvious.

Listen for: Two typical versions of the Black-throated Green Warbler's song are *zee-zee-zee-zee-zee-zoo-zeee!* and *zooo-zeee-zoozoo-zeee!* Both are buzzy and both jump between two distinct notes. Call note is a thin, sharp *tchet!*

Remember: Many birders refer to this species by the nickname BTG. This is far easier to say than the bird's entire name.

▶ *A Black-throated Green Warbler searches for insects on leaves in a foraging technique known as gleaning.*

Find it: Black-throated Green Warblers nest in coniferous forests and at higher elevations in eastern mountains. This species is a common spring and fall migrant throughout the East, occurring in any wooded habitat.

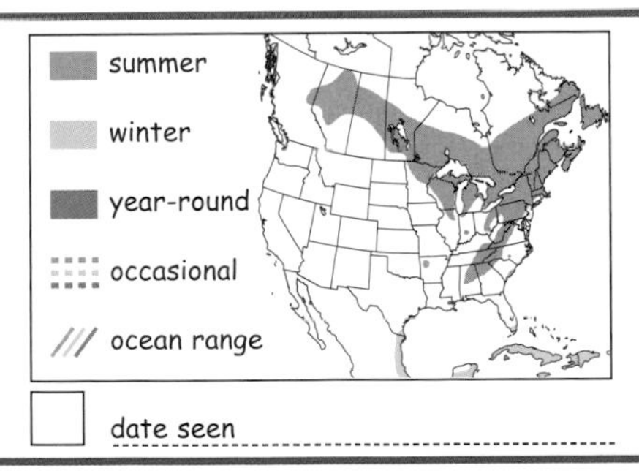

PRAIRIE WARBLER

Dendroica discolor **Length: 5½"**

Look for: The Prairie Warbler is a small warbler with a lemon yellow breast and bold black side streaks. When foraging, Prairie Warblers pump and flit their tails, and in flight their tails show obvious white outer edges. Females wear a duller version of the male's plumage.

Listen for: The Prairie Warbler's song is a distinctive series of buzzy notes rising up the scale: *zree-zree-zee-zee-zee-zee!* The song can be variable but usually speeds up toward the end. Call note is an emphatic *chek!*

Remember: The dark semicircle below the Prairie Warbler's eyes (most obvious on males) is a distinctive field mark. The tail twitching is another good field mark for this species.

WOW!
Prairie Warblers may nest in loose colonies of several breeding pairs—very unusual for warblers. Some lucky males in these colonies may have multiple female mates.

▶ *Male Prairie Warblers often sing their rising song from the top of a short tree.*

Find it: Not typically found on the prairie, Prairie Warblers prefer young woods, shrubby woodland edges, and low thickets. They often forage low, flicking their tails and flashing their white tail edges.

summer
winter
year-round
occasional
/// ocean range

☐ date seen ____________

PALM WARBLER

Dendroica palmarum **Length: 5"**

Look for: The Palm Warbler's rusty cap and pale (often yellow) eyebrow in breeding plumage are great field marks for this species. An even better field mark is the Palm's constant tail pumping. In all seasons, as a Palm Warbler pumps its tail, yellow undertail feathers are visible, even on the dullest fall and winter birds.

Listen for: Palm Warblers sing a weak trill: *tre-tre-tre-tre-tre-tre!* Call note is a crisp *tchit!* that sounds more metallic and sharper than a Yellow-rumped Warbler's chip.

Remember: Palms can range in color from very yellow to very pale, so it's best to rely on the tail-pumping behavior and yellow undertail to identify this species.

WOW!
Two folk names for the Palm Warbler, which refer to its behavior, are more accurate names: Wagtail Warbler and Tip-up Warbler.

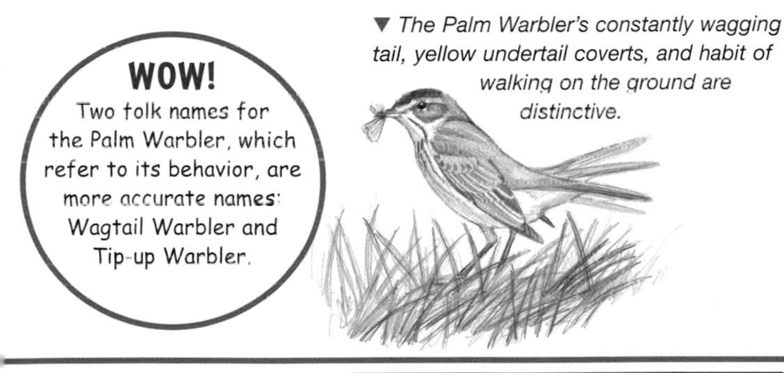

▼ *The Palm Warbler's constantly wagging tail, yellow undertail coverts, and habit of walking on the ground are distinctive.*

Find it: Rarely associated with palm trees, the Palm Warbler nests in far northern bogs. Palms are a more common sight during migration, when they forage in low vegetation and on open ground.

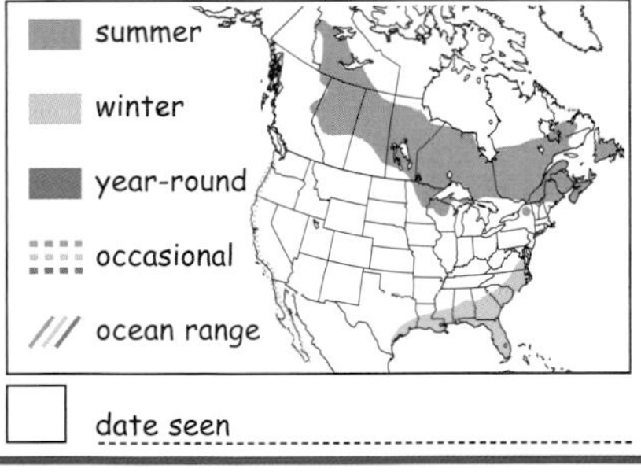

☐ date seen ______

BLACK-AND-WHITE WARBLER

Mniotilta varia **Length: 5¼"**

Male

Female

Look for: Crawling like a zebra-striped nuthatch along a tree branch or trunk, the Black-and-white Warbler makes its living gleaning insects from bark. It does not change its streaky black-and-white plumage noticeably between seasons. Its bill is long and decurved compared to other warblers.

Listen for: The song sounds like a squeaky wheel turning: *weecy-weecy-weecy-weecy*. Call note is a sharp, thick *chik!*

Remember: The Blackpoll Warbler and the Black-throated Gray Warbler look similar to the Black-and-white. In spring and summer, the Blackpoll male has a bold black cap and more subtle streaking than the Black-and-white. The Black-throated Gray Warbler has a bold facial pattern and a clear white belly.

WOW!

Two other names for this species are Pied Warbler (for its black-and-white appearance) and Black-and-white Creeper (for its foraging behavior).

▶ *Extra long and strong hind toes allow the Black-and-white Warbler to hang head-down like a nuthatch.*

Find it: Creeping along branches and trunks of large trees in old mixed woodlands, the Black-and-white Warbler can be less obvious than many other warblers. Its regular singing and constant motion are helpful in locating this species.

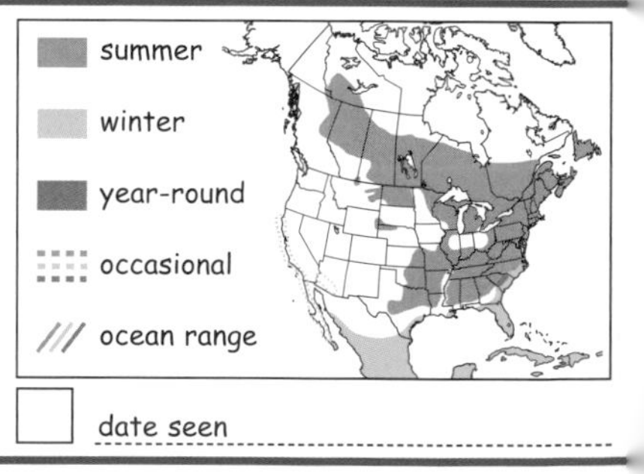

AMERICAN REDSTART

Setophaga ruticilla **Length: 5¼"**

Look for: It's hard to mistake the male American Redstart for anything else. Females and young (first-year) males are gray-green overall with bright yellow patches at the shoulders, wings, and tailbases. Like many other warblers, the redstart is active when foraging, flitting from branch to branch, often opening its wings and tail.

WOW!
In Latin America, where the American Redstart spends its winters, this species is called Candelita, the little torch.

Listen for: American Redstarts have a highly variable song. Sometimes it's *see-see-see-me-up here!* Other times it's *weeta-weeta-weeta-weet-zeet!* Often they sing one song and then the other. Call note is a clear, sweet *tchip!*

Remember: You might almost confuse the male American Redstart with a miniature version of an Orchard or Baltimore Oriole, but those birds are much larger than the redstart.

▲ *An American Redstart vaults into the air after a leafhopper.*

Find it: Redstarts prefer young woods and scrubby woodland edges for breeding and foraging. They are active singers and foragers, often zipping out from the treetops to catch a flying insect.

summer
winter
year-round
occasional
/// ocean range

☐ date seen ____________

OVENBIRD

Seiurus aurocapillus **Length: 6"**

Look for: This large ground-loving warbler is named for the domed nest (shaped like an outdoor oven) it builds on the forest floor. Males and females look alike and retain their coloration all year.

Listen for: Among our most distinctive warbler songs, the Ovenbird's song is *tea-CHUR, tea-CHUR, tea-CHUR, tea-CHUR*, given in a loud, ringing voice that starts softly and gets louder at the end. Call note is a loud *chep!*

Remember: Four other large brownish warblers share the Ovenbird's general habit of living near the ground. The Northern and Louisiana Waterthrushes, Worm-eating Warbler, and Swainson's Warbler all have eye lines rather than the Ovenbird's white eye-ring.

▼ *The Ovenbird's domed nest is often constructed under a preexisting tent of vegetation, such as Christmas fern.*

WOW!

Territorial male Ovenbirds sing a nighttime flight song during spring and summer. He flies above the treetops, singing loudly, then drops back into the dark woods.

Find it: Although its song is loud and obvious, this earth-colored bird can be hard to locate. Usually seen walking or foraging with its tail pointed up. Singing males may be perched near the trunk of a tree, just below the canopy.

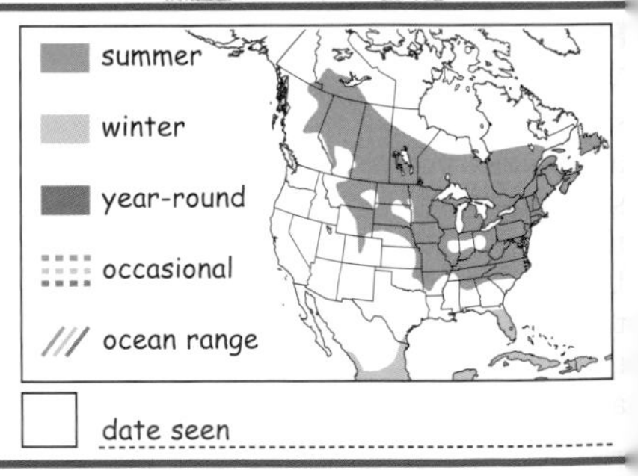

☐ date seen ______________

LOUISIANA and NORTHERN WATERTHRUSHES

Seiurus motacilla, Seiurus noveboracensis **Length: 6"**

Look for: Our two waterthrushes are confusingly similar, and both look more like tiny thrushes than warblers. For both, the constant tail bobbing as they walk is an excellent field mark. Males and females are similar, and they do not change appearance seasonally.

WOW!
Lots of experienced birders struggle to tell the waterthrushes apart. Fortunately, the two species have very different songs.

Listen for: The distinctive song of the Louisiana Waterthrush starts out with two or three down-slurred notes and ends in a sputtery jumble: *tee-yew, tee-yew, tee-yew, chicky-chick-a-chur-wow-chik!* Also gives a loud and sharp call note—*chink!*

Remember: To determine *which* waterthrush you may be watching, examine the head. The Louisiana has a bold white eye line and plain unstreaked throat. The Northern has a finely streaked throat and narrow, often buffy (not white) eye line.

▼ *The Louisiana Waterthrush fearlessly wades in swift-running streams, tossing leaves and rocks to find insects and crustaceans.*

Find it: The Louisiana Waterthrush spends spring and summer along small flowing streams and ponds throughout the eastern U.S. Breeding territories are formed in a narrow corridor along streams.

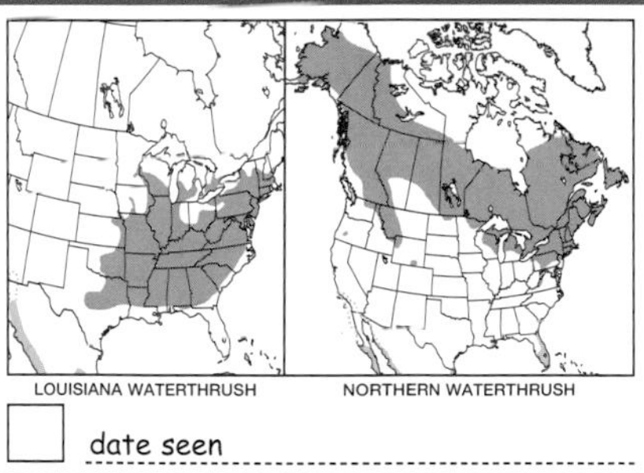

☐ date seen ________

PROTHONOTARY WARBLER

Protonotaria citrea **Length: 5½"**

Look for: The Prothonotary Warbler (*pro-THON-oh-tary*) is a bird of wooded swamps, where its loud song rings out. The male's bright golden yellow head and breast contrast sharply with the large black bill and eyes, blue-gray wings, and white undertail. Females are duller overall.

Listen for: A clear, loud song on a single tone: *sweet-sweet-sweet-sweet-sweet!* Call note is a metallic *tchit!*

Remember: The plain yellow face and head and large bill are good field marks for identifying the Prothonotary Warbler. Fall and winter birds may show a paler (not black) bill.

WOW!
The Prothonotary Warbler is named for the golden hood traditionally worn by the notary officer of the Roman Catholic Church.

▶ *A male Prothonotary Warbler wintering in a mangrove swamp.*

Find it: In its swampy woodland habitat, the Prothonotary Warbler is often heard before it is seen. Males will sing from the treetops, but both sexes will forage quite low. Nests in tree cavities or nest boxes in trees standing in water.

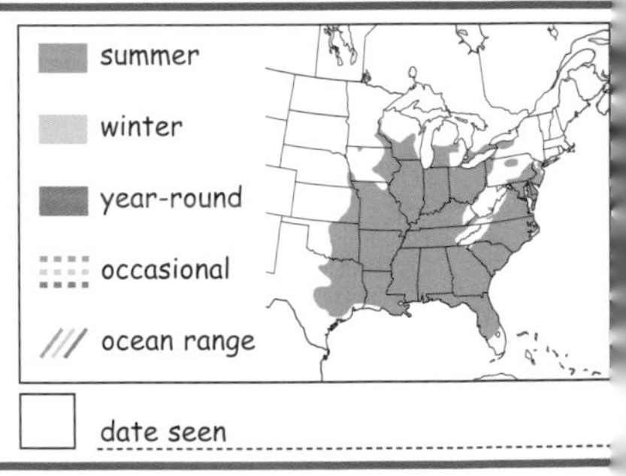

date seen

COMMON YELLOWTHROAT

Geothlypis trichas **Length: 5"**

Look for: With their black masks and the bright yellow throats for which they are named, adult males are easy to identify. Adult females and young birds are less clearly marked, but still offer clues in the obvious yellow throat patch, darkish face, and plain tan (not yellow) belly.

Listen for: Yellowthroats are very vocal birds, singing a clear, ringing *witchety, witchety, witchety*. Also utters several scold notes and rattles. Most common call note is buzzy and nasal: *cherk!*

Remember: Several other warbler species might be confused with the Common Yellowthroat, but no other warbler shows this exact pattern.

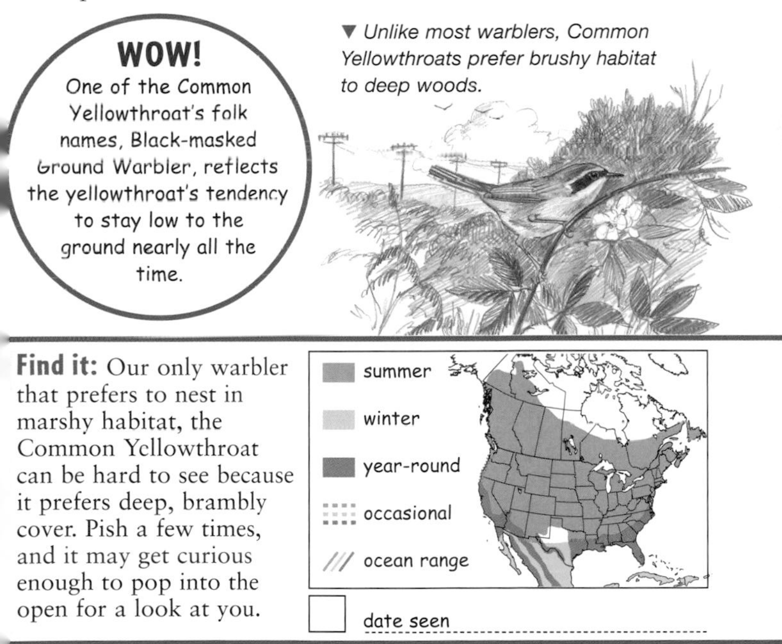

WOW!

One of the Common Yellowthroat's folk names, Black-masked Ground Warbler, reflects the yellowthroat's tendency to stay low to the ground nearly all the time.

▼ *Unlike most warblers, Common Yellowthroats prefer brushy habitat to deep woods.*

Find it: Our only warbler that prefers to nest in marshy habitat, the Common Yellowthroat can be hard to see because it prefers deep, brambly cover. Pish a few times, and it may get curious enough to pop into the open for a look at you.

☐ date seen ____________

WILSON'S WARBLER

Wilsonia pusilla **Length: 4¾"**

Look for: Adult male's round black cap, plain yellow face, and unmarked yellow underparts are diagnostic field marks. Females have yellow-olive cap, yellow eyebrow, and plain yellow face. Back is olive-yellow in both sexes and tail is dark and unmarked.

Listen for: Song is a loud, ringing series of harsh chips dropping in tone and speed toward the end: *chi-chi-chi-chi-CHETCHETCHET*. Call note is a loud *CHET!*

Remember: Wilson's Warblers have a "beady-eyed" look that few other warblers show. Yellow Warbler is similar but bright yellow overall, including the back.

WOW!
The Wilson's Warbler is named for Alexander Wilson (1766–1813), one of North America's first ornithologists.

◀ *Another folk name for the Wilson's Warbler is Black-capped Flycatching Warbler, for its habit of catching flying insects.*

Find it: More common in the West than East. Common low to ground in thick, brushy woods and alder and willow thickets, especially along streams and near other water. Abundant spring migrant in the West.

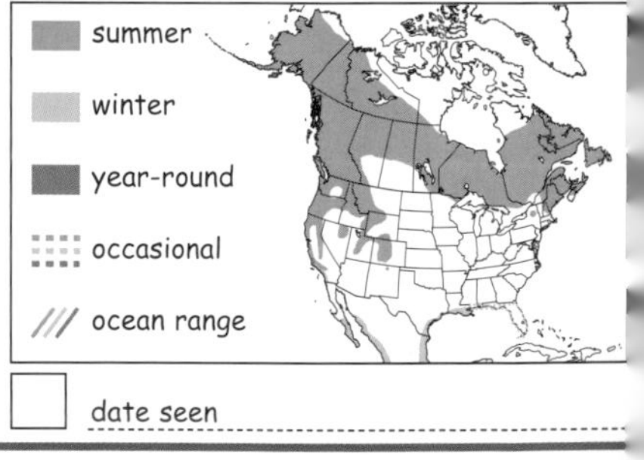

date seen

YELLOW-BREASTED CHAT

Icteria virens **Length: 7"**

Look for: The Yellow-breasted Chat is our largest warbler, though it does not really look, act, or sound like a warbler. It has a large head, a stout black bill, and a long tail. Sexes are similar, and plumage does not vary by season.

Listen for: Chats make a huge variety of weird noises, including whistles, clucks, chucks, and harsh scolding notes delivered in a series, each call separate and distinct. It sounds like this: *cherrk eeeep! woo-woo-woo! chek! wok! ank-ank-ank!*

Remember: The chat's size and bright yellow throat should be enough to identify this species. No other songbird has the clear yellow throat and breast and white spectacles (connected eye-rings).

WOW!

In his flight display, a male Yellow-breasted Chat flies across an open space, calling loudly and flapping his wings awkwardly like a wind-up toy that is freaking out.

◀ *A Yellow-breasted Chat launches into a flight song by the light of a full moon. Chats are insomniacs in spring.*

Find it: Yellow-breasted Chats often sing from the deep cover of brushy tangles along woodland edges, old overgrown fields, and streamside thickets, but they can be coaxed out by pishing or by imitating their odd sounds.

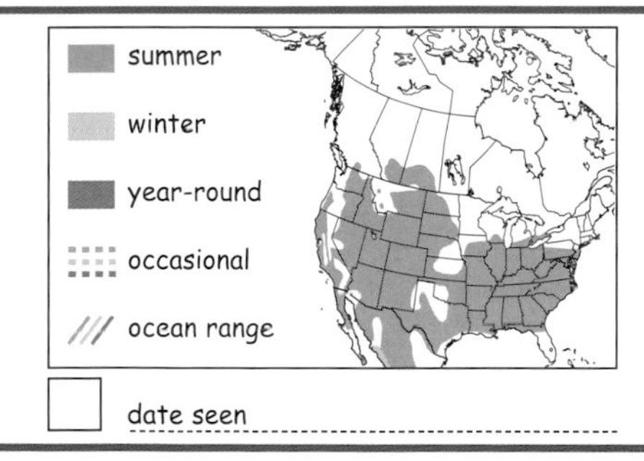

☐ date seen ______

SUMMER TANAGER

Piranga rubra **Length: 7¾"**

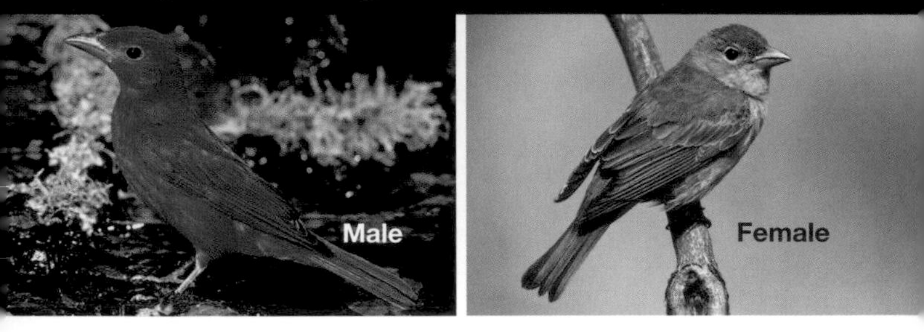

Look for: The male Summer Tanager is a bright rosy red all over and can at times appear to have a slight crest on its head. The other most noticeable feature of this species is its large dark bill, much larger than the bill of the Scarlet Tanager. The yellow-green females may show a hint of the male's reddish coloration.

Listen for: Song is similar to the Scarlet Tanager's, but richer—less harsh and more liquid. More distinctive is the Summer Tanager's explosive call: *perky-tuck!* or *perky-tucky-tuck!*

Remember: Female Summer Tanagers are more uniformly yellow than female Scarlet Tanagers. If you're confused, look at the bill. If it's long and thick at the base, you've got a Summer Tanager.

WOW!

One of the folk names for this species is Beebird, for its habit of eating bees and wasps. Its long stout bill is an ideal tool for capturing and subduing these stinging insects.

◀ *The Summer Tanager's strong, toothed bill is more than a match for the sting of a yellow jacket.*

Find it: In summer, look for Summer Tanagers in open oak or pine woods in the Southeast and in streamside cottonwoods in the Southwest. Finding one in the treetops can take patience, as they are not super active when foraging.

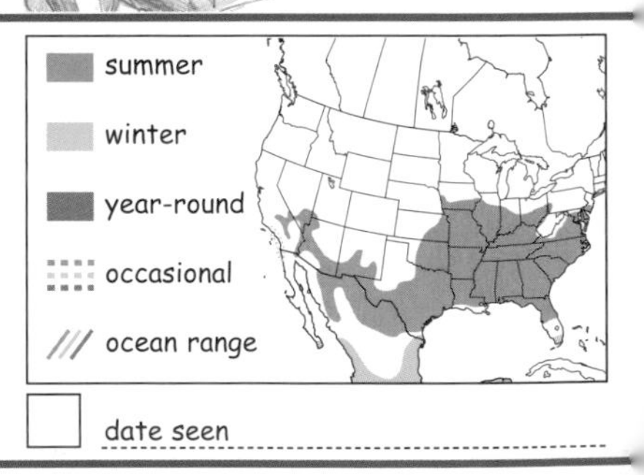

date seen

SCARLET TANAGER

Piranga olivacea **Length: 7"**

Look for: The spring and summer plumage of the male Scarlet Tanager is stunning and unmistakable. In fall, males lose all their red color and turn a dull yellow-olive (though they retain the black wings and tails). Females are olive, yellow, and gray overall in all seasons.

Listen for: A Scarlet Tanager sounds like an American Robin singing with a sore throat: *cheer-ree, chee-rear, cheer-ree, cheer-wow!* Another commonly given call is *chip-burr!*

Remember: The Western Tanager and the female Summer Tanager look similar to the female Scarlet Tanager, but their backs are darker in the center.

WOW!
No other bird in North America has the male Scarlet Tanager's combination of scarlet red body and black wings. It's too bad he has to molt into his duller nonbreeding plumage each fall.

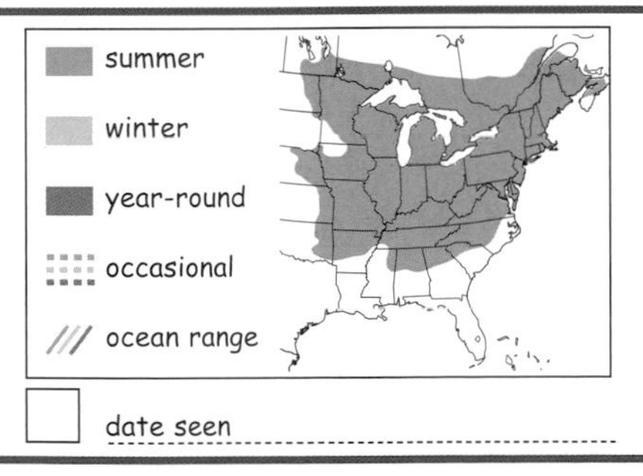

▲ *Both adult Scarlet Tanagers feed nestlings. This is a male.*

Find it: Common in leafy eastern woodlands, Scarlet Tanagers seem to prefer oak-dominated forests with large trees. Despite the male's bright colors, Scarlet Tanagers can be difficult to spot in thick foliage high in the treetops.

summer
winter
year-round
occasional
ocean range

☐ date seen --

WESTERN TANAGER

Piranga ludoviciana **Length: 7¼"**

Look for: Male in breeding plumage is unmistakable: red head, lemon yellow body, black back and wings, with both yellow and white wing bars and yellow rump. Female is similar to other tanagers but shows a gray back and obvious wing bars.

Listen for: Short, hoarse, whistled phrases that are very robinlike in quality: *brr-eet, burry, burry, brr-eet!* Calls include a soft, rising *wheee?* and a percussive *pretty-tick!*

Remember: This is our only common tanager with bold wing bars. The stout bill helps separate the Western Tanager from similar oriole species.

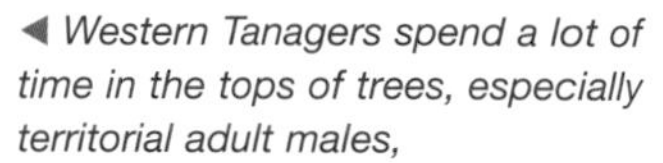

◀ *Western Tanagers spend a lot of time in the tops of trees, especially territorial adult males, singing to claim their territories.*

WOW!
The Western Tanager nests farther north than any of our other tanagers.

Find it: A treetop specialist that prefers open pine or mixed forest during the breeding season, often at higher elevations. Visits many other habitats in migration. Less active than some songbird species, so familiarity with the song is useful in finding the Western Tanager.

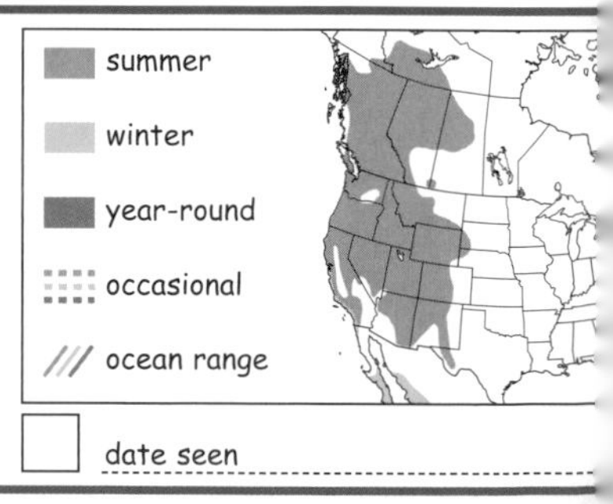

date seen

GREEN-TAILED TOWHEE

Pipilo chlorurus **Length: 7¼"**

Look for: Named for its olive green upperparts and tail, the Green-tailed Towhee's most noticeable features are its rusty cap and white throat, contrasting with the gray head and breast. This is our smallest towhee.

Listen for: Song is loud and ringing: *chert, wee-wee, CHEEE churr.* Song often compared to song of Fox Sparrow. Call is catlike: *myeeew.*

Remember: This ground-loving bird is curious and can be lured into view by *spish*ing. But when alarmed it may run away rather than fly.

WOW!
Singing male Green-tailed Towhees may incorporate the songs of other nearby species into their own songs. Copycats!

▶ *If you don't hear the Green-tailed Towhee's call or song first, you might hear the noise created as a towhee scratches through leaf litter while foraging.*

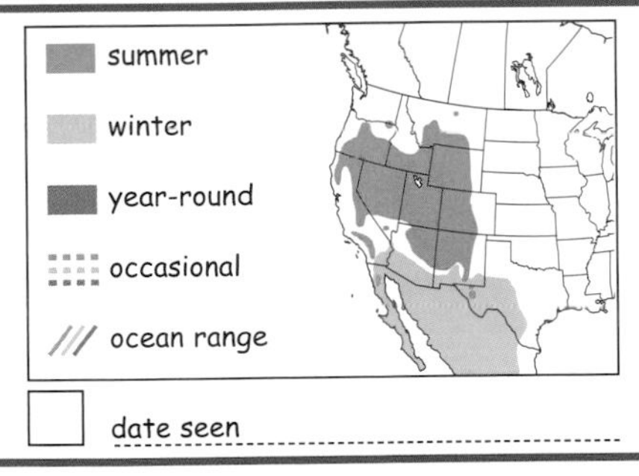

Find it: Common in a variety of dry, brushy habitats in the western mountains, including chaparral, sagebrush, riparian woods, and thickets. Often heard before it is seen.

summer
winter
year-round
occasional
/// ocean range

☐ date seen ______

EASTERN TOWHEE and SPOTTED TOWHEE

Pipilo erythrophthalmus, Pipilo maculatus **Length: 8"**

Eastern Towhee, male

Spotted Towhee, male

Look for: The former name of this species described the bird better: Rufous-sided Towhee, named for the rusty rufous sides. Spotted: wings and back of both male and female are spotted with white. Longer tailed and larger than our other sparrow species.

Listen for: One of the easiest bird songs to learn and remember, the Eastern Towee sings *drink your teeeaa!* Or simply *drink teeeaa!* Call is a similar-sounding *chew-ink!* They also use a buzzy, rising *tzeeeee!* as a flight call and alarm call. Spotted: song is *sweet-sweet teeeeaaa!* But it's less musical than Eastern's song.

Remember: Both Eastern and Spotted Towhees are larger and more colorful than the Dark-eyed Junco, which also has a black hood and back.

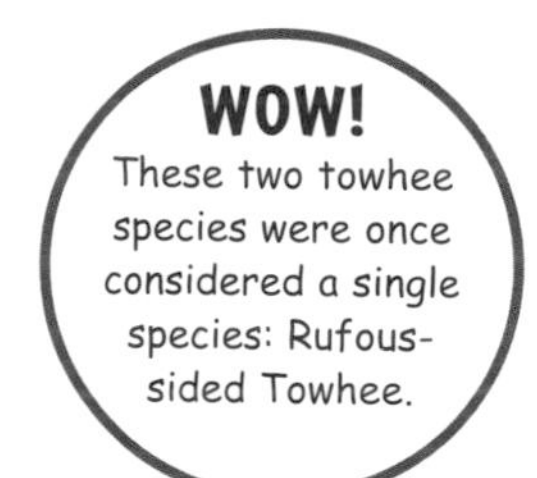

▼ *The Eastern Towhee leaps forward, kicking leaf litter back. In this way, it uncovers hidden seeds and insects.*

Find it: Eastern: common in thick undergrowth and along brambly woodland edges and old fields. Spotted: in willows, sagebrush, chaparral, and brushy woods. Both will visit backyard feeders near appropriate habitat for seed bits scattered on the ground.

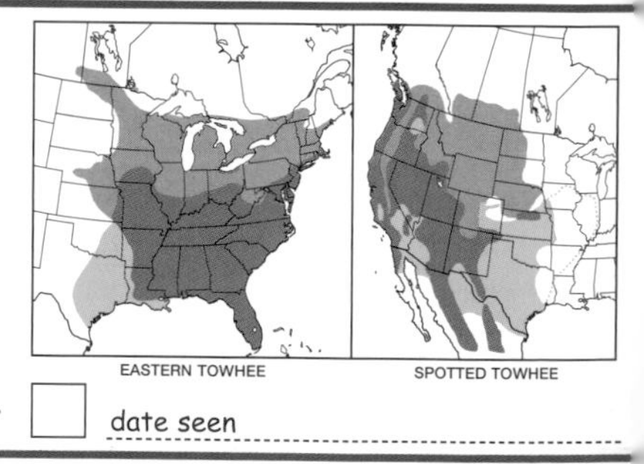

☐ date seen ____________________

CANYON TOWHEE and CALIFORNIA TOWHEE

Pipilo fuscus, Pipilo crissalis **Length: 8¾", 9"**

Look for: Both of these plain brownish towhees look like what they are: giant sparrows. Canyon Towhee has a faint rusty cap and is slightly paler overall than California Towhee, which shows rust on the throat. California is plain breasted and dark bellied. Canyon usually shows a spotty necklace, central breast spot, and whitish belly.

Listen for: Canyon: song is a loud and ringing series of notes: *chee-chee-chee-chee-chee-chee-chee*. Call is a woodpecker-like, nasal *kyerr*. California: song starts with metallic chips, speeds up, then drops in tone: *chip-chip-chichichichi-drrdrrdrr*. Call: metallic *chip!*

WOW!
These two species were once lumped together as the Brown Towhee, which is a descriptive name but kind of boring compared to Canyon and California.

Remember: These species can be told apart by range and by song. Canyon can be distinguished from similar Abert's Towhee (not shown) by Abert's partially black face.

▶ *Being ground-loving birds, Canyon (shown) and California Towhees run away from danger rather than fly.*

Find it: Both species prefer open arid scrub with scattered brush. Canyon Towhee inhabits the desert Southwest east of California. California Towhee is found in California and southwestern Oregon.

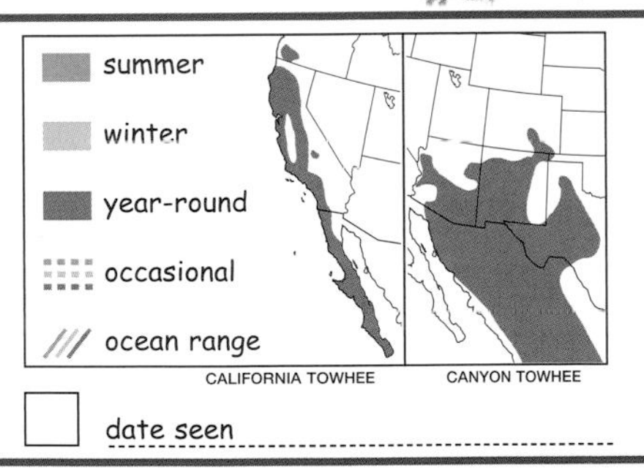

LARK SPARROW

Chondestes grammacus **Length: 6½"**

Look for: This large sparrow's bold facial markings of rufous, black, and white are diagnostic. Also note the dark central breast spot on white underparts. White sides and tip of dark tail are especially obvious in flight.

Listen for: Song is a rich, musical blend of whistles, trills, and buzzes and is fairly long in duration. Call is *tsink* and is often given in flight.

Remember: No other North American sparrow has the Lark Sparrow's combination of a bold, colorful facial pattern and central breast spot.

▶ *Lark Sparrows love to hang out with their friends in small flocks while foraging.*

Find it: A sparrow of open country with scattered bushes, trees, and fencerows; plus farmyards, pastures, and grassland with bare ground. Often forages on the ground in the open. Nearly always in small flocks, especially in migration and winter.

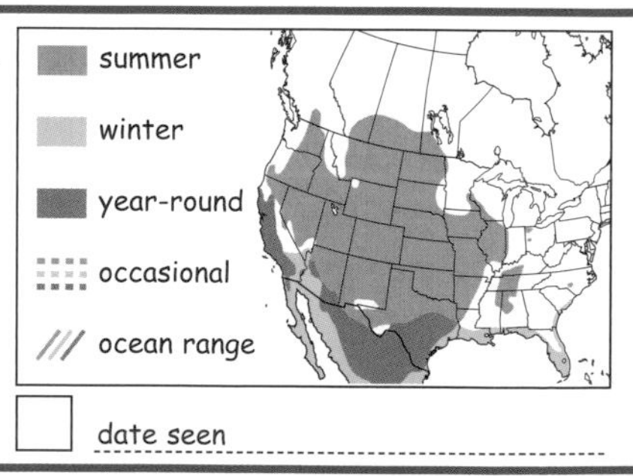

☐ date seen ____________

CHIPPING SPARROW

Spizella passerina **Length: 5½"**

Look for: In breeding plumage the adult has a rusty cap over a white eyebrow and a black eye line. Adults in nonbreeding plumage look like faded versions of their summer selves. First-year birds can be confusing, with finely streaked heads and pale faces. All Chippies look flat-headed and have a grayish rump.

Listen for: Chipping Sparrows sing a long dry trill on a single note that sounds like a cross between an insect and a sewing machine. They also utter single *chips*, for which they are named.

Remember: Though the Clay-colored Sparrow looks similar, it prefers brushy, treeless areas. Clay-coloreds have brownish tan rumps instead of gray.

◄ *Chipping Sparrows often nest low in landscape shrubbery near buildings.*

Find it: Chipping Sparrows prefer open woodlands, wooded parks, and even suburban neighborhoods, where they will nest in the most landscaped of habitat. They come to bird feeders for mixed seed and cracked corn.

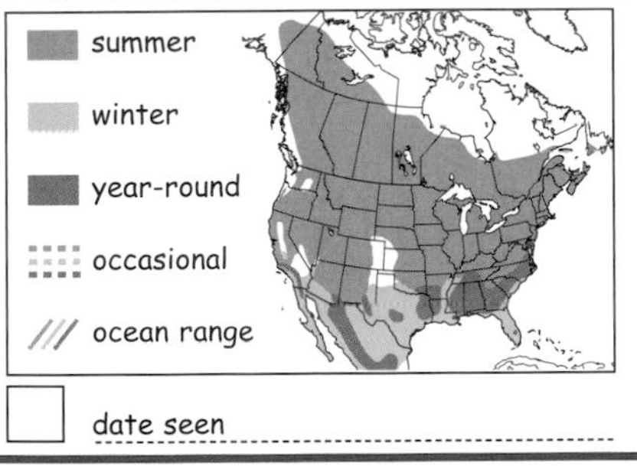

☐ date seen ----------------

CLAY-COLORED SPARROW

Spizella pallida **Length: 5½"**

Look for: Breeding adult has a dark crown, white eye stripe, and a subtle dark mustache. Very similar to a nonbreeding-plumaged Chipping Sparrow. The clear gray nape and brownish rump on the Clay-colored help separate the two species.

Listen for: One of the least musical of all sparrow songs: a series of nasal buzzes: *tzeee-tzeee-tzeee*. Call is a soft *tsip*.

Remember: Clay-colored Sparrows have a slate gray nape and a *brownish rump*. Nonbreeding Chipping Sparrows have the grayish nape and a *grayish rump*.

▼ *The Clay-colored Sparrow is a frequent victim of nest parasitism by the Brown-headed Cowbird.*

WOW!

Male Clay-colored Sparrows defend their breeding turf vigorously, excluding both Song and Chipping Sparrows where their ranges overlap.

Find it: Common in summer grassland and prairie habitat with scattered shrubs or bushes. Males perch on top of shrubs and sing their insectlike song. Often heard before they are seen, but easy to find thereafter.

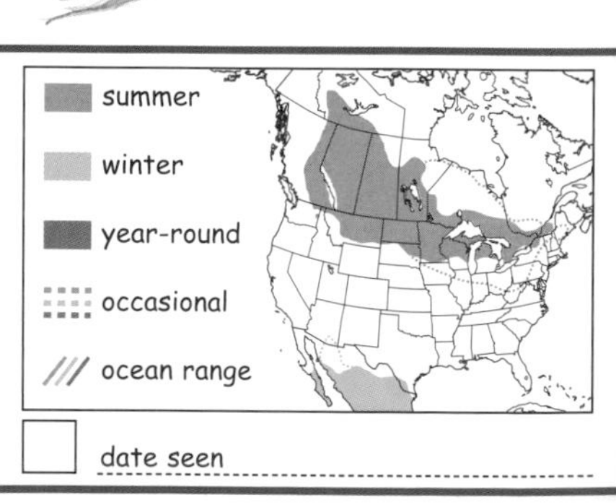

☐ date seen ----------------

FIELD SPARROW

Spizella pusilla **Length: 5¾"**

Look for: All the Field Sparrow's distinctive field marks are on the head and breast. The plain-faced look is emphasized by a white ring around the dark eyes. Adults are similar and do not change plumage seasonally.

Listen for: A series of sweet, whistled notes that speeds up into a trill and drops in tone toward the end: *too-too-too-too-tootootoottitititititititi*. The pattern is similar to the rhythm of a dropped Ping-Pong ball, which bounces faster until it stops.

Remember: The Field Sparrow's pink bill, plain face (with white eye-ring), and unmarked breast help to separate this species from the similar but less common American Tree Sparrow.

WOW!
The young Field Sparrow looks very different from its parents. It can be very confusing to identify, at least until one of the parents comes to feed it.

▼ *It's safer for Field Sparrow chicks to hide separately in deep cover than to huddle together in the nest, where their scent could attract a predator.*

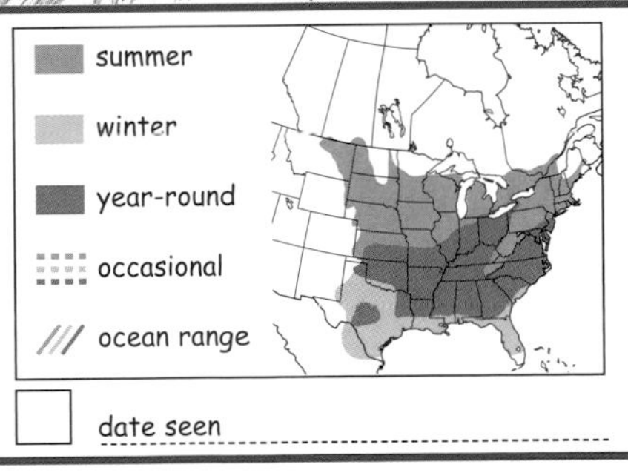

Find it: In spring and summer, males will sing while perched on a weed stem or sapling above the grass of old-fields (scrubby overgrown meadows). Field Sparrows will visit bird feeders in rural settings for mixed seed and cracked corn.

summer
winter
year-round
occasional
ocean range

☐ date seen ______________

FOX SPARROW

Passerella iliaca **Length: 7"**

Look for: This large, boldly marked bird is named for its fox red coloration, but not all Fox Sparrows are reddish—some western birds are dark brown or even gray. Sexes are similar, and coloration does not vary seasonally.

Listen for: Fox Sparrows have a beautiful song that starts with several sweet whistled notes, then a trill, and ends with a short warble: *sweet-sweet, chee-chee-chee-tititititi-chew-wee!* Call note is a loud *chip!* or *smak!*

Remember: Though similar in size, Fox Sparrows appear stockier than Song Sparrows. The fox red coloration is brighter than the Song Sparrow's earth tones, and the breast of the Fox Sparrow is spotted with rusty brown (the Song Sparrow's breast is streaked with brown).

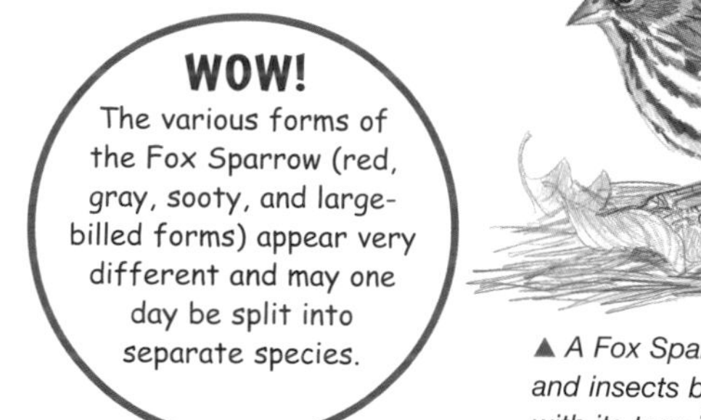

▲ *A Fox Sparrow uncovers seeds and insects by grabbing leaf litter with its toes in a two-footed shuffle and kicking it back behind.*

Find it: Fox Sparrows prefer to forage on the ground, often in dense, brushy cover, scratching for food like a towhee. They do not normally winter in flocks, but they will join other birds in a thicket or feeding below a bird feeder.

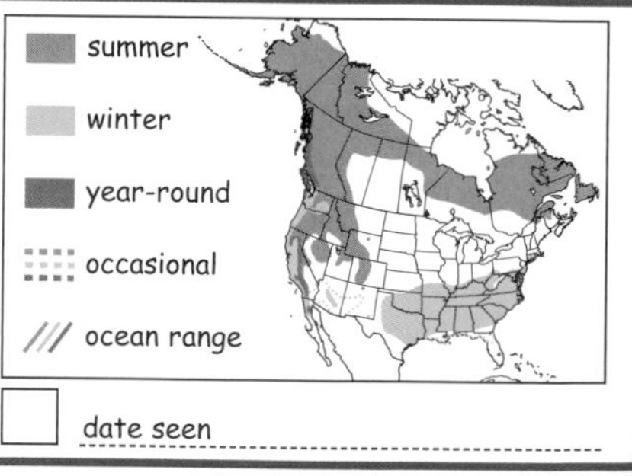

☐ date seen ______________

SONG SPARROW

Melospiza melodia **Length: 6¼"**

Look for: The chunky, streaky Song Sparrow's best-known field mark is the central breast spot, where the breast streaks come together to form a noticeable splotch. Its heavy streaking overall gives it a dark, dusky look.

Listen for: The highly variable song usually starts out with three clear, slow notes, followed by a trill, some short buzzes, and a few more single notes, speeding up as it goes: *sweet, sweet, sweet, brzzt, tititititititi, brrzzt, tee-tee-teer.*

Remember: Learn the Song Sparrow's field marks well. Then, when you encounter an unfamiliar sparrow, ask yourself, "What makes this bird different from a Song Sparrow?" The answers will be the new bird's important field marks.

WOW!
Not all Song Sparrows look alike. There may be as many as 30 different forms of this species across the continent. Fortunately, all these Song Sparrows sing the same beautiful song.

◀ *Song Sparrows are very sneaky and secretive near their nests.*

Find it: Our most widespread sparrow in North America, the Song Sparrow prefers dense cover such as brushy field edges, hedgerows, and brambles, but it's also common in backyards, parks, and cemeteries.

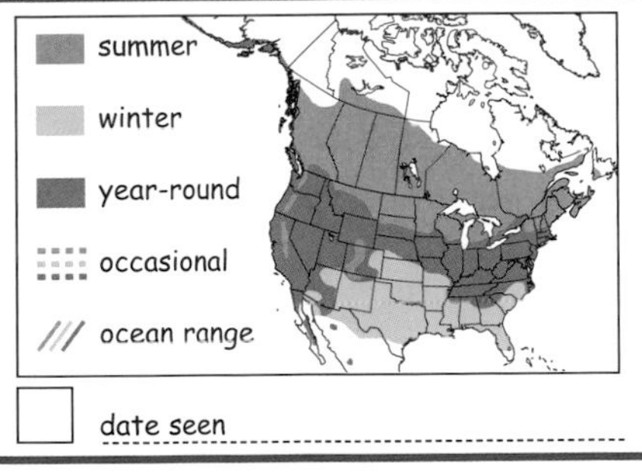

date seen

LINCOLN'S SPARROW

Melospiza lincolnii **Length: 5¾"**

Look for: A warm, colorful sparrow similar to Song Sparrow, but smaller and more finely marked. Note fine streaks on breast, gray eyebrow, buffy mustache, and buffy eye-ring.

Listen for: Song is a bright, clear jumble of phrases, rising in volume, pitch, and intensity in the middle: *burrburr ZEEEEEEEE chrup*. Sometimes compared to Purple Finch and House Wren. Calls include *chup* and *zeeet*.

Remember: The Lincoln's Sparrow looks like a Song Sparrow that got a makeover.

WOW!
This species is not named for Abe Lincoln; it's named for Thomas Lincoln, who accompanied John James Audubon on a bird-finding trip in 1833.

◄ *Where the two species' ranges overlap, the Song Sparrow outcompetes the Lincoln's Sparrow.*

Find it: Nests in thickets of alder and willow and in brushy habitat near water. In winter found in brushy woodland-edge habitat, usually near water. The Lincoln's Sparrow is a shy bird and often goes undetected.

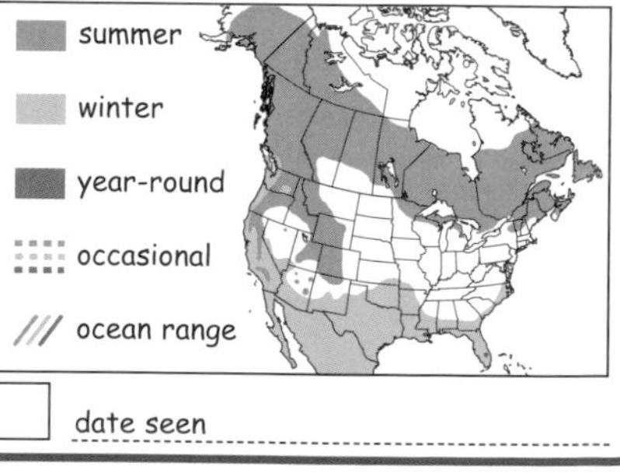

☐ date seen

WHITE-THROATED SPARROW

Zonotrichia albicollis **Length: 6¾"**

Look for: One of several species of crowned sparrows in North America, the White-throated Sparrow comes in two color varieties, or morphs: birds with white-striped crowns and birds with tan-striped crowns. Sexes are similar, and coloration does not change seasonally.

Listen for: *Old sam peabody, peabody, peabody* is the sweetly whistled, almost sad-sounding song. Two call notes are *tseeet!* and a loud, metallic *chink!*

Remember: The White-throated Sparrow's white throat sets it apart from the similar but larger White-crowned Sparrow, which has a boldly striped head but a plain gray throat.

◀ *White-throated Sparrows will sing at almost any time of year, even in winter if it's sunny.*

WOW!

It was once thought that the tan-striped morph of this species was the juvenal plumage of the white-striped adults. Studies have shown that white-striped adults usually mate with tan-striped birds.

Find it: The White-throated Sparrow prefers scrubby undergrowth along woodland edges and weedy fields but visits bird feeders for cracked corn, mixed seed, and sunflower seed bits.

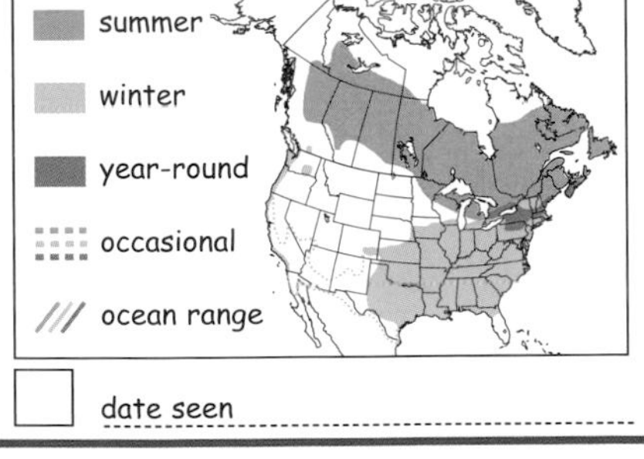

☐ date seen ____________

WHITE-CROWNED SPARROW

Zonotrichia leucophrys **Length: 7"**

Adult

Juvenile, first-winter

Look for: The White-crowned Sparrow's plain gray throat and unmarked breast help set it apart from other sparrows. Sexes are similar, and color of adults does not change seasonally. First-winter birds have brown-and-gray-striped heads until the following spring.

Listen for: White-crowned Sparrows start their song with one or two clear notes followed by a series of burry, buzzy warbles that rise up the scale: *you-can-be-so-cheez-ee!* Call note is a bright *seep!*

Remember: Your first impression of the White-crowned Sparrow may be of a large gray sparrow with an erect posture. One look at the bold head stripes and clear gray throat, and you'll know it can only be a White-crowned Sparrow. Don't forget about the young birds' tan-striped heads!

WOW!

In a 1962 experiment, several hundred White-crowned Sparrows were trapped in California and released in Maryland. One year later, eight of them had found their way back.

▶ *An adult White-crowned Sparrow forages for seeds on a stalk of lamb's quarters.*

Find it: In the East the White-crowned Sparrow is most often seen in winter and during migration. Preferred winter habitat is in scrubby hedgerows along fields, woodland edges, and thickets, where it can be found in small flocks.

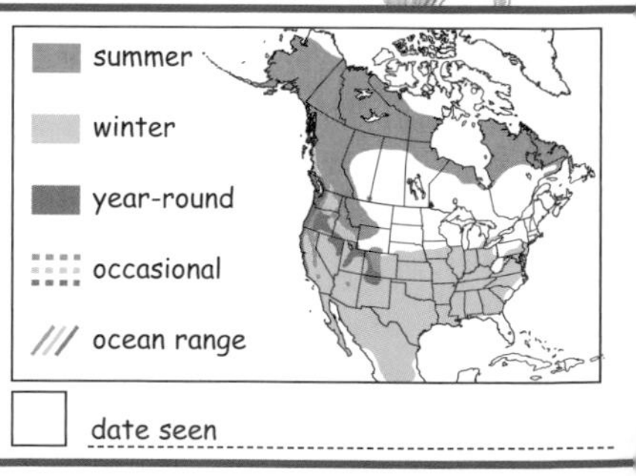

date seen

GOLDEN-CROWNED SPARROW

Zonotrichia atricapilla Length: 7¼"

Look for: Adult Golden-crowned Sparrow has a golden crown stripe surrounded by black. Slightly larger than, but very similar to, the White-crowned Sparrow. Immature birds may have only a hint of the yellow crown and lack the black head stripes.

Listen for: Song is a sad-sounding whistle: *oh dear meeee!* Or *oh lonely meee!* Call is a clear, soft *chew!*

Remember: Subadult Golden-crowned Sparrows can be identified by their plain gray face and grayish bill. Otherwise they are very similar to young White-crowned Sparrows.

◀ *The male Golden-crowned Sparrow is known to bring food to his mate while she is incubating eggs.*

Find it: Nests in stunted boreal woods and tundra scrub as far north as the tree line. Winters in dense thickets and underbrush, chaparral, parks, and gardens as far south as the Mexico border. Often found in mixed flocks with White crowned Sparrows. Will visit feeding stations for seed scattered on the ground.

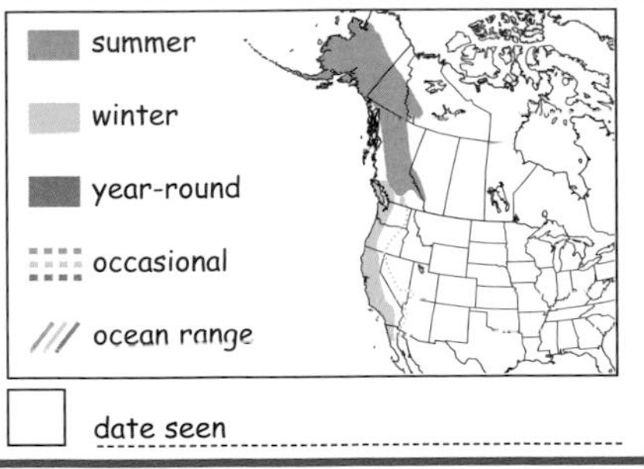

date seen

HARRIS'S SPARROW

Zonotrichia querula **Length: 7½"**

Look for: Harris's Sparrow is our largest sparrow, and its combination of a black face and pink bill (in adult birds) makes it relatively easy to identify. First-winter birds have a pale brown head and a white throat.

Listen for: If you want to hear a Harris's Sparrow sing, you'll have to head north to its breeding habitat (see below). Song is two or three clear whistles followed by two more on a lower tone: *see-see-see* [pause] *soo-sooo*. Call is *check*.

Remember: The Harris's Sparrow's large size is a great clue to its identity, especially when compared with less boldly marked young birds.

WOW!

The Harris's Sparrow was named by John James Audubon for Edward Harris, his friend and travel companion in the mid-1840s.

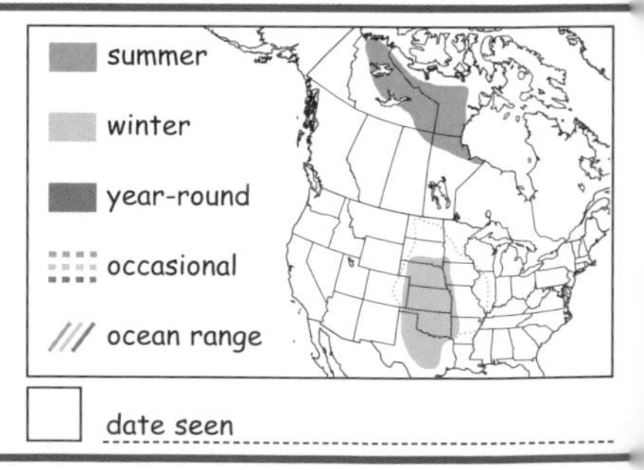

▼ *Feeder watchers in the Midwest feel extra lucky when a Harris's Sparrow visits their feeders in the winter.*

Find it: Uncommon in winter in brushy areas, hedgerows, and open woodland. Summers in far North in boreal forest and tundra scrub. May mix with White-crowned Sparrows in winter feeding flocks. Sometimes visits feeders in the Midwest.

summer
winter
year-round
occasional
/// ocean range

☐ date seen ____________________

DARK-EYED JUNCO

Junco hyemalis **Length: 6¼"**

Look for: The Dark-eyed Junco is a well-known bird even among nonbirders. Several different forms exist throughout the West, and all were once considered separate species. In flight, all Dark-eyed Juncos show obvious white outer tail feathers.

Listen for: Juncos are members of the sparrow family, and they sing a long ringing trill—*tiitiitiitiitiitii*—that sounds similar to a Chipping Sparrow's song but is slower and more musical. They also give a number of chip notes *(tick!)* and short buzzes in all seasons.

Remember: East of the Great Plains, the slate-colored Dark-eyed Junco is most common. In the West, five other forms of the species are present.

WOW!
The folk name of Snowbird is for their winter-weather coloration (gray skies above, snow on the ground) and because they appear at feeders at the first snowfall.

◀ *Flocking Dark-eyed Juncos bicker and show aggressive behavior often, their angry twitters sounding like skates on ice.*

Find it: Flocks of juncos forage along woodland edges and hedgerows, hopping over the ground in search of seeds, often chipping at each other. At bird feeders they eat cracked corn and mixed seed.

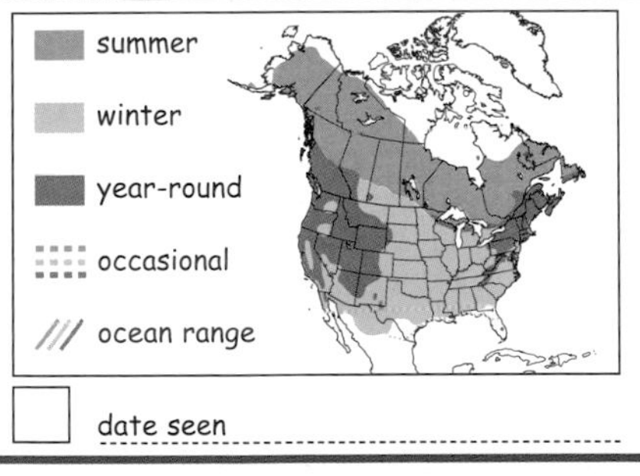

☐ date seen ____________

LAPLAND LONGSPUR

Calcarius lapponicus **Length: 6"**

Look for: Few of us get to see the male Lapland Longspur in his full breeding glory, with his black face outlined in white and the chestnut nape and collar. Winter male retains a partial collar of rust, a tan cheek outlined in dark, and heavy dark streaks on the sides. All plumages show narrow white sides to the tail in flight.

Listen for: Song in display flight is a rich, throaty warble, similar in tone to Bobolink. More commonly heard are calls: a whistled *tyew* and a dry, buzzy note.

Remember: Though it nests the farthest north, in winter this is our most common longspur species and the easiest to add to your life list.

WOW!
Lapland Longspurs returning to the Arctic tundra to breed retain extra body fat to help them survive late-spring cold and snow.

◀ *Lapland Longspurs are often found in mixed winter flocks with Horned Larks and Snow Buntings.*

Find it: An abundant breeding bird on the tundra and wet meadows of the far North. In winter, found in open grassland, plowed or planted fields, prairies, and shorelines. Large winter flocks may number in the thousands.

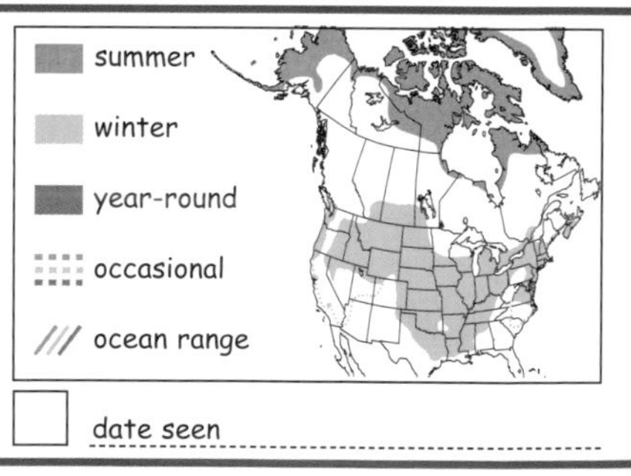

☐ date seen ____________

SNOW BUNTING

Plectrophenax nivalis **Length: 6¾"**

Look for: Most of us don't get to see the male Snow Bunting in its snowiest attire—its pure white-and-black breeding plumage worn when the birds are on the Arctic tundra. In winter, when they are present across much of northern North America, both males and females are rusty and white with black wingtips.

Listen for: The Snow Bunting's calls are a whistled descending *cheew!* and a harsh, nasal *brzzzt!* given in winter flocks. The song is a musical warble full of *tyeew* notes.

Remember: Snow Buntings like to mix with larks and longspurs. When a mixed flock takes flight, the birds flashing white wings with black tips are Snow Buntings.

WOW!
Snow Buntings can handle extreme cold—as low as 58 degrees below zero. They may burrow into the snow to shelter from the freezing wind and stay warm.

◀ *Snow buntings flock in dunes and on flats along the coasts in winter, their white wings twinkling when they take flight.*

Find it: Summers are spent in the far North on the tundra. In winter, flocks move south, where they prefer open areas such as beaches and farm fields (especially ones with freshly spread manure).

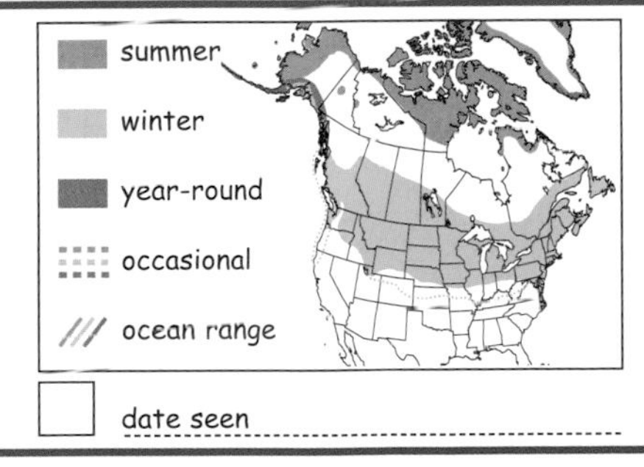

☐ date seen ____________

ROSE-BREASTED GROSBEAK

Pheucticus ludovicianus **Length: 8"**

Look for: The Rose-breasted Grosbeak is a big-headed, large-billed bird (its name means "large bill") of the treetops. Females and young males are streaky. The fall and winter adult male wears a faded, splotchy version of the breeding plumage.

Listen for: An American Robin that's had singing lessons describes the rich, musical warble: *chewee-churweo, chewee-turleo, chewee-churweo.* Call note is a sharp squeak: *eek!*

Remember: Female Rose-breasted Grosbeak and female Purple Finches look similar. The grosbeak is much larger overall and larger headed, with a broad white eye line, a finely streaked breast, and a pale, pinkish bill. The finch has a messy eye line, broadly streaked breast, and a gray bill.

WOW!
In spring migration, Rose-breasted Grosbeaks often show up at bird feeders where sunflower seed is offered, sometimes prompting a "Wow! What's that bird?"

▶ *A male Rose-breasted Grosbeak in flight is a pinwheel of black, white, rose, and carmine.*

Find it: Prefers young, open deciduous woods during spring and summer. Often forages in thick foliage near the treetops and can be best located by song or call note. In migration, it can appear almost anywhere.

summer
winter
year-round
occasional
ocean range

☐ date seen

BLACK-HEADED GROSBEAK

Pheucticus melanocephalus **Length: 8¼"**

Look for: Breeding-plumaged adult male has a black head and orange collar, underparts, and rump. Black wings are boldly marked with white. Black back shows vertical orange stripes. Adult female is butterscotch-colored below with finely streaked sides. Face is boldly patterned with dark and light stripes. Stout bill is dark above, light gray below.

Listen for: Song is a robinlike series of warbled and whistled phrases that rise and fall in tone. *Up here! Way up here! I'm singing this beautiful song for you!* Call is a sharp, spitting *tick!*

Remember: Where ranges overlap, the similar-looking female Rose-breasted Grosbeak can be identified by her heavily streaked breast. Female Black-headeds have a warmer-colored breast unstreaked in the center.

WOW!
The Black-headed Grosbeak is one of the few bird species that will eat monarch butterflies, which are toxic to most birds.

▶ *Male Black-headed Grosbeaks perform impressive song flights above perched females during courtship.*

Find it: Spends summers in deciduous woods, especially oaks and mixed oak-pine forests. Also in cottonwoods and willow groves along streams. Usually found in upper levels of trees. Winters in the tropics.

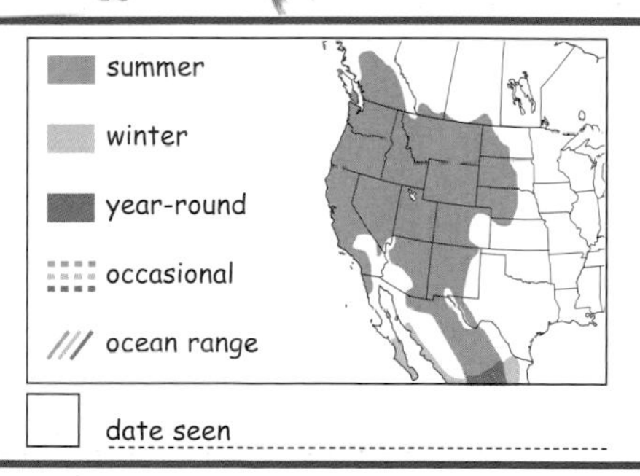

☐ date seen ______

NORTHERN CARDINAL

Cardinalis cardinalis **Length: 9"**

Look for: The male Northern Cardinal is our only crested red bird, and it's hard to mistake it for another species. The male's black facemask and the massive seed-crushing reddish bill are easy to see. Females and juvenile birds are crested and reddish brown overall, but the female has a reddish bill and subtle black face while the young bird has a plain face and a black bill.

Listen for: Northern Cardinals have a loud, ringing song that can vary: *purty-purty-purty, what-cheer! what-cheer! tee-tee-tee-tee-tee!* Also gives a loud *pik!* call.

Remember: Our red tanager species are similar in color but not in shape: they lack the cardinal's obvious crest.

WOW!

Oh no—a bald cardinal! When Northern Cardinals get infested with feather mites on the head, where they cannot preen, they can lose their head feathers. Once mite-free feathers grow in, the bird looks normal again.

▼ *A male Northern Cardinal sings and displays for a female, leaning from side to side and fluttering its wings.*

Find it: A southern bird that is expanding its range northward, the Northern Cardinal favors brushy habitat along woodland edges, open woods, parks, and backyards with dense cover, where it will visit bird feeders for sunflower seed.

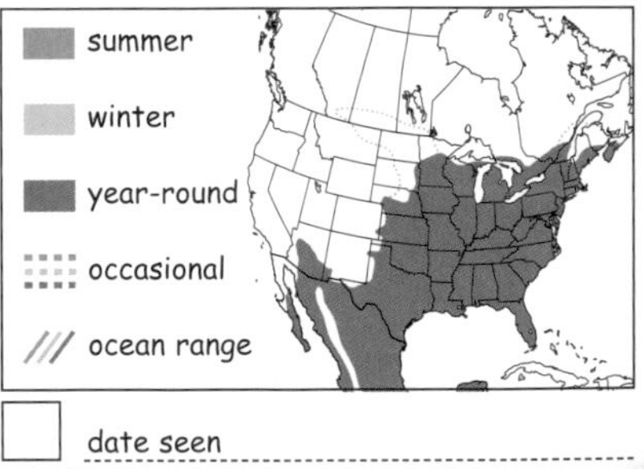

☐ date seen ____________

PYRRHULOXIA

Cardinalis sinuatus Length: 8¾"

Look for: The "Desert Cardinal" looks like a Northern Cardinal that has gone gray. Male Pyrrhuloxia is light gray overall with a red face and a red central stripe from the chin to the lower belly. Long crest is tipped in red. Stubby, conical bill is pale yellow (Cardinal bills are red). Female is gray-brown overall with a pale face and patches of red in wings, tail, and tip of slender crest.

Listen for: Song is very Cardinal-like but higher pitched, sweeter, and more drawn out in tempo. Call is a metallic *chink*.

Remember: It's pronounced *peer-uh-LOX-ee-uh*. Use this knowledge to impress your friends.

WOW!
One of the other folk names for the Pyrrhuloxia is Parrot-billed Cardinal, for the bird's stubby and curved yellow bill.

◀ *Pyrrhuloxias prefer thorny brushy habitat—much drier habitat than that preferred by Northern Cardinals.*

Find it: A resident of brushy desert habitat: mesquite thickets, thorn scrub, streamside brush. Usually found in small flocks, which move to more open and wooded habitat in winter.

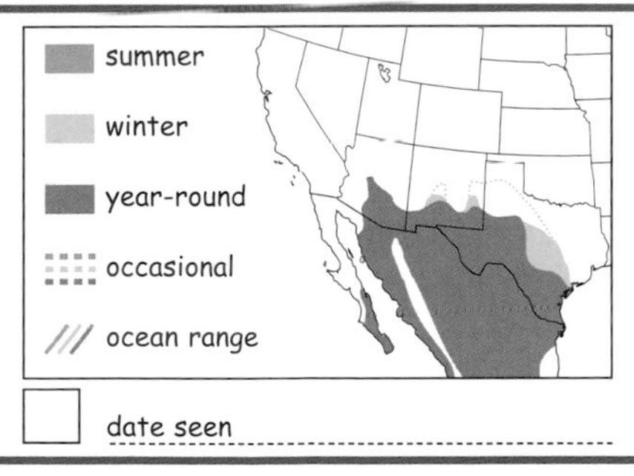

☐ date seen

BLUE GROSBEAK

Passerina caerulea **Length: 6¾"**

Look for: The Blue Grosbeak looks like a supersized Indigo Bunting with ketchup on its shoulders. Blue Grosbeaks regularly flick and spread their tails—a good field mark when light conditions obscure the birds' color (they can appear black in poor light).

Listen for: The song is a rich, burry warble with a few buzzy notes in the middle, delivered in a single unbroken phrase. The call note is a loud, emphatic *chink!*

Remember: The male Blue Grosbeak looks like a chunky Indigo Bunting, but one look at its rusty shoulder patches and heavy bill, and you know you've got a Blue Grosbeak.

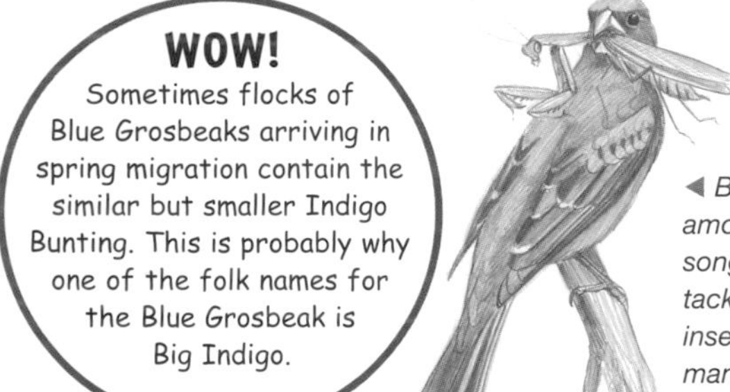

◀ *Blue Grosbeaks are among the few songbirds that will tackle large, aggressive insects such as praying mantises.*

Find it: Blue Grosbeaks often stay in thick cover in brambly thickets and along brushy field edges, but males will sing from treetop perches and fence and power lines in spring and summer. In migration, Blue Grosbeaks gather in loose feeding flocks.

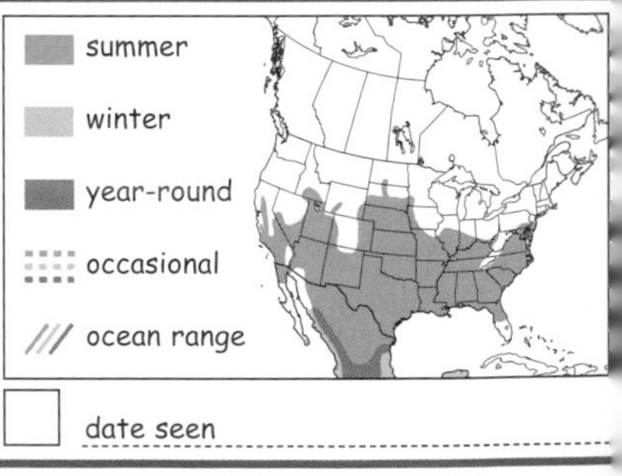

INDIGO BUNTING

Passerina cyanea **Length: 5"**

Look for: In fall the deep blue plumage of the adult male Indigo Bunting changes into the same drab brown plumage that the female wears all year. Young males may also be splotched with blue on the breast.

Listen for: Choosing a treetop perch, male Indigo Buntings sing a loud, enthusiastic song of paired notes, often descending in tone: *fire! fire! where? where? there! there! put it out! put it out!* Call notes, frequently given, are a sharp *spick!* and a short, dry *bzzt!* that sounds like an electric shock.

Remember: In very bright or very low light, the male Indigo Bunting appears all black, so relying solely on color to make an ID can be risky. Use size, shape, behavior, and sound to identify this species.

WOW!
A study with caged buntings inside a planetarium, recording the direction of their flight attempts as the star pattern above them was changed, proved that Indigo Buntings use the stars to navigate during migration.

▶ *Drab brown female and young Indigo Buntings can be identified by the way they flick their tails out to the side and back.*

Find it: The Indigo Bunting can be heard singing along woodland edges and roadsides and in overgrown farm fields and other brushy habitat. Males are most conspicuous in spring and summer while females tend the nest.

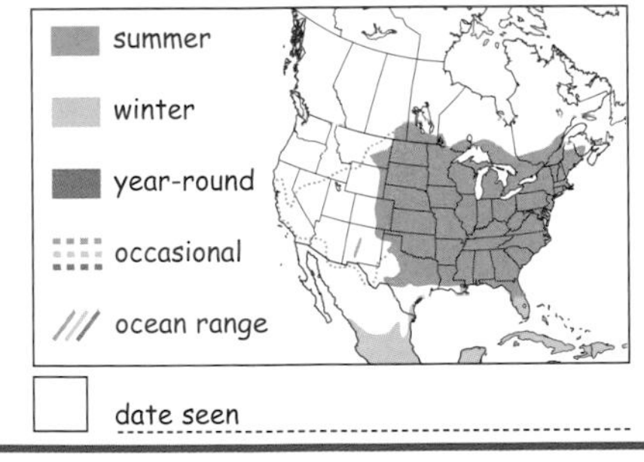

☐ date seen ____________

LAZULI BUNTING

Passerina amoena **Length: 5½"**

Look for: The adult male in breeding plumage is a bluebird-like mix of blue head, back, and wings; rusty breast; and white belly. The main difference is the bold white wing bars on the Lazuli Bunting. Female Lazuli is plain brown overall with a buffy unstreaked breast and two buffy wing bars.

Listen for: Song is a goldfinch-like jumble of sweet whistled phrases: *twee-chiddledee-twee-twee two*. Call is a dry *spik*.

Remember: If that blue bird singing from the top of a shrub has white wing bars, it's not a bluebird, it's a Lazuli Bunting.

Where Indigo and Lazuli Buntings nest in close proximity, they interbreed, and male Lazulis sometimes start to sing like male Indigos.

◀ *As the human population has expanded and subdivisions have sprung up throughout the West, the Lazuli Bunting population has declined, along with those of many other species.*

Find it: Common in summer in open brushy areas, weedy hillsides, and in shrubs along streams. Males sing from a conspicuous perch during the breeding season. In migration, found in flocks in more open habitats, such as weedy fields and brushy meadows.

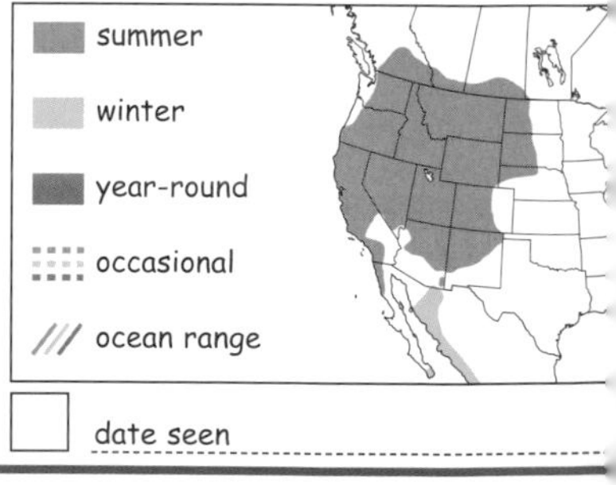

PAINTED BUNTING

Passerina ciris **Length: 5½"**

Look for: Of all the birds in North America, the male Painted Bunting is the most vividly colored. Though not as colorful, the female is distinctive too, with a color combination no other bird has—lime green above and otherwise unmarked.

Listen for: The Painted Bunting's song is a long, sweet warbling phrase. It has the musical quality of the Indigo Bunting's song, but the notes are more slurred together. Call note is a sharp, metallic *vit!*

WOW!
Another name for the Painted Bunting is Nonpareil, which is French for "having no equal." On their tropical wintering grounds they are often illegally captured and kept as pet birds.

Remember: Other small greenish birds (such as the vireos) do not have the Painted Bunting's large bill and overall unmarked plumage. Vireos have thinner bills and other obvious field marks (wing bars, eye lines, spectacles).

▶ The female Painted Bunting cares for the young all by herself. Her mate may feed the fledglings when she starts a second brood.

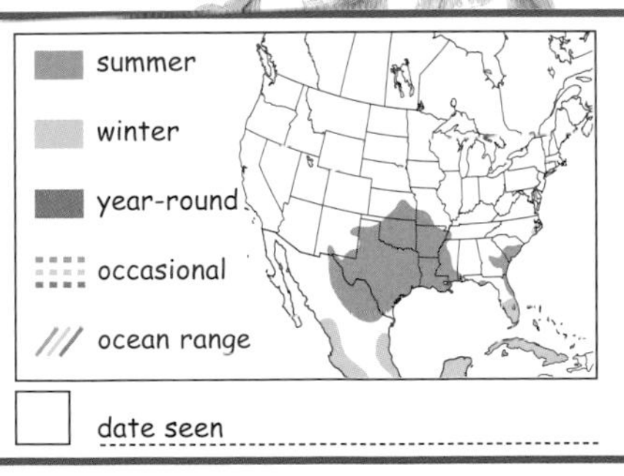

Find it: Despite the male's brilliant plumage, Painted Buntings can be hard to find. Their range is more limited than those of other buntings. They are shy birds that prefer thickets and brushy cover during the breeding season.

summer
winter
year-round
occasional
ocean range

date seen

LARK BUNTING

Calamospiza melanocorys **Length: 7"**

Look for: Breeding-plumaged adult male is nearly all black with a large white wing patch and tail tips. Adult females and nonbreeding males are brown and streaky overall with a hint of the white wing patch and bold dark mustache lines on the sides of the throat. Heavy bill is blue-gray.

Listen for: Male Lark Buntings usually give their song during display flights near the breeding territory. Song is a melody of sweet whistles, trills, and buzzes. Many notes sound similar to song of Northern Cardinal. Call is a soft-noted *hyew*.

Remember: Female and nonbreeding male Lark Buntings look like large streaky sparrows, but the super-heavy blue-gray bill is a great clue to their true identity.

WOW!
Because there are few song perches in their preferred habitat, male Lark Buntings perform their courtship and territorial songs while flying.

◀ *In flight, this mostly black bird has large white wing patches that really stand out.*

Find it: Common in summer on short-grass prairies and treeless open grasslands and prairies. During migration can be found in large flocks in any grassy, open habitat. Spends the winter in large flocks.

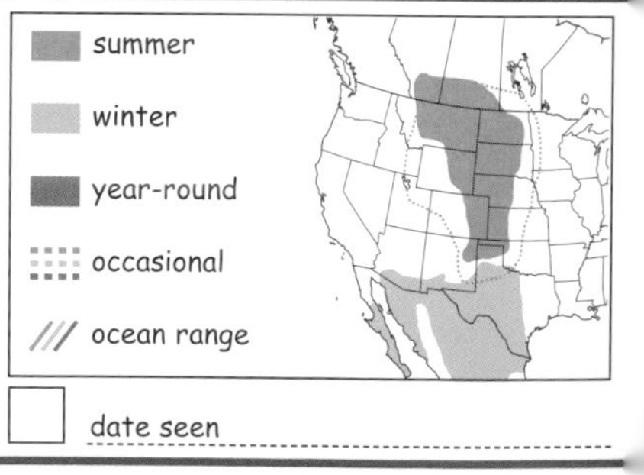

date seen

DICKCISSEL

Spiza americana **Length: 6¼"**

Look for: Looking like a miniature Meadowlark with its black V on a yellow breast, the breeding-plumaged male Dickcissel has a conical pale gray bill and a chestnut shoulder patch. Winter male is faded overall. Female looks like female House Sparrow but shows the rusty shoulder and a faint yellow wash on the breast.

Listen for: Song is a series of percussive notes for which the bird is named: *dick-dick-dick-ciss-ciss-ell.* Call, usually given in flight, is a loud, burry *bzzrt!*

Remember: The conical gray bill and rusty shoulder patch are field marks that set the Dickcissel apart from similar species in all seasons.

WOW!

Dickcissels, like some other grassland-breeding birds, have a breeding range that expands and contracts from year to year, depending on the amount of local rainfall.

► *If you are in Dickcissel habitat in summer, you are likely to hear a male singing—they are very vocal birds.*

Find it: A common summer-breeding bird in grasslands, meadows, farm fields, and weedy pastures and fencerows. Dickcissels are very vocal birds both during the breeding season and in migration, when their flight call can be heard even when the birds are flying too high to be seen.

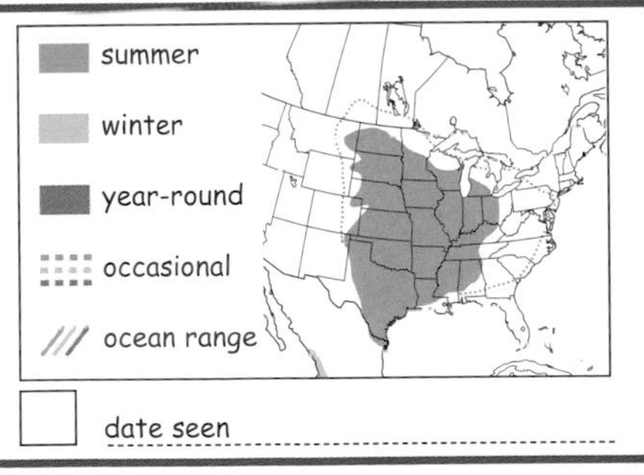

☐ date seen

EASTERN and WESTERN MEADOWLARK

Sturnella magna, Sturnella neglecta **Length: 9½"**

Look for: The lemon yellow breast with a black V sets the Eastern Meadowlark apart from all other blackbirds (except for the look-alike Western Meadowlark). Males and females look alike.

WOW!
Eastern Meadowlarks build an elaborate nest on the ground, woven out of grass. The nest often has a woven dome over it and an entrance on the side.

Listen for: Eastern Meadowlarks sing a clear pattern of downward-slurring whistles: *see I SING clear!* Or *spring of THE year!* They utter several calls, including a harsh, electrical *jrrt!* and a sputtering rattle: *brtbrtttttttt!*

Remember: It might be impossible to separate Eastern and Western Meadowlarks visually. If you hear them sing, it's easier to make a positive identification. Westerns have a lower-pitched, less musical song that has a burbly ending. Eastern's song is high, slurred whistles.

▶ *A male Eastern Meadowlark sends his clear song over a rolling hay meadow.*

Find it: Common in spring and summer in grasslands, meadows, and prairies where males sing from a prominent perch. In winter, small flocks can be found in any open habitat, including cultivated fields and the grassy edges of airfields.

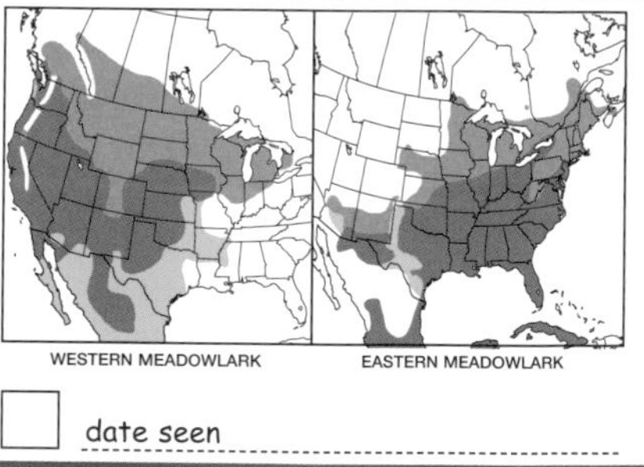

☐ date seen ______________________

RED-WINGED BLACKBIRD

Agelaius phoeniceus **Length: 8¾"**

Look for: The adult male Red-winged Blackbird looks just like its name suggests—an all-black bird with obvious red (and yellow) shoulder patches. Males flash these patches as they fly and as they sing. Females are dark and streaky, with rusty backs and buffy eyebrows and throats.

Listen for: The blackbird's song is a harsh, rising *conk-a-ree!* Females give an explosive, sputtering call: *bee-bee-bee-prrrrrttt!* Flight call is *chack!* Alarm call of male is down-slurred *tyeer!*

Remember: The female Red-winged Blackbird looks confusingly similar to a streaky sparrow, so check the bill. The blackbird's dark bill is thinner and much longer than the typical sparrow bill.

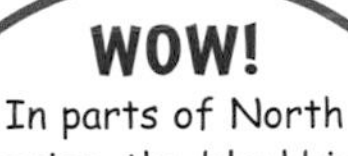

In parts of North America, the blackbird's **CONK-A-REE!** is one of the first sounds of spring's return. Males will begin singing on territory before the last snow has melted.

▼ *When singing on territory, male Red-winged Blackbirds spread their wings to show off their red shoulder patches.*

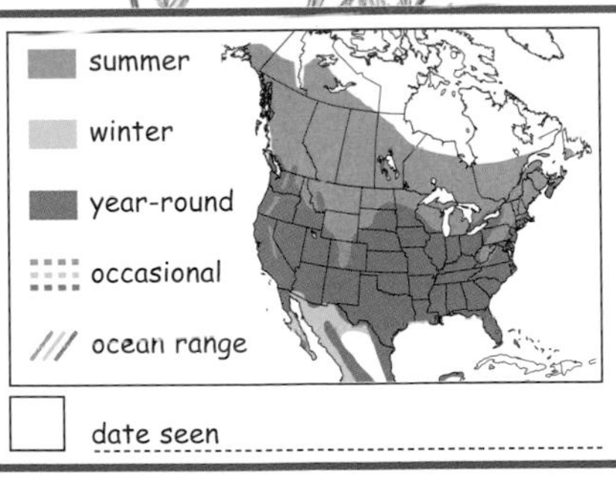

Find it: Red-winged Blackbirds are found in spring and summer anywhere there is a bit of water and some tall grass, cattails, or other vegetation: marshes, lakes, ponds. Can be found in winter almost anywhere.

summer
winter
year-round
occasional
ocean range

☐ date seen

YELLOW-HEADED BLACKBIRD

Xanthocephalus xanthocephalus **Length: 9½"**

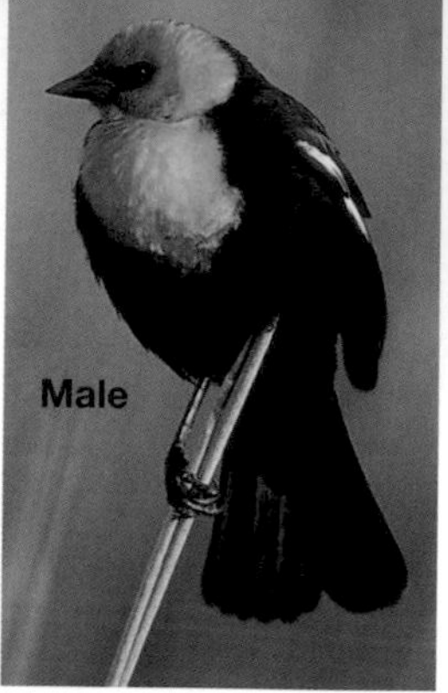

Look for: The male Yellow-headed Blackbird is a large, obvious black bird with a mustard yellow head and breast. His large white wing patches flash in flight. The female is drab brown with yellow on the face and chest.

Listen for: The male's song sounds like someone is throwing up. He points his bill upward and tilts his head to the side, showing off his colors. He does manage a few more musical notes now and then. Calls include a harsh, nasal *raar-raar-raar-raar!* and a loud, dry *chek!*

Remember: This relative of the meadowlarks shares their yellow coloration but not their habitat. Blackbirds prefer marshy settings with cattails. Meadowlarks prefer grassy meadows.

WOW!

More common in the Midwest and West, Yellow-headed Blackbirds d show up in winter blackbir flocks in the East. Sharp-ey birders can pick them out o flocks of Red-winged Blackbirds and Common Grackles.

► *The Yellow-headed Blackbird's love song ends in a bray that sounds like a dying donkey.*

Find it: Common in reedy marshes, the Yellow-headed Blackbird spends summers in noisy nesting colonies. Males perch on the tops of cattails and deliver their retching songs. Females dart to and from the nests.

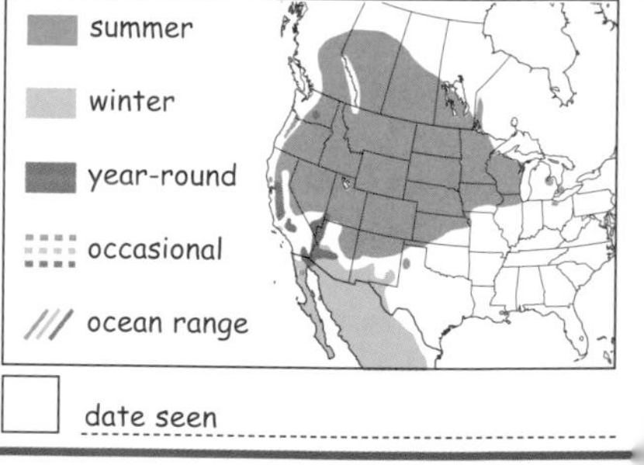

☐ date seen ____________

BOBOLINK

Dolichonyx oryzivorus Length: 7"

Look for: Male Bobolinks go through a dramatic plumage change from spring to fall. In summer, the male is a handsome blend of black, white, and butterscotch. In fall, he changes to buffy tans and browns, similar to what the female wears year-round.

Listen for: Male Bobolinks sing a burbling series of warbles, buzzes, and whistles in a long series that sounds like the start of a dial-up Internet connection or like R2D2 from *Star Wars*. Call note is a whistly, rising *wink?* Migrating Bobolinks often give this call in flight.

Remember: Fall male and female Bobolinks may look like sparrows, but they are actually blackbirds. They are larger than our sparrows and have plainer-looking faces.

WOW!

Other names for the Bobolink (a name derived from the song) are Meadow Wink (for its habitat and call note), Skunkhead Blackbird (plumage), and Butter-bird (for the yellow head patch).

◀ *A male Bobolink in full song-flight over his nesting territory.*

Find it: Locally common in grassy meadows and hay fields in spring and summer, the Bobolink is often located when the male does a song-flight over his territory. Females hide in the tall grass so are harder to see.

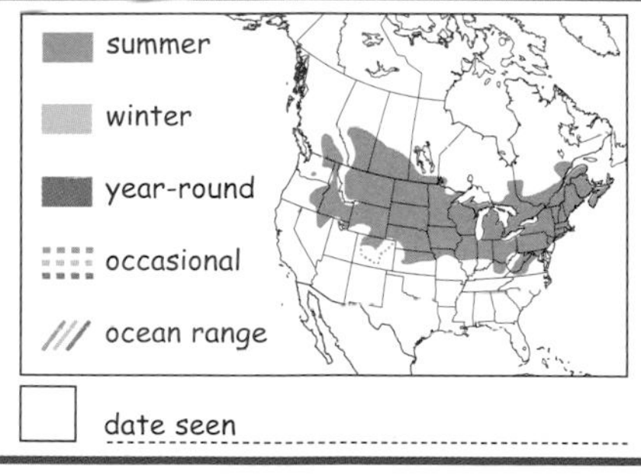

date seen

COMMON GRACKLE

Quiscalus quiscula **Length: 12½"**

Look for: In bright sunlight, the Common Grackle's feathers shine with tones of green, purple, gold, and blue. It's a dark bird that is larger than our other blackbirds (longer tail, heavier bill) but smaller than our other grackles. Females are slightly less colorful than males. All adults have pale yellow eyes (young birds have black eyes). In flight, the Common Grackle holds its tail in a V shape.

Listen for: Common Grackles make a variety of weird sounds. Their song is a series of squeaks, whistles, and grating scrapes: *krr-zheee! zhrrrt!* Call note, often given in flight, is a deep, sharp *cack!*

Remember: Common Grackles fly straighter than blackbirds, which undulate like waves as they fly.

WOW!

Common Grackles will capture bats in the air and eat them. They also ambush House Sparrows in parks and near bird feeders, knock them on the head, and eat them. Eeewwww!

▲ *Acting more like a shrike than a blackbird, a Common Grackle flies off with a House Sparrow it took by surprise.*

Find it: The Common Grackle is found from suburban backyards and city parks to farm fields and wetlands. Nesting colonies are often built in large evergreen trees. Forages on the ground and visits bird feeders.

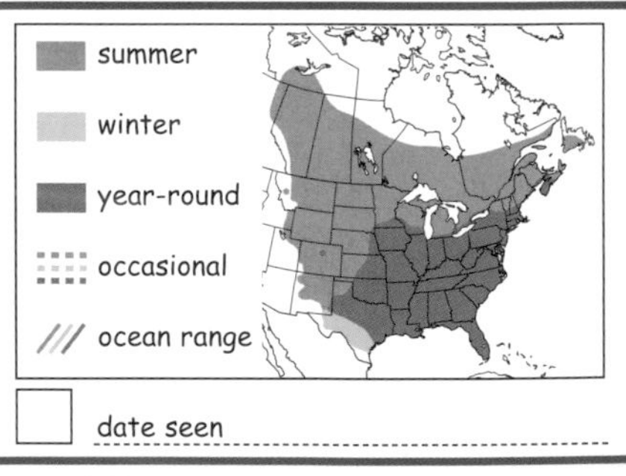

date seen

BOAT-TAILED GRACKLE

Quiscalus major **Length: 14½" (female) to 16½" (male)**

Look for: This big dark bird is like a Common Grackle with a tail extension. Males are dark bluish green in bright sunlight. Females are dark brown overall with black wings and are slightly smaller than males.

Listen for: Voice is a harsh, repetitive series of buzzing trills, squeaks, and whistles: *krssshh-krssshh-krsssh, kweet-kweet, chaak-chaak-chaak!* Many of the Boat-tailed Grackle's sounds are very unbirdlike. Call note is a deep *chuck!*

Remember: This species was formerly considered a single species with the Great-tailed Grackle, which lives in the Southwest. Great-taileds are slightly larger and not as closely associated with water as Boat-taileds.

▼ *Boat-tailed Grackles can be a nuisance, stealing pet food right off the porch. Females look like a different species from the males.*

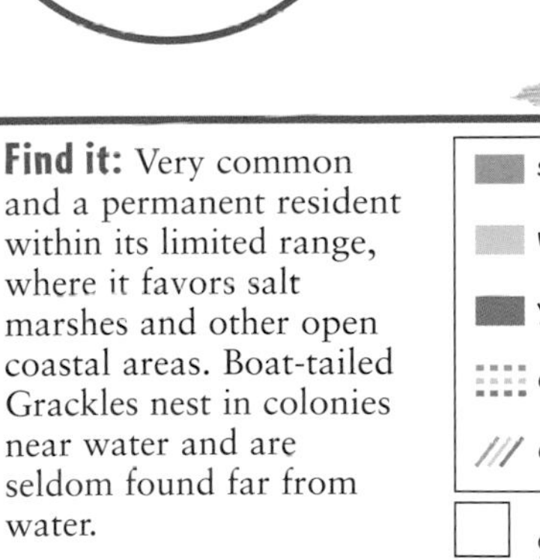

Find it: Very common and a permanent resident within its limited range, where it favors salt marshes and other open coastal areas. Boat-tailed Grackles nest in colonies near water and are seldom found far from water.

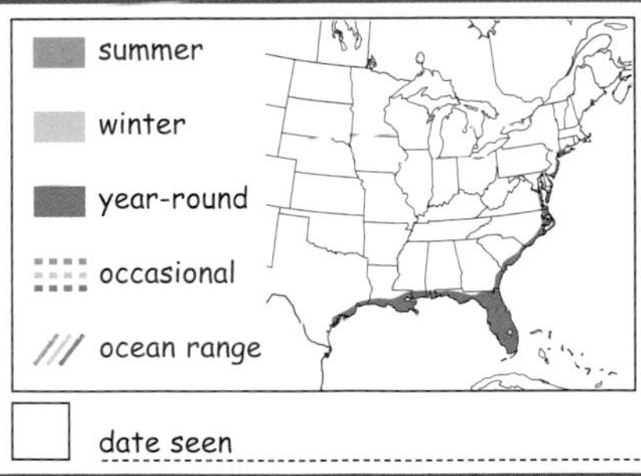

date seen

GREAT-TAILED GRACKLE

Quiscalus mexicanus **Length: 18"**

Look for: Adult male is a large glossy black bird with a long black tail, for which it is named. Females and young birds are smaller and pale brown below, dark gray above, with a pale line over the eye. This grackle is noticeably larger, taller, and longer tailed than Common Grackle.

Listen for: Song is a repeated series of loud shrieks, whistles, rattles, and harsh notes: *weet-weet-weet. Chack-chack-chack. Boit-boit-boit. Annnkannkannk.*

Remember: The similar Boat-tailed Grackle of the Southeast is never found far from the coast.

WOW!

Great-tailed Grackles roost communally, often in towns where parks and shopping centers provide both habitat and easy access to food. These roosts remain noisy all night long.

▶ *Male Great-tailed Grackles fluff up their feathers, quiver their wings, point their bills skyward, and utter their weird songs, all to charm the grackle gals.*

Find it: Common and expanding its range, the Great-tailed Grackle is impossible to miss wherever it occurs because of its large size and loud voice. Nearly always found in flocks. Forages in farmland, open groves, and feedlots. Roosts and nests in thick cover often near water.

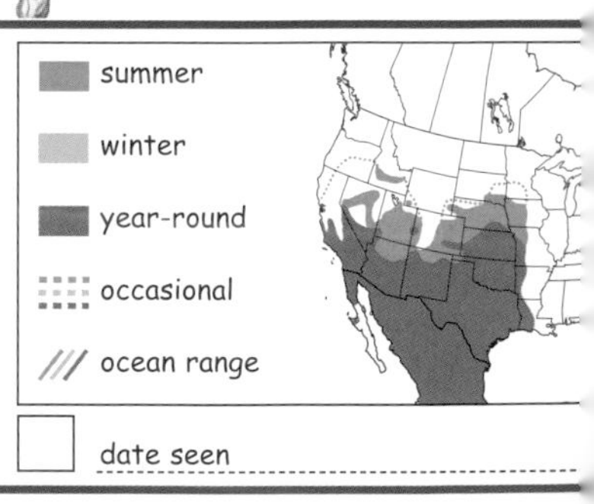

☐ date seen ____________

BREWER'S BLACKBIRD

Euphagus cyanocephalus **Length: 9"**

Look for: Adult male is all black with a pale eye. Plumage can appear glossy purple and green in bright sunlight. Tail is square ended and much shorter than those of grackles. Female Brewer's is plain grayish brown overall with a dark eye (female Rusty Blackbird has a pale eye).

Listen for: Song is shrill and unmusical: *ksh-ree!* Call is a loud *chak!*

Remember: Telling breeding-season male Rusty and Brewer's Blackbirds apart can be tricky. Adult male Brewer's never takes on any rusty color and has a slightly thicker, shorter bill than the Rusty.

WOW!
This species was discovered by John James Audubon on his final expedition in 1843 and named for his friend Dr. Thomas Brewer, a Boston ornithologist, who probably never saw a live Brewer's Blackbird.

◀ *The Brewer's Blackbird has adjusted very well to human-altered habitats. "Would you like fries with that?"*

Find it: Very common in a variety of habitats in the West, from parking lots to farm fields and mountain meadow edges. Often seen on ground, walking, jerking its head forward with each step, much like a chicken. Joins huge mixed species blackbird flocks in winter.

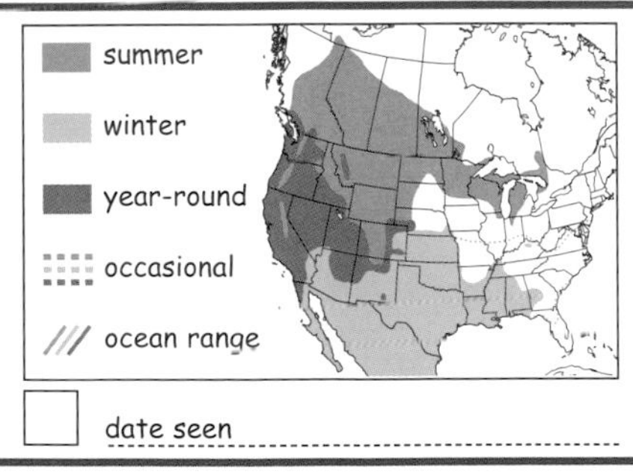

☐ date seen ____________

BROWN-HEADED COWBIRD

Molothrus ater **Length: 7½"**

Look for: This smallish blackbird has a short, stout bill and a short tail. The bill shape is a good field mark to separate this species from other blackbirds. The name *cowbird* comes from its habit of following and foraging around herds of cattle.

Listen for: The song is a weird mix of low gurgles and high, squeaky whistles. Also gives a long sputtery trill: *pt-pt-pprrrrrrrtttt!* Flight call is a high, thin whistle: *tsee-tseeeet!*

Remember: The European Starling is often seen in the same habitat as the cowbird but differs by having a long, thin bill and a very short tail.

WOW!

Cowbirds are notorious for laying their eggs in the nests of other birds. The nestling cowbirds often outcompete smaller nestmates. This has caused a decline in many songbird populations.

◀ *A female Brown-headed Cowbird removes a Wood Thrush egg and will lay one of her own. She may lay 30 eggs in a single season.*

Find it: The cowbird prefers open habitat in all seasons: grasslands, farm fields, woodland edges, parks, and backyards, where it will visit bird feeders for mixed seed scattered on the ground.

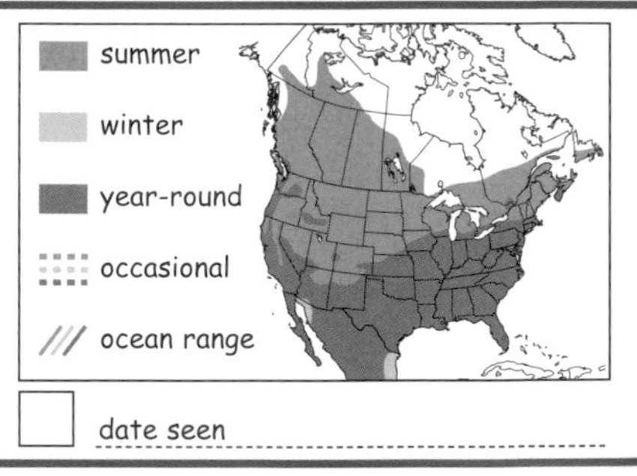

☐ date seen ____________

ORCHARD ORIOLE

Icterus spurius **Length: 7¼"**

Look for: The Orchard Oriole is small enough to be mistaken for a warbler, but one look at the long, fine-pointed bill should convince you that this is no warbler. The male Orchard is a deep rusty orange and black. Females are plain yellow-green overall.

Listen for: Male Orchard Orioles sing a rapid, variable song that is sweetly musical with a few buzzy notes: *look here! up here! see me? how's it going! chh-chh! hey you!* Call note is a soft *chuck!* or *twee-ohh!*

Remember: Female Orchard and Baltimore Orioles can be hard to tell apart. Female Orchards are smaller and greener overall; female Baltimores are washed with orange on chest and belly.

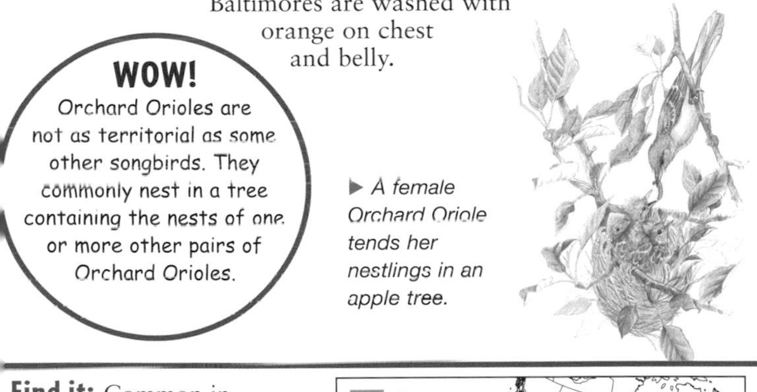

WOW!

Orchard Orioles are not as territorial as some other songbirds. They commonly nest in a tree containing the nests of one or more other pairs of Orchard Orioles.

▶ *A female Orchard Oriole tends her nestlings in an apple tree.*

Find it: Common in summer in young woods and woodland-edge habitat, Orchard Orioles prefer open, brushy settings and avoid deep woods. Males are avid treetop singers.

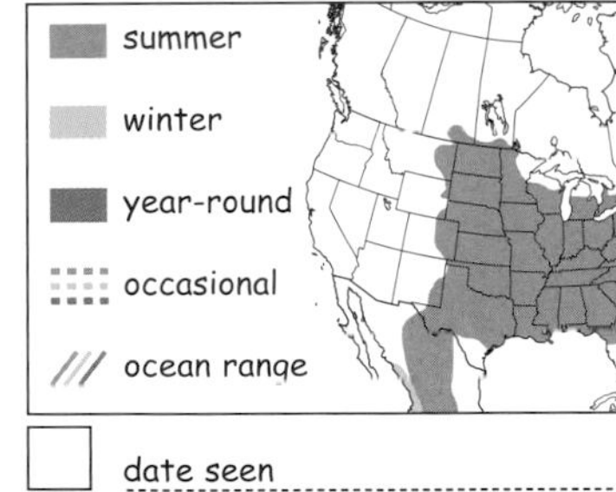

☐ date seen ____________

BALTIMORE ORIOLE

Icterus galbula **Length: 8¾"**

Look for: The classic orange and black colors of the male Baltimore Oriole are familiar even to nonbirders. No other eastern bird species has the male Baltimore's combination of size and coloration.

Listen for: The Baltimore Oriole's song is a series of clear, slurred whistles. It varies in pattern: *hey! hey you! see me up here sing-ing!* Call is a scolding chatter: *chrt-trr-rrrrrr!*

Remember: The Baltimore Oriole was once lumped together with the Bullock's Oriole (a western species) as the single-species Northern Oriole. These birds overlap in range and do interbreed, but the males are easy to separate visually (the male Bullock's Oriole has an orange face).

WOW!
The Baltimore Oriole was named for the royal colors of Lord Baltimore. Today we have a beloved bird and a major-league baseball team that proudly bear this name.

▶ *Baltimore Orioles will visit feeders for fruit, such as orange halves.*

Find it: The Baltimore Oriole prefers open woodland settings with tall trees for nesting, such as in city parks and farmland groves. Its baglike nest is often built near the end of a horizontal branch overhanging a road or water.

summer
winter
year-round
occasional
/// ocean range

☐ date seen ______________

BULLOCK'S ORIOLE

Icterus bullockii **Length: 8¼–8½"**

Look for: Adult male is bright orange below with an orange face, black throat and crown, and black eye line. Huge white wing patch (on the coverts) stands out on black wings and is obvious in flight, as is black-tipped tail of male. Female is pale orange on the head and gray below, with a gray back and subtle white wing bars.

Listen for: Song is a series of paired notes, some musical, some harsh: *cha-chacha-toowee-trickatrickatricka-reeet!* Call is a long scolding chatter: *ch-ch-ch-ch-ch-ch*.

Remember: The male Bullock's Oriole has a black crown and eye line. The similar Hooded Oriole has an orange crown and nape.

WOW!

A group of ornithologists lumped ogether the Bullock's Oriole d the Baltimore Oriole into a gle species: Northern Oriole. Then some other ornithologists split them into two species again. Stay tuned!

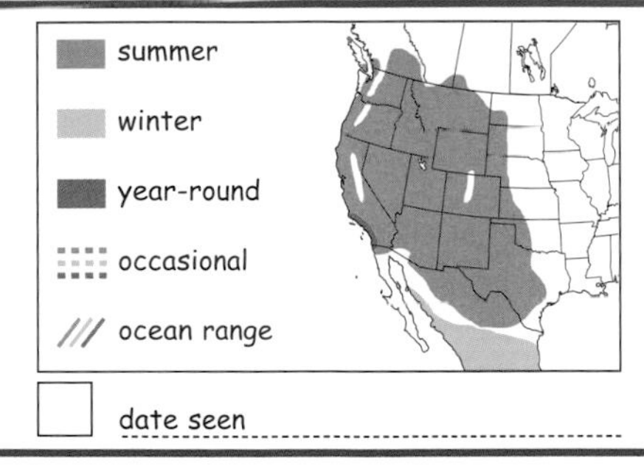

▲ *Once they identify it as a food source, Bullock's Orioles will visit a hummingbird feeder for nectar.*

Find it: Common in summer in deciduous woods, especially in cottonwoods along streams, where Bullock's Orioles sometimes form loose nesting colonies. Males are very vocal on breeding territory.

summer

winter

year-round

occasional

/// ocean range

☐ date seen

HOODED ORIOLE

Icterus cucullatus **Length: 7½–8"**

Look for: Adult male is orange and black with an orange crown and black face and throat. Bill is thinner and down-curved. White coverts on black wings are less extensive than those of Bullock's Oriole. Female Hooded Oriole is olive above, yellow below.

Listen for: Song is a jumble of whistles, squeaks, and chattering notes. Calls include a rising *wheet!* and a short, soft chatter.

Remember: The "hood" on the Hooded Oriole is golden orange, not black as in many of our other orioles.

WOW!
Some Hooded Orioles spend the winter in the U.S. in places where nectar feeders guarantee a steady food supply.

▶ *Hooded Orioles nearly always choose to nest in palm trees, particularly fan palms.*

Find it: Uncommon in the desert Southwest in summer in open woods, palm and streamside groves, parks, and wooded suburban neighborhoods. Usually associated with palm trees. A common visitor to backyard feeders for nectar and fruit.

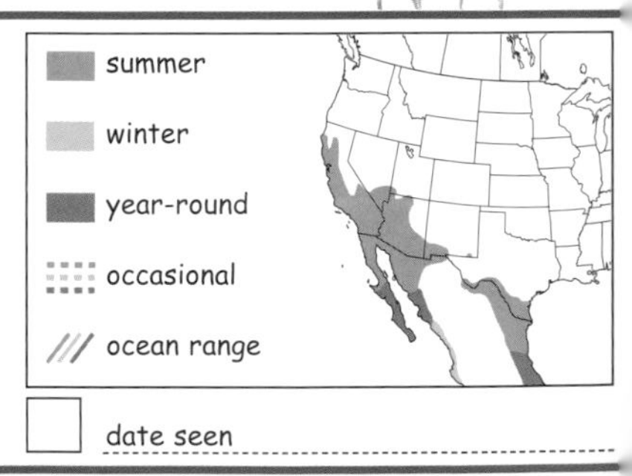

☐ date seen ____________

GRAY-CROWNED ROSY-FINCH

Leucosticte tephrocotis **Length: 6–8"**

Look for: Adult has gray head, black forehead and throat, and varying amounts of rose in the body and wing feathers. Breeding male has a blacker face, a black bill, and bright rosy patches on wings and flanks.

Listen for: For such a beautiful finch, the Gray-crowned Rosy-finch's song is nothing special: loud down-slurred *tyeew-tyeew* notes in a series, similar to the call of the House Sparrow. Flight call is a softer *tchew!* Flocks are very vocal.

Remember: This species is highly variable geographically and often flocks with other rosy-finches in winter, so it pays to look at each bird individually.

WOW!
Some high-altitude feeding stations, such as the one at Sandia Crest, near Albuquerque, New Mexico, attract all three species of rosy-finch in winter: Gray-crowned, Black, and Brown-capped!

▶ *Most birders get their life looks at Gray-crowned Rosy-finches that are visiting a feeding station.*

Find it: This is our most widespread rosy-finch, but its habitat preferences make it difficult to find: high mountain meadows, tundra, rocky summits, and snowfields. Usually in flocks. Often visits feeders at ski resorts and in mountain towns.

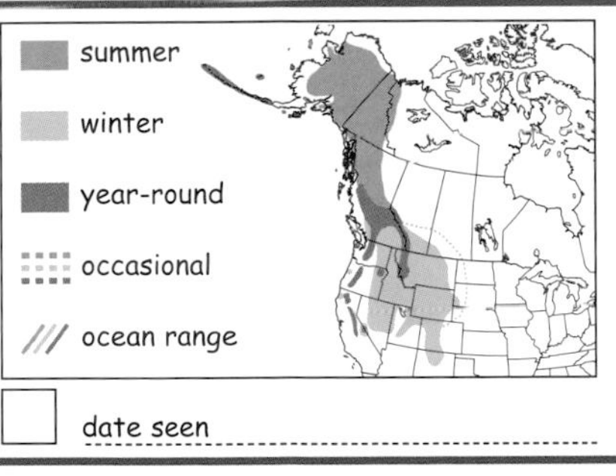

date seen

WHITE-WINGED CROSSBILL

Loxia leucoptera **Length: 6½"**

Look for: Adult male is pinkish red overall with bold white wing bars on black wings. Females and young birds are brownish and very streaky overall; they also have white wing bars.

Listen for: Song is a long unmusical rattle that slows near the end. More commonly heard is the call, given in short phrases: *chi-dit, chi-dit, chi-dit!*

Remember: White-winged Crossbills are named for their best field mark (white wing bars). Red Crossbills lack any white in the wings and are more brick red in color.

▼ Flocks of foraging White-winged Crossbills may contain hundreds of individuals. Sometimes they are joined by Red Crossbills and other finch species.

WOW!
White-winged Crossbills wander constantly in search of abundant spruce cones. When they find them, they will often stop to nest, even in midwinter!

Find it: A vagabond species of the far North that is highly reliant on the variable crop of spruce cones. Always found in flocks in spruce, hemlock, and fir forests.

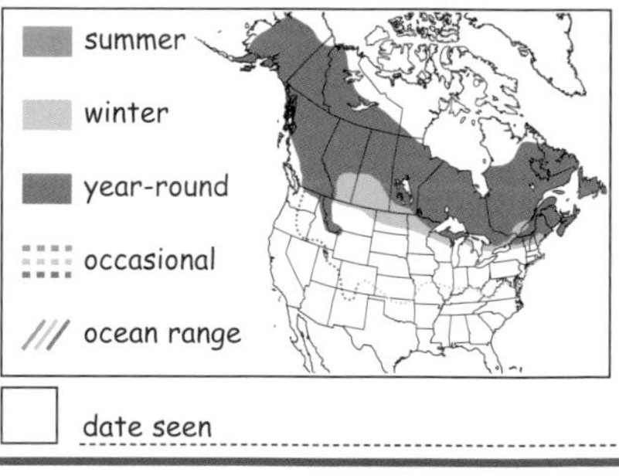

☐ date seen ____________

RED CROSSBILL

Loxia curvirostra **Length: 6¼"**

Look for: This bird's name is very descriptive: The male is a red bird with a weird crossed bill that is perfectly designed to pry the seeds out of evergreen cones. Females are dull yellow-green and have dark wings like the males. The Red Crossbill does not have wing bars, unlike its smaller-billed cousin the White-winged Crossbill.

Listen for: A short series of thin, buzzy chips passes for the song of the Red Crossbill: *twit-twit-twit-twit, jwee-zit*. Call note is a loud, hard *klip-klip!*

Remember: The crossed bill can be difficult to see from a distance. But the way crossbills feed—often hanging upside down from a pinecone, digging with the bill to pry seeds loose—is distinctive.

WOW!
There are at least nine distinct types of Red Crossbill in North America. Some have very large bills and feed on large pinecones. Others have smaller bills and eat seeds of smaller cones.

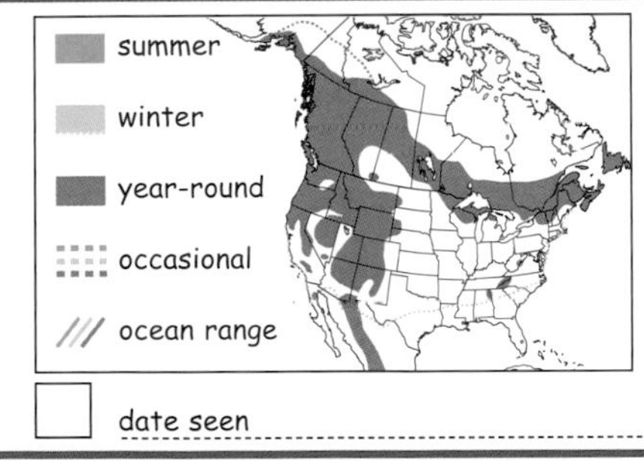

▶ *A Red Crossbill uses its scissorlike mandibles to extract spruce seeds.*

Find it: Red Crossbills are rarely found far from conifers (pines, hemlocks, spruces, firs) and are best located by the sounds made by feeding flocks.

summer
winter
year-round
occasional
ocean range

☐ date seen

COMMON REDPOLL

Carduelis flammea **Length: 5"**

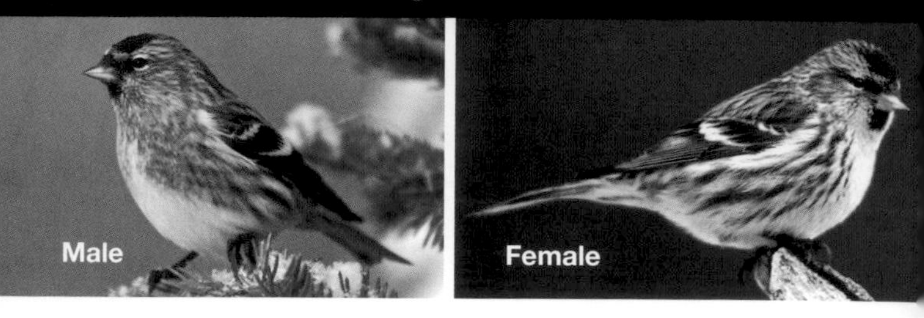

Look for: The smallest of our red finches, the Common Redpoll looks like a Pine Siskin with a red cap on its head. Streaky on backs and sides, with a tiny yellow bill surrounded by a black chin, the Common Redpoll is a very active, vocal bird. The male may have a completely pink-washed chest, while the female has a white chest.

Listen for: Song is a long series of twitters and rising buzzy trills: *chit-chit-chit-chewee, tu-tu-tu-tseeet, chit-chit-chit-zeeeet!* Common call is a chattering *ch-ch-ch-chweee!*, rising in tone on the last, longer note.

Remember: The Common Redpoll's small size and red cap make it easy to separate from other red or streaky finches.

WOW!
The polar bears of the bird world, redpolls can survive colder temperatures better than any other songbird. Extra food, stored in their crops, is digested during the night, keeping the redpoll warm.

◀ *Common Redpolls forage hastily, filling special pouches in their upper gut. Then they regurgitate the seeds, shell them, and eat them.*

Find it: In winters when food is scarce in the North, redpoll flocks, sometimes mixed with goldfinches and siskins, forage actively in weedy fields, trees, and shrubs. They will visit bird feeders for thistle seed and sunflower seed bits.

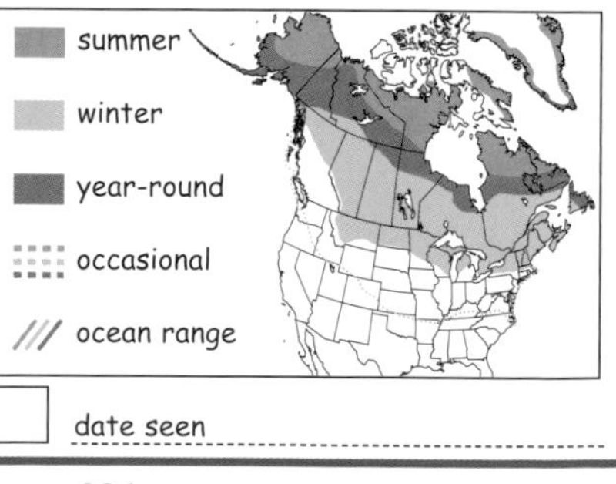

date seen ______________

HOUSE FINCH

Carpodacus mexicanus **Length: 5¾"**

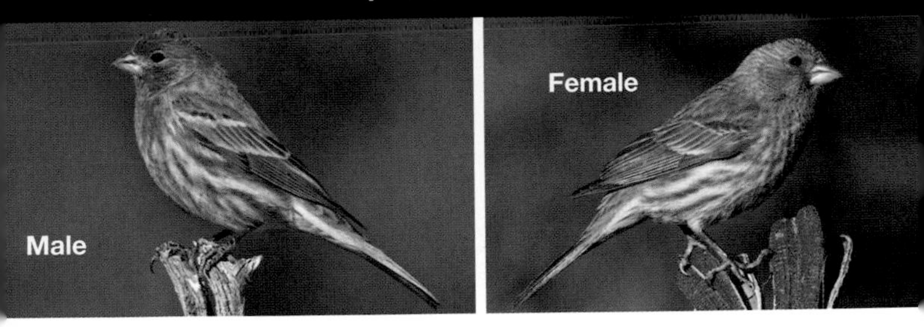

Look for: The male House Finch is washed with brick red on head and breast and streaked with brown everywhere else, especially on the flanks (sides). Females and young birds are brownish gray overall and covered with blurry streaks.

Listen for: The song is a rich series of whistled phrases ending with a few buzzy notes. Also utters a variety of call notes, from sparrowlike chirps to a rising *kweet!*

Remember: Male House Finches can vary in color from brick red to orange, but they always have blurry dark streaks on their bellies. They look dingy compared to the cleaner, brighter male Purple Finch (which has an unstreaked white belly).

WOW!

In 1940, a small number of House Finches being sold illegally as caged birds were released near New York City. All the House Finches east of the Great Plains are descended from them.

▶ *A pair of House Finches enjoy the seeds they find in birch cones.*

Find it: A native of the West, with the help of humans the House Finch has colonized the East. They are adaptable to nearly any habitat except deep woods and open grasslands, and most common near human-affected areas.

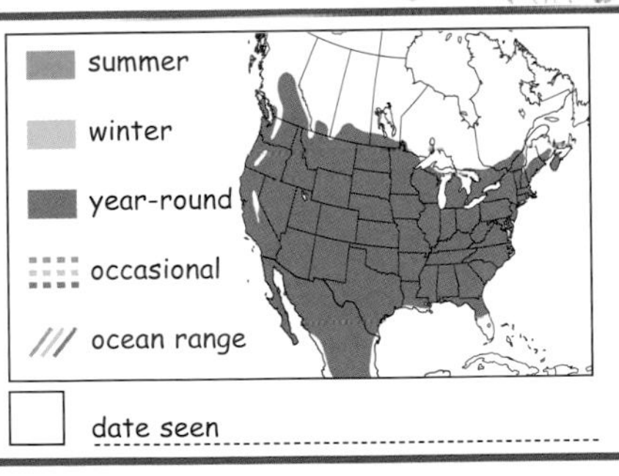

date seen

PURPLE FINCH

Carpodacus purpureus **Length: 6"**

Male

Female

Look for: The male Purple Finch is washed with a raspberry red (not purple), as if he'd been dipped upside down in raspberry juice. Unstreaked white flanks, belly, and wings help separate him from the similar male House Finch. The female is covered in short dark brown streaks.

Listen for: The bright, cheery song is a fast series of hoarse whistles: *treedle-wheedle-treedle-turtle-wheedle-breer!!*

Remember: House Finches just don't look as clean as Purple Finches do. The Purple Finch also has a larger bill, which gives it a larger-looking head. Check out the female Purple Finch's white eye line and dark bandit mask. By contrast, the female House Finch looks plain faced.

WOW!

Huge influxes of Purple Finches move south in fall and winter when the natural food crop (seeds of pines, spruces, and other trees) is poor in the North.

▼ *A male Purple Finch pulls out all the stops, singing for a female. At the height of his display, he may almost tip over backward.*

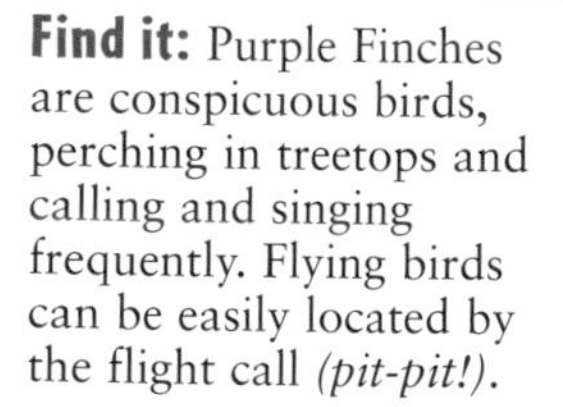

Find it: Purple Finches are conspicuous birds, perching in treetops and calling and singing frequently. Flying birds can be easily located by the flight call *(pit-pit!)*.

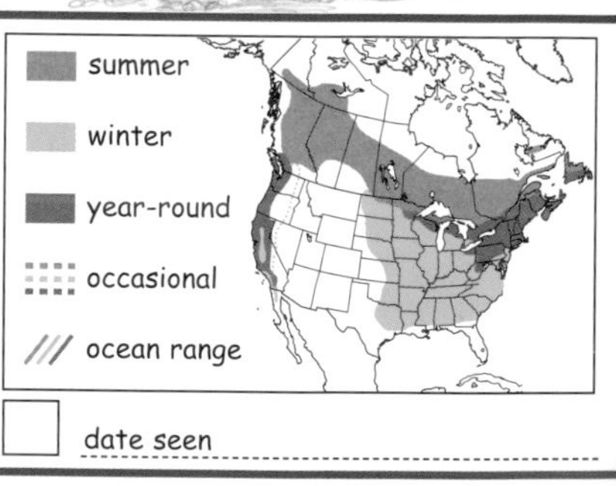

☐ date seen

CASSIN'S FINCH

Carpodacus cassinii Length: 6¼"

WOW!

The Cassin's Finch (and the uncommon Cassin's Sparrow) were named for John Cassin, a nineteenth-century ornithologist from Philadelphia, a place where Cassin's Finches never occur!

Look for: Adult male has a crown that is brighter red than the rest of its head and neck, giving it a peaked-headed appearance. Pinkish red on breast is paler than extensive red of male Purple Finch. Female Cassin's is finely streaked with brown overall, especially on the breast and belly. Cassin's Finch has a longer, more pointed bill than either Purple or House Finches.

Listen for: Song is a rich musical warble, softer in tone than song of Purple Finch. Call is a rising *giddy-up!*

Remember: Red finches in the western mountains should be examined closely. The longer, more pointed bill and the red cap of the male Cassin's Finch set it apart from both Purple and House Finches

▶ *After nesting, Cassin's Finches like to roam around in flocks, often mixing with crossbills and Evening Grosbeaks.*

Find it: A resident of mountain coniferous forest of the West and usually found in flocks, except during actual nesting. Found at high altitudes in summer, and some flocks move to lower elevations in winter.

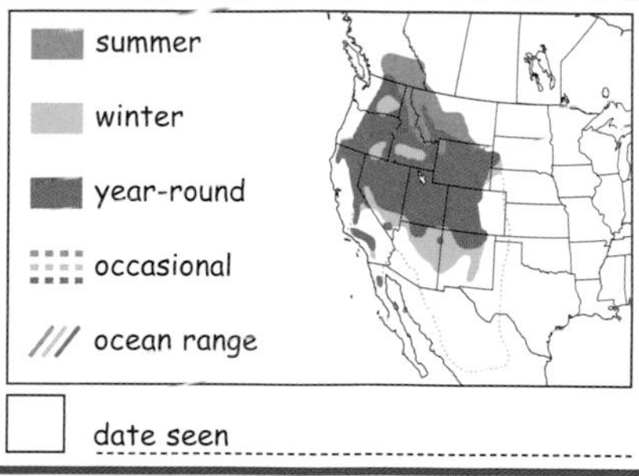

EVENING GROSBEAK

Coccothraustes vespertinus **Length: 8"**

Look for: The striking male Evening Grosbeak is a large mustard yellow finch with a big pale bill and a gold eyebrow. Huge white patches cover the bird's back and flash in flight on the black wings. Females are mostly gray with light yellow patches and less white in the wings. Coloration does not change seasonally.

Listen for: Loud and social, Evening Grosbeaks announce their presence with constant calling. Common call is a loud, descending *pyeer!* Also utters a raspy, tuneless whistle: *pirrrt!*

Remember: Like other winter finches, the Evening Grosbeak has a swooping flight style. But only the Evening Grosbeak flashes large white wing patches in flight.

WOW!
The Evening Grosbeak was named for the mistaken impression that it sang only at dusk.

▲ *The Evening Grosbeak ca exert as much as 125 pound of pressure with its bill and can crack cherry pits.*

Find it: Listen for the Evening Grosbeak's call note to find this unpredictable wanderer. Outside of nesting season, flocks move to where the food is—maples and box elders for their seeds and bird feeders for sunflower seed.

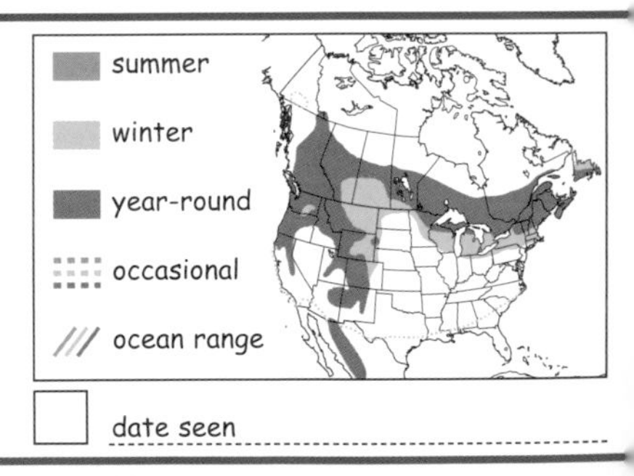

date seen

AMERICAN GOLDFINCH

Carduelis tristis **Length: 5"**

Look for: The American Goldfinch in spring and summer is familiar even to nonbirders. Bill color is yellow in spring and summer, dark gray in fall and winter. Winter-plumaged males resemble females, fooling many feeder watchers into thinking their goldfinches have left.

Listen for: The goldfinch's song is a variable mix of rich warbles and twitters, often given from a treetop perch. Twittering *potato-chip, potato-chip* call is commonly given in flight.

Remember: Dull-colored winter goldfinches are often overlooked as they forage with other species. Similar in appearance to yellow-colored warblers, the American Goldfinch has a much heavier bill.

WOW!
The American Goldfinch is the only songbird that feeds its ung a diet of seeds. Brown-headed Cowbirds hatched from eggs laid in goldfinch nests cannot survive the all-seed diet.

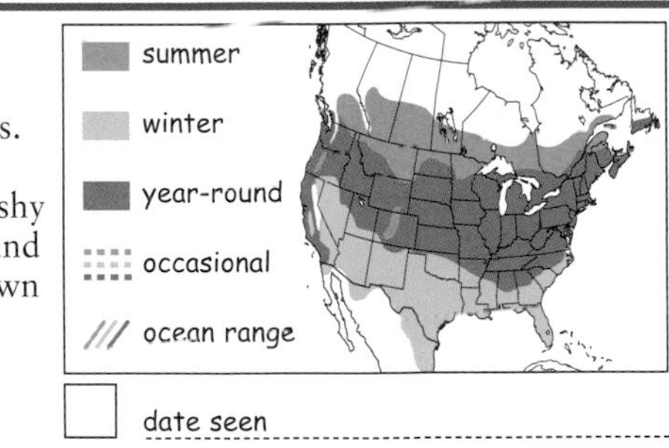

▶ *American Goldfinches eat thistle seeds and line their nests with thistle fluff.*

Find it: American Goldfinches forage in noisy, twittering flocks. They are common continent-wide in brushy habitat, along woodland edges, and in overgrown meadows. They visit feeders for sunflower and thistle seed.

summer
winter
year-round
occasional
ocean range

☐ date seen

LESSER GOLDFINCH

Carduelis psaltria **Length: 4½"**

Look for: The Lesser Goldfinch is so named because it is our smallest goldfinch. Adult male is dark olive above with a black cap and yellow below. Female is pale olive above and yellow below, very similar to female American Goldfinch.

Listen for: Song is a long rapid series of sweet notes and whistles, often repeated: *sweee-cheecheechee-tvee-tvee-tvee-tootootoo-sweee!* Often imitates phrases of other birds' songs.

WOW!
The back color of adult male Lesser Goldfinches varies geographically. Males in the southern Rocky Mountains have black backs, while most others have olive green backs.

Remember: Breeding-plumaged male American Goldfinch has a canary yellow back and white rump. Lesser Goldfinch has a dark back and rump.

▶ *The diet of the Lesser Goldfinch is mostly plant seeds. They seem to prefer thistle and sunflower seeds above all others.*

Find it: Common, but sometimes overlooked, in open brushy habitat, weedy fields, streamsides, parks, and gardens. Usually found in flocks actively foraging for seeds, uttering soft twittering vocalizations.

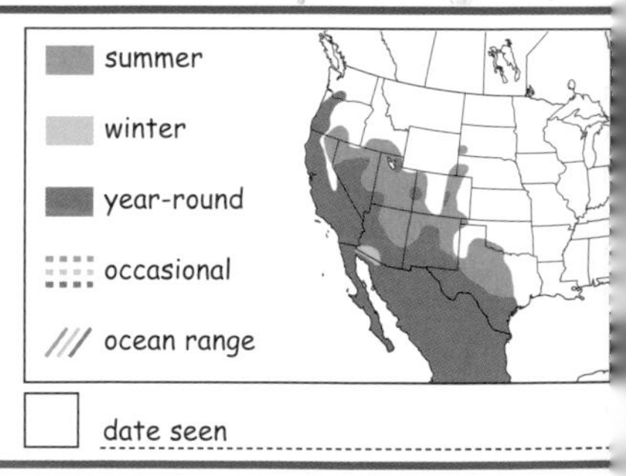

☐ date seen

LAWRENCE'S GOLDFINCH

Carduelis lawrencei **Length: 4¾"**

Look for: This handsome bird is mostly grayish. Adult male has a black face, yellow breast, and large yellow patches in the wings and on the rump. Female has an all-gray head and less yellow elsewhere. In flight, both sexes show white underwing linings and a yellow rump.

Listen for: Lawrence's Goldfinch has one of the more interesting finch songs, a long melody of musical phrases, most of which are imitations of other birds' songs. Call is a slightly harsh, rising *too-vreee!*

Remember: Flocks of goldfinches in the desert Southwest may include this species. It's worth carefully sorting through these flocks.

WOW!
This uncommon species can often be found near sources of water in remote arid areas.

◀ *Lawrence's Goldfinches are quite fond of salt and will visit a reliable source of this mineral.*

Find it: Uncommon in mixed oak-pine woods, chaparral, weedy areas usually near water. Found almost exclusively in California, except in winter, when flocks may range farther to the east, though never leaving the southwestern desert.

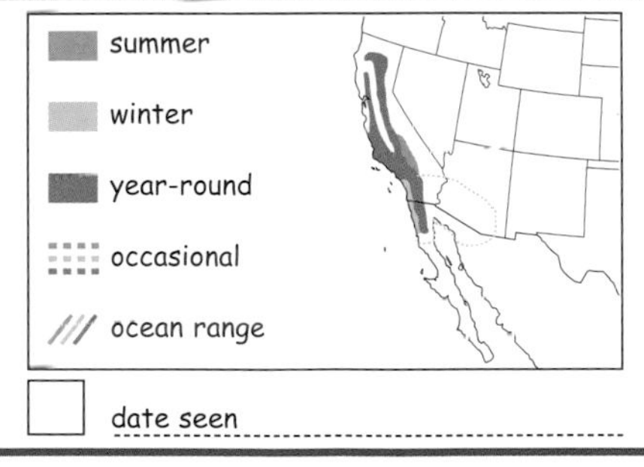

date seen

PINE SISKIN

Carduelis pinus **Length: 5"**

Look for: This streaky brown finch has a narrow, fine-tipped bill and shows yellow flashes in wings and tail in flight. They are often overlooked when mixed in with winter flocks of American Goldfinches, which are drab yellow (but not streaky) in nonbreeding plumage. The siskin's heavy streaking also makes it resemble a female House Finch, but note the siskin's smaller, finer bill.

Listen for: Pine Siskins are very vocal birds and commonly utter a harsh, rising call: *zzzrrreeeee?* Also a variety of thin, high twitters, often in paired notes: *twee-twee, jee-jee, twee-twee!*

Remember: Pine Siskins have been called "goldfinches in camouflage." They act, fly, and forage like American Goldfinches.

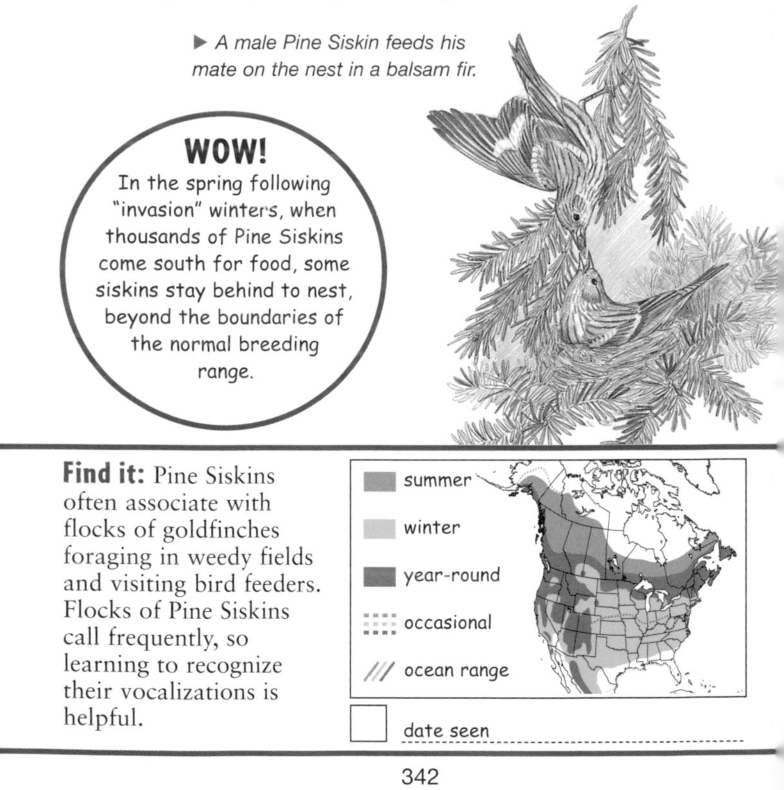

▶ *A male Pine Siskin feeds his mate on the nest in a balsam fir.*

WOW!

In the spring following "invasion" winters, when thousands of Pine Siskins come south for food, some siskins stay behind to nest, beyond the boundaries of the normal breeding range.

Find it: Pine Siskins often associate with flocks of goldfinches foraging in weedy fields and visiting bird feeders. Flocks of Pine Siskins call frequently, so learning to recognize their vocalizations is helpful.

HOUSE SPARROW

Passer domesticus **Length: 6¼"**

Look for: This chunky, big-headed sparrow is a common sight in small flocks in cities, backyards, and farmyards. In winter, the male's colors are less distinct than in spring and summer. Females are drab gray-brown year-round. Flight is direct and swift with rapid wingbeats.

Listen for: House Sparrows do not really sing a song. Instead, they constantly utter a series of husky calls: *cheerp! cha-deep!* They also give harsh rattles and short whistles near the nest.

Remember: The House Sparrow thrives in cities where other sparrows cannot survive. It is never found far from human habitation, so a sparrow you see in natural, unaltered habitat is almost certainly not a House Sparrow.

WOW!
House Sparrows were imported from Europe and introduced in New York in 1850 to control insects on city streets. By the early 1900s, they had spread across the continent to California.

▶ *House Sparrow nests are very messy, with lots of long grass stems and even bits of trash mixed in.*

Find it: Common in any human-affected landscape, House Sparrows nest in cavities (woodpecker holes, nest boxes) and often compete with native songbirds. House Sparrows visit bird feeders, eating almost anything that's served.

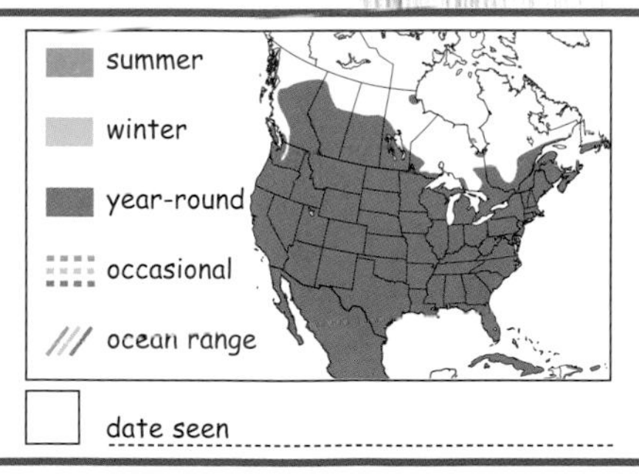

☐ date seen ______________________

RESOURCES

NATIONAL ORGANIZATIONS FOR BIRD WATCHERS

American Bird Conservancy
PO Box 249
The Plains, VA 20198
540-253-5780
www.abcbirds.org

American Birding Association
4945 N. 30th Street
Suite 200
Colorado Springs, CO 80919
800-850-2473
www.americanbirding.org

Cornell Laboratory of Ornithology
159 Sapsucker Woods Road
Ithaca, NY 14850
800-843-2473
www.birds.cornell.edu

National Audubon Society
700 Broadway
New York, NY 10003
212-979-3000
www.audubon.org

National Wildlife Federation
11100 Wildlife Center Drive
Reston, VA 20190
800-822-9919
www.nwf.org

The Nature Conservancy
4245 N. Fairfax Drive
Suite 100
Arlington, VA 22203
800-628-6860
www.tnc.org

FIELD GUIDES TO BIRDS

Brinkley, Edward S. *National Wildlife Federation Field Guide to Birds of North America.* New York: Sterling Publishing, 2007.

Crossley, Richard. *The Crossley ID Guide.* Princeton, NJ: Princeton University Press, 2011.

Dunn, Jon L., and Jonathan Alderfer. *National Geographic Field Guide to the Birds of North America.* 5th ed. Washington, D.C.: National Geographic, 2006.

Floyd, Ted. *Smithsonian Field Guide to the Birds of North America.* New York: HarperCollins, 2008.

Griggs, Jack L. *All the Birds of North America*. New York: HarperCollins, 1997.

Kaufman, Kenn. *Kaufman Field Guide to Birds of North America*. Boston: Houghton Mifflin Co., 2000.

Peterson, Roger Tory. *Peterson Field Guide to Birds of North America*. Boston: Houghton Mifflin Co., 2008.

Robbins, Chandler S., et al. *Birds of North America: A Guide to Field Identification*. Revised ed. New York: Golden Press, 1983.

Sibley, David Allen. *The Sibley Guide to Birds*. New York: Knopf, 2000.

Sterry, Paul, and Brian E. Small. *Birds of Eastern North America: A Photographic Guide/Birds of Western North America: A Photographic Guide*. Princeton: Princeton University Press, 2009.

Stokes, Donald, and Lillian Stokes. *Stokes Field Guide to Birds: Eastern Region*. Boston: Little, Brown and Co., 1996.

Thompson, Bill, III. *The Young Birder's Guide to Birds of Eastern North America*. Boston: Houghton Mifflin Harcourt, 2008.

Vuilleumier, Francois, ed. American Museum of Natural History. *Birds of North America, Eastern Region/Western Region*. New York: Dorling Kindersley, 2009.

AUDIO GUIDES TO BIRDS

Elliott, Lang. *Know Your Bird Sounds*. Vols. 1 and 2. Mechanicsburg, PA: Stackpole Books, 2004.

Elliott, Lang, and Donald and Lillian Stokes. *Stokes Field Guide to Bird Songs: Eastern Region*. New York: Time Warner AudioBooks, 1997.

Peterson, Roger Tory, ed. *Peterson Field Guide to Bird Songs: Eastern and Central*. Revised ed. Boston: Houghton Mifflin Co., 2002.

Walton, Richard K., and Robert W. Lawson. *Birding by Ear: Eastern/Central*. Boston: Houghton Mifflin Co., 1989.

——. *More Birding by Ear*. Boston: Houghton Mifflin Co., 1994.

FIELD GUIDE AND BIRD SONG APPS FOR BIRD WATCHERS

Audubon Birds: A Field Guide to North American Birds (Green Mountain Digital). www.audubonguides.com.

BirdJam HeadsUp Warblers/HeadsUp Sparrows (MightyJams LLC). www.birdjam.com.

BirdTunes (BirdTunes). www.birdtunesapp.com.

iBird Explorer (Mitch Waite Group). www.ibird.com.

National Geographic's Handheld Birds (National Geographic Society). www.nationalgeographic.com.

Peterson Birds of North America (Appweavers, Inc.). www.petersonguides.com.

Sibley eGuide to Birds of North America (Cool Ideas LLC). www.sibleyguides.com.

PERIODICALS FOR BIRD WATCHERS

The Backyard Bird Newsletter, PO Box 110, Marietta, OH 45750 800-879-2473 www.birdwatchersdigest.com

Bird Watcher's Digest, PO Box 110, Marietta, OH 45750 800-879-2473 www.birdwatchersdigest.com

Birder's World, Kalmbach Publishing Co., 21027 Crossroads Circle, PO Box 1612, Waukesha, WI 53187 800-533-6644

Living Bird, Cornell Laboratory of Ornithology, 159 Sapsucker Woods Road, Ithaca, NY 14850 800-843-2473 www.birds.cornell.edu

WildBird, PO Box 57900, Los Angeles, CA 90057 213-385-2222; (fax) 213-385-8565

GLOSSARY

Bib: the area below a bird's bill covering the throat and upper breast. Refers to same area covered by a bib worn by a human.

Binocs: abbreviation used by birders for binoculars.

Breeding plumage: the set of feathers worn during the breeding season (spring and summer). This is often the most colorful plumage, especially among male songbirds. See *nonbreeding plumage.*

Brood parasitism: a behavioral habit characterized by birds laying their eggs in the nests of other birds.

Cache: a place where food is stored or hidden for later consumption.

Call note: brief, relatively simple sound uttered by birds in various social contexts (for example, location calls, food calls).

Cavity nester: a bird that nests inside an enclosed area, such as a hollow tree, an old woodpecker hole, or a nest box.

Central Flyway: the migration route used by birds through the central portion of North America from central Canada, across the Great Plains, and southward to the Gulf of Mexico.

Checklist: a list of bird species compiled from records in a specific geographic area.

Coverts: the small contour feathers on the upper part of a bird's wing that overlap the flight feathers.

Crest: a tuft of long feathers on a bird's head that may be held erect.

Crissum: the feathers covering the undertail of a bird, as in the rust-colored crissum of the Gray Catbird.

Crown: the top of a bird's head. (See "Parts of a Bird," page 16.)

Dabbling: refers to certain duck species (such as the Mallard) that forage in shallow water, sometimes tipping forward to reach underwater food.

Decurved: curved downward (refers to the bill).

Dihedral: describes the upward-angled or V-shaped position (rather than flat or horizontal) in which certain birds hold their wings.

Diurnal: active during daylight hours.

Eclipse/Eclipse plumage: a brief period in late summer when waterfowl are molting from breeding plumage to nonbreeding plumage. Some ducks in eclipse plumage are very drab looking and cannot fly until their new feathers grow in.

Edge habitat: a place where two or more habitats overlap, such as an old meadow near woodlands. Edge habitat typically offers a rich diversity of birds.

Extinct: no longer existing.

Extirpated: no longer present in a given area (though still existing in others).

Eye line: the line over or through a bird's eye, often used as a field mark for identification. (See "Parts of a Bird," page 16.)

Eye patch: an area of feather (usually dark) surrounding a bird's eyes.

Eye-ring: a ring of color that encircles a bird's eye. A broken eye-ring is one that is not continuous, or does not completely encircle the eye.

Field mark: an obvious visual clue to a bird's identification, such as bill shape or plumage.

Flank: a bird's sides below the wings on either side of the belly. (See "Parts of a Bird," page 16.)

Fledgling: a young bird that has left the nest but may still be receiving care and feeding from a parent.

Flight call: a short, often distinctive call given by birds in flight.

Flock: a gathering of birds for purposes of feeding, resting, nesting, or migration. Winter feeding flocks of small woodland songbirds often contain several different species.

Forage: to look for food. Where and how birds forage can offer clues to their identities.

Habitat: the area or environment where a bird lives. Certain birds prefer specific types of habitat.

Hotspot: a location or habitat that is particularly good for bird watching on a regular basis.

Hybrid: the offspring produced from the mating of a male and a female from two distinct bird species.

Juvenal: plumage of a juvenile bird.

Juvenile: a bird that has not yet reached breeding age.

Life bird: a bird seen by a bird watcher for the first time, often recorded on a life list.

Life list: a record of all the birds a birder has seen at least once.

Local: refers to a species' abundance. A locally common species is one that is present in its appropriate habitat but not widespread and abundant.

Lores: the area between a bird's bill and its eyes. (See "Parts of a Bird," page 16.)

Malar: the area on the side of a bird's face below the bill and eye, sometimes referred to as the cheek.

Malar stripes: stripes in the malar or cheek area, often referred to as the mustache.

Mandible: the lower half of a bird's bill. The upper half is known as the maxilla.

Mantle: the upper back just behind the nape. (See "Parts of a Bird," page 16.)

Migrant: a bird that travels from one region to another in response to changes of season, breeding cycles, food availability, or extreme weather.

Mimic: a bird that imitates other birds' sounds and songs.

Molt: the periodic shedding of old feathers and their replacement by new ones.

Morph: a genetically fixed color variation within a species, such as the blue morph of Snow Goose. (The term is correct only when both color variations occur in the same population. "Blue" morph Snow Geese breed side by side with white birds. The word *morph* is not applied to differently colored subspecies.)

Nape: the back of a bird's neck.

Neotropical migrant: refers to migratory birds of the New World, primarily those that travel seasonally between North, Central, and South America.

Nest parasite: a bird that lays its eggs in the nest of another bird or species, forcing that nesting bird or pair to raise the parasite's young. The most common North American example of a nest parasite is the Brown-headed Cowbird.

Nestling: a bird that has hatched from its egg but is still being cared for in the nest.

Nonbreeding plumage: the set of feathers worn during the fall and winter months. Many songbirds molt from breeding plumage into nonbreeding plumage in the fall. Sometimes called winter plumage or alternate plumage.

Nonmigratory: a bird that does not migrate with the change of seasons. Sometimes referred to as a resident bird.

Peeps: a generic term for confusingly similar small sandpipers.

Pelagic: birds of the ocean, rarely seen from land.

Pishing (or **spishing**): a sound made by bird watchers to attract curious birds into the open, made by repeating the sounds *spshhh* or *pshhh* through clenched teeth.

Plumage: collective reference to a bird's feathers, which change color and shape during seasonal molts. Breeding plumage is often more colorful than nonbreeding or winter plumage, worn during fall and winter.

Plumes: long showy feathers that are part of the high-breeding plumage of many herons and egrets. These feathers were once used to decorate women's hats. The collecting of these feathers decimated wading-bird populations.

Primary feathers (or **primaries**): the nine or more long flight feathers at the end of a bird's wing. (See "Parts of a Bird," page 16.)

Range: the area in which a bird may be seen during each season of the year. Breeding range refers to the area the species occupies during the breeding season; wintering range refers to the area occupied during the winter.

Raptors: birds of prey (hawks, eagles, owls).

Recurved: curved upward (refers to the bill).

Resident: a nonmigratory species that is present in the same region all year.

Rump patch: a patch of color located above the point at which a bird's tail connects to the body. (See "Parts of a Bird," page 16.)

Scapulars: the row of feathers lying just above a bird's folded wing; the lowest group of feathers on the mantle. (See "Parts of a Bird," page 16.)

Secondary feathers (or **secondaries**): the medium-length flight feathers located on the wing between the primaries and tertials. (See "Parts of a Bird," page 16.)

Shorebirds: refers to sandpipers, plovers, and related birds (but does not include herons, gulls, terns, and other birds found in coastal areas).

Skulker: a bird that does not make itself obvious, but keeps hidden in deep cover. Many sparrow species are referred to as skulkers.

Soaring: a flight style in which a bird holds its wings steady and flies without flapping. Red-tailed Hawks are experts at soaring.

Song: a complex series of sounds, with elaborate note patterns, usually associated with courtship or territoriality.

Song-flight: sometimes called flight songs. Performed by birds (usually males) during courtship, when they sing while flying high about their territories. Many grassland nesters perform song-flights, but some woodland species do too, such as the American Woodcock and a variety of warblers.

Speculum: the patch of inner secondary feathers on the wings of waterfowl. Mallards have a bright blue speculum.

Subadult: birds that are not yet adults but are more than one year old.

Supercilium: the area above a bird's eye, sometimes called the eyebrow. (See "Parts of a Bird," page 16.)

Tail spots: spots of contrasting color (usually white) on a bird's tail, often used as a field mark.

Territoriality: behavior associated with the aggressive defense of a particular area or territory.

Territory: the piece of habitat a bird claims for its own and defends against others of its species. Birds may be most territorial during the breeding season, but some birds (such as the Northern Mockingbird) will also defend food-rich winter territories.

Tertial feathers (or **tertials**): the innermost flight feathers on a bird's wing (the closest feathers to the bird's body), which form a stack atop the rear border of the folded wing. (See "Parts of a Bird," page 16.)

Underparts: the lower half of a bird (breast, belly, undertail), often used as a field mark.

Underwing: the bottom side of a wing.

Upperparts: the upper half of a bird (crown, back, top of tail), often used as a field mark.

Upperwing: the top side of a wing.

Vagrant: a bird that wanders far from its normal range.

Vent: the feathered area under the tail and below the legs, sometimes used as a field mark. (See "Parts of a Bird," page 16.)

Waders: herons, egrets, and related birds, including storks.

Wing bars: obvious areas of contrasting color, usually white, across the central portion ("shoulder") of a bird's wings. (See "Parts of a Bird," page 16.)

Wing lining: the inner portion of a bird's underwing.

Wingpit: the area on the underside of the wing, where it connects to the body.

Wintering grounds: the range over which a bird species spends the winter.

ACKNOWLEDGMENTS

The idea for this book came out of the field trips our daughter Phoebe's class takes to our southeastern Ohio farm each year. The fascination all kids have with birds is almost limitless, and though I know most of these kids will not become avid bird watchers, a few of them might, given the right encouragement, information, and tools.

This book is for those young people between the ages of 8 and 12 who are already interested enough in birds to want to know more. I was one of those kids many years ago, and I owe a debt of gratitude to the adults who helped me along the path to becoming a bird watcher. In my younger days, Pat Murphy was a birding mentor to the Thompson family. My mom, Elsa, and much later, my dad, Bill Jr., and my brother, Andy, encouraged my interest in birds. Mom got us involved in the forays of the Brooks Bird Club in West Virginia, and today I owe much of my knowledge of birds to those kind and generous BBC members. After my parents started *Bird Watcher's Digest*, I realized that I could *watch birds for a living!* This was quite a revelation, and I really never considered another career path.

This book is made immeasurably better by the beautiful illustrations by Julie Zickefoose. Julie also helped set me straight on a lot of natural history information, and she crafted many of the clever captions in the species profiles. I could not have completed this project without you, Jules.

In expanding the text of the original *Young Birder's Guide* to include the 100 additional western species, I was fortunate to have the illustration talents of Michael DiGiorgio, one of North America's best bird artists (and an excellent banjo picker). Mike's black-and-white illustrations accompany each of the new species accounts.

Many wonderful photographers contributed their images to this book, an essential element for a birding guide since this activity is so visual.

Countless people have had a hand in molding me as a writer, editor, birder, and person. Naming everyone here is impossible, but I'd like to highlight just a few (in no particular order): Laura Thompson Fulton, Mary Beacom Bowers, Eirik A. T. Blom, Jeffrey A. Gordon, Steve McCarthy, Shila Wilson, Chuck

Bernstein, the Whipple Bird Club, and all of my friends in the Ohio Ornithological Society.

The faculty and staff of Salem-Liberty Elementary School permitted me to work with several classes of students on this book's development. I'd especially like to thank Phoebe's fourth-, fifth-, and sixth-grade classes at Salem-Liberty (by the time this book is published, Phoebe and her classmates will be nearly finished with tenth grade!): Mackayla Erb, Gregory Hill, Austin Klintworth, Drew Layton, Josh Lent, Hannah Lewis, Rikki Lockhart, Kristen Long, Brady Lowe, Alison Miller, Dana Moss, Tyson Niceswanger, Shannon O'Dell, Charity Roberts, Chelsey Schott, Amanda Seevers, Kaylee Sparks, Tara Thomas, Phoebe Thompson, Abbey Tornes, Dakota West, Paul Westfall, Jessica White, and their teachers: Mrs. Huck, Mrs. Baker, and Mrs. Biehl.

In Liam's fifth-grade class, my helpers for the update of the *Young Birder's Guide* included teachers Kelly Hendrix and Erica Schneider and students Jeffrey Bigley, Madison Binegar, Trae Dalton, Destiny Frye, Owne Gage, Hanna Grady, Nicholas Grosklos, Derik Hesson, Andrew Morganstern, Jesse Sloan, Kiersten Taylor, Morgan Wears, Kaylie Wittekind, Gage Treadway, Brenton Carpenter, Airetta Eales, Chiana Eddy, Barry Erb, Brooke Haas, Katelyn Jenks, Tyson Miller, Courtney Roberts, Tavion Rummer, Cody Sparks, Liam Thompson, Megan Tornes, Logan Wears, MacKenzie White, and Josh Zwick.

I'd like to thank the good folks at Houghton Mifflin Harcourt. My gratitude goes to Shelley Berg, Mimi Assad, Anne Chalmers, Beth Fuller, Nancy Grant, Katrina Kruse, Brian Moore, Tim Mudie, Taryn Roeder, Patrice Taddonio, Clare O'Keeffe, and a heaping helping of thanks to my (world-class) editor, Lisa A. White, for her guidance and indulgence. My agent, Russell Galen, kept me firmly grounded in the early stages of this project, which he believed in from the very start. While I was writing this book, my colleagues at *Bird Watcher's Digest* kept things running, especially (Eastern guide): Debbie Griffith, Amy Wells, Ashley Bills, Ann Kerenyi, Linda Brejwo, Helen Neuberger, Susan Hill, Katherine Koch, Claire Mullen, and Josh Schlicher. And *BWD*ers of more recent vintage: Jim Cirigliano, Kimi Carroll, Chris Blondel, Jessica Moyer, and Susie Stacy.

And last of all, thanks to you for reading this book and all of these acknowledgments. Now go outside and find some birds!

INDEX

Dea

"Jim Thompson mee... supernatural riff on the ... collections racket. This gripping supernatural adventure gives a whole new meaning to 'possession is nine-tenths of the law.'"

Ed...

Ch...

"Chri... watch... riveti... mana... one b...

An...

Hil...

"Get a... of *De...* meets... eleme... create...

Fra...

"With ... few au... action-... thriller... talent ...

Stua...

Awa...

"Sam S... finely ...

Mike...

"Holm's touch is deft and his language surefooted, a rare feat in the realm of dark fantasy. *Dead Harvest* blends back-story just wrenching enough, victims just pitiable enough, villains just ambiguous enough to keep everything on the thrummingly interesting side of noir, never bogging down in cliché. And the white-knuckle action is an indulgent pleasure – this book practically turns its own pages. The best books combine the smart with the careening, and Holm does that so well."

Sophie Littlefield, award-winning author of Aftertime

"The fight between heaven and hell takes a turn for the hardboiled in Chris F. Holm's fantastic debut novel, *Dead Harvest,* where he's created a character as pulpy and tough as anything Chandler or Hammett dreamed up in his doomed Soul Collector. Holm's writing is sharp, powerful, and packs a wallop."

Stephen Blackmoore, author of City of the Lost

CHRIS F. HOLM

Dead Harvest

The Collector Book One

ANGRY ROBOT
A member of the Osprey Group

Lace Market House,
54-56 High Pavement,
Nottingham,
NG1 1HW, UK

www.angryrobotbooks.com
Collectorville

An Angry Robot paperback original 2012

Cover art by Amazing 15

Distributed in the United States by Random House, Inc., New York.

ISBN 978-0-85766-218-7
eBOOK ISBN 978-0-85766-219-4

Printed in the United States of America

9 8 7 6 5 4 3 2 1

For Katrina. For always.

There is no greater sorrow
Than to be mindful of the happy time
In misery.

DANTE ALIGHIERI

1.

Light spilled through the window of the pub as I watched them, casting patches of yellow across the darkened street but conveying no warmth. It had been three rounds now, maybe four, and Gardner had yet to pay for a drink; his reading tonight went well, and they were falling over themselves to share a pint with Britain's Greatest Living Author.

I fished another Dunhill from the pack, lighting it with the dwindling ember of the one that preceded it. The ground around me was littered with cigarette butts – I'd been standing there a while. But the moon was high overhead, and the night was getting on. I wouldn't have to wait much longer.

Finally, midnight rolled around, and the last straggling patrons were ushered out into the chill spring air, the barkeep locking up behind them. Gardner headed up St Giles, listing slightly. I took a last long drag off my cigarette, and then pitched it into the street, falling in behind him. I kept some distance between us, in case he looked back.

He didn't.

A few blocks later, he ducked into an alley to take a leak. I gave him a minute, and then followed. He was leaning one-handed against a wall, pissing behind a dumpster. The toast of Oxford, or so I'd been told. From here, it was hard to see.

He turned toward me, zipping up his fly. When he spotted me, he started, and damn near tipped over. "Who the bloody hell are you?" he asked. "What are you doing here?"

I stepped toward him. My hand found his chest and reached inside. He knew then. Who I was. What I was doing there.

"Sorry," I told him. "It's nothing personal."

I yanked it free then; that light, that life. Gray-black and swirling, it cast long shadows across the alley, and its song rang bittersweet in my ears. Of course, if anyone had happened by, they'd have seen nothing, heard nothing. No, this show was just for me. For Gardner, too, perhaps, though even then I couldn't be sure.

Gardner's body crumpled to the ground, whimpering as it hit the pavement. I paid it no mind. It was already dead, or near enough. Sometimes it takes a minute for the meat to get the message.

I removed from my pocket a bit of worn cloth and a small length of twine, wrapping my prize in the former and binding it tight with the latter. The whole package was scarcely larger than an acorn. I slipped it into my inside coat pocket and then set off down the street, whistling quietly to myself as I disappeared into the night.

2.

Sorry – it's nothing personal.

I wish I could tell you I have no idea how many times I've uttered that phrase. That I have no idea how many bodies I've left crumpled and inanimate in my wake. I wish I could tell you that, but I can't.

The truth is, there've been thousands. Some, like Gardner, are so damn surprised, they never even see it coming. Some spend their lives in fear of the moment, and catch my scent a mile away; they beg, they plead, they scream. In the end, it doesn't matter – I always get what I came for. And I remember each and every one of them. Every face. Every name.

I collect souls. The souls of the damned, to be precise. Not the most rewarding gig, I'll admit, but I didn't choose it – it chose me. Once upon a time, I was a man named Sam Thornton. I paid my taxes. I went to church. I didn't litter. I was a model fucking citizen, and then it all went to shit. That business with Gardner? Sixty-odd years ago that was me, and believe me, my collection was nowhere near as pretty.

The River Cherwell glimmered in the morning sun as I strolled along its bank, the path before me empty but for the occasional enterprising Oxford student out for a pre-class jog. By noon the place would be packed with folks eager to exorcise the demon winter – couples strolling hand in hand through gardens rife with fresh buds, tourists poling rented punts up and down the river – all manner of lively good cheer I'd just as well avoid. Now, though, I'd done my deed, burying Gardner's soul deep beneath a patch of dog's-tooth violet still weeks from flowering, and I thought that for a moment, at least, I could wander in peace. I should have known better. That's the bitch about being damned – things rarely shake out your way.

"Collector!"

Her call came from behind me, carried like a song on the breeze. "Morning, Lily," I said, turning. She was a few paces back on the path, her red hair cascading down over a whisper of a summer dress, her bare feet leaving no prints on the dirt path as she approached. "Aren't you up a little early?"

"When I rise is no concern of yours, Collector. And I've asked you not to call me that."

"Right," I replied. "Must've slipped my mind."

She cast an appraising glance my way, the faintest of smiles playing across her face, and despite myself, I flushed. "You look like shit," she said. "Why you persist in eschewing the living in favor of these rotting meatsuits, I'll never know."

"The living give me a headache."

"That *is* what they do best."

"This a social call?" I asked, shaking a Dunhill from the pack and striking a match.

"Hardly. Are you going to offer me one of those?"

"No," I replied, taking a long drag and slowly exhaling. "So who's the job?"

"Her name is Kate MacNeil."

"Contract or freelance?"

"She struck no bargain. Her actions are to blame."

"What'd she do?"

"As I understand it, she slaughtered her family."

"Christ," I said, noting her disdainful glare. "Where is she now?"

"Manhattan," she said. "I trust that's not a problem?"

"It's a place like any other," I replied.

"Of course it is. But as you well know, failure is not an option. I simply thought that, given your history there..."

"I'll get the job done."

"Yes," she said, "I expect you will. You should know that there's a timeline on this. It seems she's caught the eye of some rather influential... people. I wouldn't dally."

"I never do."

"No," she said. "You never do." She caressed my cheek, a teasing gesture, and then strolled northward past me up the footpath. A warm breeze kicked up from the south, and her sundress clung to her beautiful frame.

"Oh, and Collector?" she called, glancing backward.

"Yes?"

"Do try to enjoy yourself, won't you?"

And suddenly she was gone, replaced by a teeming swarm of butterflies, left to scatter on the warm southern wind.

A few streets from the river garden, I found myself a news stand. I managed to buy a copy of the *New York Times*, and tipped the guy a twenty pound note – after all, I wasn't going to need it. Under the shade of a massive oak, I lit another Dunhill, savoring the richness of the tobacco. I'll tell you, their food might be for shit, but the Brits sure as hell know how to make a cigarette. My pack was still half full, and I didn't relish the thought of leaving it behind, but I had a job to do.

MacNeil, it turns out, made the front page. My guess is it was the Park Avenue address as much as the three dead bodies that landed her there. Hell, a couple blocks further north, she might have been above the fold. I skimmed the article. Seems some neighbors heard screaming and called the police. By the time they arrived, Kate's brother and father were dead; the cops got there just in time to watch her slit her mother's throat. Took six officers to bring her down, and by the time they did, she was unconscious. Now they had her under guard at Bellevue Hospital, at least until she woke up. With a little luck, I thought, I could have this wrapped up before she even does.

I tossed aside the A-section and flipped through the

paper until I found the obits. Papers like the *Times*, the obituaries are always a crapshoot. More days than not, they're stuffed front to back with octogenarians of some historical import – a touching gesture for friends and family, I'm sure, but it doesn't do *me* a load of good. Today, though, I was in luck. A playwright, thirty-five. Pills and booze, an apparent suicide. Not half-bad looking, either. It didn't get much better than that.

I closed my eyes and focused. My limbs grew heavy and ungainly as I pulled away. The jasmine scent of spring retreated, replaced by hollow nothingness.

Somewhere behind me, a body convulsed, thrashing about on the grass as a thousand synapses misfired. Then the world lurched, and it was gone.

The first thing I noticed was the smell – a harsh ammonia reek that burned my sinuses and caught in the back of my throat, making me gag. My stomach clenched and I doubled over, or tried to. My head clanged against something just a couple feet above, a muffled thud. I pressed against the liquid darkness. Cold vinyl pressed back, slick and unpleasant. Clumsy fingers fumbled in the darkness as I followed the line of the zipper. It ended just overhead. I forced a finger through, metal teeth digging flesh, and then pushed the zipper open.

I kicked free of the body bag. The chill of the morgue drawer stung my naked skin. My heart raced – the useless panic response of a fledgling meat-suit. I took a deep breath and closed my eyes, and the flutter slowed.

Hands against the back wall, I pushed, and the morgue drawer slid open. The room beyond was dimly lit, but after the absolute black of the drawer, I squinted still. I stumbled to a large utility sink, clumsy as a newborn foal. Bile rose in my throat, and I retched. It happens every time. A reflex, I suppose – just the body's way of trying to get rid of me. I try not to take it personally.

The water ran cool from the tap. I drank from cupped hands. Whiskey and pills and sick swirled toward the drain. The water calmed my stomach, and the act of drinking was an anchor, fixing me in place. The body no longer fought my movements, no longer coursed with fear. I stretched my limbs, testing each in turn. Not a bad fit, really. Possession can be a tricky thing, particularly with the dead. You've got to find one in decent working order, for one – if you don't get to them quick enough, they tend to run a little rough. And they've all got their quirks. The guy I left in Oxford, for example: bum hip, lousy stomach, and apparently scared shitless of bugs. You get something *that* ingrained, there's no stopping it, and considering he'd been on his floor a couple days before I found him, I was lucky he didn't have a fucking heart attack. I had to shower for an hour before his skin stopped crawling. Still, it beats taking the living – no thoughts, no memories, no baggage. Their constant yammering is enough to make you want to take a header off a bridge and bail on the way down.

I glanced at my wrist, a useless gesture. The watch

I was looking for was a continent away, adorning the arm of a corpse. I looked around. The clock on the wall read 5am. If I were a betting man, I'd have said this place'd be deserted for least another hour. Of course, I'm *not* a betting man – in my line of work, I've seen my share of wagers, and believe me, the house always wins. Still, it couldn't hurt to poke around a bit.

I peered through the gloom at the bank of morgue drawers behind me. They gleamed faintly in the pale glow of the exit signs. Numbers, no names. I padded naked toward the door at the far end of the room. Beside the door hung a clipboard – a list of names, arranged by drawer. Three of them MacNeil.

I crossed the room and slid one out. As I unzipped the bag, the copper tang of blood prickled in my sinuses, and I went a little woozy. Great – New Guy had a thing about blood. *That* was gonna be a treat.

He was a boy of maybe twelve, with straw-colored hair and a smattering of freckles across his face. His feet were bare, his pajamas in tatters, and there was so much fucking blood, it was impossible to tell what color they were. His hands were nicked and scraped, his face mostly spared – it was clear he'd tried to protest, to protect himself. His chest was a tattered mess – bone protruding, soft tissue visible beneath. I zipped him up and slid him back.

The father was a mess as well. Well over six feet and not a slight man at that, he looked as though he'd been tossed about like a rag doll. He had at least a dozen fractures that I could see, arms kinked at improbable angles, legs a twisted wreck. His chest, too, was riddled with holes – knife wounds, like the boy – some flecked with chips of bone from the force of entry.

The mother, though – she was something else entirely. With her chestnut hair and her elegant features, she was beautiful once, no doubt, but now her body was a maze of tiny cuts – thousands of them, each no longer than an inch, marking her skin like some unholy etching. And there was something else, too. A familiar scent, mingled with the metallic bite of blood.

Alcohol.

Jesus – these cuts, they weren't intended to kill. They were meant to hurt like hell. To make this woman scream. I wondered how long it took before the neighbors took notice and called the cops. From the look of Kate's mom here, it could have been hours. And Kate just kept on cutting, waiting patiently for her audience to arrive before she slit her mother's throat.

I was suddenly glad Kate MacNeil would be cuffed and unconscious when I came to collect her. She wasn't to be trifled with, it seemed, and borrowed body or not, the pain's the same.

I slid the drawer closed, eyeing the gooseflesh on my arms as I did. It *was* cold in here, I realized, noting the tension in my muscles, the ache in my joints. I left the autopsy suite, snatching a lab coat from a line of hooks in the anteroom beyond. Pressing through a set of swinging double doors, I found myself in a hallway ablaze in fluorescent light. The hall was empty, its walls scarred with the scuff-marks of countless carelessly piloted stretchers, and I crept quietly down it, mindful of the doors on either side.

At the end of the hall was a locker room. A set of utility shelves stood along one wall, stacked high with clean scrubs, all neatly folded and arranged according to size. I took a set and slid them on, admiring myself in the mirror. I was a little pale, a little thin, but already my face showed signs of color, and for a dead guy, I cut a dashing figure in the powder blue scrubs. You could hardly even call it theft – in a couple hours I'd leave this body behind, and both it and the clothes I'd pilfered would wind up right back here.

I put the lab coat back on and headed for the door. An elderly woman pushed a mop bucket past me in the hall, but she paid me no mind. Between the lab coat and the few days' stubble that graced my cheeks, I looked like I'd just pulled a double shift.

I pushed through a set of glass doors and stepped out into the pre-dawn half-light. It was cold – bitterly so – as though the first kiss of spring I'd felt in Oxford was still some weeks away from warming the dead gray of New York's steel and concrete. From where I stood, First Avenue was pretty quiet – just the odd commuter among a dozen or so delivery trucks rumbling northward from the East Village. Bellevue lay a few blocks to the south. I pulled my lab coat tight around me and set off walking, my bare feet aching as the chill of the sidewalk leeched upward through my soles.

3.

It'd been sixty-five years since I last laid eyes on Bellevue. Sixty-five years, four months, and seventeen days. Since then, it had changed plenty, with its modern glass atrium jutting skyward and glinting in the morning sun, I almost didn't recognize it. But the cold, impassive stone face I remembered all too well stared outward from behind the glass, and my own new face twisted into a smile of grim remembrance. Try though we might, we never can quite deny who we once were.

The hospital itself was a massive structure, occupying twenty-five floors and two city blocks. In the nearly three centuries of its existence, its halls had spread and shifted and wound among themselves like vines on a trellis. The result was a tangled labyrinth of wrong turns and dead-end corridors, peppered with the occasional brightly colored map in what I can only assume was a fit of architectural sarcasm.

Of course, it would help if I knew what I was looking for; all I had to go on was what I read in the paper.

Killing spree, coma – the girl could be anywhere. Prison and psych wards make for tricky collections – they've got armed security, locked rooms, the whole nine – but in most hospitals, they're also overflowing. My hope was they were keeping Kate somewhere a little less secure. I played the odds and headed for the ICU.

As the elevator doors opened, I knew I'd struck pay dirt. The ICU was a sleek, modern affair, all glass and light – the better to see you with, my dear. A few rooms in, a uniformed cop sat slouched beside an open door, his nose buried in a Scudder novel. I strode past him down the hall. He didn't spare me a second glance. Through the glass-paned walls, I caught a glimpse of the room's sole occupant. She looked so tiny and so frail as she lay still in her bed, surrounded by the blip and whir of medical equipment. But her wrist was cuffed to the bedrail, and her hair was flecked with blood – no doubt about it, she was my mark.

I continued without pause down the hall, flashing the nurse at the station a smile as I passed. She flushed and returned the favor. In my line of work, I don't get looked at that way often. Almost a shame the assignment was so easy; it'd be a waste to ditch this skin-suit so soon.

As I neared the end of the hall, I glanced back toward the nurses' station. The nurse was clacking away at a computer terminal, her back to me. I ducked into the nearest room. In the bed was an elderly gentleman – his eyes closed, his pallor gray. A tube snaked from his mouth to a machine beside the bed that accordioned up and down, pumping breath into his lungs.

I approached the bed, my bare feet silent on the tiled floor. The only sounds in the room were the blip of his heart monitor and the grim, mechanical hiss of the respirator. I took the man's hand in mine. It was cold and dry. At the end of one finger was a small white clip, a wire running from it to the tangle of machinery beside him.

I grabbed the wire and yanked free the clip, letting it fall to the floor as I strode out of the room. A shrill monotone pierced the air as the heart monitor flatlined. Alarms sounded at the nurses' station, and I was buffeted by medical personnel as they rushed past me down the hall.

As diversions go, they don't get any easier than that. Time was I'd have had to almost kill the guy to get that kind of rise out of everybody. Now all you have to do is unhook a wire. I only hoped Kate's guard would be as easily distracted.

I snatched a chart at random from the nurses' station and set out for Kate's room. I strode with purpose toward the door, thinking doctorly thoughts. You'd be surprised how often that sort of thing works.

This time, no such luck. The cop stood as I approached, sidestepping in front of me as I tried to shoulder past.

"Where the hell you think you're going?"

"I'm here to see the patient," I replied, brandishing the chart by way of evidence.

He scowled. "You ain't her usual doctor."

"I'm from Neurology. They called me in for a consult."

The cop looked me up and down, eyes lingering on my bare feet. His hand crept toward the gun on his belt. "I'm gonna need to see some ID."

I lunged forward, slamming him against the door-jamb. His hand found the gun. Steel scraped leather as it slid free of its holster. I pressed my hand to his chest and reached inside. His eyes went wide as I clenched tight his soul.

"David," I said. "She knows. She knows what you did to him." Somewhere, an eternity from the swirling blackness where we stood, a gun clattered to the floor.

I withdrew my grasp, and David crumpled. He was shaking, whimpering. Tears streaked his pallid face.

"No," David whispered.

"You know it's true, or *I* would not."

"No," he repeated. "No no no no no!" He scurried backward along the wall, his gun and his assignment forgotten. He scrambled to his feet and took off at a dead run, not looking back.

Inside, the room was quiet. Just the steady blip of the heart monitor, the soft tap of the IV drip, and the gentle sigh of Kate's breathing. She was younger than I expected – she couldn't have been more than seventeen. And Kate was beautiful. Her auburn hair spilled across the pillow like a thousand adolescent fantasies, and though her eyes fluttered in dream, her face carried no hint of worry or concern.

I'd gladly give a limb to have dreams like hers, I thought. But then, these limbs weren't mine to give.

I approached the bed, caressing her cheek a moment before resting my hand on her breastbone. "Sorry," I said. "It's nothing personal."

I reached inside. My head was suddenly filled with light – blinding, beautiful. I clenched shut my eyes against it, but it wasn't any use. Still it streamed in, the purest white. Not devoid of color – full of it. And with it her song. So beautiful. So sweet. I staggered backward, blind and helpless. My hand pulled free, and the light and song were gone. I collapsed to the floor, tears streaming down my borrowed cheeks, whether from the beauty of what I'd seen or the sudden horrible absence of it, I didn't know.

I looked around. I'd so lost myself in that light, that sound, I was unsure of where I was. The lines and angles of the hospital room seemed suddenly harsher, somehow. Colder. My heart thudded in my chest. I climbed trembling to my feet, my body drenched in a cold, acrid sweat. I knew that scent. I'd smelled it a thousand times in the moment before I tore soul from flesh.

It was fear.

It was fear, and it was mine.

I approached the bed again. With shaking hands, I reached toward her. I hesitated, my fingers scant inches from her breastbone. I wasn't sure if I could do it. I knew I couldn't not. I closed my eyes, steeling myself for what was to come.

That's when she started screaming.

My eyes flew open. Kate was staring back at me, her eyes wide with fear. She thrashed against her restraints, cuffed wrists clanging violently against the bedrail. Her screams echoed through the tiny room, blotting out all thought.

"My God, is she all right?"

A nurse, in the doorway. I forced myself to focus. "She's seizing!" I replied. "Give her something to calm her down."

The nurse hurried to Kate's bedside, snatching a needle from the cart beside the bed. "Pushing four of Ativan." Kate's thrashing slowed, and her cries died down to little more than a whimper. Her eyes met mine. Terrified, pleading. Then the spark within them guttered and died, and her lids came crashing down. Kate MacNeil was once again asleep.

I, unfortunately, had no such luxury. My mind was reeling. Adrenaline coursed through my veins, urging me to flee. I knew I had a job to do. But that light, that song – in all my years, I'd never seen anything like that. Something wasn't right here.

"Her wrists," I said, embarrassed by the sudden quaver in my voice. "They've been abraded by the cuffs. I'd like to have a look at them. Do you have a key?"

"I'm not supposed to unlock her," the nurse replied.

I nodded toward Kate's sleeping form. "You think she's going anywhere?"

She hesitated a moment, and then fished a small set of keys from her pocket, unlocking first one set of cuffs, and then the other. "The police should really be here when she wakes up," the nurse said. "They're going to want to talk to her."

I nodded my agreement. "The officer that was stationed at her door was asleep when I arrived. When I woke him, he said he was gonna head down to the cafeteria for a cup of coffee. If I had to guess, I'd say they're not going to be too happy with him if they find out he was gone when she came to. You could catch him if you hurry. Tell him to call in, let them know that she's been stirring."

"And her?"

"I'll keep an eye on her until you get back."

She gave me a curt nod, and took off at a jog. A good kid, I thought – the trusting sort. It almost made me feel bad for what I was about to do.

Beside the bed was a wheelchair, folded and propped against the wall. I yanked it open. Then, with a glance over my shoulder to ensure I wasn't being watched, I slid the IV from Kate's arm. Blood welled red in its wake. I blotted it with the bed sheet, and replaced the tape that had held the IV in place. Then I lifted her into the wheelchair. Her eyes fluttered, but she didn't stir.

Outside Kate's room, the hall bustled with activity. The nurse had headed left, so I went right. No one gave me a second glance as I wheeled her down the hall, her head lolling to one side.

What the hell was I supposed to do now? It would only be a matter of minutes before they discovered she was missing, and this girl was a hot commodity. I knew what I *should* do was make the collection and be on my way. I also knew that wasn't going to happen – not until I figured out what the hell was going on.

"Hey! Hey, you!"

The call echoed down the length of the crowded hallway. I pretended not to hear – just kept on pushing Kate down the hall like I hadn't a care in the world. As soon as we were around the corner and out of view, I broke into a run. The wheelchair rattled and shimmied beneath my sweat-slick hands – any moment I expected her to spill out of the chair and onto the floor. But she stayed put, and I kept running.

Some fucking plan *this* was.

There was a clatter of footfalls behind us, a bevy of shouts. We reached a bank of elevators, and I pressed the call button. My lungs and legs were burning, and my heart thudded in my chest. Kate, for her part, seemed content to sit and drool on the shoulder of her hospital gown. At least it beat the screaming.

The elevator door pinged open. Two uniformed cops rounded the corner, guns drawn. I rolled Kate into the elevator and chose a floor at random. Then I hit three more below it, just to keep them guessing.

The cops were rapidly approaching, and still the door was open. I pounded on the button to close it, and slowly, it began to move. One of the cops made a leap for the door, arm extended in desperate attempt to halt the door's progress. A second more, and he might have made it, but he was too slow, too late. The door slid shut. A bang reverberated through the elevator shaft as he pounded on the door in frustration. The sound filled the elevator car, and then receded as we lurched downward.

And just like that, we were gone.

4.

Shit.

My hope was that I'd've had a couple of minutes before they noticed she was missing. And when they did, they'd be looking for a dangerous, psychotic, *awake* little girl; I could just wheel her out the front door unnoticed. But now they'd seen me with her, and that was going to complicate matters a bit.

Of course, the fact that they'd seen *me* wasn't the end of the world. I could always leave this body behind and find another; around here, there were no shortage of volunteers. But whatever I looked like, I was still going to have to get her out, too. I couldn't afford to chance it in some broken-down old body. It just wasn't worth the cost.

So what, then? My mother used to tell me when God closes a door, he opens a window, but I was guessing God didn't give two shits about me. Not to mention we were standing in a tin box with one door and no fucking windows. One door that was soon to open onto God knows what.

The elevator jerked to a halt. I left Kate in the center of the car and pressed myself tight to the wall beside the door. As hiding places go, it wasn't much of one, but it would buy me a second or two, and I wasn't about to go out without a fight.

When the doors slid open, I heaved a sigh. There was no one waiting, gun in hand, to reclaim Kate or evict me from this body. Of course, that didn't mean much – the hospital was too big to put a cop on every floor at the drop of a hat, but you can be damn sure they were gonna have the exits covered. The best I could hope for was to buy a little time, maybe come up with a game plan.

I poked my head out into the hall. The plaque on the wall read Radiology. At the end of the hall, a couple of orderlies were wrangling an elderly woman out of a stretcher and into a wheelchair, but otherwise, the place was deserted. I wheeled Kate's sleeping form out of the elevator and down the hall. The old woman met my gaze as we approached. The orderlies paid us no mind. I flashed her a smile of reassurance, and she smiled back, wan and tired and grateful. Then the three of them disappeared into the imaging suite, leaving the empty stretcher behind.

I said a prayer for her. It was the least I could do.

Thanks to her, we just might get out of here alive.

Though it was just past 8am, Bellevue's ER was already bustling. The waiting room was crammed full of the sick and injured: children, hacking away on their mothers' laps; junkies, gray and shaking from withdrawal; a man in chef's whites, bleeding into a dish towel. The staff were bustling, too, with the cold efficiency of folks who've seen worse than this more times than they could count. The only indication they'd seen us at all was the slight change in their trajectory as they strode past us in the hallway. They had a job to do, and we didn't concern them in the slightest, so long as we were out of the way.

To tell you the truth, they didn't concern me much, either. What *did* concern me was the cop manning the entrance. I'd been watching him for going on ten minutes through the criss-crossed panes of safety glass set into the doors that separate the waiting room from the ER. In that time, he hadn't budged, hadn't yawned – hell, he'd barely even *blinked*. Getting past him wasn't going to be easy. Lucky for me, an ER is a diversion waiting to happen; all I had to do was bide my time.

Turns out, I didn't have to wait long. My pulse quickened in anticipation as I heard the squeal of approaching sirens. And I wasn't the only one. You could see it in their drawn faces; you could hear it in their clipped, efficient tones as they relayed details and called out orders. It was an accident. A bad one. Three ambulances en route, with more to follow. Patients in critical condition. They readied examination rooms and operating suites, and I readied myself as well. There were maybe thirty yards and one cop between me and freedom, and however this went down, I was only gonna get one shot.

A dozen doctors, nurses, and orderlies pushed past me through the double doors as the first of the ambulances rocked to a halt outside the ER. The ambulance doors burst open. Inside, the EMTs hovered over a stretchered form barely recognizable as human. The stretcher was unloaded, no small feat since one of the techs knelt atop it, straddling the patient. A woman stumbled out of the ambulance cab, her face scraped, her hair matted with blood. When she saw the man in the stretcher, she began to scream.

Two more ambulances screeched to a stop beside the first. Stretchers were unloaded amidst a sea of shouted instructions. The whole place was swarming with people – staff, the wounded, a throng of onlookers, pressing close. Still, the cop held fast. I jumped as the double doors banged open beside me, the first of the stretchers rolling past, at the center of a medical maelstrom.

Outside the doors, the crowd of onlookers pressed closer, eager to absorb every lurid detail. Over the din, I heard a shouted *"Do something!"* – and reluctantly, the cop abandoned his post. He herded the crowd backward, trying desperately to give the doctors room to work.

That was as good a chance as I was going to get. I gripped tight my borrowed stretcher and pushed it through the double doors. Kate, strapped down atop it, didn't stir. There'd been a set of folded blankets stacked on the shelf beneath the woman's stretcher, and one of them now covered Kate to her chin. Her head I'd bandaged with supplies stolen from a nurse's treatment cart left unattended in the hallway. The effect was less professional than I'd intended. It wasn't going to stand up to any kind of scrutiny, but from a distance, it did the trick.

The ER waiting room had become a triage center; dozens of people not injured enough to require an ambulance were being sorted through by doctors too hurried to spare a second glance at me. Stretcher after stretcher careened past me, headed toward the operating suites. As I pressed toward the entrance, two more ambulances arrived, and were then abandoned as their crews wheeled their respective payloads inside. In his hurry, one of the drivers left his ambulance running. I swear I could have kissed him.

Though the sky was gray overhead, and the air was thick with exhaust, the cool morning breeze was like a balm to my frayed nerves. The sidewalk was cold and rough beneath my feet. The wheels of the stretcher folded upward as I shoved it into the empty, waiting ambulance.

I slammed shut the rear door and headed for the cab. A hand grabbed my shoulder – not gently. I turned to find the cop, dark eyes glowering at me from beneath a furrowed brow.

"Where you think you're going?" he asked. Skeptical, but not yet hostile. That was OK. Skeptical I could work with.

"Patient transfer."

"Where you taking her?" he asked.

"Him," I replied. "Got a suite waiting at Beth Israel."

"Where's your badge?"

"Sorry?"

"Your badge? You know staff's supposed to display it at all times."

"Of course," I replied. I patted the pockets of my lab coat, a smile of contrition pasted on my face. My left hand dipped into a pocket, and his eyes followed. He never saw my right hand coming.

The punch connected with the bridge of his nose. A crunch of bone, a spray of blood, and he went down. Not dead, just sleepy. Parlor tricks like the one I'd pulled on the guard upstairs are all well and good, but sometimes, you just gotta go the direct route. Besides, subtle's never been my strong suit.

I climbed into the cab and threw the ambulance into gear. I gave it a little gas, and it lurched forward. Through the side mirror, I saw a kid of maybe ten staring slack-jawed back at me. He was tugging at his mother's arm and pointing toward the cop sprawled across the pavement, but she ignored him. The show outside the ER was still going strong, and she wasn't about to miss it.

I gave the kid a wink and a lazy mock-salute, and then pulled out of the hospital drive, disappearing into the early morning traffic.

5.

"City and state?"

Between the din of the nearby traffic and the work crew drilling through the sidewalk just a half a block away, I couldn't hear a word she said. I pressed the handset tighter to my ear and huddled closer to the payphone. "I'm sorry?"

"City and state?" the woman repeated.

"Uh, Manhattan," I replied. "Manhattan, New York."

"What listing?"

"Jonah Friedlander."

The line hissed and clicked, and the woman was replaced by an automated voice that spat out the requested address. I was hard-pressed to tell the two apart.

I dropped the handset back onto its cradle and hunched across the lot to the waiting ambulance. Across the intersection from where I stood, a police cruiser sat idling at a light. I watched him from the corner of my eye, thinking inconspicuous thoughts. The gas station was packed three deep with cabs waiting for a crack at the pumps, and droves of pedestrians filtered through for a paper or a cup of morning coffee – no way the cop had made me. Of course, the meat-suit didn't want to hear it. His heart was pounding a mile a minute; his palms were sweating; his mouth was dry as dust. Just once, I'd like to possess me a Mob enforcer or something. These peaceful, law-abiding sorts make this job of mine a bitch.

Inside the ambulance, Kate was still unconscious. The question was, for how long? My stomach roiled as I recalled the bitter tang of blood and alcohol that clung to her mother's mangled corpse, and I gave her restraints a tug to ensure they were secure. Then I thumbed the ignition, and the ambulance sprang to life. I pulled out of the station and onto the crowded city street, disappearing into the swell of traffic.

Friedlander's apartment was a third-floor walk-up in Chelsea, the kind of place a realtor might charitably call a quaint Manhattan brownstone. It was brown, true enough, but its façade was faded and crumbling, and the paint on the sills had blistered and peeled, revealing rotten wood beneath. The whole building had the look of a musty old sweater – one well-placed tug and the whole thing might come tumbling down.

The front door was propped open with a rolled-up newspaper, and thick bacon-scented smoke poured skyward from the hallway beyond. From somewhere inside, a smoke alarm cried. I nudged the door open with my foot and carried Kate's sleeping form across the threshold. The ambulance I'd left in an alley a block north. They'd find it soon enough, but I didn't much mind – Penn Station lay just a couple blocks to the east, and they'd be expecting us to run. Either I was too smart to run, or too stupid, but either way, we'd be safe here a while.

I was huffing pretty good by the time I got her up the stairs. My muscles burned in protest, and my eyes stung from sweat and smoke. Friedlander's door was cordoned off with police tape. I ran a fingernail along the jamb, breaking the seal, and then I tried the knob. Locked. I set Kate down and looked around, to be sure I didn't have an audience, and then I shouldered the door, hard.

White-hot pain radiated outward from the point of impact, but nothing happened. I tried again. More of the same. Just my luck, I thought: the building is a fucking dump, but the one thing the landlord didn't cheap out on was the locks.

Again, I slammed into the door. There was a sickening crunch as the doorjamb splintered, and then I spilled into the apartment, tumbling gracelessly to the floor. I lay there a moment, waiting for the pounding of my pulse to subside. Then I dragged my ass up off the floor and carried Kate inside. I dropped her into a threadbare old armchair, and then went back and closed the door, throwing the bolt and setting the chain.

I stifled a yawn. My shoulder ached like hell, and I felt like I'd just run a fucking marathon. Some collection this was turning out to be. I'd botched the job, snatched the girl, and in all likelihood become the target of a city-wide manhunt. All of which paled in comparison to the world of shit I'd be in when word got out I'd disobeyed an order. Failure was bad enough; insubordination was... I didn't even *know* what. Far as I knew, I was the first.

So the clock was ticking. I had to sort this shit out fast – the last thing I needed was to be made an example of. I shook my head as I recalled Lily's parting words: *Do try to enjoy yourself, won't you?*

Enjoy myself. Right. Well, I thought, no time like the present.

After all, there's gotta be *something* to drink in this place.

"Hey!"

The voice drifted toward me from someplace far away. It seemed faint and unimportant, and whatever it was, it could wait.

"Hey! Wake up!"

The voice was louder now, more insistent. I did my best to ignore it.

"Wake *up*, you sick son of a bitch!"

I opened my eyes. I wished I hadn't. Sunlight streamed in through the windows like an ice pick to my brain. I raised a hand to shade my eyes. It didn't help.

Kate was awake in the armchair, struggling against her makeshift restraints. They were just a couple of bed sheets, really, twisted into ropes and tied behind the chair back, but it was the best I could do on short notice. I was happy to see the knots had held – I was pretty drunk when I tied them. I'd found a half a handle of Maker's Mark by the bathtub, along with a smattering of multicolored pills scattered across the linoleum. I'd taken a couple at random, in the hopes they'd help the throbbing in my shoulder. The results were mixed – the shoulder was doing pretty good, really, but the throbbing in my head made it a lateral move at best.

"I see you're feeling better," I mumbled.

"Feeling better?" Kate said. "I'm tied to a fucking chair!"

"Yeah," I said. "Sorry about that. I hear tell you put on quite a show the other day, and I wasn't wild at the prospect of being included in the encore."

"I don't know what you're talking about."

"Anybody ever tell you you're a lousy liar?"

"Anybody ever tell you it's impolite to hold people hostage?"

I laughed. "You think *that's* what this is? Sweetie, I'm trying to *help* you here."

"Help me. Right. I bet you say that to all the girls."

"Just the ones I tie to chairs," I replied with a smile. "So tell me, what'd they do?"

"What? Who?"

"Your family – what'd they do? Your mother cut your allowance? Your little brother read your diary? Maybe Daddy wouldn't let you drive the Bentley?"

"Don't you talk about my family."

"Suit yourself," I replied. I rose stiffly from the couch and padded into the kitchen. "You hungry?"

"What?"

"I asked if you were hungry."

"I… I don't think so."

"Well, I'm starving." I cracked open the fridge. Not much there – just a half a bell pepper, a few eggs, a hunk of cheddar cheese. "Tell you what – I'm gonna make myself an omelet. You want some, you're welcome to it."

Kate eyed me quizzically for a moment, but said nothing. I busied myself in the kitchen, chopping and whisking and grating. I found a skillet in the cupboard. A pat of butter and I was off and running. My stomach rumbled in anticipation.

"So," she said finally, "you some kind of doctor?"

"No," I replied.

"Oh. I thought – I mean the clothes and all…"

"I stole them. And so far, you're pretty much the only one I fooled."

"You got a name?"

"His name was Jonah. I guess that's as good as any."

"What's that supposed to mean?"

"Nothing," I replied. "Omelet's up."

Plate in hand, I dragged a chair from Friedlander's dining set over to Kate's armchair and sat down beside her. The omelet was steaming, and the mingling scents of sautéed pepper and melted cheddar were intoxicating. Kate tried her best to look disinterested. I split the omelet in two with my fork and scooped up a goodly bite, offering it to her.

She shook her head. "You think I'm eating that, you're nuts."

"Fine by me," I replied. I stuffed the forkful into my mouth. It wasn't half bad. I chased it with another, and then another. Soon, I'd polished off half the omelet. I was about to start in on the other half when she finally caved.

"Wait," Kate said. "Maybe just a bite." I gathered up a forkful and held it out to her. She frowned a moment, still doubtful, and then took the bite. Her eyes went wide. "It's good," she mumbled grudgingly as she chewed.

"You ask me, it coulda used a little Tabasco, but it turned out OK. When's the last time you had anything to eat?"

She shrugged against the restraints. "Dunno." Kate wolfed down a couple more bites, just as fast as I could feed her. Soon, fork hit plate, and I set both aside. "Water," she said. No please or anything, but still, it was progress. I filled a glass from the sink and tipped it to her lips. She lapped it up greedily, water dribbling down her chin.

"Easy," I said. "You're not careful, it's gonna come right back up."

"Thought you weren't a doctor," Kate said.

"I'm not, but I'm also not an idiot. You've been out a couple days – it's gonna take your stomach a little time to adjust."

"A couple days? What in hell did you *do* to me?"

"Hey, don't blame me – you were unconscious when I found you."

"Then how–"

"Wait," I said, "you're telling me you really don't know?"

"Know *what*?"

I ignored her question. "Kate, before waking up here, what's the last thing you remember?"

Her face twisted into a scowl. "I – I'm not sure. I remember coming down for breakfast. Mom was in the kitchen, packing lunch for me and Connor. Dad was on the phone in his study. Connor was at the piano playing 'Chopsticks' – Dad yelled at him to keep it down, said he couldn't hear himself think. Then things went a little fuzzy. I must have hit my head or something, because I remember smelling blood. After that, it's just fragments. My brother, crying. The scent of alcohol. Sirens, wailing in the distance. I think I might have spent some time in a hospital. I remember a bright light. Someone was screaming – I think it was me. Then I woke up here, tied to this chair."

"That's all you remember?"

"That's it."

I don't know why, but I believed her. "Kate, there's something you should know."

"What?"

I snatched the remote up from the coffee table and clicked on NY1. No surprise, we were the top story.

"... the hunt continues for seventeen year-old Katherine MacNeil, prime suspect in the brutal slayings of her mother, Patricia Cressey-MacNeil, her father, real estate magnate Charles MacNeil, and her eleven year-old brother, Connor.

MacNeil was under guard at Bellevue Hospital Center Tuesday when she escaped with the help of an unidentified white male, age unknown. Anyone with information regarding MacNeil is urged to call..."

The anchor was replaced with a picture of a smiling Kate, clearly taken from her family's apartment, and a sketch of me that could've been any skinny white guy in the Tri-State area. Beneath us ran a number. I turned off the TV.

Kate stared at the blank screen for a while. Not blinking, not speaking. When she finally did speak, her voice was thin and frail.

"I... I don't understand."

The look on her face said otherwise. "Yes, you do."

"There has to be some sort of a mistake."

"There's not."

"Why would they think I'd done such a thing?"

"The cops found you at the scene, Kate. They saw you..." *slit your mother's throat,* I thought. "They saw enough."

"But why? *How?*"

I remembered the light that enveloped me as I'd clutched tight her soul. I remembered her song ringing loudly in my ears as I crumpled to the ground. "I don't know," I replied.

"That's why you tied me up," she said. "You were *afraid* of me."

"Yes."

"Then why'd you help me escape?"

"That's complicated."

Kate eyed me a moment. "Yes," she said, "I imagine it would be."

She fell silent for a while. I let her sit in peace. What could I say to her, really? Her family was dead. Dead by her hand. Words weren't going to change that.

I set about cleaning up the mess from breakfast. I was halfway through the dishes when she found her voice.

"This place," Kate said. "It's not yours, is it?"

"What makes you say that?"

"Doesn't seem like you, is all."

I smiled. "It belongs to a friend of mine. He wasn't using it, and we needed a place to stay. I figured we'd be safe here for a while, while I sorted things out."

"And have you? Sorted things out, I mean."

"I'm working on it," I said.

"Yeah," she replied. "Me too."

6.

"I need to use the bathroom."

Kate hadn't said a word in hours – she'd just sat and stared at nothing. Of course, it's not like she had a lot of other options, being tied to a chair and all.

"I'm not sure that's such a good idea," I replied.

"I'm serious. I've really got to go."

"Then I'll get you a trash can."

"What's the matter – you scared to untie me?"

"Something like that."

"Well, it's got to happen sometime," she said. "You can't keep me here forever."

She had a point. Of course, she'd left three solid counterpoints cooling at the morgue. Still, I'd snatched her for a reason. There was something sour about this collection, and I sure as hell wasn't going to figure out what by playing babysitter all day long. Besides, maybe I untie her and Kate tips her hand. She goes postal and my little moral dilemma gets resolved in a hurry. That happens, I finish the job, and to hell with the light show.

Man, I hope she tries to kill me, I thought. I could use a happy ending.

"All right," I said, fetching a chef's knife from the kitchen, "I'll let you go. But you're gonna get out of that chair and head straight to the bathroom. When you're done, you're to get back in the chair – no argument, no complaint. Those are my terms. You break them, things are gonna get unpleasant. We got a deal?"

"Yeah," she said. "We got a deal."

I knelt behind the chair back and cut through the makeshift restraints. As they fell to the floor, I took a big step back. The knife I kept at my side, all kinds of casual, like I wasn't figuring on how fast I could put it between us should things get ugly. Just because you're thinking about stabbing somebody doesn't mean you have to be a dick about it.

Kate flexed first one limb, and then another. When she was sure all four still worked, she rose, unsteady, from the chair. She limped the length of the floor to the bathroom. If I had to guess, I'd say the cops were a little less than gentle when they finally took her down.

"Three minutes," I told her as she reached the door. "Not a second more."

She nodded, and shut the door behind her. I let out a breath I hadn't realized I was holding. Then I set the knife on the coffee table and collapsed onto the couch. The clock on the cable box read one-fifteen. I kept an eye on the bathroom door and wondered for about the hundredth time just what the hell I'd been thinking taking the girl.

Three minutes passed, and no Kate. I figured I'd give her a break – she'd been in that chair the better part of a day, and before that, she'd been cuffed to a stretcher. Besides, she was still decked out in a hospital gown; it's not like she was wearing a watch.

When four minutes had gone by, I got a little irritated. Then five ticked past, and I was downright pissed. By the time six minutes rolled around, I was banging on the bathroom door.

"C'mon, Kate, you've had your fun. Time to get back in the chair."

No response. I tried the knob. Locked. "I'm not fucking around here, Kate! Open this door or I swear I'll break it down!" Still nothing. I put an ear to the door. I heard the sound of running water, and beneath it, something else. A low, wet gurgle. Like someone choking. Like someone dying.

Shit.

I slammed against the door, and rebounded hard, sprawling across the living room floor. Pain radiated outward from my shoulder in nauseating waves. I regrouped and tried again. I managed to stay up this time, but it still hurt like hell, and the door didn't give an inch. She must have barricaded it somehow. Didn't want me ruining her big exit.

Inside the bathroom, Kate's ragged breathing ceased. I was out of options. The problem was, this body of mine was exhausted, and I didn't know if it had the juice for me to make the jump. Still, I had to try.

I clenched shut my eyes and focused on her,

choking on the other side of the door. Blood trickled from my nose at the sudden strain, and my mouth filled with the taste of pennies. The world went dark as I pulled away. Friedlander crumpled, his head slamming into the floorboards with a sickening *thwack*. Then, for a moment, there was nothing.

When I opened my eyes, I was staring at the bathroom ceiling. I couldn't breathe. Somewhere inside my head, Kate was shrieking. It's like that with the living – damn near impossible to concentrate with them always carrying on.

I tried to roll over, but Kate's limbs were like lead. Pills, I'd guess. I was such a fucking idiot. After the way Friedlander checked out, I should have thought to check the medicine cabinet. Now I hoped I wasn't too late.

The nausea hit me like a freight train. Kate's entire body clenched. I struggled with her sluggish limbs, and managed to tip us over. Cheek met tile, and my vision swam. Beside me lay a smattering of empty bottles. Prescription, the lot of them. I tried in vain to read the labels, but my eyes wouldn't cooperate. Whatever she'd taken, she hadn't been fucking around.

Acid scorched my throat as Kate's body tried to purge itself of me. I stayed put. A couple dozen pills weren't so lucky. Her stomach heaved again. Pills and sick spilled across the tile.

Control came by degrees as her body relented to my demands. Still, it was blunted by the drugs. I didn't have much time.

On quivering limbs, I forced myself to my hands and knees. Still the sickness came. I glanced toward the bathroom door. I could barely keep my head up. She'd barricaded it, all right. A set of wooden shelves, wedged between the door and tub. I clawed at them with clumsy hands. The shelves were jammed tight, and my grip was weak.

My strength faltered, and again I hit the floor. I grabbed the shelves and yanked.

Nothing.

Kate's lids buckled under the weight of narcotic slumber. The rest of her wasn't far behind. I mustered all my failing strength, pulling the shelves as hard as I could. Then the world went dark, and Kate was gone.

I was back in the Friedlander body, sitting in the recliner and holding a towel full of ice to the knot on my head, when Kate finally came to. She'd been out for nearly a day. I'd left her lying on the couch, her head turned aside in case she wasn't done throwing up. She hadn't been. For a while I thought this headache was for nothing – she'd grown sicker and paler with every passing hour. Around midnight, though, she'd turned a corner. She'd stopped throwing up, and a little color returned to her cheeks.

Now, her eyes fluttered open. Kate looked around a moment, confused. I saw a flicker of remembrance as her gaze met mine.

"What…" she rasped. "What *happened*?"

"You took some pills," I replied. "You tried to check out. Just lie still a bit – you're going to be OK."

"Pills," Kate repeated, casting her gaze toward the open bathroom door. "Of course."

"How'd you know you'd find them there? The pills, I mean."

"I didn't – I just got lucky. But every bathroom's got a mirror. Figured I'd slit my wrists and just fade to black. I guess I wasn't quite as lucky as I thought." Her face was clouded with suspicion. "What the hell did you *do* to me?"

"I did what I had to. You could have *died,* Kate."

"I wish I were dead."

"Yeah, well, I'm glad you're not."

She snorted. "You're *glad*? You were so scared of me, you tied me to a chair."

"You tied up now?"

Kate lifted her head. The effort caused her to wince. "No," she replied. "But I ought to be. I'm not to be trusted. I killed my family." A single tear slid down her cheek.

"No, Kate," I said. "I don't think you did."

"I don't understand. You said they saw–"

"I don't doubt what they saw. I just don't think it was you that killed them."

"You're not making any sense."

I flashed her a wan smile. "Maybe not," I replied. "Or maybe you and me just have different ideas about what makes sense."

She clenched shut her eyes and pinched the bridge of her nose between thumb and forefinger. "My head is killing me," Kate said. "If you plan on talking around in circles all day, I'm going to need a couple aspirin."

"I think you've had enough pills for one day."

"Then how about you start talking straight?"

"Believe me, Kate – you're better off not knowing."

She let out a barking, humorless laugh. "Better off? You think I'm *better off*? My family is dead. I would be too, if I'd had my way. But you went and stopped me – God knows how, but you did. Now I'm holed up in some shitty apartment, a fugitive from justice, and I feel like I'm going out of my head. So to hell with what you think. I need answers, Jonah. I need the truth."

I looked at her a long, appraising moment. Kate looked back, angry and expectant. To hell with it, I thought. "For starters," I said, "my name isn't Jonah."

And then I told her. What I was. Why I came. I expected shock, anger, disbelief. But she just listened, without comment, without interruption. It wasn't till I finished that the questions came.

"So in the hospital, that was you?"

"Yes."

"You'd come to collect my soul."

I repeated, "Yes."

"I thought I'd *dreamt* it. I remember a sudden pain – pain and fear – and then this, this *light*..." Her hands found her chest. "But I haven't any mark. Any scar."

"The invasion isn't physical."

"So why didn't you take me?"

"I just... *couldn't*," I replied. "When I make a collection, there's this moment – this beautiful, terrible moment when my hand closes around the soul, and I see everything. *Experience* everything. A lifetime of beauty, and of happiness, and of sorrow. I see every kindness. Every slight. Every moment that's led them to my grasp. But the souls of those that I collect are just hollow echoes of their better selves – they're occluded by the darkness within. Yours was different. Pure. Unfettered."

"You make me sound like some kind of saint."

I smiled. "I wouldn't know anything about that. What I *do* know is evil changes a person, tainting everything until no memory is untouched. Only in your case, there was no stain."

"But how can that be? I mean, my family–"

"Kate, that wasn't your fault."

"But I have these flashes. These memories. Horrible reminders of the things I've done."

"I know."

"Then how could it not be my fault?"

"Tell me," I said, "earlier, when you were in the bathroom, how did you get out?"

"*You* got me out."

"Yes, but how?"

Kate's brow furrowed as she struggled to remember. "I was groggy. Sleepy. Then all of the sudden, you were in my head. I threw up. You rolled me over, so I wouldn't choke."

"Then what happened?"

"I'd barricaded the door," she replied. "You clawed at it, I think. I don't know – I was so groggy, all I wanted was to sleep."

"Did you want to do those things?"

She shook her head. "All I wanted was to die."

"And yet here you are."

I let the sentence hang in the air for a minute. She was slow getting there, but eventually, realization dawned. "You're saying someone else was in my head? That *they* killed my family?"

"Not someone," I replied. "Some thing."

"Some *thing*?"

"Kate, there aren't many folks like me out there, and we're kept on a pretty short leash. We never take what isn't ours to take; we just do our jobs – no argument, no deviation. Not to mention, I read the news coverage – there's no way someone like me could've mustered the kind of strength they're talking about. No, whatever did that wasn't human."

"Which leaves what, exactly?"

"A demon, most likely."

"A demon."

"Yes."

"But that's *insane*."

"Any more insane than what happened in the bathroom? Demonic possession is far from unprecedented, Kate. Most possessions go unnoticed; the body chosen is simply a conveyance, a means to an end – when the task at hand is done, the possessor leaves, and no one's the wiser. Seems like your guy had other plans."

"How can you be so sure? How can you be sure I didn't just suffer some psychotic break and kill them myself?"

"Because possession is by nature a violent act. You're forcing an unfamiliar body to succumb to your will. When you possess the living, you're also fighting the impulses of their conscious mind. That kind of struggle is sure to leave a sign."

Kate's brow furrowed. "What kind of sign?"

"It's hard to describe. You ever lend out a sweater, and when you get it back, it just doesn't fit right?"

"I guess."

"It's kind of like that." Kate seemed to accept that, which was fine by me. She didn't need to know the rest. That whatever had done this had violated her with such fury I'm surprised she'd even survived. That it had gouged and splintered her mind like nails against a coffin lid. That I'd been so terrified by what I'd seen, when I returned to this body, I hadn't stopped trembling for hours. No, she didn't need to know any of that. Which was fine, because I sure as hell wasn't going to tell her.

Kate said, "So where does that leave us, then? I mean, if I'm innocent, you'll be on your merry way, right? No harm, no foul. And I what – spend the rest of my days in a loony bin? And that's if I seem nuts enough to keep me out of prison. I mean, as far as the rest of the world is concerned, I'm still the one who killed them, right? Forgive me if I sound ungrateful – I'm glad I'm not damned and all, but this still pretty much sucks."

I sighed. "It's not that simple. You're marked for collection, Kate. And once you're marked, you're collected – it's as simple as that."

"Can't you talk to your boss or something – explain there's been some kind of mistake?"

I shook my head. "Lilith's not exactly the understanding type, Kate, and even if she were, she's not the one calling the shots."

"Then who is?"

"The short answer is, I don't know. The longer answer is, I don't know 'cause they don't want me to. Lilith is my handler, and she's the only one I ever deal with – I couldn't go around her if I tried. But she's made it very clear that babysitting me is nothing but a chore to her, something passed down from on high – or on low, I guess you'd say. Besides, I doubt an end run around Lilith would even do us any good. These are the denizens of hell we're talking about, Kate – I've got no reason to believe her bosses would be any more receptive than she would. No, I think the best thing we can do is stay off the radar for a bit, while we figure out what's going on."

"What happens if they find out that you're helping me?"

"I don't know," I replied. "As far as I know, no Collector's ever willfully disobeyed an order before. But what we're talking about is mutiny – insubordination against the authority of hell. I'm pretty sure I don't want to find out."

"Why not just take my soul, then? It's not like I have anything left to live for."

"I can't. Whatever's going on here, your soul's not mine to take. My job is to collect the wicked, the corrupt. The taking of a pure soul is forbidden – the results would be catastrophic."

"Catastrophic how?"

"We're talking some serious End of Days shit here, Kate."

"Oh," she said. Her eyes no longer met mine; she seemed suddenly fascinated with a spot between us on the floor. "OK, then. But if I'm marked for collection and you can't collect me, where does that leave us?"

"I don't know. Being marked isn't something you can easily fake – whoever did this has got clout, to say the least. Which means this wasn't just some demon on a joyride – whoever did this had an agenda. The way I figure it, our best bet is to figure out who's behind this before they get wise to the fact that you're not in the ground and send someone to finish us both off. That is, if we can keep clear of the cops for long enough."

She surprised me with a laugh, full and throaty and beautiful. "That's our *best* bet?"

"Near as I can tell."

"Well, shit," she said, and despite myself, I smiled.

"Yeah," I replied. "Shit."

7.

"So," Kate asked, "what now?"

I shrugged, chasing a mouthful of pastrami sandwich with a long pull of Brooklyn Lager. It had been a few hours since Kate woke from her little chemical nap – she'd polished off her sandwich in record time, and I was pleased to see some color returning to her cheeks. I'd stuck around until I was pretty sure she wasn't going to make another go of it, but eventually hunger got the best of me. I swapped my scrubs for a pair of jeans, a T-shirt, and some battered Chuck Taylors, and hiked down to the bodega on the corner for a pack of smokes and a bite to eat. The cigarettes tasted like shit, but the sandwiches weren't half bad, and after a day of traipsing all over town barefoot, I was happy for the wardrobe upgrade. Friedlander might've lived in a dump, but at least I knew the clothes fit.

"I don't know," I said, finishing my sandwich and tapping a cigarette from the pack. "I've got a contact in the demon-world who might have some idea who's behind this – I thought I'd pay him a visit, see what I can see. Only I'm not exactly relishing the idea."

"Is he – I mean, do you have to go..." she stammered. "Is he in Hell?"

I laughed. "Near enough – he's in Staten Island."

"Oh," she replied. "But you've been? To hell, I mean?"

"Have I *been*? Sweetheart, I'm sitting in it."

"I don't understand."

"Hell isn't some faraway land, Kate. It's right here – in this world, in this room. Heaven, too, as near as I can tell. They're just, I don't know, set at an angle or something, so that they can see your world, but you can't quite see them. Occasionally, the boundaries break down, and the result is either an act of horrible savagery or of astonishing grace. But make no mistake, they're always here."

Kate's brow furrowed as she looked around the room. "I guess I always imagined hell to be all fire and brimstone."

I lit my cigarette and took a long, slow drag. "You ask me, I'd guess heaven and hell look pretty much the same," I replied. "Only in hell, everything is just a little out of reach."

There was a long pause before Kate spoke again. "You don't seem so bad to me," she said.

I laughed. "Thanks, I think."

"So how'd you wind up here, doing what you do?"

"That," I replied, "is a story for another time."

• • • •

The summer of 1944 was one of the hottest the city had ever seen. The streets of Manhattan seemed to ripple in the midday sun, and the bitter stink of sweat and garbage clung heavy to anyone who dared to venture outside. Even the breeze off of the harbor offered no relief from the oppressive heat. Every night as I made my way back home, I watched as passengers crowded three deep at the bow of the ferry, eager to feel the wind on their faces. But the air was still and thick with diesel fumes, and all they got for their trouble was a sheen of sweat atop their brows and angry glares from those they jostled.

Home back then was a tenement in the New Brighton neighborhood of Staten Island, about twenty minutes' walk from the ferry terminal. The place was ramshackle and overcrowded, and the racket from the munitions factory across the street was as constant as it was maddening. Still, as I hobbled up the stairs, I was greeted by the heavenly aroma of garlic and onion, so I couldn't much complain.

Inside, Elizabeth was standing by the stove, her back to me. A Benny Goodman number drifted across the room from the radio in the corner, and she tapped her foot in time. When I closed the door, she started, and then smiled. I crossed the room and gave her a kiss.

"Sam," she said, blushing, "you know the doctors said you shouldn't do that!"

"To hell with them. You're my wife – I'll kiss you if I damn well please."

"How'd it go today?"

I shrugged off my suit jacket and yanked the tie from my collar, tossing both across a chair. "Same old story. They said

I'm more than qualified, that my references are sound, but there's just no way a gimp like me is gonna keep up with the demands of the job."

"They actually said that to you?"

"No, of course not – they said a man in my condition."

"Ah," she said, as if confirming something she had already known.

"What do you mean, ah*?" I snapped. "Just because the words they use are flowerier doesn't make 'em any likelier to hire me, now does it?"*

Tears shone in Elizabeth's eyes. She blinked them back and looked away.

"Liz, I'm sorry," I said. "I'm just frustrated, is all. I'll find something eventually, and then we'll get you better – you just wait and see."

I put a hand on her shoulder. She shrugged me off and returned to the stove.

"Whatever you're making smells fantastic," I said. Though her back was still to me, I could see her posture relax.

"It's braciola" she replied. It was my favorite, and she knew it. I felt like an ass for snapping at her – God knows it was the last thing she needed right now.

"How're you feeling today?" I asked.

She flashed me a smile over her shoulder. "Well," she said. "I think the tincture Annie got for me is working."

"Liz, that's great! You'll beat this yet, you wait and see."

She dropped her gaze and said nothing for a moment, then: "Oh, I forgot to tell you – Johnnie Morhaim stopped by to see you. Third time this week, I think."

"Yeah, I'll bet he did. He comes around again, you just let him knock, OK? I don't like the thought of the two of you here alone together."

"Honestly, Sam, he's always been perfectly polite to me. Don't you think you're overreacting a little?" I shot her a look that made it clear that I thought no such thing.

The timer on the stovetop buzzed. Elizabeth took the pan off of the heat and transferred its contents to a serving plate. "Go wash up," she said. "Dinner's ready."

I kissed her neck and headed down the hall to the bathroom. The water ran rusty from the tap, and I waited for it to run clear before splashing my face and washing my hands. I heard the familiar patter of water against tile, and cursed softly to myself – the fittings must be loose again, I thought. And as I ducked my head beneath the vanity to reach the pipes beneath, something in the trash can caught my eye.

It was one of Elizabeth's handkerchiefs, crumpled and discarded; I could just make out the delicate stitching of her initials peeking out over the rim of the can. Despite the heat, my skin went cold, and my heart thudded in my chest. I fished it from the trash, certain of what I'd find.

The ivory surface of the kerchief was flecked with blood. Elizabeth's blood.

Whatever lies she told me, we were running out of time.

The wind ripped across the harbor as I leaned against the deck rail of the ferry, savoring the bite of the chill salt air against my face. Behind me, an unfamiliar Manhattan skyline receded in the distance. So much had changed since I'd last been back, but as the low-slung buildings of the Staten Island waterfront swung into view, a shiver of remembrance traced its way along my spine. I guess the past is never quite as far behind us as it seems.

The sun dipped below the horizon as I wandered away from the terminal, blanketing the streets of the island in shadow. I pulled Friedlander's pea coat tight around me, my hands thrust deep into its pockets.

The old tenement was just as I remembered it. The first floor now housed an adult bookstore, its storefront windows papered over from within and its sign declaring XXX VIDEOS BOUGHT AND SOLD, but otherwise the years had failed to leave their mark. The same couldn't be said of the rest of the street. Most of the storefronts sat vacant. The old munitions factory was bricked up and abandoned. On a stoop two doors down, a bedraggled old man slouched unconscious and mouth agape, a bottle of Mad Dog dangling precariously from his hand.

"Hey, sweet thing, you lookin' for a little company?"

I turned around. Behind me stood a working girl, shivering in a hot pink tube-top, a fake leather miniskirt, and a rack to match. Track marks traced the veins of her forearms.

"Maybe," I told her. "But I'm not from around here. You got somewhere we could go?"

She looked me up and down. "For you, sailor, I'd lay down right here."

"I was thinking someplace a little more private."

"I know a spot a couple blocks from here, long as you don't mind the hike."

I didn't. She led me by the hand to a decrepit row house, nibbling on my ear all the while. I pretended not to notice. Inside, the place was a mess. The paint on the walls was discolored and flaking. The floor was littered with newspaper, empty bottles, and God knows what else. A smattering of stained and filthy mattresses were scattered throughout the front room. A few of them were occupied: junkies, mostly, sprawled amidst their needles, lighters, and scorched bits of tinfoil.

My date dragged me toward the stairwell. I followed. At the foot of the stairs, a man was slouched against the wall. His sleeve was rolled up, and his arm was tied off with a length of rubber tubing. A hypodermic needle jutted from his arm. His eyes fluttered as we stepped over him, but he didn't stir.

"Nice place," I said as we reached the landing.

"I think the time for talking's passed," she replied, pushing me up against the wall. She kissed me, then. Her breath reeked of latex and menthol cigarettes. Involuntarily, I pulled back.

"Whatsa matter, sport, you rather get right to it?" Her hand found the zipper of my jeans. I pushed it away. Her face read hurt and angry, but the emotion never registered in her blank addict's stare. Then her eyes filled with black fire, and her hurt expression disappeared. That's when I knew I'd found my mark.

Quick as death, her hand found my throat. Her grip was like iron, crushing my windpipe as she lifted me off the ground. My teeth rattled as my head connected with the wall. She held me there, pinned, as my feet tried in vain to reach the floor.

"This body isn't yours," she said. Her voice was suddenly raspy and hoarse, nothing like the treacly croon she employed out on the street.

"I could say the same of you," I squeaked.

"She gives it freely."

"I'm sure she does." My feet kicked against the wall. My vision went a little gray around the edges. I hoped to hell we got to the point before I passed out.

"Who are you?"

"An old friend."

"Most of my old friends would rather see me dead."

"Can't imagine why," I replied. My face had passed red and was headed toward purple. Spots swam before my eyes.

"Why are you here?" the creature speaking through her asked.

"Because I need your help."

She released her grip. I crumpled to the floor, gasping. By the time I'd regained my wits, the *other* had gone, and the girl was glaring at me with glassy-eyed disdain.

"The boss'd like to see you," she said.

"Yeah, I thought he might." I rose unsteadily to my feet, a hand on the wall for support. Without another word, she headed back down the stairs and out of sight. I stumbled after.

She led me through the front room to a grimy kitchen, its broken, gaping window doing little to alleviate the stench of rot that emanated from the open refrigerator. In the kitchen was a door. The girl opened it, revealing a set of rickety stairs that led down to the basement. She descended. I followed.

The basement was close, fetid. The only illumination was from a series of bare light bulbs dangling from the ceiling at irregular intervals. Many were out, and all were so covered in grime they did little to dispel the murk. At the edges of my vision, half-seen figures writhed and moaned and wailed, in pleasure or pain I wasn't sure. There were people strewn everywhere, some shooting up, some grinding against each other in varying states of undress. One man, withered by drugs or disease or both, rocked back and forth, his knees tight to his chest. He'd scratched his forearms raw, and he clawed at them still, nails furrowing flesh. As I passed, I heard him muttering "I'm sorry I'm sorry I'm sorry," again and again, to no one.

My escort led me through this sea of human detritus to the far corner of the basement. The light was warmer here, brighter – the result of dozens of candles, casting tiny halos of light from every surface. A lush Oriental rug occupied the space, and the walls were lined with shelves, cobbled together from scrap wood and cinder blocks and adorned with thousands upon thousands of books. Also on the shelf was an ancient record player, which crackled with the sounds of some old jazz standard – Billie Holiday, I'd guess. And at the center of it all was a man, clad in a pale blue suit and a hat to match, his diamond tie tack catching the candlelight and casting tiny rainbows across his black silk tie. He was draped casually over a high-backed leather chair, a glint in his eye and a smile on his cold, handsome face.

"Sam Thornton, as I live and breathe," he said. "Well, *live*, anyway. I didn't expect I'd be seeing you again."

"Merihem," I replied.

"You know, Sam, there was a time you called me Johnnie."

"There was a time I didn't know any better."

Merihem gave the girl beside me a nod, and she disappeared into the darkness. "That was some stunt you pulled, walking in here like that. I could've killed you. Woulda been a shame, really – that body suits you."

"I'm surprised you recognize me."

Merihem laughed. "I didn't, at first. Your meat-suit fooled her eyes just fine, but my own eyes are another matter."

"And you," I said, taking him in. "You haven't changed a bit."

"I'd like to think I've mellowed," he said, a grin playing on his face. "But that's not exactly what you meant, is it? My kind are too dignified to trawl among the monkeys; the body you see is a projection, nothing more. I gather you didn't drop by to catch up on old times – why don't you tell me just what the hell you're doing here?"

"It's about a girl."

"Isn't it always?"

"I suppose it is," I said, "but this one I was sent to collect."

"Ah, a little on-the-job romance! So what – you figured you could stash her in a black market body and buy you two some time? Maybe jet off to Cabo for a week or two before you do the deed? You've got stones, my friend, I'll give you that – but believe me, it's more trouble than it's worth. Your handlers will see through you just as surely as I did, and they won't find the situation half as amusing, I assure you. My suggestion is you finish the job and move on. Afterward, bring her body by if you like – I'll pop one of my girls in there, and you can have yourself a go."

"Much as I appreciate the offer, I think I'm gonna have to pass. See, I *tried* to collect this girl, only it didn't take. Her soul – it knocked me back. So I panicked and I snatched her."

Merihem guffawed. "This the chick that offed her family? Man, I've been reading about you – you walked her ass right out of the goddamn hospital! You know, that sketch doesn't do you justice."

"Thanks. But here's the thing – I'm pretty sure she didn't do it."

He shook his head. "Not possible. If they sent you, she did it – end of story."

"Yeah, only I've got reason to believe someone else was driving."

He squinted at me. "OK, the Sam I knew, he wasn't stupid, which means you probably know how nuts that sounds. I mean, any demon coulda taken this chick out for a spin, but she'd be lit up like a Christmas tree for anyone who knew to look. No *way* she gets marked for collection. No, a con of that magnitude would take some serious clout – not to mention one hell of a death wish."

"Death wish? Death wish how?"

"You think either side wants a war?" Merihem spat, and any hint of Staten Island disappeared from his voice, an affectation easily discarded. "When last it happened, one-third our number fell – and all because a son of fire refused to kneel before a son of clay. You couldn't *begin* to understand the world of shit that would rain down upon us all if one of our kind was caught damning an innocent soul to rot in hell for all eternity. You're not the only one who's duty-bound, Collector. We all have our roles to play. We do them, and do them well, because the alternative is unthinkable."

"For you, perhaps. Maybe not for everyone."

"OK. Say you're right – which you're *not* – and your girl's been set up. That means whoever's responsible acted against the explicit wishes of the Maker and the Adversary both – and is powerful enough to have done so undetected. If that's the case, what the hell do you expect that *you* are going to do to stop them?"

"I don't know. But I have to try."

"You're pissing in the wind, Sam. If you came here for my counsel, I say keep your head down and do your fucking job."

"I didn't come here for your counsel – I *came* here for your help."

"Did you now?" He smiled. "I'm surprised at you,

Sam – I would've thought you'd learned better than to seek favors from my kind. The price is often steeper than you think."

"The way I figure it, you owe me one."

He laughed then, a big, roaring laugh that rebounded like a chorus off the concrete walls of the basement. "I owe you one! Ha! That's why I've always liked you – there aren't many who'd dare march in here and speak to me that way."

"Yeah, well, it's not like I've got anything left to lose."

"We *all* have something to lose, Sam. Most of us just can't see it till it's gone."

Merihem fetched from his pocket a small leather case, from which he selected a cigar. He clipped the end with a brass-plated guillotine and struck a light with a matching lighter, rolling the tip back and forth within the flame for a moment before placing the cigar in his mouth and taking a long, slow drag.

"I'll help you," he said finally, loosing a heady cloud of smoke that hung thick around him like a shroud. "I'll ask around, see if there's anything to this theory of yours. Just keep your nose clean for a couple of days while I do my thing, OK? Try not to do anything stupid."

"Thanks, Merihem. I appreciate it."

"You understand I still think your theory's full of shit. But better I do the asking than you – the folks who hold your leash don't take kindly to sedition."

"I'm sure that's true."

"If I find nothing, you'll take the girl?"

"I'll consider it."

"You'll *consider* it."

"That's right."

Merihem sighed. "No matter," he said. "By you or someone else, she'll be taken soon enough. Is there someplace I can reach you?"

I smiled. "You don't really think I'd tell you where we're staying, do you?"

He returned the smile. "No, but one does have to try." He recited a number. "Can you remember that?"

I repeated it back to him. He nodded his assent. "Call me there in two days' time. And Sam?"

"Yes?"

"In the interim, try not to get yourself killed, would you?"

8.

Charcoal-smudged clouds scudded westward across the Manhattan skyline as the first faint rays of sunlight peeked over the eastern horizon. The city was quiet as I left the ferry terminal and strolled west toward Battery Park. I'd walked the streets of Staten Island all night, trying to process what Merihem had told me. But no matter how I looked at it, it just didn't make sense.

I mean, I was sure of what I'd seen. *Something* had been inside Kate's head. Something powerful. Something vicious. Something certainly capable of the horrors I'd seen in the morgue. Not to mention, if I was wrong and Kate was to blame, then why torture the mother? Why wait for an audience to arrive before slitting her throat?

No, whatever killed Kate's family had been putting on a show. It wanted no doubt in anyone's mind that Kate had done the deed. Why? That, at least was simple. It wanted to ensure I'd finish the job – no fuss, no mess, no questions. It *wanted* her taken, and that's the part I couldn't square. I mean, in a war between heaven and hell, who wins?

As I wandered, lost in thought, across State Street, I tapped out a cigarette, cursing as it slipped from my cold-clumsy hands. I bent to retrieve it. Only then did I hear the roar of the engine. Loud and low and approaching fast. I looked up. An old Crown Vic skittered around the corner off of Pearl, tires squealing. It leveled the yellowy gaze of its headlights on me and bore down hard. Thirty feet. Twenty. Ten. I was running out of time.

I leapt aside. Not fast enough. My hip exploded in pain as bumper met flesh.

The impact spun me end over end. I tumbled to the pavement like a rag doll, cracking my head against the centerline. The driver laid on the brakes and the Crown Vic came to a screeching, crooked halt amidst a cloud of thick blue smoke that reeked of melted rubber. I tried to move. It didn't take. My left leg felt like it was full of hot lead. My head didn't feel much better. Then the car clunked into gear, and the reverse lights came on.

I was beginning to think these guys didn't like me.

The engine whined as the car swerved backward toward me. Close and coming fast. With all I had, I threw myself aside, or tried. With my leg still not cooperating, I barely moved a couple feet. As I glanced toward the car, I caught a glimpse of my own frightened stare, reflected in the chrome of the bumper. But in an instant it was gone, replaced by a blur of fender as the Crown Vic whizzed past, scant inches from my face. I collapsed backward onto the pavement. My chest heaved with every ragged breath as I stared, spent, at the gray morning sky. Two for two, I thought – not too shabby. I was out of gas, though, and I knew it. If they came at me a third time, I was toast, and this body was heading right back where I found it. I wondered queasily whether the docs would even recognize poor Jonah once that Crown Vic had its way. It wasn't a comforting line of thought.

But they didn't take another pass. Instead, the engine cut out. Four doors opened, and then slammed shut. Four sets of shoes clattered across the pavement. Three stopped well short of where I lay – they spoke in hushed tones, their words lost to me on the breeze. The fourth approached me, blotting out the morning sky as he hunched over my crumpled form. He was fuzzy, hard to see – as if lit from within. I was pretty sure that wasn't just because of the crack I took to the noggin. My breath caught in my chest. My vision dimmed. I tried in vain to stretch my consciousness, to find myself another vessel, but the effort was too great – all I got for my trouble was a searing pain between my temples and the copper scent of blood prickling in my sinuses. Sirens, faint as hope, echoed in the distance. In that moment, I didn't care I was a fugitive – I just prayed they'd be in time. Whatever these guys wanted with me, it wasn't good, and it's not like I was gonna go down swinging.

"Is it dead?" called one of the stragglers.

"No," replied the one above me. "It lives."

"Come, Ahadiel. We have to go. Perhaps next time, we will finish him."

And then, sirens drawing closer, they fled.

I woke by degrees. The first thing I was aware of was my leg, which throbbed in time with the beating of my heart. Next came the sirens. They were everywhere, reverberating off the walls around me. I opened my eyes. Light flooded in, and my head erupted in white-hot pain. I clenched them shut again and retched. That meant concussion. Explained the fuzziness.

Again I opened my eyes, slowly this time. My stomach clenched, but I didn't vomit. It was progress. I looked around. I was lying in a broad trash-strewn alley, tucked between a dumpster and a loading dock.

And I wasn't alone.

By instinct, I tried to find my feet, but my hip felt heavy and out of joint, and my leg couldn't take the weight. I got to one knee before collapsing to the ground with a scream.

"Quiet," said the young man who sat beside me, nodding toward the mouth of the alley – toward the source of the sirens. "They'll hear you."

He was a wiry kid of maybe twenty-three, in a tattered army surplus jacket and dirt-smeared jeans. His pallor was gray, his face gaunt, his black hair was longish and matted. His eyes darted this way and that, looking anywhere it seemed but at mine. His frame and clothes suggested homeless. His furtive gaze suggested crazy. In his hand he held a knife, matte brown with rust and filth.

Christ, I thought – this day keeps getting better and better.

"What makes you think I don't want them to hear?"

"You told me. In my head."

I eyed him, suspicious. "I did."

He nodded. "In my head, I heard you calling. Afraid. Trying to escape. So I came to help."

"Look, about that – I appreciate the help, but I really gotta go."

"You are not who you are."

My heart skipped a beat. "Come again?"

"You are not who you are," he repeated. "Your body – it fits you funny, like borrowed clothes. And the voice you used to call me is not the voice you use now."

The kid rocked back and forth as he spoke, and still his gaze avoided mine. It was clear he wasn't quite right in the head – but could he really see me?

I rested my weight against the loading dock and stretched my consciousness toward him – probing, testing. The pain in my head redoubled as I struggled to focus. My body went slack as I pulled away. My vision dimmed.

I brushed against his mind, and he flinched as if stung. I settled back into the Friedlander body. The kid stared at me with wide-eyed terror.

"That isn't very nice," he said, shaking his head, his knife held ready between us. "My head is crowded enough already."

"I'm sorry." My hands were raised palm-out, my tone placating. "It's just that most people, they can't see me. What I am. Their minds won't let them."

He scowled. "You thought I was crazy."

"Of course not!"

"Everyone thinks I'm crazy. I guess maybe I am. But the pills, they dull everything. The tastes, the smells, the sounds. They reduce it all to ash. You ask me, I think crazy seems the saner option."

"Listen, kid, you got a name?"

"My mother called me Anders."

"Nice to meet you, Anders. Mine called me Sam. You think maybe we could do without the knife?"

He looked down at the knife in his hand as if seeing it for the first time, and then at me. From his jacket Anders produced a makeshift scabbard of duct tape; he slid the blade into the scabbard, and both disappeared into his jacket.

"Sorry," he said. "I was worried they'd come back. The ones who hurt you."

"Did you see them?"

"Yes. They were not like you. They were fuzzy. Hard to see. Like looking at the sun."

Shit – angels. That's what I was afraid of. What they wanted with me, I had no idea, but it was clear it wasn't good.

I pushed myself up off the ground and clambered awkwardly to my feet, careful to keep my weight on my good leg. "Anders," I said, "I have to go. I don't think I can walk, so you'll have to help me. You think you can do that?"

Anders nodded. "Is this about the girl?"

"What do *you* know about the girl?"

"Before, in my head, when you were trying to escape – you said she was in danger. That you had to save her. That everything depended on it."

"I did?"

"Yes."

I eyed him appraisingly. "So you in?"

Anders shrugged. "I guess," he said. "I mean, I'm not busy."

I laughed.

Anders added, "You said something else, too, you know."

"Yeah? What's that?"

"You said you thought *she* might save *you*."

I smiled and shook my head. I didn't doubt what the kid said, but I'd been a fool to even think it. After all, I was lost a long time ago.

9.

"Are you all right?" Anders asked. "You don't look well."

"I'm fine," I lied. Truth was, my head was fucking killing me.

"You're slurring. You need to sit down."

I opened my mouth to argue, and then closed it again. Anders was right. We'd been hobbling along for what seemed like hours, and I was exhausted. My leg was throbbing, my mouth was dry as dust, and my head felt like it was full of angry bees.

I looked around. The world lurched – my vision was slow to respond. We were heading north on Church, a few blocks south of City Hall. At the corner was a mounted cop, lazily scanning the crowd from atop his steed. I looked away. Beside us was a family of tourists, decked out head to toe in New York gear, and walking hand in hand. Their youngest, a girl of maybe six, caught my eye as they passed. Her eyes flickered with black fire as she spotted me, and her smile faltered, replaced by a look of pure hatred. As soon as it appeared, though, it was gone. She shot me a quizzical glance as though I was to blame, and then she smiled again, turning her attention once more to the sights of the city.

"I think maybe I *should* sit down," I said, "but not here. We need to get off the street."

Anders led me through a narrow parking lot to a side street. Beside a rusted metal door marked as the service entrance for the deli around the corner sat a battered dining-room chair, curlicues of green vinyl arching skyward from its cracked and peeling seat. Anders dropped me into the chair and plopped down onto a milk crate beside it.

I closed my eyes and willed the throbbing in my head to stop. It seemed my head had other plans. But at least sitting down, my leg was tolerable, and after a couple dozen blocks serving as a human crutch, I'm sure Anders was grateful for the rest. Crazy or not, he sure as hell never signed on for this.

We sat in silence a while: me stock-still as I waited for my head to clear, and Anders rocking gently back and forth, his gaze fixed at a spot just in front of his shoes. Eventually, though, his curiosity got the better of him.

"The men who attacked you," he said. "They were cops?"

"Not exactly."

"Then who?"

"That, I'd rather not say."

Anders nodded, as though that were answer enough for now. "But you're not fond of the cops –

I've seen the way you look at them. Watchful. Wary. Always quick to look away before they see you."

The kid was nuts, maybe, but not stupid. "I guess I like them fine," I said. "Only right now, they're not too fond of me."

"Why?"

"I took something that didn't belong to me."

"So you're a thief."

I smiled. "I guess you could say that."

"And the others?"

"What others?"

"The lady throwing bread to the pigeons. The man at the window in the coffee shop. The little girl, just now. All like you – like someone else behind the eyes – but only for a moment. They've been watching you. They've been watching you, and you've been terrified."

"Not like me," I said. "Not themselves, but not like me."

"Then what?"

Ah, hell, I thought. If he can see them – Anders deserves to know. "They call themselves the Fallen. But demons, devils, djinn – you can call them what you like."

He fell silent a moment, as if processing what I'd told him. "These demons – they're looking for you? Hunting you?"

"I don't think so," I replied. "These creatures, they're powerful, and clever as well. Any of them could've taken me if they wanted. No, I think they wanted me to see them. I think they wanted me to know that they were watching me."

"Watching you – why?"

I thought back to what Merihem had said to me. *You think either side wants a war? When last it happened, one-third our number fell – and all because a son of fire refused to kneel before a son of clay. You couldn't* begin *to understand the world of shit that would rain down upon us all if one of our kind was caught damning an innocent soul to rot in hell for an eternity.* My guess was, whoever Merihem had been leaning on had got to talking. Not that I should be surprised – if this morning was any indication, my days of flying under the radar were over. "It's complicated," I said.

"The men who attacked you – were they demons, too?"

"No."

I could have told him, I guess. That they were angels. I told myself then that he wouldn't have believed me, but I'm pretty sure that's crap. I think I was worried that he *would* have. I mean, Anders was a little off-kilter, yeah, but he seemed like a good kid. Who's to say he wouldn't have taken the angels' side? The way I figured it, the shape I was in, I needed all the help I could get. If that meant keeping the knife-wielding crazy person in the dark, then so be it.

He shook his head. "You don't seem very popular."

"It's been a rough couple of days," I agreed. "You didn't ask for any of this, you know. You wanna walk away, now is the time."

"The pills they gave me, they said they'd make it better. The fear. The worry. The things I thought I'd seen. They told me it was all in my head. But that wasn't entirely true, was it?"

"No, I suppose it wasn't."

"Closing your eyes won't make the world go away. I'm in if you'll have me. Besides," Anders added, looking me up and down, "you seem to be doing pretty lousy on your own."

By the time we made it back to Chelsea, day had evened into dusk and the lights of the city reflected amber in the overcast sky. It felt like we'd been walking for days.

Though this time there was no fire, no billowing bacon-scented smoke, the front door of Friedlander's building was unlocked. In retrospect, I should've seen that stroke of luck for the warning sign it was. At the time, I was so damn tired, all I wanted was to get upstairs and get some sleep.

The stairs themselves were tricky. With one hand on Anders' shoulder and the other on the banister, I half-hopped, half-hoisted myself to the top. By the time we reached the third floor, my lungs were burning, my face and neck were slick with sweat, and my chest and good leg ached from exertion. I collapsed to the floor beside Friedlander's door, exhausted. From somewhere down the hall, a dog yapped, driving into my temples like a furry little ice pick. I wished to hell it'd shut up.

Anders jiggled the doorknob. "Locked," he said. "You got a key?"

I shook my head. Anders shrugged and took a knee. From his jacket, he produced a small screwdriver and a scrap of metal wire. A bit of fiddling, and the lock clicked home. I pushed myself up off the floor and limped over to the door. This time, the knob turned fine. I pushed open the door and threw an arm around Anders. Together, we shambled across the threshold into the darkened apartment.

Inside, the place seemed deserted. The lights were off, the curtains drawn; the only illumination was the wedge of light that spilled into the apartment from the open door. My heart fluttered in panic as I opened my mouth to call for Kate, but the word died on my lips as the darkness was pierced by an animal scream. I was peripherally aware of a flash of movement, a glint of metal, and then I was falling. I slammed into the floorboards and skittered across the room, watching as Anders dove for the open door, his arms thrown up to shield his head. Our assailant followed, a cry of raw fury escaping her lips.

It was Kate, I realized. And as she drew her hands high above her head, I realized the glint I'd seen was a knife.

"Kate?" My voice had abandoned me, and all I could muster was a hoarse whisper. Anders was backed against the doorjamb – his eyes pleading, his hands raised in defense. Kate brought down the knife. *"KATE!"*

At the sound of her name, she wheeled. Too late to stop the knife, but not too late to deflect it. Anders rolled sideways, and Kate drove the knife into floorboards instead of flesh. Her eyes went wide with horror and she released the blade, backing slowly away from it as though it were an animal poised to strike. "Sam?" she said. She sounded suddenly small and afraid.

"Yeah, kid, it's me."

"But I thought – I mean, you were gone for *hours*, and then the door was rattling... I figured they'd gotten you – that they'd gotten you and come for me." She looked me up and down. "God, Sam, you look like shit!"

I laughed. The effort made me wince. "Lay off the funny, kid – laughing makes my everything hurt."

"Who the hell is this?" Kate jerked her head at Anders, who was staring up at her from the floor with a mixture of awe and terror.

"Long story. Why don't you close the door, and I'll tell you all about it."

She closed the door and helped me up. Together, we made our way to the couch. Anders collected himself from off of the floor and headed to the kitchen. He got a glass of water from the tap and handed it to me with shaking hands before taking a seat on the armchair, as far away from Kate as he could manage.

I took a sip of water and began to talk. I told Kate of my meeting with Merihem, and about the run-in with our friends in the Crown Vic. I told her of my rescue by Anders, and our subsequent trek across Manhattan. I left out the fact that Merihem claimed there was nothing I could do to save her, the identities of the folks who tried to run me down, and the attention my little field trip had garnered from the demon realm. The way I figured it, she'd had a bad enough week already.

Through it all, Anders sat listening quietly. When I finished, he spoke. "I know you," he said to Kate. "You're the girl on the TVs. Ten of you in every storefront. They say you killed your family."

"Sam here thinks I was framed."

Anders' gaze settled on the knife still jutting from the hardwood floor.

"Yeah," Kate said, following his gaze. "I'm really sorry about that. It's just that Sam had been gone so long, I was worried he'd been caught or something, and then things got really creepy here–"

"Creepy?" I interrupted. "Creepy how?"

"I don't know – just creepy. I mean, there was all kinds of commotion next door earlier, and I swore I heard a scratching in the walls. Then that damn dog started barking for no reason…"

Scratching in the walls. I leapt to my feet and hobbled to the wall that abutted the apartment next door, gritting my teeth against the pain. "Which wall – this one?" I asked.

"Yeah, how'd you know?"

"The others are either exterior or they face the hall." I scanned along the wall until I found what I was looking for. A heating vent, nestled in the far corner between wall and ceiling. My stomach dropped as I caught a flicker of motion like a snake receding into its hole, only this snake glinted like glass, like metal. Like the kind of camera a SWAT team would use to monitor a room.

"That dog wasn't barking for no reason," I said. "It's time to go."

But I was too late. As I hobbled toward the couch, the lights cut out, and the apartment was plunged into darkness. Anders found his feet and wandered over to the window, pulling aside the curtains and peeking out.

"It doesn't look like an outage," he said. "The rest of the block is fine."

"Anders," I said, "get away from the window."

"What? Why?"

"Get away from the window now!"

Anders must've heard something in my tone that rattled him; he leapt back from the window as if stung. In that moment, the window imploded, spraying glass and wooden splinters through the darkened apartment. Something clattered to the floor, and the room began to fill with thick noxious smoke, ghostly white by the reflected glow of the street lights. The hall outside the apartment echoed with a chorus of shouts. The floor resounded with the force of approaching footfalls, coming toward us from down the hall and up the stairs.

I realize now that someone must've tipped 'em to our presence – our faces had been plastered all over the news, after all, and with me in scrubs, carrying Kate's robed form down the street, we weren't exactly subtle getting here. No doubt some busybody neighbor spotted us and called it in. Cops were probably camped out all damn day, keeping an eye on Kate and waiting for her accomplice to return so they could spring their trap and snatch her back.

Like I said, *now* I get it. *Then*, though, all I knew was they were coming. They were coming, and I couldn't let them take her.

My leg erupted in pain as I sprinted across the darkened room. I paid it no mind. The gas was thicker here – it burned my eyes and clawed at my throat and sinuses like a rabid animal. All I wanted was to curl up on the floor and wait for the pain to go away. Of course, that didn't seem like much of a plan. So instead, I grabbed Anders and Kate by the arm and dragged them through the darkness toward the bedroom, slamming the door behind us.

The air in the bedroom was a little better. My eyes and throat still burned, but I felt a little more human – a little more in control. I pulled them close, shouting over the din of the raid. "Listen very closely. They're coming in, and if I don't do something to stop them, they're going to take us all. I can't let that happen. I'm going to need to create a diversion. You two stay in here and count to fifty. Then you go out the window and down the fire escape. Don't stop for *anything*, you hear me?"

"Won't they be watching for us?" Anders said to me.

"Not if I do my job."

"No!" Kate shook her head. "We're not going to just leave you here!"

"Kate, there isn't any other choice. My leg's shot – I ain't going anywhere. And without a diversion, you wouldn't make it five steps."

In the other room, the front door thudded. The jamb held, but it wouldn't for long. We were running out of time. Kate looked at me a moment, her eyes red and streaming from the gas, and then she leaned toward me, planting a kiss on my cheek. "Be careful, OK?"

I smiled. "You just worry about staying alive, all right? Once you're safe, I'll follow, I swear. There's a park at the corner of Ninth and Twenty-eighth – do you know it?" She nodded. "Good – I'll meet you there. And Anders?"

"Yeah?"

"You keep her safe."

The front door splintered inward with a sickening crack. It was time. I closed my eyes and concentrated, my lips moving in a silent prayer that this would work. Swapping bodies takes strength, strength and focus, and the shape I was in, both were in short supply. Not to mention the fact that possessing the living is not without its price. Still, my only alternatives were capture and death. If I were captured before I did my thing, then they'd get Kate, and she was as good as damned. As for death? Just because in my case it isn't permanent doesn't make it any more of a picnic. If ever I were gonna dig deep, now was the time.

They stormed the apartment. From my hiding place in the bedroom, I touched each of them in turn. The rookie, all fear and nerves – no use to me. The jaded old-timer, just looking to get through this so he could get back to banging his wife's sister. Ditto with him. The commanding officer who knows deep down he's thought of as an officious prick. Nope. But the one who was first through the door? Quiet. Competent. The one they all trusted. He was exactly what I was looking for.

I threw my mind at him with all I had. The Friedlander body convulsed around me as I struggled to pull away. Every muscle clenched as one. Tendons snapped like rubber bands. I shrieked in agony, but still I pressed on. My nose erupted in a torrent of blood, and for a fleeting moment everything went red as a vessel in my eye burst under the strain. Then, suddenly, the pain evaporated, and all went dark.

Friedlander was gone.

My mind slammed into the cop's like a freight train. He buckled, but kept his feet. His stomach clenched, threatened to purge. By force of will, I kept it down.

I wheeled around. Just the three of them inside with me, armored up like they were heading off to war. A lot of effort for such a little girl. My earpiece crackled with static and shouted commands, but I ignored it. Instead I raised my firearm, a mean looking fully automatic assault rifle that looked to weigh about a ton. This guy handled like a dream, his muscle memory doing all the work. He let out a panicked wail inside my head as I pulled the trigger, three quick bursts. Just like that, the advance team went down. My guy had decent aim – one of 'em took a stray bullet in the shoulder, but the rest hit them square in the breadbasket. If the vests had done their jobs, breathing was gonna hurt like hell for a while, but all three ought to live.

I approached the open doorway to the hall. A thousand shouted questions in my ear. I considered yanking the earpiece, but then I thought better of it. The better to hear you with, my dear.

A rustling to my right. One of my teammates was scrambling to get to his knees, his gas mask clouded with condensation from his labored breathing. His rifle lay useless halfway across the room. I watched him as he groped for the piece strapped to his ankle. Not on my watch. I cracked him hard in the face with the butt of my gun, and he fell limp to the floor.

I took a moment to check the others. They were both out. Best not to disturb them, I thought – they look so peaceful when they're sleeping.

The front door lay in the center of the floor, the hinges a splintered mess. I pressed my back tight to the wall beside what was left of the door frame and listened. If anyone was right outside, I didn't hear them. I rolled along the wall onto my belly, gun at the ready, and sprayed a few rounds into the darkened, fog-laden hall.

Again, the radio squawked. *"Jesus Christ, what the hell is going on up there? Flynn! Jenkins! Skala! Fischer! Anybody – report!"*

"We've got shots fired, and three men down," I replied, injecting what I hoped was the appropriate amount of panic into my voice. "They got past us, sir. Send all units to the front entrance – suspects are armed, and I think they mean to shoot their way out!"

I let off a few bursts into the hall to punctuate my point. From somewhere below me, I heard the *pop, pop* of return fire. The radio filled with chatter as cops were redeployed. I hoped that Kate and Anders were on the move – they were never going to get a better shot. I fought the urge to fall back and join them – for this to work, I was gonna have to keep the pressure on.

I crawled into the hall, pausing at the top of the stairs. If anyone had seen me, they didn't let on, and anyway, they had no reason to shoot at me if they had – I looked like one of them. Still, bullets *hurt,* so you can never be too careful.

The stairwell wound around a central shaft that cut clear down to the first floor. I rested the barrel of my gun between the wooden balusters and squeezed off a few shots toward ground-level. No response this time – they were either waiting me out, or they were already on the move. I slinked down the stairs to the next landing and tried again. Still no response.

The second-floor hallway was bathed in eerie white light, streaming in through the transom above the front door from the spotlight they'd trained on it from their position on the street. I steered clear of the beams, hugging tight to the shadow-clad floorboards. From where I lay, I had a clear shot at the front door. Gritting my teeth against the possibility of actually *hitting* anyone, I took it. Shafts of white light poured through the holes I'd punched through the door and swirled ghost-like with the settling remains of the tear gas. It was oddly beautiful.

I lay there a while, occasionally loosing a round or two on the poor innocent door to keep this standoff going. I wanted desperately to retreat and check on Kate and Anders, but they couldn't have been taken or I would've heard it over the radio. No, the best thing I could do for them was to stay put and give them time to run. When this was over all I had to do was find a quiet corner while they stormed the place and walk right out that front door. No one would be the wiser.

It was a decent plan. A solid plan. And all it took was a creaky floorboard to let me know it was never gonna happen.

The floorboard in question was about five feet to my right, just three steps up from my second-floor perch. By instinct, I rolled away from it, bringing around my gun – incessant yammering aside, this guy sure beat the last meat-suit for handling – but I was too late. It was the rookie, his face stripped of his gas mask, his eyes wide and frightened. He had his 9mm trained on me, the barrel bobbing between my head and chest in his shaky, unsure grip.

"Drop it, Mike."

I did what he said, setting the rifle on the floor beside me. I wasn't wild about my odds, lying flat on my back as I was, so I rose slowly to my knees, my hands raised in what I hoped was a placating gesture.

The rookie said, "Stay put, Mike – I don't want to have to use this."

"And I don't want to make you. Why don't we talk about this?"

I stepped toward him. He retreated.

I reached for the rookie's name. It wasn't hard to find – old Mike here was shouting to him at the top of his imaginary lungs. I said, "C'mon, Owen, it's *me* – why don't you put that thing down, and we'll walk out of this together."

"But you – you *attacked* us!"

"I'm sorry. I wigged. I thought they were behind us. This is all just a big misunderstanding."

Owen looked incredulous. "You *wigged*?"

"That's right."

"You wigged and took out your *team*?"

"Look, it was an accident. I said I was sorry." Again I stepped closer. This time, he didn't back away. "Just put down the gun. I mean, you're not really going to *shoot* me..."

I took another step, made a play for the gun. Owen screamed and backed away.

The last thing I remembered was a flash of white light, and the thunder of gunfire.

And then falling.

And then nothing.

10.

"All right, Mike. Why don't you walk me through this again?"

I was sitting chained to a table in a Tenth Precinct interrogation room. The fluorescent light overhead was making my head throb, and my chest was fucking killing me. Of course, it could've been worse – the way that rookie's hands were shaking, I'm lucky he didn't put a bullet in my head instead of my vest.

"I've been through this all a dozen times, lieu," I said, affecting a tone of weary resignation. "When we took the door, the room was quiet. I entered first. The gas was so thick, I couldn't see a goddamn thing. Something musta gone weird with my earpiece, 'cause I swore I heard movement behind me. I thought we'd been outflanked, and I panicked."

"You panicked."

"That's right."

The lieutenant gave me a look like I was something unpleasant he'd just stepped in. We'd been going around like this for hours, he and I. At first, I figured

I could wait him out – after all, this particular meat-suit was a cop in good standing; they had no reason to suspect he was involved. But as the night wore on, it seemed less and less like they were just gonna cut me loose. Of course, I could've just pulled a little body-swap and left poor Mike sitting here while I walked right out the front door, but that plan came with a big fucking catch. See, a demon takes a body for a ride, all the vessel's left with is a blur of disconnected fragments and images; the demon's thoughts remain occluded. Me? I don't have that kind of power. Just one more reason I prefer the dead: I jump ship now and Mike starts singing. They'd mostly think he'd gone off his nut, I'm sure, but they'd probably send a couple cruisers to the park regardless. My guess is they'd have Kate in custody before I could get within ten blocks of her. So for now, at least, there was nothing I could do but wait.

"Listen, Flynn, I want to believe you, but honestly, I don't know what the fuck to think. I got a kid out there who swears up and down you turned around and popped your team just as cool as can be. I got a body on the scene that matches the description of the perp who marched the MacNeil girl out of the hospital two days ago, and I got a coroner who tells me he collected the same body damn near a *week* ago from the same goddamn apartment. I got a little girl who butchered her goddamn family slipping past the best-trained unit in the country. And in the middle of it all, I've got you, telling me it was all just a big fucking misunderstanding."

"So where does that leave us?" I said.

The lieutenant rubbed absently at the back of his neck, a pained look playing across his face. "I don't have a fucking clue. And I hope to God this shakes out your way, Mike, but until I get some answers, I'm afraid you ain't going anywhere."

The thing about a deal with the devil is you don't always know you've made one till it's too late. I'd like to think I didn't. Then again, looking back, I'm not sure knowing would've changed a thing.

I found Johnnie Morhaim on the corner of Franklin Avenue and Van Buren Street, shooting craps out on the sidewalk with a pack of drunks and kids. Every town's got a guy like Johnnie Morhaim: quick to smile with a temper to match, Johnnie had a hand in every bum racket and crooked deal from Edgewater to Rockaway Beach. I'd met him a few months before, when Elizabeth and I had just moved to New Brighton; he'd been putting a crew together for some job or another, and he'd heard I needed work. It didn't take me too much poking around to find I didn't want the kind of work he was offering, but he never seemed to get the message – every week or so he'd happen by and ask me how the hunt was going. Maybe I should've caught the twinkle in his eye, the swagger in his step when he stopped by. Maybe I should've realized the guy had juice, and if he wanted to keep me desperate, all he had to do was put out the word and not a soul in town would hire me. Maybe I should've seen the setup for what it was, but I swear to God I didn't. Nope, instead I cursed my lousy luck and hobbled my way right back to Johnnie, just like he knew I would.

Johnnie scooped the dice up off the sidewalk amidst a chorus of shouts and jeers, pausing just long enough to take a swig from the bottle of rye that sat brown-bagged between his knees. If anybody else saw him swap the dice for a pair within the bag, they sure as hell didn't let on.

"Johnnie," I called, "you got a minute?"

He never even looked at me. "Can it wait?"

"Not long."

He tossed the dice across the sun-bleached sidewalk. The crowd erupted. "Elevens again, boys! Guess today's my lucky day!" Johnnie snatched up the loaded dice and pocketed them in one swift motion. Another pull off the bottle and the straight dice came back out to play. He handed them to a kid on his right and rose stiffly to his feet. "Your roll, sport – me and Sammy got some business to discuss. And don't think I won't be back for my money, hear?"

We strolled down the street a ways, Johnnie strutting along like he owned the whole damn town, me limping just a couple steps behind. He fetched a cigarette from behind his ear and struck a match; I tapped a fresh one from my pack and lit it as well. "So, Sammy," he said, smiling, "any luck on the job front?"

"That's kind of why I'm here."

"Yeah? You reconsider my proposition?"

"I'm coming around."

"That girl of yours – how's she feelin'?"

There was no point lying – the answer was written all over my face. "Not good. Something's gotta give, and quick. You said you know a guy could use a little help?"

"That's right," Johnnie said. "He's gonna hafta meet you first, of course. A nice, upstanding fella like you is just the kind a guy he's lookin' for, though, so you don't got nothin' to worry about. Your old lady's gonna be just fine – you wait and see."

"Set up the meeting – I'll be there. Just tell me where and when."

For a moment, I thought I saw a flicker of black fire dancing in his eyes, but it was gone just as quickly as it appeared. "All right, Sammy," he said, extending his hand to me. It hung in the air between us for a moment, and then I took it. His grip was cold and hard as stone. Johnnie shook my hand like we'd just concluded some high-powered business meeting, no trace of humor or irony in his eyes. "Looks like you got yourself a deal."

It turns out when was 3pm Tuesday. Where was Mulgheney's, a tacky little gin joint on the Upper East Side, just a block north of Midtown. Mulgheney's was the kind of place that sprung up three to a block across the whole city in the years after Repeal, all chrome and neon and drunken good cheer. Problem was, at Mulgheney's, the chrome was just a touch too gaudy, and the neon lights a hair too bright, their harsh glare revealing that what appeared to be drunken good cheer was a perhaps a little desperate, painted-on. The cumulative effect was a place too classy for the guys who worked the loading docks across the street, and too coarse for the moneyed set that populated the surrounding blocks. All of which sounded just about right for a cohort of Johnnie's.

The place was quiet when I arrived: a couple old-timers, nursing drinks at the end of the bar. A working girl, dividing *her time between sipping her gin and tonic and nibbling on the ear of her john, whose suit – a little loose on his frame, but well-made, and only slightly out of style – suggested banker, and whose glassy eyes read well past drunk. And in a booth in the back, a large, red-faced man in a dusty brown suit and a fedora to match sat flirting with the barmaid, a buxom brunette in a skirt so high and a neckline so low they damn near met in the middle. A shock of red tie hung around the man's neck, and the woman fingered it playfully as she laughed at whatever it was he'd just said. But then he spotted me standing in the doorway, blinking in the sudden gloom of the bar after the brilliant glare of the afternoon sun, and he waved me over, his massive chins bobbing up and down. I shuffled toward him, clenching my jaw against the pain in my knee and willing the limp out of my gait.*

"Sam?" he asked, once I reached his booth. "Sam Thornton?"

"That's me."

"Good to meet you," he said. "Name's Dumas. Walter Dumas."

He extended a hand. I shook it. Up close, I saw his bloodshot eyes, the gin blossoms that spread across his massive cheeks. It was pretty clear the guy was a few drinks to the good. He told me to have a seat, asked what I was having. I slid into the booth and said I wasn't thirsty. Dumas just shook his head and laughed.

"Nonsense! Dinah, bring the boy a shot o' whiskey and a beer, and what the hell, the same for me as well!"

"You got it, sugar," she said. She tapped Dumas playfully on the nose, leaning in as she did so he could better ogle the vast expanse of cleavage that pressed upward from her blouse

in brazen defiance of gravity and decency both. Up close, her perfume was dizzying, and the apples of her cheeks were pricked with red, from rouge or drink I didn't know. She flashed me a wink as she turned to fetch our drinks, and then retreated to the bar, Dumas eyeing her all the while.

"Fine piece a tail on that one," he said. "Got a husband, of course, but then that's no concern o' mine."

I said nothing. Dumas just smiled.

"So, Japs or Krauts?" he said.

"Excuse me?" I said.

"The limp – Japs or Krauts?"

"Actually, neither. I've never served, though not for lack of trying. I enlisted back in '42, but they bounced me on account of my wife's condition."

"Yeah, Johnnie mentioned she's a lunger." At that last, I flinched and cast my eyes around the bar to see if anyone had heard. Once tuberculosis moved to the lungs, it was both deadly and highly infectious – if word got out about Elizabeth, they'd surely lock her away in some horrid sanitarium where she'd slowly waste away to nothing. I refused to let that happen. Lucky for me, not a soul in the place was paying us any mind.

Dumas said, "You seem healthy enough, though."

"Docs say I'm doing fine." Of course, that was only half of what the doctors said. The whole of it was I'm doing fine *for now*. That living with Elizabeth, it was just a matter of time. The first few times, it didn't bother me – I mean, docs'll tell you all kinds of shit about eating your vegetables and laying off the drink, and that doesn't mean you listen to the letter. But you hear it enough, and eventually, it gets to you.

I'd be lying if I said I didn't break out in a cold sweat every time I stifled a cough, wondering – is this the time my hand comes back flecked with blood? I'd be lying if I said I wasn't terrified. But I needed this job, and saying all that wasn't gonna help me none. Besides, the way Dumas was looking at me, I got the sense he already knew it.

Our drinks arrived, and Dumas clanked his shot glass against mine, sloshing whiskey across the table, before tossing it back and chasing it with a swig of beer. I followed suit. My stomach roiled when the whiskey hit. Dumas held up his shot glass, signaled for two more.

"So if it wasn't in the war, where'd you get the gimpy leg?"

"Bad bit of business back in San Francisco. Back then, I worked the night shift at a foundry – at least, until we struck, that is. The owner of the place didn't take kindly to the idea, hired some boys to break it up. Some of us got a little more broke up than others."

"Ah, so you're a union man," Dumas declared, beaming. "No wonder Johnnie sent you my way!"

"I don't follow," I said.

"Don't tell me Johnnie didn't tell you! You're among friends, brother! I run the International Longshoremen's Union, Local 1566. Christ, that Johnnie's quite a card, setting up a meeting like this and not telling you what it's all about – you musta thought we were the Cosa Nostra or some shit!"

The barmaid brought our next round, her ruby lips parting in a smile as she leaned in close to set down my whiskey, the warmth of her breasts pressing against my arm. My face was flushed with embarrassment, though I wasn't sure why, and my head was fuzzy from the whiskey, from the barmaid's scent, from this weird-ass meeting. I tossed back the shot, and set fire to a cigarette. Neither helped to quell my unease.

"So that's what this is all about?" I asked. "A union job?"

His massive head bobbed up and down from behind his bottle of beer. "The union's always on the lookout for guys we can trust – guys who ain't afraid of a little hard work. Johnnie says you're good people, and his eccentricities aside, he ain't never steered me wrong yet. So whaddya say – you in?"

"I don't even know what the job is *yet."*

Dumas shrugged. "A little of this, a little of that. Errands, and the like. Nothin' you can't handle, I'm sure."

Another shot appeared in front of me. I downed it without a second thought.

"Johnnie said you knew a guy could help my wife."

"That's right. I know a group o' docs at Bellevue say they're running some kind of trial. A miracle drug, to hear them tell it. They think that it's a cure."

"And they're willing to treat Elizabeth?"

Dumas nodded. "Ever since we got into this goddamn war, most of the medical equipment and supplies in this country have been diverted to the front, which means that stateside they're in short supply. Now, I'm all for supporting our boys overseas, but the way I figure it, we gotta keep the home fires burning too. Now, nothing comes into or out of the harbor that my guys don't have a hand in – we just make sure some of it stays *here, and finds its way into some suitably appreciative hands. Workin' the docks ain't easy – we see our share of cuts and scrapes and broken bones. But you keep the sawbones happy, and they're more than willing to return the favor. We'll get that little missus of yours into that trial just as easy as you please, and soon she'll be right as rain."*

"You can seriously do that?"

"You have my word."

"Then just tell me what I have to do."

"Nothin' yet, 'cept to go home and tell your wife the good news. The work you'll be doin', it ain't steady, but it pays well when it pays, so don't you worry about that. We'll call you when we need you."

"I look forward to hearing from you."

"Excellent. Now if you'll excuse me," Dumas said, nodding toward the bar, "I've got me a barmaid to attend to."

Eleven hours.

Eleven hours they'd left me here, sitting alone in this holding cell without so much as a word. In fact, these past two hours I hadn't even warranted a glance from the officer standing watch. Not that I was surprised – it had been written all over their faces as they led me back here: I was a crooked cop. A traitor. I guess they figured they could leave me to stew awhile, see if maybe it loosened my tongue a bit.

Well, if they wanted me to stew, they sure as hell got what they wanted.

I sat there in that dank fucking cell, my meeting with Merihem playing over and over in my mind. *Any demon coulda taken this chick out for a spin,* he'd said, *but she'd be lit up like a Christmas tree for anyone who knew to look. No way she gets marked for collection. No, a con of that magnitude would take some serious clout – not to mention one hell of a death wish. You couldn't* begin *to understand the world of shit that would rain down upon us all if one of our kind was caught damning an innocent soul to rot in hell for an eternity.* So assuming I was right, why set up the girl? And who the hell had that kind of power?

More importantly, if someone was going to all this trouble, what was going to happen once word spread that I'd failed to collect her?

I had more questions than answers, but there was one thing I *did* know – I had to get out of this cell, and fast. Whoever or whatever was after Kate, they'd come too far now not to give chase, and I meant to be there when they found her. The problem was, this skin-suit wasn't apt to play nice – he'd roll on me the minute I let him up off the mat, and my little plan to save the world would be over before it had even begun. Of course, there *was* one other option, but it didn't exactly fill me with warm fuzzies.

But on the balance, what's one innocent life, when weighed against the Apocalypse?

Truth to tell, I'd known for hours that there wasn't any other way, but it took a while to find the nerve. Just the thought of it set my hands shaking, and filled my stomach with angry, crawling things. I mean – yeah, I take lives every day, but only those that are mine to take. This, though, this was something else entirely.

This was murder.

Still, it wasn't like I was taking his *soul.* Just extinguishing his mortal flame. He'd be better off without it, really. He'd be free to, I don't know, frolic through the fields of heaven or whatever. That's what I told myself, at least.

From the screaming in my head, I'd say neither of us much believed it.

The bed frame creaked in protest as I tipped it on its end and wedged it against the wall beside the toilet. They'd taken my belt and laces, of course, but my uniform shirt looked strong enough, and the sleeves were more than long enough to do the job. I stripped to my undershirt and knotted one sleeve of my button-down around the top of the bed frame. Then I climbed atop the toilet and tied the other sleeve around my neck.

Death, as a Collector, is a strange experience. For one, it hurts like hell. I mean, I suppose dying is never all that pleasant, but we Collectors seem to get a little extra in that regard. Whether it's a header off a bridge or a handful of pills, the agony is always the same. Kind of a stupidity bonus, I suppose. Still, we all try it a time or two before we catch on. The first time you take a soul, the experience is a little rough – most rookie Collectors think death the better option. And every once and a while, you see something that you just can't shake, and you get to thinking maybe this time it won't be so bad – maybe this time, they'll just let me fade to black.

Believe me, they never do.

Then there's the simple inconvenience of it all. See, a Collector's not like a demon – we can't exist outside a vessel. And when a vessel dies, any invading soul is expelled. So when we die, we get automatically reseeded somewhere else. If there's a rhyme or reason to where we end up, I sure as hell can't figure it. It could be around the block; it could be around the world. Both of which, I was forced to admit, would be better than my present accommodations.

Still I hesitated, whether from guilt or some nagging sense of self-preservation, I knew not which. I caught a glimpse of my vessel's reflection in the polished steel mirror bolted to the wall beside me: though his hair had silvered at the temples, and his face was well-lined, he couldn't be more than forty – a baby, by my reckoning. His eyes, a piercing blue, seemed to beseech me not to do this. I wondered if I even could.

Then I pictured Kate, so small and frightened and alone, and my hesitation evaporated.

I stepped off of the toilet.

I stepped off of the toilet, and nothing happened.

At first, I thought I'd just miscalculated – that I'd left too much slack in the shirt, and wound up just standing here, tied to the bed frame like an idiot. Then I looked down. My feet scrambled for purchase a good six inches off the floor. Just the sight of them swinging there made me break out in a cold sweat. And yet somehow, I was still breathing.

Whatever the hell was going on, I was sure of one thing: this was not my fucking day. I couldn't even manage to kill myself properly.

"You'll forgive my interference, I trust, but I found your chosen method of egress a touch… drastic."

The voice came from somewhere to my right, its

honeyed tones resonating off the cold masonry of the cell walls. Hanging there as I was, I couldn't see who the voice was coming from. I opened my mouth to reply. All that came out was a hoarse squeak.

"So sorry," continued the voice. "Where are my manners?" The sleeve around my neck abruptly slackened, and I tumbled to the floor.

He was a tall, slender man, and he was standing in the far corner of my cell. Though I was looking right at him, he remained fuzzy and indistinct, like something half-glimpsed out of the corner of my eye. His hair was neither light nor dark; his eyes were neither brown nor green nor blue. In fact, I could scarcely be certain he was a *he* at all: he was more the *impression* of a man, a collection of vague, impassive features, imbued with an odd internal light and clad in a suit of charcoal gray. Black gloves of supple leather graced his hands. He extended one by way of assistance, and I took it, climbing to my feet.

"What are you doing here?" I asked. His eyes seemed lit from within, his every movement suffused with preternatural grace. It was all I could do not to look away.

"Why, Collector, I would have thought that you'd be grateful – after all, I just spared you no small measure of suffering, did I not?"

"But you – you're a seraph, aren't you? An angel of the highest order. It seems odd you'd deign to meddle in the affairs of Man – or stoop to rescuing a lowly Collector from hanging himself."

The angel smiled. "It seems you know your angelic hierarchy. But tell me, Collector, how well do you know yourself? Your given name, for example, is from the Hebrew for 'heard by God'. Perhaps it is by God's grace that I've come to rescue you. Then again, perhaps I simply wish to save this vessel of yours from prematurely shuffling off this mortal coil. After all, this man is a warrior for good – he deserves better than to be discarded once his usefulness to you is at an end."

"So which is it? Did you come here to spare me or to save him?"

"It is a fallacy of your human perspective that it must be one or the other. Can it not be both? Or, failing that, can it not just be?"

"You're telling me mine is not to wonder why."

"I'm telling you to have faith in the will of God," the angel amended.

"Faith is belief in the absence of proof. As far as proof goes, I've seen my share. The way I figure it, that means faith for me is no longer an option."

"I speak not of faith that God exists, but of faith that grace lies not beyond your reach."

"I made my choice a long time ago. Save your talk of redemption for someone who deserves it."

His eyes danced with mischievous cheer. "Like, perhaps, the MacNeil girl?"

"So *that's* what this is about."

"Again you persist in this fruitless quest for *understanding.*"

"Yeah," I said, "I'm funny that way." Then my brain played a little connect-the-dots and I flashed the angel a rueful smile. "The guys in the Crown Vic this morning – they were *your* boys, weren't they?"

"An unfortunate misunderstanding," the angel replied. "I was laboring under the misapprehension that you were willingly subverting the ancient balance, and I reacted accordingly. Now I understand that your intentions are pure, and that you've simply been misled."

"So what – you're here to scare me straight?"

"I'm not here to *scare* anybody, Collector. I'm simply here to remind you that this détente of ours has lasted for millennia, and it has done so because the balance has always been carefully maintained – by those like me, as well as those like you. I would be loathe to see anything disrupt that balance – the results would be catastrophic."

"And if the girl is innocent?"

"Not a soul among us is innocent," he replied, "but of course that is not what you mean. You might be surprised to know your concerns have not fallen on deaf ears. I've looked into the matter myself, and I've been assured that she is anything but. To put it plainly, she's been deceiving you."

"I don't accept that."

"Whether you accept it or not is immaterial. The girl's collection is inevitable. If you truly care for her, the best thing you could do is collect her yourself. If you fail, they'll send another, and I doubt that Collector will share in your compunctions. You could spare her a world of pain with a simple act of mercy – and in the process, spare this world a war the likes of which it's never seen."

The angel gestured toward the cell door. It slid open as if of its own accord.

"So you're just going to let me go?" I asked.

"Yes."

"And what about the cops? They're going to wonder where the hell I went."

"I assure you, they'll remember nothing of this. It's best that way, don't you think?"

"You have to know I still mean not to take her."

"I have faith that when the time comes, you'll do what's right."

What's right – sure. I untied my shirtsleeve from the bed frame and slipped on the shirt. The angel conjured a business card from thin air, extending it to me. "If ever you need assistance," he said, "don't hesitate to give me a call."

I glanced at the card. It was a white so bright it seemed illuminated from within. On it was no number, no address, just a single embossed word, printed black as moonless night: *So'enel.*

"Thanks," I said, tucking the card into my pocket. Then I shuffled out of the cell block and through the oddly silent precinct house, fetching back my belt and laces from the abandoned guard's desk along the way. Outside, the sidewalk was flush with foot traffic, folks in business suits headed home from work.

With a glance back to be sure I wasn't followed, I descended the steps of the precinct house, disappearing into the crowd.

11.

Night had settled over the city by the time I made my way to the park. I was relieved for the anonymity the darkness afforded, but I didn't relish the prospect of tracking Kate and Anders down in it. At just a single city block, Chelsea Park wasn't a ton of ground to cover, but when you've got an angry horde of demons on your tail, you don't feel too compelled to stray from the cold comfort of the sodium-vapor lights and into the shadows beyond – missing girl or no.

Twice I wandered the perimeter of the grounds – up Ninth to Twenty-eighth, then over to Tenth and back down to Twenty-seventh – but Kate was nowhere to be seen. I hopped the low metal fence-rail and cut across the grounds. At this late hour, the park was devoid of patrons, with the exception of the derelicts who took refuge beneath her trees and sought comfort on her benches. As I wandered the footpaths beneath the canopy of leaves, I shivered. Sheltered as it was from the stone and brick and glass of the city, which seemed to radiate the sun's heat for hours into the night, it was colder here – achingly so. I shoved my hands into my pockets and pressed on, hoping against hope that I would turn the corner and find them there, waiting.

Eventually, my head caught on to what my gut had known all along: Kate and Anders were gone. The thought of Kate wandering the city with just a mental case with a bowie knife to protect her made my stomach lurch. I mean, Anders was a good kid, but what the hell was he gonna do if they came across another Collector, sent to do what I wouldn't? And if she *were* taken, what then? Apocalypse?

All of which meant there was no plan B: I had to find them first.

"Hey, pal, you got a smoke?"

He was huddled under a tree at the edge of a basketball court. With his matted gray beard and his ratty, timeworn clothes, he nearly disappeared into the gloom.

I patted my pockets reflexively, but of course I didn't have any. Whatever Flynn here had in his pockets when I snatched him had been confiscated before I ever came to.

"Sorry," I replied. "I wish I did."

"How 'bout a little cash, then?"

The second voice was lower, raspier, and dripped with Bond-villain menace. All of which was secondary to the fact that it was coming from about six inches behind me.

I said, "Listen, friend, you don't wanna to do this –

I've got nothing you could possibly want, and believe me when I tell you I'm more trouble than I'm worth."

"I think we'll be the judge of that, *friend*." Something cold and hard jabbed into the small of my back as if to punctuate his point. By the look of his cohort, I doubted it was a gun; more likely than not, I was being held up with an empty bottle of Night Train.

This day just kept getting better and better.

Guy One found his feet and clambered over to me, a look of demented glee pasted on his face. Guy Two had a death grip on my shoulder and continued to jab the not-gun into my back like if he pretended hard enough, maybe this time it'd go bang-bang for real. "Check his pockets," he called over my shoulder. His breath reeked of garbage and decay. His buddy didn't smell much better.

Guy One's fingers found my pants pocket and dipped inside. I saw my chance and took it. I slammed my head into his nose and he went down screaming. Blood spattered across the concrete. I grabbed Guy Two's wrist and twisted, hard. Something snapped, and he folded like a cot. My knee connected hard with his throat as he went down. He crumpled into a writhing, wheezing mess, his precious bottle shattering on the ground beside him. I stood at ready between them, my feet straddling the three-point line of the ball court, but they were all out of fight. Damn shame, I thought – I was just getting warmed up.

"Now, boys, I hope you don't mind if I ask you a few questions."

"Fuck you," said Guy One. Of course, with his nose a twisted wreck, it sounded more like *fug-OOH*. Still, you had to give him points for trying.

"I'm looking for a girl. Sixteenish, pretty. She would've been traveling with a guy about her age. Either of you gentlemen see her?"

"Ead shid ad eye."

"Sorry," I said, "didn't catch that one. Wanna give it another try?"

"Ead shid ad eye. *Eadshidadeye*!"

"Ah – eat shit and die. Charming. But I'm done playing."

I hunched over him and plunged my hand into his chest. He shrieked like a frightened child. Then I wrapped my fingers around his soul, and his shrieks died down to a whimper.

"Now," I said, bathed in the black light of his soul, "I'm going to ask you again. *Did you see her*?"

His eyes were wide with terror. Guy One said nothing. Then I gave his soul a tug and he started singing, his voice thick and nasal, his broken nose mangling his consonants.

"Y-yeah, I s-s-saw her. They l-left a coupla hours ago, when the cops came through to shake us out."

"Any idea where they went?"

"N-n-no!" The Ns like Ds.

I released him. He crumpled to the ground, crying like a newborn. "W-w-wha...what did you *do* to me?"

"Gave you a taste of what your eternity's gonna look like if you're not careful. You're gonna get the hell out of here and get yourself clean, you hear me? Stay off the drink, get yourself a job, and if ever you end up running this racket again, I'll be back for you. We clear?"

Guy One nodded, his face full of fear and awe. I was full of shit, of course, but what's the harm of a little white lie every now and again in the service of a good deed?

I snagged a handful of crumpled bills from the man's pocket – his take of the night's spoils, no doubt – and left him shaking on the pavement as I headed back toward Tenth. My head was reeling from the glimpse into his withered soul, and what little information he'd given me was ringing in my head. So Kate and Anders had made it this far, and they fled before the cops had seen them. That meant I still had a shot. But if I was gonna find them, I was going to need some help.

And so I set out to find me a payphone, oblivious to the eyes that tracked me through the darkness, watching.

I found a bank of payphones on the corner of Ninth and Twenty-sixth. One of them was missing entirely, and the second's handset was nowhere to be seen. I snatched the third off of its cradle and pressed it to my ear. It was dead. I muttered a silent prayer, to which side I wasn't sure, and punched in the number Merihem had given me. For a second, nothing happened. Then, somewhere in the city, the other phone began to ring – an odd, queasy, *reluctant* sort of ring. Still, I coulda done a jig.

After three rings, Merihem answered.

"I was beginning to think I wasn't going to hear from you, Sam." The voice was breathy and feminine, but there was no mistaking Merihem's tone. If I had to guess, I'd say he camped one of his girls out by a random payphone somewhere in the city in anticipation of this call. Locked up as I'd been, I wondered how long I'd left her standing there. I decided that I didn't really care.

"We need to talk," I said.

"I'm not sure that's such a good idea."

"Yeah, well, I ran out of *good* ideas a few days back, so it'll have to do. If you'd like, I can come to you."

"*No*!" Merihem's voice quavered for a moment – panic? fear? – but then he caught himself, and his composure returned. "That won't be necessary."

"Where, then?"

"The corner of Eleventh and Sixth. One hour. Don't be late."

"I'll be there," I replied, but there wasn't any use. I was speaking into a dead receiver. Merihem was gone.

12.

My muscles ached beneath the thin fabric of my uniform shirt, whether from my recent exertion or the chill spring air, I knew not which. I popped into a Duane Reade to buy a lighter and a pack of smokes, and then I struck out south toward my meeting with Merihem.

Though the night was cold, the streets bustled with people, and the air was redolent with an intoxicating mix of meat and spice and cooking oil from the sidewalk carts I passed, which mingled oddly with the scent of subway exhaust pouring upward from the ventilation grates beneath my feet. For a while, I wandered the streets at random, ducking down side streets, doubling back the way I came, but if anyone was following me, I didn't see them. For a time, I thought I caught a pair of eyes watching me through the crowd, but it was just a young boy begging for change, his face streaked with dirt, his jacket three sizes too big. I tossed him a couple bills from my would-be assailant's stash and kept on walking.

The corner of Eleventh and Sixth was quiet – aside from the Chinese place down the block, the place was mostly residential, all red brick and white trim and Woody Allen charm. Why Merihem would have chosen here to meet was beyond me. And speaking of, he was nowhere to be seen. I lit another cigarette and waited.

Three cigarettes later, I was getting antsy. I began to pace. I strolled up and down the length of the block, watching for Merihem all the while. Looking back, I must've passed the place a dozen times before I spotted it.

It was a low stone wall, wedged between two buildings and discolored with age. Hidden in the shadows as it was, it's no wonder I nearly missed it. I approached it cautiously, wary once more of being watched. Atop the wall, a wrought-iron fence stretched skyward. At the center of the wall was a gate, a lock dangling open from its hasp. I touched the gate and it swung aside.

"I was wondering when you'd come."

I squinted into the darkness. Eventually, an image resolved: Merihem, sitting propped against a tree amidst a sea of clinging ivy, a large obelisk headstone jutting skyward beside him. The graveyard itself was small, just a handful of weathered old headstones sticking improbably out of the ground and surrounded by buildings of towering brick.

"You could have told me where to find you. Speaking of, what's with the digs? You got something against meeting someplace we could get a drink?"

Merihem smiled, teeth flashing white in the darkness. "This cemetery was intended as a resting place for the sick. For nearly a quarter-century, those riddled with disease were interred here, in this soil. In 1830 city planners put a halt to that, insisting they be buried elsewhere; it seems the living have a limited tolerance for pestilence and plague so near to where they lead their desperate, fruitless lives."

"Look, Merihem, as fun as it is for me to reminisce about your salad days, we've got business to attend to."

"Hold your tongue, Collector. You think I selected this place so that I could regale you with tales of times gone by? I am the *bringer* of pestilence – this place is hallowed ground for me. Here, I cannot be harmed."

"What do you mean, *harmed*? Harmed by who? Merihem, what the hell is going on?"

"I did as you asked. I looked into this girl of yours."

Merihem fell silent, as if unsure what to say next.

I didn't have time for this. "*And*? What did you find?"

"A world of shit is what I found! This girl, she's caught the attention of some higher-ups – it seems they like her style. The way they tell it, she's destined for great and terrible things, Sam, only here you are, fucking it up for all of us."

"What do you mean *all of us*? All of us *who*?"

"You. Me. *Everybody*. Since word got out you've gone off the reservation, the angelic world is in an uproar. They've been leaning pretty hard on their Fallen brethren, convinced your little rebellion here is the first volley in some sort of insurrection. Now the demon-world is *pissed* – pissed at *you*."

I thought back to the black stares from the passers-by on my way back to Friedlander's apartment. "Yeah," I replied, "I got that feeling."

"Did you now? Well, believe me when I tell you, Sam, the folks we're talking about, it isn't a far cry between pissed and murderous. We may be lowly creatures in the eyes of God and Man, but a good many of us enjoy our little fiefdoms in this world, and would take personally any attempt, perceived or otherwise, to wrest them from our grasp. If they come for you, I'm not going to stand in their way – I'm pariah enough just for asking around. We go back a ways, you and I, but I'm not about to die for you. You go down, you're going down alone."

"Then what am I supposed to do?"

"There is no *supposed to* – *supposed to* implies options. I hate to rain on your parade, Sam, but that whole free-will thing? Kind of the dominion of the living. That isn't you anymore. You're nothing, now. Carrion. You just collect the fucking girl – period. If you're very, very lucky, that will be enough to spare your soul. There are worlds besides your own, Collector, and trust me when I tell you your hell is Paradise in comparison."

I hesitated, suddenly unwilling to tell him what I came to tell him. But as he said, I was out of options. "Listen, Merihem – even if I wanted to collect her, I couldn't."

"What are you talking about?"

"She's gone."

"I don't understand."

"Yes, you do."

"Are you telling me you *lost* her?" Fear crept into Merihem's tone. It didn't exactly fill me with warm fuzzies. If Merihem was this spooked, things were even worse than I thought.

"Look, the cops musta tracked us to where we were staying – they were waiting for me when I got back. I was able to keep her out of custody, but we were supposed to meet up after, and she never showed."

Merihem looked me up and down. "I guess that explains the new vessel. Police issue, no doubt?"

"Not that it matters, but yeah."

"And your girl – she just up and disappears? Sounds like the actions of an innocent to *me*." His tone dripped sarcasm.

"There were extenuating circumstances."

"Of course there were," he said.

"Merihem, I have to find her."

"I should say so."

"I kind of called you here to help me."

"That's funny – I thought you called me here so I could report on the *last* favor you asked of me. It seems our friendship is a costly one, Collector. Costly and dangerous."

I ignored the jibe. "Are you gonna help me or not?"

"Do you truly mean to take her?"

"I don't know," I admitted.

"That's simply not good enough."

"Damn it, Merihem, what if I'm right? What if this girl isn't meant to be taken? Am I supposed to just ignore what I've seen? To collect the girl like nothing ever happened and go on about my merry way?"

"What's the alternative? The balance must be maintained. If you're wrong, then this girl's fate is sealed. Refusing to take her would be seen as an act of war. Are you really willing to risk all that because Sam fucking Thornton had an *idea*?"

"I guess I am."

"Such hubris your species suffers from. No matter – if you fail to collect the girl, I'm sure they'll send another."

At that, I bristled. "Let them."

"Ah, yes – ever the protector. Good to see you haven't changed. And who knows? Perhaps you'll get lucky and dispatch the first they send her way. The second, even. But the third? The tenth? The thousandth? This game can't last forever. In the end, they'll get what they came for, and you'll get what you deserve."

"Then I guess we're done here." I turned on my heel and headed back toward the open gate.

"Sam, wait." I hesitated, not turning around. Merihem continued. "There's a man in Chinatown named Wai-Sun. He runs an antique shop on Eldridge."

"And?"

"Wai-Sun specializes in arcane objects – items of singular power. Weapons, talismans, and the like. He may be able to help you find what you're looking for."

"Thank you, Merihem."

"You understand the position you're putting me in by even meeting with you – I can't be seen as party to your sedition. If I see you again, I'll kill you myself. And Sam?"

"Yes?"

"Be certain that I don't."

13.

The bell above the door jangled as I stepped inside Shangdi Antiques on Eldridge. My sinuses prickled with the spicy scent of old wood and the dust of times gone by. The shop itself was tiny, and its wares were stacked atop each other at random, creating an accidental labyrinth whose walls remained standing in sheer defiance of the laws of physics and common sense. The sign in the window read "Rare Objects Our Specialty!" I hoped that it was true. Eldridge, it turned out, had no shortage of antique shops. This was the third place I'd visited today, and so far, I hadn't found any Wai-Sun. I couldn't help but think that Kate was running out of time.

"Can I help you?"

The call came from somewhere deep within the stacks, the English unaccented but nevertheless spoken with the melodic tones of one for whom Mandarin is his native tongue. I traced the voice back through the narrow winding aisle, nearly toppling an ancient bamboo birdcage in the process. I emerged to find a man standing behind the cluttered antique desk that served as the store's counter and polishing a small lacquered box with an oiled rag. He was short and stout, clad in a worn blue button-down and a dusty pair of suit pants. Thin wisps of white hair lay across his pate in a halfhearted comb-over. As I approached, he set aside the box and smiled.

"I hope so," I replied. "I'm looking for a man named Wai-Sun."

His smile faltered. "And what, pray tell, do you want of this Wai-Sun?"

"I've lost something, and I was hoping he could help me find it."

He gestured toward the piles of antiques surrounding us. "As you can see, we carry here a great many things – I am certain whatever it is you're looking for, we can find for you a suitable replacement."

"What I'm looking for is a girl."

Something flickered in his eyes. Fear? Suspicion? "I don't understand," he said.

"I think you do."

"Who are you? What are you doing here?" His hand crept toward the register. His eyes never left mine. If this was indeed my guy, I didn't want any part in whatever he was reaching for.

I raised my hands in what I hoped was a placating gesture. "My name is Sam Thornton. I'm here because a girl has gone missing, and it's important that I find her. I spoke to Merihem, and he told me you may be able to help."

The man broke into a smile, his hand no longer creeping toward the register. "Merihem sent you, did he? That bastard owes me fifty bucks. Sorry about all the subterfuge, but when one deals in items such as mine, one must be careful of the company one keeps. So you say you've lost a girl, eh? Let's see if we can find her, shall we?"

He removed from a desk drawer a worn wooden top and a creased map of the city, setting both on the desktop. I eyed them with suspicion. "*That's* what's going to help me find her?"

Again, Wai-Sun smiled. "Mystical objects need not be as elaborate as one might think. After all, appearances can be deceiving. So your girl – do you have anything of hers? A lock of hair, perhaps, or an article of clothing?"

I shook my head, and he frowned.

"No matter," he said. "I think I have something in the back that might do the trick."

He brushed aside the curtain that separated the front room from the back, and disappeared into the murk beyond. "So, this girl, she is of some importance, is she not?"

"She's my mother's sister's girl," I lied. "I was supposed to have her for the week, and she ditched me so she could meet up with her boyfriend. If I don't find her, Mom's gonna have a fit."

"Come now," he said, "there's no need to bore me with your falsehoods – I am merely making conversation. Your secrets are your own." Behind the curtain, something clattered to the floor, and Wai-Sun cursed softly under his breath.

"You need a hand back there?"

"No cause for alarm – I'll be out in a moment!"

There was something about his tone that didn't ring true. It was too cheery. Too earnest. Too at odds with the whispered epithet I'd heard him utter mere seconds before.

Something wasn't right here.

Silent as death, I ducked behind the desk and approached the curtain. The racket in the back room continued. Gingerly, I pushed the curtain aside.

Wai-Sun lay in the center of the storeroom, glassy eyes staring upward toward the ceiling. The floor around him was thick with congealing blood, glistening in the lamplight. His face was twisted into a rictus of pain, and he looked as if his throat had been ripped clean from his body. Well, anything but *clean* – tattered shreds of flesh clung to the ruined remains of his neck, exposing pink-white glimmers of bone beneath.

My Wai-Sun was standing, his back to me, in the far corner of the room, ransacking a set of small wooden drawers mounted above a rough-hewn workbench. His clothes, his hair, his *everything*, were identical to the man who lay lifeless on the floor beside him.

Too late, I realized what happened: that piece of shit Merihem had set me up.

Suddenly, my Wai-Sun straightened and turned.

"I really wish you hadn't done that," he said. Seeing him there, hearing him speak while two feet away he lay dead in a pool of his own blood, set my head and stomach reeling. "If you'd simply given us the girl's location, I might have let you live." His eyes flickered with black fire, and his features became suddenly vague – a mere *suggestion* of the Wai-Sun that lay ravaged at my feet. He seemed somehow to expand, his small frame suddenly filling the room. All around him was a halo of shimmering, liquid blackness, like silk fluttering weightless in an underwater current.

"No," I said. "You wouldn't have."

"Sounds nice, though, doesn't it? Merciful. Of course, I've never been much for mercy." The darkness pressed against my mind, obliterating all thought. I tried to tell my legs to run. They weren't listening.

"Who *are* you?"

"I think you misunderstand the situation, Collector. I'm the one who'll ask the questions. Now tell me – where the fuck is the girl?"

"You don't listen well, do you? If I knew where the girl was I wouldn't *be* here. Of course, Wai-Sun could've probably found her for you, if you hadn't gone and torn out his throat."

"You expect me to take criticism from a *monkey*? Wai-Sun was useless. He might as well have thrown open a window and shouted for her, for all the good he did. No, to find her I need someone with a *connection* to the girl – which, for the record, is the only reason you're still standing."

"If you think I'm going to deliver her to you, you're out of your fucking mind."

I didn't even see him move. One moment, he was standing half a room away. The next, his hand was on my throat. His eyes met mine, and I was plunged into darkness so complete, for a moment, I thought I'd ceased to be. Then he threw me across the room, and the darkness lifted.

I crashed into a stack of half-assembled wooden chairs. He was on me in a flash. He yanked me from the rubble by my arm. Something in my shoulder snapped. "I think with the proper encouragement, Collector, you'll tell me everything I need to know." He let me go, and I tumbled to the floor. Then he kicked me so hard my vision went dim and my mouth filled with the copper tang of blood.

The kick lifted me up off the floor and sent me sailing across the room. I slammed into a bank of shelves and crumpled to the floor, the shelves crashing to the ground atop me. Pain blossomed in my head and in my chest – exquisite, clarifying – and the world snapped back into focus. I clambered to my feet, shrugging aside the splintered wood and shards of glass that used to be the contents of the shelves.

I flashed him a half-crazed smile of defiance. "So tell me, demon, do you have a name?"

Again he struck. Just a momentary blur, and then darkness enveloped me, and I saw nothing. Great claws dug into my chest and I was lifted skyward, slamming into the ceiling before falling back to the floor, the storeroom rubble scratching and piercing my skin. I coughed and tasted blood.

"Are you the one who did this to her? Killed her family, set her up?"

The blow came from behind this time. It was like a fucking bus. I ricocheted off the workbench and smacked head-first into the wall before tumbling to the floor. A close one, I thought – if I hadn't gotten my arms up in time, that woulda been curtains for this meat-suit. Two in two days – it might have been some kind of record.

Then again, if I had died, I would have missed out on all this fun.

"You can make all of this stop, you know," the demon said to me. "Just help me find the girl, and I've no further quarrel with you. I promise I'll dispatch this vessel of yours quickly and you'll be free to go about your wasted, scavenging existence."

"That's a lovely offer, really." I lay prostrate on the floor, and drew breath in ragged, hitching gasps. "And after careful consideration, I've decided you can go fuck yourself."

The gap between us disappeared. A hand, cold and unyielding as marble, closed around my neck. My ears filled with the sickening noise of my own strangled gurgles; my legs pistoned in the rubble. I was running out of time.

"Wait!" I squeaked, and the grip slackened, just a shade. "Wait. I'll help you find her." The demon released my neck, instead grabbing me by the collar and dragging me out into the front room. He dropped me to the floor, and, once again Wai-Sun, wiped blood –

my blood – from his hand onto the wooden top, smearing the rest onto the map.

"You have made a prudent choice, Collector. Once I have the girl in hand, you have my oath that I shall kill you quickly."

I nodded, and spat blood onto the painted concrete floor.

"Now – clear your thoughts. Think of nothing but the girl. If you attempt to deceive the map, I will find out, and when I do, your suffering to date will be nothing compared to what you have in store. Are we clear?"

"Clear," I rasped.

The false Wai-Sun closed his eyes. I didn't. Instead I watched him as he descended into trance, my grip tightening around the dagger I'd snatched up off the floor of the storeroom. It was an odd little thing – pure silver by the look of it, with an ornate filigreed handle and a series of characters etched along the blade, in what to my eyes looked like Aramaic. I didn't know for sure if it could hurt a demon, but Wai-Sun's talents were acknowledged by Merihem and this creature both – the way I figured it, this was the only shot I had. All I could think was I'd better not miss.

The demon began to hum – a low, atonal, guttural tone, which was soon accompanied by a second higher one, and then one higher still. The top righted of its own accord and began to spin. At first, it skittered wildly around the table, and then it settled into an elliptical orbit. I tried to force any thought of Kate from my mind, which was about as useful as, I don't know, something not so useful. The top's orbit began to decay – it spun in ever smaller ovals, until it had centered on an area of maybe six by nine blocks. At least she was still somewhere on the island, I thought, but this had gone on long enough – any longer, and I'd be giving up the farm.

I dove toward the false Wai-Sun, drawing the dagger high overhead and plunging it deep into his chest. His eyes snapped open, and he staggered backward. The humming ceased, and the top skittered off the desktop and across the floor. The demon's eyes registered shock and surprise; he backed into a cherry end table and stumbled. His mouth opened, and closed, and opened again, emitting a dry, whistling rasp that built upon itself like waves capping against the shore. Tears sprung up in his eyes and spilled down his face. Soon his whole body was shaking, and he doubled over, bracing himself against the corner of the desk.

The demon, I realized, was laughing.

He said, "You fool. Did you really think that pitiful blade would hurt me? I'm a fucking *demon*. But don't worry – that's one mistake you won't have long to regret."

He approached, slowly this time, as if savoring the moment. I backed away. My hip connected with a mahogany buffet, and I tried too late to scramble over it. He backhanded me, and I sailed across the room, toppling a pile of furniture and sending a half-dozen vases shattering to the floor.

I made for the front door of the shop, but my way was blocked. The demon just smiled. I clawed at the mound of junk that barred my path, tossing anything and everything toward my assailant in a desperate attempt to slow him down long enough to make my escape. I bounced a pearl inlay music box off his temple, but it left no mark, and he just laughed – that horrible, wheezing laugh, like dry leaves on pavement. I heaved a wooden chest to the floor between us, but he simply gestured, and it moved aside. It was clear he was enjoying this.

I flung myself atop the pile as the demon closed the gap. As I clawed my way to the summit, he grabbed my leg in an iron grip. I kicked at him with my free leg, connecting with his jaw. It was like kicking a fucking tree. But daylight was so close, the shop door just a few feet beyond the mound of junk I lay atop – surely he wouldn't chase me into a crowded street?

I was pretty sure I knew the answer to that question, but still, I had to try.

Despite my efforts, he dragged me backward, daylight dwindling to nothing as I slid backward down the pile, loosing a small avalanche of timeworn junk. I grabbed whatever I could and winged it at him – a wind-up clock emblazoned with Mao's wizened face, a cane in the shape of a serpent – but still backward I slid. As he dragged me down to face him, my hands closed on a small ceramic Lucky Cat, the kind you'd find in Asian restaurants the world over, this one chipped and faded and ugly. But I was too late: his eyes, black as starless night, bored into my own, until nothing left of me remained, it seemed. His brittle cackle filled my head as I tumbled toward oblivion. In one last frantic act of rebellion, I brought the cat down hard onto his face. The way I figured it, if I was going down, I was going down swinging.

Something happened then, or rather several somethings, in such rapid succession it's not clear just what happened when. The darkness lifted, and consciousness returned, streaming in pure and true like first morning light. The demon released his grip, and I fell limp to the floor at his feet. A horrible, piercing shriek filled the air, rattling windows in their casements and setting off car alarms for a dozen blocks around. And, as I watched him stagger backward, the demon grew pale, indistinct – his insubstantial hands clawing helplessly at his torn and shattered face, the sharp edges of the broken figurine slicing through his flesh like so much Jell-O.

I skittered backward on the floor away from him, pure animal instinct urging me to flee. The demon fell to his knees, and then toppled to the floor – now charred black beneath him as if from fire, though just feet away, I felt no warmth. The shriek died to a whimper, and then fell silent. A voice – no longer connected to the transparent waif of a body that lay before me, but instead comprised of the myriad creaks and roars and scratches and whispers of the buildings and traffic and scuffing shoes and whooshing fabric that surrounded me as I lay on the floor of the dead

Wai-Sun's store – called to me, full of hatred and menace and fear:

You have no idea what you've just done. You've sealed your fate, and the girl's as well. You cannot kill us all, Collector, for we are Legion – and you cannot keep her from us forever. My brethren shall dine on the tender flesh of her soul.

Then the body before me burst – thousands of horrid, nameless, mewling things pouring forth from it and scattering to all corners of the store, disappearing into the murk. After a moment, their unnatural squeaking had ceased, but still my skin crawled from the sight of them, and my teeth were set on edge. I pushed aside furniture, sure they were still there – watching, waiting – but whatever they were, they were gone now.

I didn't have a fucking clue what had just gone down, but of one thing I was sure: whatever just happened, I was suddenly alone.

14.

The morning sun ducked behind a passing cloud, and I wrapped my arms tight around my chest to defend against the sudden chill. The signal changed, and I stepped out into the street, the ceramic shards in my pocket jangling as I hit the crosswalk on Morton, headed northwest toward Seventh Avenue on Bleecker Street. Since I left Wai-Sun's, I'd been wandering for hours, taking refuge in the quiet chaos of the Village. A far cry from the rigid grid of streets and avenues that traversed the rest of Manhattan, the tangled streets of Greenwich Village seemed as good a place as any to get lost – which was fine by me, since beaten and bloodied as I was, the last thing I needed was to be found.

I still wasn't sure just what in the hell happened back there, but one thing was certain – I was lucky I'd gotten out of Wai-Sun's alive. After I'd dispatched the false Wai-Sun, I'd collected up the shattered remains of the ceramic cat and stuffed them in my pocket. I'm not sure what kind of mojo that cat had, or whether it would work again, but I figured it couldn't hurt. Of all the things the demon had told me, at least one of them was true: *Mystical objects need not be as elaborate as one might think.*

After sweeping up the remains of the cat, I'd drawn the blinds, flipped the sign to Closed, and gotten the hell out of there, locking the door behind me. It was only a matter of time before Wai-Sun was found, but I wanted to be well away from there when he was. Besides, the longer it took for word to spread I'd killed a member of the Fallen, the better. The last thing I needed now was a pack of demons with a vendetta on my tail.

Once I'd left Wai-Sun's, I set out walking toward the neighborhood the top had circled in its last lazy arcs before skittering off the table and across the room. Of course, the top had only narrowed it down to an area of maybe fifty blocks, and it wasn't like I could just go around knocking on doors. Policeman-suit or not, that was liable to arouse exactly the sort of suspicions I could really do without. Still, the top was all I had, and one way or another, I simply had to track Kate down.

Fun as all that sounded, though, it was gonna have to wait. Right now, I had to deal with whatever it was that was following me.

I'd first spotted him last night on the way to my meeting with Merihem – a dirt-streaked kid in a jacket a few sizes too big, sitting at a busy corner and begging for change. I wouldn't have given him another thought, except I spotted his reflection in the window of a Korean take-out joint earlier this morning, and then again a couple minutes ago, when he got chased off from a news stand a half a block ahead of me for loitering. The kid didn't look to be more than eleven, and he was thin as a rail, but I didn't let that fool me – plenty of demons like to take a spin in the little ones, and tiny frames or not, demonic strength is all the same.

I lagged back a while to make sure he caught sight of me, and then ducked into a narrow service alley beside a dingy neighborhood pub. The stained brick walls were a scant three feet apart, blotting out the morning light, and the alley smelled of rotting garbage and piss. I held my breath and soldiered on.

The alley intersected with a haphazard courtyard, just a couple of picnic tables and a pair of withered birch trees overlooked by three buildings' worth of windows; the rear of the bar and the dry cleaner's next door made up the windowless fourth wall, bisected by the alley I'd just cut through. Clotheslines criss-crossed the sky above.

Yeah, I thought – this'll do fine.

Other than the alley, the only way out of the courtyard was through one of the three buildings. I checked the doors – two were locked, but the third was propped open with a dented Folgers can, filled with sand and littered with cigarette butts. I glanced back the way I came. There was no sign yet of my pursuer. Good – that meant I still had time. I dragged one of the picnic tables over to the far wall, and climbed atop it. After a minute or two of wild, flailing leaps, I managed to snag the fire escape ladder. It extended downward, rattling like a rusty chain, and then slammed into the tabletop with a satisfying *thunk*.

I hopped down from the table and retreated to the propped courtyard door. I set the can aside and stepped into the building, shutting the door behind me. I'd done my job well – through the narrow pane of safety glass set high into the door, I had an eyeline to the ladder and the alley as well. Now, all I had to do was wait.

Turns out, I didn't have to wait long. Maybe a half a cigarette after I'd assumed my post, I saw the kid's head duck around the corner of the alley. He was a cautious one, I'd give him that – he stuck to the shadows, his tattered, down jacket pressed tight to the dingy alley wall. He paused there a moment until he was sure there was no sign of me, and then he trotted over to the picnic table, circling it a time or two as though unsure what to make of it.

"Come on, you son of a bitch," I muttered, "take the bait."

After what seemed like forever, he did. I watched him scamper up the ladder, haul himself up onto the first landing, and continue on up the stairs toward the roof and out of sight.

It occurred to me then that I could run – just head out the way I came, and be rid of this tail, maybe for good. But I needed answers, and running wasn't going to get them for me. So instead, I forced myself to sit and finish my cigarette, allowing him ample time to reach the roof, and then, stubbing out the butt on the heel of my SWAT-issue boot, I slipped out the door and followed.

The pebbled roof bit into my tender stocking feet as I slinked across it, ceramic shard in hand. My boots were tied together at the laces and draped across one shoulder; I'd taken them off so I could ascend the fire escape unheard. But six stories of rusting waffled iron had bit into my soles and left me raw and hobbling, and now the kid was nowhere to be seen.

The rooftop was dotted with massive air conditioning units, and the odd pyramidal structure that housed the stairwell entrance jutted upward from the center of the building, blocking my view of the roof beyond. I clung tight to one of the air conditioners and crept toward the edge, painfully aware that, should I suddenly have to run, my chances were nil. The best laid plans and all that, I guess.

I wheeled around the corner of the AC unit, shard at the ready, but there was no one there. I approached the next, and crouched behind it, wary of remaining too exposed. Slowly, I circled, the seconds stretching on for hours it seemed, but again I came up empty.

Ahead lay the shed that allowed access to the stairwell. The roof behind me was hidden from sight by the hulking mass of the air conditioners. I let out a breath I hadn't known I'd been holding, and approached the stairwell door.

It was locked, as I'd expected, which meant he had to be beyond the shed. I crept around it, my thumb stroking the smooth surface of the ceramic shard for reassurance. My foot came down on something hard and sharp – a bottle cap, left over from some rooftop party, no doubt – and I stumbled forward. It was then that I saw him: leaning over the edge of the building, a hand on the handrail that curved upward over the low stone wall and provided access to the fire escape below. This fire escape was street-side, opposite the one we'd come up on – he must've assumed I fled down it, eager to be rid of my irksome little companion. But I had other plans. I stepped clear of my hiding place and strode toward him, the cat-shard brandished before me like a knife.

"Lose something?" I said.

The kid spun around, eyes wide with fright. His mouth opened and closed, but no words came out. He tried to back away, but his thighs connected with the rooftop wall – had he not been holding the rail of the ladder in a vise-grip, he would have surely gone over.

"Who are you?" I asked. "Why are you following me? Are you one of *them*?"

Still, he said nothing.

I stepped closer, shard held at ready. "One way or another, you *will* answer me."

Again I stepped toward him. He flinched but held his ground. Then my head snapped back as someone behind me grabbed a fistful of hair and yanked. I staggered backward. The tender flesh of my neck dimpled as a knife blade pressed tight against my windpipe.

"Easy, pal," said a voice into my ear, "the kid's with me."

15.

The hand yanked my head back. I struggled in vain against it. Knife parted flesh, and blood, warm and slick, dribbled down my neck.

"Stop fighting," said the voice. "I'll kill you if I have to."

I fought against the panic and stopped struggling. Instead I reached out with my mind toward my assailant's – probing, searching. If it was human, I could grab a hold, try to get it to release this body, and be back inside my policeman-suit before its owner got three steps. The only snag to that plan was this possession stuff is a little unpredictable – I had no way of knowing whether Stabby here was gonna clench up and dispatch my little cop-friend before I got a chance to play the hero. Between the real Wai-Sun, and the replacement I'd dispatched, I was pretty sure I'd already seen enough death for one day.

Turns out, fate had other plans. As I grazed his mind with mine, my assailant flinched as if stung. The knife clattered to the rooftop, and he released his grip on my hair. I wheeled on him, my face a tug-of-war of confused and surprised.

"Anders?"

"*Sam*? Jesus, you scared the *shit* out of me! I could tell this body didn't belong to whoever was driving, but I had no idea it was *you*!"

"I left the old one in the apartment," I said. "The place was his, anyway." I pocketed my cat-shard and dabbed at my neck with the palm of my hand. It came back streaked with blood.

"Sorry about that," Anders said, his furtive gaze regretful. "I thought you were one of *them*. A dark-eyed one, tired of simply watching." The kid I'd been following had yet to relinquish his grip on the handrail – he was just staring at me and Anders with a mixture of bewilderment and fear. Anders shot him a reassuring smile; it looked out of place on his gaunt, worry-lined face. "It's all right, Pinch. This is Sam – he's one of the good guys."

That characterization was a dubious one at best, but I wasn't in the mood to correct him. "Anders, what the hell is going on here? Is Kate all right? Who the hell is *this*?"

"Kate's fine – I'll take you to her. We tried to wait for you at the park like you said, but things got dicey quick. A bunch of guys were going door to door flashing Kate's picture around, asking if anybody'd seen her. They wore the skin of cops, but I knew better – their eyes shone black as night. I grabbed Kate and we got the hell out of there. Pinch here offered to stay in case you showed, but when you *didn't...*" he swallowed hard. "We thought you might be dead."

"Truth be told, you weren't too far off." I looked the new kid up and down, then, not bothering to hide my suspicion. Pinch let go of the ladder, and took a couple tentative steps toward me. "Pleased to meet ya," he said. He extended a hand. I ignored it. It hung there between us for a moment, and then he let it drop.

"Anders, what the hell were you *thinking* bringing someone else into this? Does he know where you're keeping Kate?"

"Relax, Sam. The kid's the best pickpocket in town – wasn't anybody gonna get the drop on him."

I said, "I just did."

"Yeah, only that almost didn't work out too well for you, did it?" Again Anders smiled. "Look, all I gave Pinch was the number to a payphone down the street. Told him if he saw anything, he should give me a call. A few minutes ago, he did. Seems he didn't like the look of your little setup, thought maybe he ought to bring along some backup."

"Still, if anyone had gotten that number out of him, it would have only been a matter of time before they tracked you down."

"I can *hear* you two, you know," said the kid.

Anders replied, "The way I saw it, without you around, we were as good as dead already. The number was a risk I was willing to take."

"I'm *standing* right here." Pinch spoke again, his voice tinged with impatience.

"Why in the hell was he following me in the first place?" I said.

"I told him if anybody else came looking for Kate, hang back and keep an eye on 'em. I hear you put on quite a show, questioning those homeless guys."

"You coulda gotten him *killed*."

The kid bristled. "I can take care of myself."

I replied, "No offense, kid, but you have no idea what you're dealing with. You're in *way* over your head."

Pinch just smiled and held a good-sized shard of ceramic up to the light and turned it over in his hand, inspecting it. My hand flew to my pocket. It was a whole lot emptier than I remembered. "Did you just almost attack me with a *cat*?" he asked.

"Don't touch that," I said, snatching back the cat-shard. "It's dangerous."

"Good thing you never tangled with my grandma, then – she had a couple dozen of these things. Coulda gotten messy."

I said nothing, settling instead for seeing if maybe I'd spontaneously developed the ability to shoot death rays from my eyes. Anders took the hint, and pulled the kid aside. "Listen, Pinch, why don't you take off? I'll catch up with you later, OK?"

"Whatever," the kid said. He trotted back toward the fire escape he'd come up on. Before Pinch disappeared from sight, Anders stopped him with a shout.

"Hey, Pinch?"

"Yeah?"

"You did good today."

The kid flashed him a smile, and disappeared behind the stairwell shed.

"You know you never should have brought him in," I said. "The kid's a liability."

"The kid's a *friend,* Sam."

"Yeah," I said, "same thing."

Dumas, it turns out, was as good as his word – two weeks after our meeting at Mulgheney's, we got a call from the research group at Bellevue. They said that they had an opening in their program, and that Elizabeth looked to be a perfect match. She couldn't believe her luck. I hadn't told her that Dumas had promised to get her in, so worried was I that he wouldn't deliver. In fact, I hadn't told her much about the meeting at all – I didn't have to. She was so over the moon I'd found a job, she didn't care much what it was. Which was fine by me, since I couldn't have told her what it was yet if I'd tried. I hadn't heard a word from Dumas since our meeting, and were it not for the call from Bellevue, it may as well have never happened. In retrospect, I'm sure that was all part of his plan. Once he had Elizabeth to use as leverage, he knew he had his hooks in me but good – there was nothing I wouldn't do to get her well.

I got my first call less than twenty-four hours after they'd admitted Elizabeth to Bellevue. The assignment was simple enough: just pick up a package and drop it in a locker at Penn Station. I was given a car, an address, a time and date. The car was a '42 Studebaker. The address was on the waterfront. The time was 4am. I guess that shoulda clued me in that something was hinky, but those were different times.

Least, that's what I like to tell myself. Sometimes, it seems to me the times haven't changed that much at all.

When I arrived at the pier, all was quiet. Though sunrise was still an hour away, the morning air was already stifling, and my clothes clung heavy to my skin. A cargo ship sat, moored and lightless, at the far end of the pier, a ramp jutting upward to her deck. I hobbled toward her, my progress tracked by a trio of crewmen who lounged smoking amidst the shipping crates that were scattered along the wharf.

By the flag flying from her mast, the ship was registered in Jamaica, but the crew mostly didn't look the part. Their appearance and the occasional snippet of Spanish that drifted to me through the still morning air led me to guess that Mexico had been this ship's last port of call. No one addressed me as I approached, nor did they object as, hesitantly, I mounted the ramp and limped upward toward the deck.

On the ship, I was greeted by a dark-skinned boy of no more than sixteen, who led me wordlessly to the captain's quarters, knocking twice on the open door before ushering me inside. The captain was a wiry man with eyes and skin of deepest brown, and an accent to match the flag atop the mast. He sat behind a massive wooden desk, scarred and pitted – and stacked high with books and charts. He didn't rise when I entered, and as I approached to shake his hand, he waved me off, instead nodding toward a worn leather suitcase standing just inside the door.

"I believe that is what you came for," he said. "Now take it and get the hell off my ship."

His tone was angry, to be sure, but the quaver in his voice belied the strength of his words. This man was afraid, I realized. Of me. Of Dumas.

Unsure how else to respond, I did as the captain said, retreating from his cabin without another word. The suitcase was heavy, and cumbersome as well. Twice as I descended the narrow ramp to the wharf, I stumbled, and nearly fell. But if the crewmen watching from behind the glowing embers of their cigarettes found my lack of grace amusing, they sure as hell didn't let on – there was nary a snicker or chiding comment to be had. It seemed the captain was not the only one who was frightened by my new employer. I was beginning to wonder if I ought to be as well.

It was just past 5am when I arrived at Penn Station, suitcase in hand. A far cry from the modern monstrosity now crammed like an afterthought beneath the hulking behemoth of Madison Square Garden, the old station was a soaring structure of glass and granite, its imposing colonnades oddly out of place alongside the deserted sidewalks of early morning. I left the car at the curb and wrestled the suitcase inside.

According to the board, the first train of the day – an overnight from St Louis – wasn't scheduled to arrive for another twenty minutes. Aside from an old man in coveralls, pushing a mop around like he didn't give a damn if the floor got clean, the concourse was deserted. A bank of lockers sat along the far wall, and I dragged my payload toward them, wincing as I heard my awkward, shuffling gait repeated back to me as it echoed through the vast empty space.

When I reached the lockers, it was clear I had a problem: with its stiff outer frame, the suitcase was just too damn big. No way was it gonna fit. But I wasn't about to blow my first assignment, so I decided to improvise. I'd just empty the contents of the suitcase into the locker, and drop the empty suitcase off when I returned the car and the key.

When I unzipped the suitcase, a sudden vinegar tang tickled my nostrils, and something else as well, earthy and unpleasant. It put me in mind of Mission Street out in San Francisco, where the hopheads used to beg for change to support their habits. The case was stuffed with paper bags, each dotted with oil spots and wrapped around something the size and shape of a brick. I took one out and looked inside. A compressed block of yellow-brown powder stared back at me, confirming what my nose had known all along.

Heroin. Musta been fifty grand's worth, maybe more. Whatever it was worth, it was more money than I'd see in a lifetime, that's for sure.

And there was something else for sure, too: no way was I gonna stand here in full view of anybody who cared to look and unload this thing into a locker. Which meant if I didn't figure out what I was gonna do with this shit and quick, I was pretty well screwed.

Footfalls echoed like gunshots through the concourse. I dropped the bag back into the open suitcase and wiped my hands off on my pants. Three bleary-eyed kids trotted past, dragged by their mother toward the platform, no doubt there to greet their father upon his return from St Louis. My eyes tracked them for a moment, but they never gave me a second glance. I zipped the suitcase and lugged it back through the station to my waiting car. I circled the terminal until I hit Eighth, and then I headed northeast toward Mulgheney's.

Dumas and I were gonna have ourselves a little chat.

The walls of the narrow corridor seemed to tilt and sway by the light of Anders' match-like reflections in a funhouse mirror. I followed behind him in the darkness, dragging one hand along the wall beside me to orient myself. The air around us reeked of moisture and rot, and the concrete beneath our feet was cracked and chipped – and littered with pots and pans and empty cans of God knows what, their labels faded to sallow obscurity.

Match burned flesh, and Anders cursed, dropping it to the floor. The match's flame guttered and died, plunging us into total darkness. My heart thudded in my chest as I remembered the eyes of the false Wai-Sun, their blackness so absolute it reduced all thought of light to the fleeting recollection of a half-remembered dream. I clenched my eyes against the panic and willed my heartbeat to slow.

We were three blocks and seven stories from the rooftop, in the basement storeroom of an abandoned restaurant. It looked like they'd ditched the place mid-renovation; the stenciled storefront window read Molly's, but the lettering was only half filled-in, and the entire storefront had been papered over with yellowed pages from the *New York Post*, the headlines eight months old. The front door was chained shut, but Anders led me around back to a secluded alley, wheeling aside a small dumpster as far as its chain would allow, to reveal a sidewalk-level service entrance, one scarred and rust-flecked corner peeled skyward just enough to get a grip. Anders grasped the corner with both hands and jerked it upward. Rusty metal squealed in protest, and then gave. Once we clambered inside, he bent the door back into place, reducing the bright afternoon sun to a mere trickle, watery and insubstantial. By the time we rounded our first corner, even that faint light disappeared, and we were reduced to traveling by match-light.

I had to give it to him – he'd stashed her someplace nice and hard to find. Wai-Sun's top coulda done a dance on the fucking roof and I *still* might've never found them.

Anders struck another match and we continued down the hall. I realized the detritus that lined the hallway was anything but random. By the light of the match, Anders zigged and zagged between makeshift walls of cans, and stacks of pots balanced precariously atop each other as if by a precocious child.

"Your work?" I asked.

"I figured if they found us, I didn't want 'em coming quietly," he said.

As we climbed the stairs, the darkness lessened. To our right was what used to be the kitchen. Once doubtless stuffed with ovens and dishwashers and stainless steel countertops, all that now remained were a series of black rubber mats and a wide double sink collecting dust on the far wall. To our left, a short hall led toward the dining room. Light trickled amber through the papered windows beyond, bathing Anders and I both in a peculiar golden light.

The light reflected yellow from a set of eyes glaring at us from a darkened corner of the kitchen. They locked on mine a moment, and then disappeared without a sound. Just a rat, I told myself. Nothing to worry about. Still, I suppressed a shiver as again I was reminded of my meeting with the demon – and of the horrid creatures he'd carried inside.

Just beside the stairs was a door. A small placard that read "Office" hung crooked at its center. Anders approached it and knocked: first twice in rapid succession, and then thrice more.

"Kate, it's me," he said.

From behind the door came the clunks and scrapes of furniture being moved. The lock disengaged with a *click*, and then the door swung inward. Kate stood in the door frame, looking haggard but beautiful as ever, a smile dying on her lips as she saw me.

"Kate, you've no idea how relieved I am to see you," I said, but she just backed away.

"Anders, who *is* this?" she asked.

"Kate, it's me – Sam!"

"Anders, he *told* you that? He told you that and you *believed* him?"

Anders was struck dumb by her response. Looked like I was on my own.

"OK, I took you from the hospital. I saved your life when you took those pills. I made you an omelet!"

"If you have Sam somewhere, you might've made him tell you all those things!"

I racked my brain for anything that might convince her. "When you were young, you used to be afraid of the man who lived downstairs. For years, you refused to take the elevator alone, and at night you'd sleep beneath your bed, your pillows under your blankets as a decoy in case he came for you."

She stared at me for a long moment, but I don't think Kate really saw me – she had a faraway look in her eye, like she was suddenly somewhere else entirely. "He had a glass eye," she said finally.

"What?" Anders said to her.

"He had a glass eye, and it didn't fit so well. Once, when we were talking in the elevator, it fell out. He popped it back in like nothing had happened, but from then on I was terrified of him. But how could you possibly *know* that?"

I flashed her a wan smile. "Comes with the job, kid." Truth was, my head was crammed full of countless such moments, every one of them but Kate's serving as a painful reminder of a soul I had dispatched. They filled my dreams in my sleep, and when sleep would not come, it was those stolen memories – those cast-off echoes of a life misspent – that robbed me of my rest. They were my punishment. My burden to bear. And they were never very far from reach.

But Kate didn't need to know any of that just now. She beamed back at me and threw her arms around my neck, squeezing until I thought I might pass out.

"Where are my manners?" she said once she released me from her grasp. "Come in, come in!"

Anders and I followed her into the office. She swung shut the door, and Anders helped her drag the scarred metal desk back in front of it. They tilted it on its side such that the desktop was wedged beneath the doorknob, bracing the door closed. The room itself was small and cramped, and flickered with the light of a dozen candles, which dripped wax on every filthy surface. Besides the desk, there was a ratty desk chair, its black vinyl cushions cracked and peeling, a hulking gray filing cabinet, and a dusty old floor lamp, its cord chewed through just inches from the base. I fingered a stack of unlabeled cans piled high atop the filing cabinet, and Kate smiled. "Pickings are kind of slim around here," she said. "We never know what we're gonna get until we open them. They're mostly just beans, but Anders swears he can tell which ones are peaches by the sound."

My eyes settled on a pile of old clothes in the corner, arranged in a sort of makeshift bed. "Church up the street is having a clothing drive," Anders said. "I snagged those off the steps last night. Figured we're as needy as anyone. We're sleeping in shifts," he added lamely, as if I might have assumed otherwise.

I tried to raise an eyebrow at that last, only to find that Flynn here couldn't manage it. "I'm just glad you two are safe," I said.

"And what about *you*?" Kate asked. "When last we saw you, you were convulsing on the floor, and now you show up here days later in a new body, only this one already looks like you put it through the wringer. Spill it, Sam – I want to hear everything!"

And so I told them. I told them how I shot my way out of the apartment, and how I'd requested all units to the front of the building, allowing them an opening to escape. I told them how the rookie got the jump on me, and put a bullet in my vest. I told them about the hours of interrogation, and my subsequent release. I told them of my meeting with Merihem, my run-in with the demon in Chinatown, and my unlikely deliverance at the hands of a small ceramic cat. They listened rapt throughout, asking only the occasional question of clarification, and I was suddenly struck by how *young* they both were – far too young, I thought, to have to deal with such unpleasantness. Then again, if life is suffering, these two were old beyond their years.

Funny, how that thought failed to comfort me.

What I didn't tell them were the *circumstances* of my release, or indeed of my meeting with the seraph at all. Even now, I'm not sure why. Maybe I didn't want to frighten Kate with the knowledge that the angels were aligned against her. Maybe I wanted to spare her the seraph's accusations of her treachery and deceit, and the fear and doubt they would instill. Maybe I didn't want to plant the notion in her head that I'd eventually betray her, as the seraph said I would.

Or maybe, just maybe, there was some small part of me that wondered if what the angel had said was true.

16.

"I'm coming with you."

Kate's statement hung in the air like a trial balloon, daring me to shoot it down. After two days of itchy, nerve-jangling wakefulness, I'd curled up on the office floor for a little shut-eye, waking just moments before to the sound of clanking pipes. Kate and Anders were busying themselves in what was left of the kitchen, their candlelight reflecting orange off the open office door. I propped myself against the wall and rubbed sleep from my eyes with bloodied knuckles. I had no idea how long I'd been asleep. Long enough for the soreness to set in. I don't know if you've ever had the experience of being tossed about like a rag doll, but I gotta tell you, I don't recommend it.

"Are you off your nut?" I called back, my voice echoing through the dark expanse of the basement kitchen. "That's *completely* out of the question!"

"Oh, come *on*, Sam, I'm not some helpless little girl. If this guy knows who set me up, I want to help you get him."

"First of all, Kate, Merihem is not a *guy* – he's a *demon*. As in powerful and evil and, whether he's involved in framing you or not, very interested in getting his hands on *you*. Or have you already forgotten why I got my ass kicked just yesterday?"

"I haven't forgotten. I just figured maybe you could use me – you know, like bait."

I said, "Bait only works when you've got yourself a trap to put it in."

"So then – what's the plan?"

"I don't know yet – but it sure as hell involves you staying *here*."

"You're being ridiculous."

"Am I? Let's forget for a second that the entire demon-world is looking to deliver your immortal soul to eternal damnation, quite possibly triggering a war of literally Biblical proportions – you're also the target of a citywide manhunt on the part of New York's Finest. You can't exactly flash that face of yours all over town."

"No?" she asked, strolling through the office door and giving me a catwalk twirl. "How 'bout *this* one?"

I had to admit, the transformation was impressive. Kate's long auburn locks were now shorn into a jagged bob that traced the line of her jaw. She'd bleached it all a platinum blonde, with a streak of blue framing her face to each side. Thick hoops graced her ears, and another wrapped around one nostril. A studded leather choker wound its way around her neck above a vintage T-shirt and tattered jeans patched with bits of plaid. A pair of work boots worn shiny from years of use finished off the outfit. She grinned at me with blue-painted lips, eyes sparkling from beneath streaks of metallic blue eyeshadow.

"Well? What do you think? The clothes are mostly from the bag we snagged – Anders ran out for the rest this morning. The nose ring is a fake, but it looks legit enough, I think."

"I gave him that money for *food*, not so you could play dress-up."

"All the food in the world isn't going to do us much good if I can't ever leave this basement."

A fair point, I had to admit. But still, going after Merihem was a far cry from simply walking the streets unnoticed. "Kate, I'm sorry, but there's just no way. You're staying here with Anders, and that's final."

Hot breath clouded the windshield of the van as I sat watching the stoop of Merihem's Staten Island lair, smoking cigarette after cigarette as much for warmth as out of boredom. The engine skipped a bit, and the van shuddered as if from a sudden chill. I knew how it felt. I'd snatched this rusty piece of shit from a parking garage over on Prospect Avenue, and swapped its plates with another just as ugly at a liquor store a couple blocks away. The way I figured it, even if anybody reported this baby missing, the cops would spend their night chasing down the wrong van. By the time they sorted out what happened, I'd be long gone. Still, if I'd known the heat was busted on this one, I might've opted for Door Number Two.

"You want to give me one of those?" Kate asked, eyeing the cigarette as she shivered inside her leather jacket.

"Not a chance."

"Come on – it's *freezing* in here."

"Hey, you're the one who wanted to come. Besides, these things'll kill you."

"I thought *you* were supposed to kill me."

"Yeah, well," I said, "the night is young."

"I still don't see why we couldn't stop off for coffee and doughnuts – I mean, this *is* a stakeout, after all."

"Maybe if you hadn't blown all our cash on that get-up of yours, we might have."

"Hey – this get-up is what got me here. Not to mention, you just stole a *car*. You can't find a way to score a couple bucks?"

"Sorry – I'll try to snatch a body with a debit card next time."

For the first time in the three hours we'd been sitting here, Kate fell silent. We watched the flophouse for a while in the sudden quiet, nothing much happening but the occasional junkie heading in, or a john coming out. Wind whipped down the street, tipping trash cans and rattling the low-slung shrubberies that clung, gray and dead, to either side of the stoop. Though the doors and windows of the van remained closed, the wind cut through them like nothing at all. My knuckles ached from it, and Kate, in the passenger seat, pulled her knees up to her chest and hugged herself for warmth.

"I don't know how you do it," she said finally.

"Do what?"

"Swap bodies like that. I mean, I changed my hair and my clothes and I feel like a different person. It's got to be hard not to lose track of who you are."

I shrugged. "It's not so hard, really."

"No?"

"I once read that nothing fixes something so intensely in your memory as the desire to forget it."

"What's *that* supposed to mean?"

"Nothing," I said. "Looks like we're on."

A figure had approached the stoop. Not an inch over four feet, and a slight four feet at that, he looked tiny and afraid in the orange glare of the sodium-vapor street lights. A filthy down jacket hung loose around his frame.

"You've got to be fucking kidding me," I said.

Kate shot me a puzzled glance. "Who the hell is that?"

"A liability," I replied.

Pinch paused at the bottom of the stoop, casting furtive glances left and right, and then he ascended the steps, knocking on the flophouse door. I stubbed out my cigarette and cracked the window. Whatever went down, I was damn sure I wanted to hear it.

After a moment, the door opened. Behind it was a chocolate-skinned woman in a leather halter and a denim miniskirt; a luxuriant head of cinnamon locks that was almost certainly a wig cascaded down over her naked shoulders. She was rail-thin, with sunken eyes and a face that could have been a young-looking fifty or a weathered thirty. My money was on the latter.

"Ain't you a little young to come 'round here, sport?" she asked. Her words dripped with condescension. A smile played across her face.

"I'm here to see Merihem," Pinch replied.

"Kid, I don't know where you heard that name, but believe me when I tell you, you'd best forget it quick, you hear? Now why don't you run along to Mommy – I'm sure she'd hate to hear what kind of trouble her baby's gettin' hisself into."

"It's about the girl."

"What girl you talking about?"

"You know what girl," Pinch said.

"Honest, baby, I don't. Maybe you could come inside and tell me?"

"I'll only talk to Merihem."

"Well, then, I guess I got no choice. Come on in, child, and I'll take you to him."

"I'm *young*, I'm not *stupid*. He wants to talk to me, he can bring his ass out here."

Her eyes flashed with anger at that last. "You'd best watch that mouth of yours, boy – you don't know who it is you're speaking of."

"I know enough," he said. "Just go get him."

The woman disappeared back into the house, and the door swung shut. Pinch shifted from foot to foot as he waited, rubbing his hands together to ward off the cold. He glanced around again, looking down the street away from us, and then directly toward the van. If he saw us inside, he didn't let on.

"I don't get it," Kate said. "She seemed pissed he wouldn't go inside, but that chick was twice his size – why didn't she just grab him?"

I smiled despite myself. "Because she *couldn't*. See, she can try to tempt him all she likes, but if he won't enter of his own accord, there's nothing she can do to make him. Sin is all about free will, which means evil has no power unless you grant it."

"Tell that to my family."

I flushed. "Kate, I'm sorry. You know I didn't mean–"

"Forget it," she said. "Something's happening."

The flophouse door swung open again, but this time, the errand girl was nowhere to be seen. Merihem looked down at the boy, a benevolent smile pasted on his face. Even from here, I could see it didn't touch his eyes. They exchanged a few words, and then Pinch beckoned Merihem to follow him. I dropped the van into gear and waited.

They stepped off of the curb and headed west across the street – Pinch leading, Merihem a couple steps behind. I floored the gas and the van lurched forward. Beside me, Kate screamed.

"Sam, what the hell are you *doing*?"

"Hold on to something," I replied.

"I thought this was a stakeout!"

"Change of plans."

The van shook like it was coming apart at the seams, and the engine whined in protest, but I kept the pedal to the floor. Merihem looked toward us, startled by the sudden noise. His eyes registered shock and surprise as they met mine. Then they registered the windshield as the van slammed full bore into him.

I hit the brakes. The van screeched to a halt. Merihem didn't. He skittered across the pavement for a moment, a tangle of limbs and tattered clothes, and then slid to a stop, leaning heavily against the curb.

I threw open the driver's side door and sprinted toward him, tire iron in hand. An acrid cloud of burnt rubber hung like fog over the roadway. Merihem shook his head as if to clear it, and tried to stand. I hit him with the tire iron, and he went down. Just stunned, I knew, and not for long, but it was all I needed. I leapt atop him and stuffed a shard from the ceramic cat into his mouth, wedging it tight such that the tip dug into the soft flesh of his palate. Merihem whimpered in sudden pain.

"Pinch, now!" I called. The kid picked himself up off the pavement and yanked a roll of duct tape from his coat pocket, tossing it to me.

"Jesus, Sam," Pinch said, "could you have cut that any closer?"

"You're still standing," I replied. I tore off a length of duct tape and pressed it tight to Merihem's mouth, wrapping it around his head a couple times for good measure. I grabbed him by the lapel and pulled him close, his nose nearly touching mine.

"The shard in your mouth – you know what it is?"

Merihem nodded, eyes wide with fear.

"Good. If I were you, I'd concentrate real hard on not biting down on it, or you might end up going bye-bye, you get me?"

Again, he nodded. I kicked him over, and grabbed his wrists, binding them tight behind his back with duct tape. Ankles, too. He grunted something unintelligible. I ignored it.

"Pinch," I called, "help me get him up! Kate, get the doors open!"

I grabbed Merihem by the arms. Pinch scooped up his ankles. Together, we hauled him to the van. Kate, who'd watched the whole affair with obvious horror through the windshield of the van, snapped out of it in time to climb in back and throw open the rear doors. We tossed in Merihem, and Pinch climbed in, too, pulling the doors closed behind him. Then I hopped into the driver's seat and punched it. The whole affair couldn't have taken more than thirty seconds, start to finish.

Son of a bitch, I thought – we just kidnapped a *demon*. I glanced back at the demon in question, noting with no small measure of fear the hatred that glinted in his eyes.

I'd better be right about the girl, I thought, because if I was wrong, the horrors of this existence were *nothing* compared to the torment I had in store.

17.

"Sam, what the hell *was* that back there?"

Kate glared at me, her face flushed from anger and cold both. The abandoned munitions factory towered overhead, its long shadow hiding us from the damning glow of the street lights and protecting us from prying eyes. The lot beside the loading docks was cracked and overgrown, maybe four decades of detritus littering seemingly every inch – beer bottles, fast-food wrappers, yellowed scraps of newspaper. At the far end of the lot, a tattered baby carriage sat on its side, one wheel spinning in the chill breeze. The chain-link fence around the property had gone up long ago, topped with barbed wire, but the padlock on the gate was rusted through, and a few good whacks with the tire iron did the trick. Anders and Pinch were inside with our guest. Kate, it seemed, had other plans.

"Look, Kate, I don't have time for this right now."

"The hell you don't. You said we were going there to watch, and instead we fucking snatch the guy? And what's with the kid? You make like you don't know what's going on, and next thing I know, he's in the goddamn van! You sent him, didn't you, you son of a bitch? You sent him, and you just decided not to tell me!"

"If I'd told you," I asked, "would you have let me do it?"

"Of course not," Kate replied. "He's just a kid, for God's sake!"

"You think I don't *know* that? You think I would've sent him if I had any other choice? If I'd gone to the door myself, I wouldn't have lasted ten seconds – they'd have dragged me in there and torn me limb from limb. That whole free-will clause doesn't apply to me – my fate was sealed a long time ago, and that means I'm fair game. No, for this to work, I needed someone human – someone *innocent*. Obviously, I couldn't send you, since you're the one they're looking for, and half the fucking demon-world saw Anders and me together when he helped me back to Friedlander's. That left the kid."

"Still – you just sat there and *deceived* me."

"I couldn't run the risk you'd wig out and botch the job. This isn't a *game* we're playing, Kate. If I let them take you, there's a good chance that this world is over. If that happens, that kid and everybody else are in for a life of suffering and agony, so if I've got to make a tough call or two, that's fine by me. My only priority is to keep you safe."

"Even if it means lying to me?" Kate asked.

"Yes."

"And Anders? Did he know?"

I paused, considering a lie – before reluctantly settling on the truth. "Yes."

"So it's just *me* that you don't trust."

"That's not it at all, Kate. Anders knows the kid. I don't. For the plan to work, I needed Anders to go talk to him, get him on our side – and someone had to prepare this place ahead of time for our arrival. If I could have left them out of this, I would have. But this I couldn't do alone."

"Hey, guys?" Anders said, poking his head out the door beside the loading dock. "This really isn't the best time. You maybe wanna come inside and talk to the angry demon?"

"Just give me a minute," I replied. Anders ducked back inside. "Listen, Kate, I appreciate your objections – really, I do. But whether you like it or not, Merihem is the closest thing we've got to a lead, which means we've got to know what he knows. Now, if that means I've got to hurt him, then so be it. If you can't be around for that, I understand. But we're too deep in this to look back now."

"You think he knows who killed my family?" Kate asked.

"He might."

"You think he's gonna talk?"

"I'm not sure."

"If he doesn't," she said, "I'll kill the bastard myself."

Candles flickered in the cold expanse of the factory, throwing shadows – of girders and machinery too cumbersome to have been removed – across the dirt-streaked windows and graffiti-tagged walls that surrounded us. Merihem sat duct-taped to a wooden chair in the center of the room, his mouth still bound. The chair – which we'd, uh, *borrowed* from the dining room of Kane and Anders' restaurant hideout – was propped against an I-beam that jutted upward from the uneven concrete floor and disappeared into the darkness above. Between the chair legs and the I-beam lay a scrap of two-by-four maybe three feet long, into which I'd wedged a half a dozen shards of ceramic, all pointing skyward. A length of nylon rope, looped around the chair's back legs at one end and clutched in Anders' closed fist at the other, spanned the seven or so feet between us. If Merihem tried anything, Anders just had to give the rope a tug and the chair would fall. If that happened, Merihem was gonna get a back full of goodbye. To his credit, he seemed to know it. Though his eyes glinted with cold, animal fury, he sat as still as death.

"Merihem," I said, "I'm going to remove the shard from your mouth, now. You so much as flinch, I swear I will end you, you hear me?" Merihem nodded once. "Good. Anders?" Anders nodded as well, and coiled the rope once more around his hand, stretching the line tight between them. Just a twitch, and it'd be curtains for Merihem.

The tape wound around Merihem's head several times, and came off reluctantly, tearing flesh and hair free as it did. He winced, but did not move. The shard was still in place – the strain on Merihem's jaw was obvious as he struggled to keep it open to prevent the sharpened tip from plunging deeper into the soft tissue of his palate and sending him to oblivion. Gripping his jaw with one hand, I reached in with the other and yanked free the shard. Beside me, Anders tensed, but Merihem just flexed his jaw a moment, and then was still.

"I take it you found Wai-Sun, then," Merihem said.

"What, *this*?" I said, holding up the shard. "No, this I got at Yankee Stadium on Kill a Fucking Demon Day." I wiped it off on my shirt and dropped it into my pocket.

"That the girl?"

"I'm sorry, am *I* the one tied to the chair? How 'bout I ask the questions for now, and maybe later we can switch."

"Cute, Sam – real cute. I'm going to kill you all, you know. I'll start with the little one," he said, nodding toward Pinch. "Then him," Merihem said, indicating Anders, "then you. I'll make the girl watch."

"Yeah, that's nice," I said. "But before we do that, why don't you tell me why the fuck you set me up?"

"It was nothing personal, Sam – you of all people should know that. It's just the girl's a hot commodity. Besides, I didn't have a choice – he got to me just after we met."

"Who? Who got to you?"

"His name is Beleth."

"Never heard of him."

"That's because he doesn't often deign to meddle in the affairs of Man." Then, addressing Kate: "You, missy, have attracted some serious attention – you should be flattered!"

"Go fuck yourself," Kate replied.

Black flames raged for a moment in Merihem's eyes. He blinked, and they disappeared. "Ooh, she's feisty – I can see why you like her so much, Sam. Maybe I'll take a go at her myself. I mean, she'll be kind of pretty once I tear that fucking ring out of her nose. Honestly, I've no idea what these kids today are thinking."

Kate fingered the nose ring. I shot her a look, and she stopped.

"So this Beleth," I said, "what's his interest in me?"

"His *interest*? You're in the *way*, Sam, it's as simple as that. That the girl will be collected is a foregone conclusion. The only one who doesn't seem to know that is *you*. You've become an embarrassment – you're making our whole damn operation look like a bunch of bumbling amateurs. The folks you're crossing don't enjoy being made fools of."

"Is Beleth the one who set Kate up?"

"Get it through your head, Sam – *nobody* set her up. It's been all her, all along. Every blow. Every slice. Every agonized scream. All of it the result of the depraved little creature scowling so adorably beside you. You understand, dear, that I mean no offense – I'm actually quite a fan. It's just time for you to come home, is all."

"Thanks for the invitation," Kate said, "but I'd really rather not."

Merihem smiled, all teeth and ill intentions – the kind of smile you feel in the pit of your stomach.

"Sweetheart, you make it sound as though you have a choice."

I interrupted. "So this Beleth – what else can you tell us about him?"

"I can tell you that he's a ways above my pay grade. Until Blondie here came into the picture, I'd never met him – I'd only heard the stories."

"Stories? What kind of stories?"

"They say he's a great monarch of the Depths. That he's most favored by the Adversary. That he's got a significant role to play in the great battle to come."

"You mean Armageddon?"

Merihem scoffed. "I sure as shit don't mean *Survivor*."

"*Who* says?" I asked. "What kind of role?"

"How the fuck should I know? *They* say, you know? This shit's all been foretold. *Beleth is a mighty and terrible king of the netherworld. His name shall bring forth the sounding of trumpets*. That sort of thing."

"That's not a lot to go on."

"Hey, they're *your* books, man. It's not my fault you people take lousy notes. I'll tell you this, though: if he had any fucking idea I was telling you this shit –"

"He doesn't."

"Not yet, maybe, but rest assured he will – and when he does, we're *both* gonna pay."

"Merihem, Beleth is dead."

At that last, his face dropped. Gone was the glimmer of fury in his eyes. For the first time, Merihem looked scared. "What the hell do you mean, dead?"

"Just what I said."

"Oh, fuck – the shards – I mean, I just figured you *escaped*!"

"I did. Right after I killed him."

"Shit, Sam, do you even realize what you've done? Nobody's killed one of my kind since the last Great War! If word gets out that Beleth is dead, the Fallen are liable to get the wrong idea, figure he's a victim of the crackdown. That happens, we've got war in the fucking streets. Not to mention, it's gonna come out eventually what *really* happened, and that's gonna lead them all to the both of us. When that happens, this little girl is gonna be the least of your worries. Man, you've fucked us but good."

"So what's the play, then?" I asked.

"Sam, you *have* to let me go – it's the only way. I can make sure nobody catches wind of what we've done. Beleth's got to have some enemies in the demon-world – our only shot's to try and put this all on them. If it looks like he's been killed by one of his own, we can maybe avoid a war."

"Avoid our asses in a sling, you mean," I said.

"That, too. You're in no position to begrudge me my motives, Sam."

"You forget, I'm already in a world of shit for taking the girl. Honestly, what's a little more heat?"

"I'm not talking a *little,* Sam. You're a fucking gnat right now – an annoyance. Word gets out you killed Beleth, they're gonna think you're trying to jump-start the End Days. That'd make you priority number one for both sides. We're all *happy* in our roles, Sam.

Comfortable. Isn't anybody on either side that wants to see the balance disrupted."

I fell silent a moment, mulling what he'd just told me. "If I let you go," I asked, "what assurance do I have you'll do as you say?"

Kate balked. "Sam, you couldn't seriously be considering letting him go?"

If Merihem heard her, though, he gave no indication. His eyes were locked on mine, his face betraying nothing. "You have my word," he said.

"Your word," Anders said. "Some fucking use *that* is."

"Yeah, Sam – let's finish this guy," Pinch chimed in.

"His word is his bond," I said, quietly. The corners of Merihem's mouth turned upward ever so slightly, almost imperceptible in the flickering candlelight. Almost.

"What?" Kate asked.

"His word is his bond," I repeated. "He's obligated to honor it. It's the way of his kind." I didn't say the rest. That his kind is disinclined to make pacts that end well for the second party – witness my day job. I didn't mention it because the way I saw it, we were *both* desperate. We *both* stood to lose. And if letting him go bought me enough time to clear Kate's name, then the deal would have been worth it, and the consequences be damned.

"So he'd *have* to help us?" Pinch asked.

"We let you walk out of here, and you leave us be, you got me? You don't come after the girl, you don't *send* anyone after her – you don't let it slip you might know where she is. Same goes for any of them. These kids are untouchable."

Merihem nodded. "All I'm worried about right now is my own ass. They tie me to Beleth's death, and it's all over. Far as I'm concerned, I never saw you."

"Anders," I said, "set down the rope."

"Are we really gonna do this?" he asked.

"I don't see we have a choice."

"This is ridiculous," Kate said.

"Anders, the rope."

Anders let go of the rope. It fell to the floor. I let out a breath I didn't even realize I'd been holding.

"Sam," said Merihem, "you're making the right choice."

I swear I never saw it coming. One minute, Kate stood fuming beside me, and the next, she'd closed the gap to Merihem. In one smooth motion, she kicked the chair out from beneath him. He teetered for a moment, his eyes wide with fear and surprise, and then he fell atop the shards. A horrid, guttural scream pierced the air and blew out windows the factory over. Candles guttered and died all around us. Anders crumpled to the floor, head in hands, and Pinch began to cry. But Kate never wavered, never flinched. As Merihem's writhing, fading form burst open, releasing the thousands of nameless scurrying things that passed for his soul, she spat on it, paying no heed to the terrible creatures that crawled, dragged, and scampered across her feet.

And under her breath, nearly lost beneath the echoing screams, she said, "That's for my family, you evil son of a bitch."

18.

Finally all was silent, and the mewling creatures gone. Anders was lying on the concrete floor, his eyes clenched shut, his face twisted in pain. He held his hands to his ears, a useless gesture. The sound he sought to keep out was in his mind: the anguished cries of those nameless, scurrying things that were once Merihem as they faded from existence. I knew, because I'd heard it twice now. Just two more things I wished I could unremember. Two among thousands.

I shambled over to where Anders lay, my borrowed body trembling, my knees threatening to buckle. I told myself that it was just a natural response to what I'd just been witness to, but I knew that wasn't completely true. Merihem's death had rattled me in a way Beleth's had not. Merihem wasn't a friend – not exactly – but we had a history, he and I, and that's not something you can easily forget. Now he was dead. Dead because of me. And it was a senseless death, at that – no honor, no dignity, no reason at all it had to happen. Demon or not, I couldn't help but think Merihem deserved better than that.

"Anders – are you all right?" He looked up at me and nodded. Anders was lying, of course, but that he was well enough to lie was a good sign. "We've got to get moving. Half of Staten Island must've heard those windows blow – we haven't got a lot of time."

I felt terrible for the kid – lacking whatever filter prevented normal people from seeing the world as it really was, only to be branded a nutcase, by them and me at first as well. Of course, if any of those so-called normal people could see the things that Anders had seen, they'd be a little twitchy, too.

I helped him to his feet, and nodded toward Pinch, who had retreated to a far corner of the room. Pinch sat with his back to the wall, rocking back and forth with his knees hugged tight to his chest. "Go help him," I said, "I'll take care of Kate."

Kate, for her part, was nowhere to be seen. Not that *that* meant much – most of the candles were extinguished during Merihem's exit, and the few that remained did little to push back the encroaching darkness. I noticed a thin rectangle of paler darkness along the far wall – a door, standing slightly ajar and leading to the night beyond. No doubt that's where she'd gone. I gave chase, and prayed she hadn't gone too far.

She hadn't. I found Kate standing with her back to me in the center of the abandoned, weed-strewn parking lot. She was shaking, I noticed, and she held her arms tight across her chest, hugging herself. It wasn't entirely from the cold, I thought. Demon or not, you couldn't just take a life and not have it rattle you a little. I once heard that it gets easier. I think they had it backwards. After a while, you just get harder.

"You wanna tell me what the hell just happened back there?" I asked.

She turned and looked at me, her eyes flashing with angry rebellion. "I ought to ask you the same thing. Did you think I was going to stand idly by as you let that bastard walk out of here?"

"You're damn right that's what I thought! Letting Merihem go was the smart play. I don't know if you've noticed, Kate, but we're kinda short on allies right about now, and thanks to you, we've got one less."

"You think he was an *ally*? I've got a newsflash for you, Sam – Merihem was a *demon*. As in evil. I did the world a favor, killing him."

"The hell you did. You wanna do the world a favor? Try dropping this bullshit vengeance trip and get on board with the whole keeping-you-alive thing."

"Bullshit?" Kate spat. "You think that this is *bullshit*? You said yourself they killed my *family*, Sam. This was just my way of trying to even the score."

"I said that *one of them* killed your family. I never said that it was Merihem."

"Does it matter? They're all the same."

"No," I said, "they're not."

"They're demons. End of story."

"You know what separates a demon from an angel? Choice. Angels are beholden to the will of God. Not a bad gig if you can get it, I guess. No doubt. No pain. No fear. No free will, either, but most don't seem to mind. There were some, though, who did – some who thought free will was worth losing everything for. They turned their backs on who they were, which meant turning their back on God. They were cast out for their impudence, forced to live a twisted, perverted existence, forever obscured from the light of God's grace."

"Why are you telling me this?"

"Because you need to understand that whoever killed your family made a choice to do so. Because back there, you just did the same. Demons aren't the only ones with free will, Kate. Be sure you use yours wisely."

"You think that Merihem was innocent," she said.

"Of this, yes."

"What makes you so sure?"

"Merihem was a corrupter of souls, a bringer of pestilence. For his line of work, this world of yours is fertile ground. He had no more interest in seeing it end than you do."

"That doesn't exactly make him sound like one of the good guys."

"I never said he was. But this is bigger than you, Kate. Bigger than what happened to your family. If they succeed in collecting you, we're talking about the end of the world. I'll take my help wherever I can get it."

Kate gazed in silence at the pavement for a moment. When she spoke, it was barely a whisper, and her eyes never left the ground. "The last time I spoke to them, it was in anger."

"What? Who?"

"My mom. My dad. My brother. I'd been planning a road trip with some friends for the summer. There's this music festival out in Washington – three days of bands and camping and whatever. It just seems so fucking silly now. Anyways, Dad said I could go, but Mom thought I was too young to go traipsing across the country by myself. I tried to tell her I wouldn't be by myself – that we'd be fine – but she wouldn't hear any of it. We ended up shouting at each other over the breakfast table, and I said some things…"

Tears spilled down her cheeks, and she was suddenly racked with sobs. "Kate," I said, "you don't have to tell me –"

"Yes, I *do*. I can't just keep carrying it around. It's too much." I nodded, and she continued. "I told her that I hated her. That my *real* mother would've let me go. That I wished that *she* was dead instead."

I was taken aback. "Your real mother?"

Kate nodded. "She died when I was very young. Complications from childbirth. And Dad… I mean, I know he missed her, but he never took it out on me. When I was three, he met Patricia. She's the only mother I've ever known. I just can't believe I *said* those things – and all over a stupid fucking *trip*!"

"I'm sure she knew you didn't mean it."

"Did she? Did she know I didn't mean it when I killed her husband right in front of her? When I killed her *son*? Did she know it while I tortured her?"

"Kate, that wasn't *you*. You have to understand that."

"How can you be sure? How can you know I didn't, I don't know, invite something in when I said what I said? That I didn't open the door for this to happen?"

"It doesn't work that way, Kate. If a moment of anger was enough to invite a possession, there wouldn't be demons enough for the demand."

"You *say* that, sure, but you aren't certain – I can see it in your eyes. You've seen what I'm capable of," she said, nodding toward the factory door. "You've seen what I can do when I get angry."

"Yeah, I have, but I've also seen your *soul*. I know you weren't responsible for your family's death, Kate, even if you don't. You've just got to trust me."

Kate brushed tears from her cheeks and looked at me, eyes rimmed with red. "And what about what I did back there? If you looked at my soul now, what would you see? Have I been tainted by what I've done? Can you just collect my soul now, and go on about your merry way?"

"It doesn't work that way, Kate. You knew full well what Merihem was when you did what you did. Besides, you're innocent in all of this – he and his kind had no business meddling in your affairs."

She laughed – a shrill, humorless bark of a laugh. "So I just get a freebie, then?"

"I wish it were that easy," I said, "but taking a life – human or not, justified or not – it eats at you. You take enough of them, it'll hollow you out from the inside, until there's nothing left but a husk of your former self. I don't want to see you head down that path."

"Is that what *you* are, Sam – a husk of your former self?"

I shook this borrowed head, shrugged these shoulders that weren't mine. "Sometimes I think I'm something even less than that." I took her hand, led her back toward the open factory door. She didn't resist – not exactly – but there was no volition to her movements; I felt like I was posing a doll. "C'mon, kid," I said, squeezing her hand in mine, "time's short. We've got to get you out of here."

The midday sun reflected off the chromed storefront of the bar, casting haloes of light across the sidewalk and causing me to squint. I took a sip of coffee from the mug in front of me, but it was cold and bitter, and seared like acid as it went down. I pushed the mug aside. Really, I shoulda stopped drinking this shit three cups ago: my eyes were dry and itchy, and felt too big for their sockets; my scalp was crawling from the caffeine and the lack of sleep. But I wasn't about to slink off to bed. Not with a fortune in heroin stuffed into the back of a borrowed car. Not without talking to Dumas.

When I left Penn Station, I headed straight to Mulgheney's, *but by the time I got there it was nearly 6am, and they'd been closed for hours. I parked the car out of sight around the block, and plopped myself down on a stoop across the street that afforded me a decent view of the entrance to the bar. I was determined to sit here for as long as it took, and anyways, what choice did I have? Dumas never gave me his number or address, so all I had to go on was that* Mulgheney's *was his favorite watering hole, and he had the look of a guy who had himself one hell of a thirst. The way I figured, it was only a matter of time before he showed.*

Eventually, though, the waiting wore on me, and I realized if I was gonna last the day, I was gonna need a little pick-me-up, and a bite to eat as well. So I moved camp to a lunch counter just a couple doors down, and ordered up a cup of coffee and a plate of steak and eggs, rare and over easy. The eggs came over hard, and the steak well, but the coffee did the trick, and the refills were free. Two hours later, though, the guy behind the counter lost his patience with me and quit topping me up, hence the cold and bitter. Didn't matter, though. Just as I was beginning to contemplate the odds on another sip being any better than the last, I spotted my mark.

Dumas was half a block away, slouching toward the bar in a sweat-stained camel-colored suit, a matching cap atop his head. I tossed a couple bills onto the counter and slid off of my stool. As I approached, he pulled the cap off of his head and mopped his brow with his sleeve. The cap blocked his view of the street. He never saw me coming.

I caught up to him just steps from the entrance of the bar, grabbing a fistful of lapel and pinning him to the wall. His face was a mask of shock and surprise, and his eyes glinted in sudden anger. Still, he made no move to stop me.

"You set me up, you son of a bitch!"

His prodigious brow furrowed. "Sammy, what is this about? Set you up how?"

"Don't play dumb with me. That package I was picking up? It was smack."

"Now how the hell would you know that? Your orders were to pick it up and drop it off, not to open it."

"Yeah, well, I did."

"Why on Earth would you go and do a thing like that?"

"Why doesn't matter – what matters is what was inside."

"Believe me when I tell you, Sam, it matters very much. That dope, it belongs to some pretty dangerous people – people who would not take kindly to you messin' with their product."

"It didn't fit," I said.

Dumas cocked his head, shot me a puzzled look. "What?"

"The suitcase. I tried to put it in the locker, but it didn't fit, so I figured I'd just take out the contents, leave 'em in the locker like you said."

To my surprise, Dumas laughed – a big boisterous full-bodied laugh that set his chins quivering. "It didn't fit? Shit, ain't that a hoot!"

"Yeah, a regular laugh riot."

"Ah, you know what they say – the best laid plans and all that. You didn't leave it there, did you? All unwrapped and everything?"

"No, I didn't leave it there," I snapped. "It's in the car."

"And where's the car?"

"It's safe."

"Good boy, good boy. So you been waiting here for me ever since?"

"That's right."

"Sounds like you've been having yourself one bitch of a day. Why don't you come inside and we'll discuss it over a drink, like civilized men? Maybe I can explain myself a bit, you'll see I ain't as bad as I might seem."

I don't know why, but I released him. Dumas straightened

his jacket, picked his cap up off the sidewalk, and gestured for me to head inside.

He led me to a booth in the back – his usual, it seemed, the one I'd met him in before – and flagged down the bartender, ordering a beer and a shot apiece. When they arrived, Dumas downed his shot and took a pull of beer. I ignored mine. He eyed me a moment, giving me a chance to reconsider, and then shrugged.

"Listen," he said, "I'm real sorry about this mornin'. You weren't meant to see that."

"That doesn't change the fact I did."

"You're right, of course. I guess I owe you an explanation."

"What good is explaining gonna do?" I said. "I'm no dope peddler."

"Nor am I, Sam – nor am I. But I am *in shipping, and if there are people willing to pay mightily for their shipments to arrive in time and unmolested, who am I to turn them away? What is in those shipments is their concern, not mine. And OK, yeah, maybe this time, I knew what was in the suitcase, but so what? These folks ain't giving this shit to schoolchildren, they're running a* business. *As in, if people wanna buy it, it's none o' mine."*

"You can't expect me to just look the other way, pretend I never saw what I saw. The world doesn't work that way."

"Believe me, Sam, people see what they choose to see every damn day of their lives. Besides, I'm not the bad guy here, and neither are my clients. You wanna blame somebody, you blame Uncle Sam. These clients o' mine, they were perfectly happy running booze across the border, and wasn't nobody complaining then. But then Repeal yanked the rug right out

from under 'em, and what do you expect 'em all to do? They got a right to make a living, after all."

"Sure, they got a right, only I don't want any part in the living they choose to make. The catch is, now I'm stuck with a car full of dope and nowhere to put it. Or rather, you are, 'cause I'm out." I slid the keys across the table toward Dumas. They came to rest against his substantial belly, which pressed tight against the table's edge.

"You're out."

"That's right."

Dumas nodded, raised his hands in acquiescence. "All right," he said. "I can see you've thought this through. I guess all that's left is the matter of your wife, then. Or had you forgotten?"

"You leave Elizabeth out of this."

"It'd be a damn shame if she got dropped from the program now – I hear she's makin' such progress, after all."

"Damn it, she hasn't done anything wrong. You wanna punish me, you go ahead, but you leave her be."

"Oh, don't worry, Sam, you'll get yours, but the deal was you work for me, your Elizabeth gets the treatment she so desperately needs. You don't work for me, she doesn't – it's as simple as that."

"You'd really do that to her? You'd really let an honest woman die?"

"Oh, no, Sam – not me. You. You go back on this deal of ours now, it's you who's letting her die. Her blood is on your hands."

I dropped my gaze then, to the shot that lay in front of me, and to the beer. I stared at them a while, not moving, not *speaking. Then I tossed the former back, and chased it with the latter, glugging away at the beer until there was nothing left but foam.*

"All right," I said. "Just tell me what I need to do."

Back in the factory, Anders sat huddled beside Pinch, one arm slung around the boy's shoulders. Pinch was shaking, and tears welled in his eyes, but he bit them back. A tough kid, I thought, but still just a kid. I felt sorry for him. I felt sorry for them all.

The sound of sirens cut through the still night air, drifting through the empty window frames and reverberating off the factory walls like an unholy orchestra. We didn't have a lot of time.

I searched the charred wreckage of the chair for the remains of the ceramic cat, but they'd been mostly ground to dust – there wasn't enough left of them to threaten a cockroach, much less a full-sized demon. That left only the shard that I'd removed from Merihem's mouth, its slight weight in my shirt pocket an uncomfortable reminder of just how tenuous a protection it was.

I ushered them out the door and into the van, slamming shut the doors behind them. A glimpse of flashing red and white through an alley, a siren's wail approaching. The van's engine didn't want to catch. Just a sputter, then nothing, over and over again. Eventually, though, it fired to life, and I dropped it into gear, lurching away from the curb without lights and screaming down the street.

"Where are we going?" Kate asked.

"Don't know. First thing is, we've got to find a spot to ditch the van – somebody might've seen us snatch Merihem, and even if they didn't, the thing's too hot to hold for long. After that, you three are gonna hafta hole up a while. I'm gonna try and get some answers."

"But with Merihem gone, aren't you kinda out of sources?"

"No," I said, my face set in a frown. "There *is* one other."

19.

"Collector," she said, a smile dancing across her luscious lips – lips painted a red so deep they looked black by the pale glow of the moon. The color of lust, I thought. Of blood. "I confess, I was surprised you called – and as you know, I don't surprise easily."

"Thanks for meeting me here, Lily."

Her smile faltered. On that face, with those lips, it was like snuffing out the sun. "You know I hate it when you call me that," she said, "and as for meeting you, it's not as if I had a choice."

She was right, about the latter, at least. I'd ditched the van in an alley off of Lafayette. Allison Park was just a couple blocks away, all old-growth forest and verdant lawns and quiet. Once an asylum for dying sailors, the park would suit my needs just fine. I'd stashed the three kids in a picnic shelter buried deep within the trees. Just a shingled roof atop a dozen rough-hewn posts, a stack of picnic tables chained together in one corner, the structure was more concealment than shelter, but it was well away from prying eyes, so for now, it'd have to do.

Once there, I'd sent Anders in search of a few supplies: a cast-off feather; the branch of a withered, dead tree; a night-blooming flower. When he returned, I'd set off east through the trees, my items in tow. I'd also taken Anders' knife, and Pinch as well, who looked a little ill at the request. But still he came, and when I told him what I'd needed, Pinch never faltered. His blood dripped black onto the makeshift altar I'd constructed out of stones at the center of the asylum's old cemetery, and he retreated to the forest's edge, dabbing at his arm with a kerchief.

Really, I don't know what the hell I was thinking, bringing her there. I'd considered taking the seraph up on his offer of calling upon him should I need anything, but I didn't know what constituted cause for summoning an angel. Besides, what I needed was the skinny on who could've framed Kate, and as far as that went, So'enel played for the wrong damn team. All of that sounded plausible enough, and all of that paled in comparison to the simple fact that the reason I hadn't summoned the angel was because he scared the shit out of me. That left me with no option but Lilith.

For all my effort, I wasn't at all sure she'd show – these sorts of invocations are more the domain of the living, and my Sumerian ain't exactly up to snuff. But show she did, strolling out from a copse of trees in a sheer nightgown that only served to amplify the graceful, feline movements of the body it was intended to conceal.

"So," she said, strolling barefoot toward me and

running a fingernail down my face, my neck, "what is it I can do for you, Collector?"

"I need some information."

"Ah. I see. Tell me, are the stories true? I hear you've grown yourself a conscience. That you've gone rogue. That you've defied the Maker and the Adversary both."

"I'm just trying to do my job."

"You *are*." She smiled again. My heart skipped.

"I am. Only I'm not sure this is my job to do."

A frown settled on her face, delicate and adorable. "I'm not sure I understand. You are a Collector. Your job is to collect. More specifically, your job is to collect those souls I *tell* you to collect. Honestly, I'm having difficulty understanding why I shouldn't just report this little revolt of yours the moment I take my leave of you, and wash my hands of this whole sordid affair."

"Lily, what if this girl isn't meant to be collected?"

She laughed then, a throaty purr I could feel in my socks. "Tell me, Collector – is she beautiful?"

"What the hell does that have to do with anything?"

"I'll take that as a yes. So what, you think if you can save the girl from hell, you'll become a real boy, and the two of you will ride off into the sunset?"

"I know damn well it's too late to save myself. But this girl's an innocent. It's not too late for her."

"Is she nearby?"

I hesitated. I wished I hadn't. "She's safe."

"Of course she is. I only asked because you could not have brought me here without the blood of an innocent willingly given, and we both know *yours* hardly qualifies..."

"I found another volunteer. This isn't exactly my first day."

"Ah," she said, "very prudent. Keep the poor unfortunate soul far away from little old me. Or is there perhaps another reason you brought someone else? Perhaps you didn't care to discover her blood was not so innocent as you'd hoped?"

"Could be I thought that. Could be I figured whoever set her up might have worked some mojo on the girl that woulda kept her blood from doing the trick. Tell me, what do you know about a demon named Beleth?"

"Of Beleth I know volumes. I know he's a demon of great influence and power. I know that he's a fierce warrior, and a fiercer lover. I know that he's taken quite an interest in you of late. But none of that is what you wanted to know, now, is it? What you really want to know is could Beleth have orchestrated the girl's collection? What you want to know is what he would stand to gain should the girl be taken? Really, Collector, you should know better than to play coy. That invocation of yours binds me to secrecy– your equivocating accomplishes nothing but the waste of both our time."

"Fine, then. Did Beleth set up the girl?"

"I haven't the faintest."

"*Could* he have?"

She thought a moment. "I suppose. I mean, obviously, it's never been done before, so I couldn't say for certain, but a being of his power could certainly make a go of it. The question, though, is why?"

"Maybe he was bored. Maybe he had grown tired of the truce. Maybe he's a fucking demon, and this is just what demons do. His motivations really don't concern me much – what concerns me is undoing what he did."

She was peering at me now, as if for the first time. A puzzled frown darkened her exquisite features. "Tell me, Collector – why the past tense?"

Fuck. Rookie mistake. Best to play dumb: "I don't follow."

"He *was* bored. He *had* grown tired. What do you know that I don't?"

"Nothing," I said, a bit too quickly. "I only meant –"

She cut me off. "He's *dead*, isn't he? He's dead and *you* killed him."

I said nothing for a moment, just squirmed beneath her withering glare. "Yes," I said. "I killed him. Merihem, too."

"Do you have any idea what you've done?"

"They would've done the same to me," I said.

"Yes, I suspect they might have – as would have been their right. But for all your talk of protecting the balance by refusing to collect an innocent, you sure have a funny way of maintaining it. This could well lead to the very thing you claim you're trying to avoid."

"Yeah, but if Beleth set up the girl–"

"You idiot – Beleth couldn't have set up the girl! For a creature of his kind, his power is inextricably linked to his being, his essence. If he had set her up as you claim, his death would have released her, and I assure you it did not." Lilith saw my face drop. Again, her smile came out to play. "You really thought he did it, didn't you?"

"I hoped he had, yes," I replied.

"Then tell me, why on Earth did you kill Merihem?"

"Merihem's death was an accident. We needed information. He was something shy of cooperative."

"Should I take that as a warning, Collector? Perhaps I should endeavor to be more forthcoming. Still, I thought the two of you were... not *friendly*, exactly. Collegial, I suppose. I'm surprised you had it in you to kill him."

"I did what I had to do."

She appraised me a moment, frowning. "You're lying. It's written all over your face. *You* didn't kill him, did you? It was the *girl*."

"One of his kind killed her family," I shot back. "She saw a chance to even the score, she took it. You can't blame her for that."

"Of course, of course. Or perhaps the girl worried that Merihem might expose her for the charlatan she is? After all," Lilith said, caressing my cheek with the back of one blood-colored nail, "how long do you think she'd last without her big, strong protector watching over her?"

"You're wrong about her," I said.

"Maybe, maybe not. It hardly matters. They sent another to collect her, you know."

"I suspected they might. Collectors I can handle."

"Don't be so sure. This Collector is one of Beleth's own. A thousand years he's walked the Earth since Beleth first sired him, and not a shred of humanity remains. He's more demon now than man."

My stomach dropped. "Bishop," I said. "They sent Bishop, didn't they?"

She raised an eyebrow. "You know him?"

"We've met."

"Ah, but of course you have! Then you know full well what the girl is in for. You, too, I'd imagine. As I understand it, he was something of a pet to Beleth. You see, he thinks of Beleth as his Savior – his one true God. What do suppose a creature such as he would do to the man that killed his God?"

I said nothing – just stood there, stunned. She approached me then, and draped one arm around my neck, pulling me close. Her body pressed against mine, and my head swam with the scent of her, all jasmine and spice and sex. I clenched shut my eyes to steady myself, but it wasn't any use. As her lips brushed against my ear, she spoke.

"This vessel suits you, Collector – we could have had such fun with it, don't you think? It's a pity they will flay it alive for what you've done. And who knows? Perhaps I'll see you then. One way or another, I think I'd like to hear this body scream."

Then, suddenly, she was gone – and with her, her warmth, her dizzying scent. I stood shivering in the darkness, alone.

• • • •

A frost had settled across the cemetery, the blades of grass crunching beneath my feet as I trudged back to the treeline, and to Pinch. He paid me no mind as I approached, instead staring at the spot from which I'd come. He stood wide-eyed and mouth agape, his forearm streaked with blood. The kerchief lay forgotten at his feet.

"Who *was* that?" he asked, his voice small and faraway.

"Nobody. We have to go." I crunched past him, into the forest. He didn't budge.

"She was *beautiful,*" Pinch said. "Bring her back."

"Maybe later. Right now, we have to move."

"I could give her more blood," he said. I watched in horror as Pinch fetched Anders' knife from his pocket and dragged the blade once more across his forearm. Fresh blood welled, glistening black in the moonlight.

I grabbed him by the wrist, trying desperately to still the blade. He struggled against my grip. That's when I hit him. A backhand blow across the face, hard enough to knock him down. Pinch glared up at me from the ground, eyes full of cold fury. At least it beat the moony stare of a moment before. I extended a hand to help him up. Reluctantly, he took it.

"Her name is Lilith," I said. "And believe me, you want nothing to do with her."

"Lilith," he repeated, in the reverent tone of the devout. "Who is she? *What* is she? Is she a god?"

"A god?" I laughed. "Pretty fucking far from. As to what she *really* is, that's complicated. Some say that she's the night. The southern wind. Some believe that she was the first woman to walk the Earth – that she was cast out of Eden because she refused to be subservient to Man. There are some who say she is the mother of demon and djinn, to incubi and succubi – to all the creatures who walk the night, and prey on your kind."

"So which is it?"

I shrugged. "Who knows? The books were written long ago, most by folks like you, struggling to make sense of things we weren't meant to know. Not a one of them is right, or maybe they *all* are, I don't know. Either way, the lot of them, Christian, Egyptian, whatever – they're all just dim reflections of the world beyond, offering nothing but distorted, funhouse images of what they attempt to explain. What I *do* know is that Lilith is powerful, a creature of great influence and even greater beauty. Which is to say she's dangerous. You felt what her presence was like, and that was from forty feet away. Up close it's even worse. You'd do well to stay away from her – she's corrupted even the bravest and truest of souls, and she'll try to do the same to you, if you give her half a chance."

"Why doesn't she affect you?"

I laughed. "Believe me, she does. But in my case, it's only incidental. See, I've got nothing left for her to take. Now come on – we've got to go."

We set off through the woods. My muscles ached from exertion and from the cold, but still, I set a brisk pace. Pinch struggled, panting, to keep up. The path was lazy and meandering. I had time for neither. I left the trail behind, plunging into the forest proper. I hoped to God I was headed in the right direction. Now was not the time for mistakes.

Sneaker scraped against wood, and Pinch yelped, tumbling. A tree root, thick and gnarled, had blocked his path, sending him to the ground. Reluctantly, I stopped and gave him time to find his feet.

"Jesus, Sam – where's the fire?"

"No fire – we just have to go, is all."

"This about that Bishop guy?"

I pondered lying. I figured – what's the point? "Yeah," I said. "It's about Bishop."

"What kind of a name is Bishop, anyway?"

"What kind of a name is Pinch?"

"Fair point," he said.

"Anyways, it's not his name, it's his title. Was, anyway. Word is, he was a powerful man in the church during the Middle Ages. Had himself a school. Problem was, his students – young boys, all – had a habit of turning up dead. He took their eyes, their tongues, their hands. Other things, too. Of course, he had the protection of the church, so there's no telling how many boys he killed, and nobody knows what he was doing with the bits he took – although if you heard the speculation, you'd likely cry yourself to sleep."

"And now he's after us?"

"Yes."

At that last, Pinch sat down hard and put his head between his knees. His face looked pale and clammy by the light of the moon, and he gulped greedily at the cold night air like he was going to be sick.

"You OK?" I asked.

"Fine," he said, raising his head after a moment and climbing unsteadily to his feet. "Just wondering what I've gotten myself into, is all. So what do we do now?"

"We get the hell out of here, for a start. Find someplace crowded. Someplace public."

"Wait a minute – I thought crowded was bad. I mean, this guy is like you, right? He hops from body to body? I mean, he could be *anyone*."

"Yeah, but he's good at his job – the best, maybe. He knows better than to cause a scene. Besides, if I'm gonna take him on, I'm gonna need some spare bodies. The last thing I need is for him to kill me and send me halfway across the fucking globe."

"Spare bodies? That's encouraging."

"I'm not here to keep your spirits up – I'm *here* to keep you alive. I should've never gotten you and Anders involved."

"If you hadn't," Pinch said, "she might be dead already."

"Yeah," I said. "That'll be some comfort if I get you killed."

"So this Bishop guy – how are you gonna see him coming?"

"I don't know. What I do know is that he's close."

"You can sense other Collectors?"

"I can sense this one."

20.

"Kate? *Kate*?"

Pinch and I had been walking for half an hour. Navigating the woods was tougher than I'd expected, and somehow we'd managed to miss the picnic shelter altogether, winding up on the wrong end of the park's long, narrow pond. By the time we got back on track, the shelter was empty. I prayed we weren't too late.

"Kate!" I shouted again, my voice hoarse with fear.

"Sam?" The call came from the darkness to our left. "Sam, thank God!" Kate broke free of the treeline and leapt into my arms, Anders trailing just a couple yards behind. As she squeezed my battered ribs, I winced, but still, I held her tight.

"Jesus, Kate, you scared the shit out of me – where the hell did you *go*?"

"You guys were gone so long," she said. "We started to worry, thought maybe you were in some kind of trouble."

"Damn it, Kate, you know better than that! The last

thing we need is for you to go running into harm's way. Besides, Pinch and I had things under control."

Kate saw the scowl on my face, and replied with one of her own. "Sam, what's wrong?"

"Nothing. We have to get moving, is all."

"I'm just glad you two made it back in one piece," Anders said, clapping Pinch on the shoulder. "You and him and a ritual and a knife – who knows what might've gone down?"

"Who, indeed?" Pinch said.

My stomach lurched. I shoved Kate aside, and lunged toward Pinch. Kate squealed in surprise, and Anders looked shocked, but Pinch didn't. He knew he'd fucked up.

After all, when's the last time you heard an eleven year-old say *indeed*?

I was fast. Pinch was faster. He plunged his blade deep into Anders' side and tossed him into my path. We collided, and tumbled to the ground in a mess of blood and limbs. As Kate stood frozen, a look of horror on her face, Pinch closed the gap between them.

How could I have been so stupid? I *knew* Bishop was somewhere nearby – I could *feel* it. I should have seen Pinch's sudden bout of nausea for what it was: Bishop taking over. But I didn't. Didn't see, didn't think. No, instead, I led him right fucking to her.

I struggled to free myself of Anders, but he was dead weight – limp and uncooperative. I didn't have time, though, to worry about him. Right now, my only thought was of Kate.

Pinch/Bishop grabbed a handful of Kate's hair and yanked. She yelped and fell to one knee. His gaze traveled up and down her trembling form, as if seeing her for the first time. "My, but you're a pretty one," he said. "Not my type, of course, but I suppose I could make an exception, just this once."

His hand plunged into her chest. Kate shrieked in pain and fear. Muscles screaming in protest, I rolled Anders off of me – he collapsed to the forest floor beside me, blood pulsing around the blade in his side and oozing black from his mouth. I scrambled toward the figures of Kate and Pinch, locked in their horrible embrace, no thought in my head but that I could not let this happen.

I don't even know where the rock came from. I must've picked it up along the way, but even now, I can't recall. Wherever I got it, I brought it down hard on Bishop's head, again and again, until finally, he released his grip on Kate's soul and collapsed to the ground.

The rock fell slick from my hand, and the night air prickled with a sudden copper tang. Still, it wasn't until Kate scampered backward away from me, tears welling in her eyes, that I realized what I'd done.

"Kate –" I began, but then stopped, unsure of what to say. "Kate, I'm sorry. I couldn't let him take you." I looked down at the body at my feet – an enemy no longer, just the bloodied remains of an innocent child. "I couldn't let him take you."

Kate continued her retreat until a tree trunk barred her way. She pressed tight against it, as if she simply could not bear to be any closer to me than she had to. "Don't talk to me," she said. "You don't get to talk to me ever again, you hear me?"

"Kate, listen to me. He was going to *kill* you. Worse, even – he was going to take your *soul*. Do you even understand what that means? An eternity of torment, and not just for you. If he had taken you, he'd have opened the floodgates. We're talking a full-scale war between heaven and hell. You think Pinch would have wanted that?"

"Don't you stand there and tell me what he would have wanted." Tears spilled down her face, a twisted mask of pain and grief. "He was a *kid*, Sam. He was a kid, and you killed him. You're no better than the rest of them. A monster."

I hung my head, squeezing shut my eyes so that I wouldn't have to see the blood that clung stickily to my hands. "I did what I had to do."

"Yeah, well, you won't have to do it anymore. I don't care what happens to me – we're through. I won't be a party to any more bloodshed. You'll just have to find another life to ruin."

Just then, a low, wet gurgle sounded in the darkness. It was accompanied by a hitching, labored breathing, arrhythmic and faint.

Anders.

I left Kate where she sat, wheeling toward the source of the noise. I didn't have long; in moments, the horrid sound of Anders' labored breathing was replaced by an even more terrible silence.

Anders lay on the ground where I had left him. His eyes were clenched, his pain evident. The blade lay beside him in the grass, slick with blood. One blood-drenched hand lay beside it in the grass, and his sleeve was slick and dark as well. It looked to me like he'd removed the blade himself. I wished to God he hadn't. The blade would have slowed the bleeding, maybe bought us a few minutes, but now that he'd removed it, there was nothing left to stanch the flow. Anders was running out of time.

"Kate!" I called, but she didn't answer. "Kate, I need your help!"

Still nothing.

"Damn it, Kate – you can hate me later. Right now, I need you over here, or Anders is going to die!"

There was a rustling in the darkness, and Kate appeared beside me. She said nothing. She didn't have to. The anger in her eyes said it all. It seemed she'd hate me now, whether she chose to help or not. So long as we didn't lose another life tonight, I figured I could live with that.

I grabbed her hand and pressed it tight to the wound in Anders' side. Kate recoiled slightly from my touch, but when I let her go, her hand stayed. "I need you to put pressure on the wound – more than you think you need, OK?"

"He's not breathing."

"I *had* noticed," I said. What I was going to do about that, though, I had no idea. I hovered over Anders' still form, unsure. I mean, I'd seen it done before – in a movie or two – but the whole CPR thing was a little after my time. Honestly, I'm usually more concerned with halting breath than with restoring it.

"Switch with me," Kate said.

"What?" I looked at her, confused.

"Oh, for God's sake, *switch with me*!" She released the wound and grabbed my hands, shoving them in place. "You always gotta be the hero, don't you?" She pressed her mouth to Anders' and exhaled twice. Then she placed her palms against his breastbone and pressed downward in a steady rhythm. I just watched in amazement.

"I took a babysitting course, a few years back," she said, and then once more blew breath into Anders' mouth. "CPR was a requirement. Of course, that doesn't mean I know what I'm doing."

"You're doing fine," I said. In truth, I had no damn idea, but I hoped to God that I was right.

Again Kate pressed her lips to his. This time, when she released him, Anders sputtered and coughed, blood spraying red across his teeth and lips. The breathing was a good sign. The blood was not. Kate might have bought us a little time, but this kid was gonna need a doctor if he was gonna live.

"That... wasn't...Pinch," Anders said, his voice a brittle whisper, his eyes clenched shut against the pain.

"No," I said, "it wasn't."

"Then who?" he asked, between panting, labored breaths.

"A Collector, like me. They call him Bishop."

"I saw... I mean, I *knew* that something was

different... that he'd *changed* somehow. I just figured it was the... the ritual. I should have said something. I should have tried to stop him..."

I took his hand in mine. "You did fine, kid. Now, though, I need you to save your strength – we're gonna get you some help. Just relax, and try not to speak."

"But Kate... is she OK?"

I looked her in the eye. Truth was, she looked anything but. "Yeah, kid – Kate's OK."

"Good," Anders said, and then promptly lost consciousness.

Kate checked his neck for a pulse. "Still beating, she said, "for now, at least. You think he's going to make it?"

"No," I said, "but if he's gonna have a shot, we have to move *now*."

"So," Kate said, the brittle, frost-laden grass crunching beneath her feet, "you knew that guy?"

We'd only been walking a few minutes, headed south through the park toward what I hoped was the nearest street. With Anders' limp and blood-slick form cradled in my arms, it felt like we'd been walking for hours. For maybe the fifth time now, I hitched him upward, trying to re-establish my grip. But the kid was heavier than he looked, and the sheen of sweat and blood that graced his arms, his neck, his back, made it tough to hold on. The going was slow, and the makeshift bandage I'd jury-rigged from the Flynn meat-suit's uniform shirt wasn't going to hold for long. We were running out of time.

"Yeah," I said, "I know him a bit."

"So what – you guys stand around the water cooler, chat about the souls you've snatched, that sort of thing?"

"Not exactly. Bishop is the one who collected me."

We trudged in silence for a moment. Finally, Kate broke it.

"I'm sorry," she said. "I didn't know."

"How could you have?"

"I don't know. I just – it's terrible, isn't it? Being taken, I mean."

This time, it was my turn to pause. "Yes. Yes, it is."

"I swear I can still *feel* him. Clawing. Tearing. Struggling to rip free my soul."

"Listen to that feeling," I said. "For the collected, it never really goes away. If you're lucky, you came close enough, and it'll stick with you, too."

"If I'm *lucky*?"

"Damn *right* if you're lucky. Bishop's not done with you yet, Kate. If you can hold on to that feeling, you might be able to sense him coming. It could give you the edge you need to escape him."

"So you can feel it, too? You can tell when he's nearby?"

"Yes," I said.

"Then how – I mean, with Pinch…"

"I didn't listen to my instincts. I got too close to the job. To Pinch. To all of you. I got too close, and you can see where it's landed us all. You can be sure I won't make that mistake again."

"So if he's done it once before, what's to stop him from doing it again? I mean, how do we know that Anders is Anders?"

"What's your gut tell you?"

Kate frowned in concentration. "I – I don't know. I'm still a little rattled, but it's fading. I mean, he *seems* like Anders. Still, that's not a lot to go on."

"It's enough," I said. "No way would Bishop have hitched a ride with Anders. The kid is badly hurt, and he might not make it. If he'd entered Anders, he might not find the strength to leave before it's curtains, and then he's fucked. Folks like me, we're happy enough with the living or the dead, but the *dying*, they're a whole 'nother matter. See, in death, the body expels any invading soul. And since a Collector can't exist without a body, that means when one of us dies, we end up reseeded somewhere else at random. Could be a freshly buried corpse half a world away. It could be a baby down the street, too weak to lift its own head, let alone give us the boost we need to jump away. So you keep listening to that gut – it's done fine by you so far."

Kate shuffled along quietly for a moment, her face set in a thoughtful scowl. "Sam?" she said, finally.

"Yeah?"

"If he'd succeeded in taking me, would I be a Collector, too?"

"Maybe. Maybe not. I don't know. It's not for me to say."

"Better than the alternative, I suppose. You know, a lake of fire or whatever."

I looked at the crumpled, dying figure I held cradled in my arms. "No," I replied, "it really isn't."

21.

As we approached the edge of the park, headlights shone through the trees – beacons of hope sweeping past us in the darkness. It was late, and the traffic was slight, but I was confident we'd find what we needed. But the slog through the park took longer than I'd expected, and the kid was fading fast. I only hoped it wasn't too late to make a difference.

With Anders' bloody, wheezing frame cradled tight to my chest, I broke from the cover of the trees, staggering out into the street. Behind me, Kate screamed, but I paid her no mind. The screech of tires pierced the night, and the air hung thick with burnt rubber. It drifted blue-black across the roadway, stinging my eyes. I blinked back tears, and squinted against the sudden glare of headlights.

Looked like I found my mark.

It was a Volvo station wagon, blue as sky beneath the streetlights, and it rocked to an awkward, diagonal halt just inches from where I stood. The driver, a woman in her fifties, was fumbling with a cell phone, her eyes wide with fright. I hoisted Anders over my shoulder, Flynn's well-muscled frame protesting under the strain, and broke for the driver's side door, yanking it open with my free hand and clawing for her phone. She was too stunned to resist. I snatched the phone from her hand, and tossed it in a lazy arc toward the woods. Her eyes flitted back and forth between the patch of woods in which it landed and me – filthy and bloodied in an undershirt and navy trousers, my only hope of passing as a cop in her eyes the uniform shirt currently pressed tight to Anders' wound – her face twisted into a rictus of terror.

"T-t-take the car," she said.

"I don't want the car," I said.

"I... I have money." She twisted in her seat, fumbling around in the back for her purse. I grabbed her wrist, and she turned, her gaze meeting mine.

"I don't want your money, either. This boy – he's hurt. What I *need* is a ride."

"I don't..." she stammered, "I mean, I *can't*–"

"Do you know where the nearest hospital is?"

She hesitated, but only for a moment. "Yes."

"Then you can."

She stared at me a moment, her face a silent plea.

"If you don't do this, he'll die."

That did the trick. She clicked the rear doors unlocked. "Get in," she said.

I dropped Anders in the back, and gestured Kate in there as well. I climbed into the passenger seat, fetching Anders' blood-streaked knife from my pocket and laying it at ready across my lap. Our Good Samaritan didn't fail to notice. The blood drained from her face, and she gripped the steering wheel hard damn near enough to break it off, her knuckles bone-white.

"*You* did this to him?"

I didn't hesitate. "Yes," I said.

"You're a *monster,*" she replied, not just a little bit of steel in her voice. "A goddamned monster."

"Lady, you have no idea how true that is. And if you don't start driving, I swear you're gonna get the same."

Again her tires squealed. This time, the car lurched forward.

"Easy!" Kate called from the back. "He's seeping through his bandages. I'm doing my best to stanch the flow, but if you rattle around too much, I won't be able to keep the pressure on."

"She slows down, it doesn't matter how careful she is – the kid's gonna die," I replied.

We squealed around a corner, rocketing through a red light. I braced myself against the door handle, the knife gripped tightly in my free hand. I didn't think our driver was gonna be a problem, but Bishop was another matter. As far as I knew, he might be halfway around the world right now, but even if he *were,* he wouldn't stay that way for long. If he'd found his way back in time to see our little traffic stunt, our new friend here'd become a liability right quick. If that happened, I had to be ready. No qualms. No hesitation.

Still, I hoped for all our sakes it wouldn't come to that.

The woman glanced at Kate in the rear-view. Her eyes narrowed. "You're that girl from the news, aren't you? The one that killed her family."

Kate said nothing.

"You think that silly punker get-up's going to fool people for long? Your picture's been on every television in the city. It's just a matter of time before they find you."

"Just shut up and drive," Kate said.

"I'm trying," the woman replied, and then, as she screamed past her intended turn: "Shit!"

Do you have any idea where we're going?" I asked her.

"Do *you*?" she shot back.

I thought a moment. "St Vincent's is close, if it's still around. But we shoulda been there by now."

"It is," she said, "although it hasn't been St Vincent's for years. And we would have been, if I weren't pointed in the wrong direction when you stopped me. You would have done better to carjack someone headed south."

"A lot's changed since the last time I was here. That's kind of why you're doing the driving," I said. "Now get us turned around, and quick."

"What's to stop me from just driving straight to the police?"

"This, for one," I said, brandishing the knife. "But more importantly, there's no time. You take the time to turn us in, the boy dies. You look like a decent person to me. I think you're gonna make the right choice."

"They'll almost certainly apprehend you when we reach the hospital," she replied.

"Then what exactly is the problem? Now if you wanna get out of this alive, you're gonna shut your mouth and get us to the hospital, you hear?"

I was kinda shocked she listened, but I guess she'd already said her piece. She just gripped the wheel and drove like it was the last lap at Indy, barreling down the street with breakneck speed – and ignoring every light, every sign, every lane marker on the way. Had her lips not been pursed in grim concentration, I'd have thought she was enjoying herself. Of course, right now, I couldn't give a shit about her motivation – all I cared was that we get Anders some help before it was too late.

It wasn't till we picked up a tail that I realized what she was doing.

He came screaming out of a Dunkin Donuts parking lot about a half a block back, siren blaring. Red and white lights strobed through the cabin of the Volvo.

"Sam," Kate said, "we've got company!"

I glanced back. The cop was gaining fast. A triumphant smirk flickered across our driver's face, and the speedometer needle began to drop as she coasted toward the shoulder.

I held the knife up to her neck, and she went rigid in her seat. I said, "You do not stop, you hear me? You just keep on driving till you get us where we're going."

The driver said, "I... I can't just *ignore* him."

"That's *exactly* what you're gonna do."

"There'll be more of them, you know, and not just behind. If they cut us off, I'll have no choice but to stop."

"If you stop this car before you get us to the hospital, I swear you'll wish you hadn't. Drive through them if you have to. This kid is not dying on my watch. Am I clear?"

She nodded. The fear in her eyes had returned. That was good. The cop was gaining, though. That was bad. The funny thing was, I didn't see any others. At the time, I didn't know why, but that fact – which should have comforted me – instead left me with a gnawing pit of worry where my stomach should have been.

Of course, it didn't help that Anders stopped breathing.

It wasn't a peaceful sort of thing, either, like drifting away in the quiet hours of the night. It was more like a flailing, writhing, drowning-on-dry-land sort of thing. Anders' limbs swung wildly through the cabin of the Volvo, one leg connecting hard with the back of the driver's head and sending the car careening onto the sidewalk toward a darkened storefront. I grabbed the wheel and jerked us back onto the street, receiving a glancing blow to the temple for my trouble. Kate was shrieking, and Anders was making a horrid, gasping noise that sounded like a pipe organ collapsing on itself.

Our driver was shouting now, too, in fear and panic, and to her credit had us more or less back on track. Things got dicey for a second as we leapt the center divider, and the sudden glare of approaching headlights made a collision seem imminent, but she yanked the wheel to the right, and sent the car sailing back into our lane in a rain of sparks and a squeal of rending metal.

And still, our pursuer remained.

We were close now, the structure of the hospital looming over the tops of the timeworn Colonials that surrounded it. Anders' flailing had died down, but it was hard to take that as a *good* thing. Not to mention, I had no fucking idea what I was gonna do about the cop. One thing's for sure – planning's never been my strong suit. Eh – if I was right, and this girl's soul really *did* hold the fate of the world in the balance, at least I'd know that God has got a sense of humor. I mean, shit, he could've sent her a savior with a *clue*.

In the distance, a backlit sign jutted from a well-manicured garden, marking the hospital entrance. I pressed the knife to our driver's side. "You don't slow down until we reach the entrance, you hear me? No signal, no warning, *nothing*." She just nodded, her eyes never leaving the road. A good little trouper, that one. I confess I was relieved – the last thing I needed was more innocent blood on my hands.

When the turn came, she didn't hit the brakes, she just yanked the wheel. The car skittered a second, and then the back tires caught, and we rocketed forward. Thank God she'd listened to what I said. If she hadn't, we would have all been dead.

The police cruiser slammed into our car with a spray of glass and the sickening crunch of metal on metal. His front end connected with our back-left fender, and we one-eightied. The car rocked hard on its shocks as we slammed into the curb, but it could have been worse. Had we slowed to take the turn, he'd have caught us dead to rights, and we'd have rolled for sure.

The cop was out of his car – which had beached itself on the hospital's now-ruined sign – in a flash. His gun was drawn, and he was running toward us, closing the gap between his wreck and ours with lightning speed. Our driver looked stunned, confused, but I wasn't – not anymore. It was clear now why he'd pursued us alone, why he'd never called for backup: this guy was no more a cop than I was. It was Bishop, back to finish what he'd started.

The bastard was good – I'd give him that. I'd hoped Pinch's death had at least bought us some time. I'd hoped we'd lost him – that he was strapped to a bed in some old folks' home in Dubai or something, never to be seen again. I'd hoped that maybe, just maybe, we'd catch a little break. Turns out, I'd barely even slowed him down.

Shows what hoping will get you.

Bishop must've been waiting for us. Listening. He knew we couldn't flee the park without causing a scene, so he camped out in the nearest cop and waited for the calls to come rolling in. If I had to guess, I'd say his meat-suit's partner was standing outside Dunkin Donuts with a handful of coffee and crullers, wondering where the hell his buddy and their cruiser went.

I looked our driver in the eye. She looked at me, and then at Bishop, clearly registering the hate and anger that strained the features of his borrowed face. "Listen, lady, we need to move."

"What?" she asked. Her voice seemed small and far-away.

"That guy's not friendly. There's no time to explain – you're just gonna hafta trust me."

"Trust you? How could I, when you hurt that boy..."

"I just told you that so you'd do as I said. It was him," I said, gesturing toward the approaching cop. "You hear me, it was *him*!"

Whether it was my words or her own instincts, something got through. She slammed the car into gear, and lurched forward, jerking the wheel toward Bishop as he raised his gun to fire. The movement caught him off-guard, and he squeezed off a few wild shots. Two slammed through the front end of the car, and the engine quit, but we just kept on rolling. The third punched through the windshield, and our driver screamed in pain.

I barely took a moment to register her injury – a spray of blood against the driver's side window, a hand clutching the meat of her shoulder – before I leapt out of the moving car and sprinted toward Bishop. He'd fallen backward onto the pavement, dodging the surging Volvo, and I threw myself atop him as he struggled to bring the gun to bear.

Cold steel pressed against my cheek. A deafening blast rocked the night. I clenched shut my eyes in anticipation of the expulsion to come. I was sure that this was curtains. Instead, a sudden warmth trickled down my ear, and the world went quiet. My face was stippled with burns from the particulates, but I was otherwise OK.

The bastard had missed.

Again, Bishop tried to aim the barrel toward me, but the report of the last shot had weakened his tenuous grip. I grabbed his wrist and slammed his hand to the ground. The gun clattered to the pavement, just out of reach. Bishop lay pinned beneath me, and I swung wildly, again and again, connecting with his cheek, his jaw, his nose. My damaged eardrum throbbed in time with the thudding of my heart, with the rhythm of my flailing blows. I forced myself to hold back – just a touch – and remember the innocent within. The last thing I needed was another death on my conscience, and anyway, *unconscious* would do just fine.

At some point, he stopped fighting. I thought it was a ploy. Then I caught his fearful gaze, leveled not at me, but at the Volvo. A feint? Maybe. But I bit, nonetheless. I hazarded a glance, and was glad I did.

The car had finally stopped rolling, coming to a rest against the far curb. As I watched, our driver pushed open her door, and doubled over, retching in the gutter.

I leapt up off of the pavement, or as near to leapt as my weary bones could manage. Our driver rose, leaning heavily on her open door as she wiped absently at her mouth with the back of her hand. She turned toward the rear door of the car, an oblivious

Kate behind it, still struggling to keep pressure on Anders' wound. Lucky for me, the driver was so focused on Kate that she never glanced back.

I sprinted toward the Volvo, desperate to stop our driver from reaching Kate. At the last moment, the woman wheeled toward me. I kicked the open driver's side door with all I had. It slammed shut hard on her, bouncing back open as she crumpled to the ground. I lay a moment, winded, willing my battered limbs to move.

The knife was a surprise. Not the happy cake-and-balloons kind, either. More like the gut-wrenching, excruciating, hope-you-don't-black-out kind. I must've left it in the cab of the car when I'd gone after the cop. Wherever I'd left the knife, driver-lady/Bishop had found it, and was kind enough to return it to me, by which I mean she planted it a good three inches into the meat of my thigh. Blade scraped bone, and for a second, everything went dark.

When the lights came back on, the nice driver-lady was standing over me, the knife – blade down – raised high above her head. A wicked smile warped her otherwise kind features. I tried to move. My legs weren't listening. Kate watched helpless through the car window – I willed her to run, but she just sat there, frozen.

The blade dropped. Actually, the whole damn *woman* dropped. Just collapsed atop me like so much rubble. I rolled her off of me. The knife fell from her hand, coming to rest in the grass just beyond the curb.

Standing just behind her former perch above me was the cop – his face swollen and bloodied, his sidearm in one hand, a small tuft of blood and hair dotting the barrel from where he'd pistol-whipped the woman. He extended his free hand to help me up. I took it.

"That *thing,*" he said, "is it unconscious, too? Or will it just grab hold of someone else?"

I could barely hear him over the ringing in my ears. I looked down at the woman. She was out cold. "Yeah," I replied, "it's out, too – but probably not for long."

"It was in my *head*. I mean, I was just sittin' in my cruiser, and next thing I knew, I was puking my guts out, and I wasn't in control. That's fucking nuts, right? I mean, I must be fucking nuts."

"No," I said. "You're not nuts."

"It wanted to *kill* you."

"It was after the girl. I was in the way."

"The girl – she's the one from the news? The one we've been looking for?"

"Yeah."

"She didn't do it, did she? Kill her family, I mean."

"No, I don't believe she did."

The cop glanced back toward the hospital. The entrance was a few hundred yards away; it looked like a crowd was gathering. I thought I heard sirens, although that could've been the ringing in my ears. As I stood shakily between the wrecks of the cruiser and the Volvo, our unconscious driver at my feet, it was hard to believe this whole fucking mess had gone down in a matter of seconds.

The cop caught my glance, and no doubt he heard the sirens better than I. "They'll be here soon," he said. "The paramedics. The cops. You should go – just take the girl and leave. I'll clean up this mess."

"There's a boy in the car. He's hurt."

"I know. I... remember, I guess. I'll see to him. What about her?" He nodded toward the woman at our feet.

"Long as we're gone when she wakes up, Bishop's got no reason to stick around."

"Bishop," the cop repeated. "Is that its name?"

"No one's left that knows his name," I replied. "Bishop's close as we can get."

"That's not how it thinks of itself," he said.

"No?"

"No."

"What, then? What does Bishop call himself?"

"God," he said, his voice catching in his throat. "That thing believes it's God."

22.

"Sam, what the hell are you *doing*?"

"Just stand back."

I peeled my blood-soaked undershirt from my frame and wrapped it tightly around my bruised and battered fist. The blood seeped between my fingers, cold and slick in the chill night air. I was painfully aware that this blood wasn't mine to shed, and the fact that Bishop had a hand in shedding it did little to assuage my guilt. Of course, if I was right about the girl, any blood shed in the cause of keeping her safe was an acceptable loss. I just couldn't help but wonder if Pinch and Anders would disagree.

I swung my arm as hard as I could, connecting with the window of the Taurus and sending a spray of glass scattering through the cabin. I winced in anticipation of an alarm – one of the most horrid inventions of the modern age, as far as I'm concerned – but there was none. I popped open the door from the inside, and snatched the duffel bag from the back seat with my unbound hand. Very slick little smash-and-grab, I thought – smooth and professional.

That's when I fell down.

We were at the far end of the parking lot from the mess we'd left behind, obscured from view of the first responders by the rambling hodge-podge buildings of the medical center itself. We'd hovered at a distance long enough to watch them intubate Anders and wheel him into the ER, and then we split. They worked quickly on him, swarming like bees on a hive. I took that to be a good sign – it meant they thought they had a chance of saving him. I hoped to God they could – I'd seen enough death for one day, and damned or not, my conscience couldn't take another.

Our driver was another story. She came to just as we'd left the scene, and her injuries appeared minor. After what she'd experienced, I was reasonably sure she wouldn't roll on us, but I couldn't swear to it. Besides, our new cop friend had his hands full explaining just what in hell went down, and when they realized his story didn't add up, you'd better believe they were gonna fan out and check the area. I didn't plan to be there when they did.

All of which sounded nice, but there was a catch, in the form of a throbbing knife wound in my thigh. Truth is, I could barely support my own weight, and I'd lost enough blood that I was feeling pretty woozy. If I couldn't stanch the flow of blood, the whole fleeing thing was kind of out of the question. Which brought me to the car.

Now, I'll admit, hightailing it to the ass-end of the campus on a skewered leg doesn't sound like the brightest of ideas, but I had my reasons. I'm sure I could've found what I was looking for a little more close by, in one of the Beemers, Land Rovers, and Audis that populated the doctors' spaces. Problem was, they were a little too visible for my taste, located close to major entrances as they were, and you can be damn sure they'd have alarms. So I had to settle for something a little more working-class, in a nice, little out-of-the way section of the lot that looked to be reserved for support staff – nurses and the like – with nary a Mercedes in sight.

When I collapsed to the pavement, Kate rushed to my side, a cry of alarm escaping her lips.

"Damn it, Kate, you've got to keep quiet!"

She shot me a look that would have stopped a charging bull. "This from the guy who just busted in a car window. Damn thing sounded like a gunshot. What in the hell are you looking for, anyway?"

I nodded toward the bag at my side. "The gym bag," I said. "Open it."

She did as I asked. Inside was a set of women's gym clothes – sports bra, T-shirt, shorts, sneakers – as well as a set of street clothes and a towel. I snatched at the latter and missed.

Kate frowned and pressed the towel into my hand. I held it tight to my bleeding thigh, clenching my eyes tight against the pain. "Sam, you're not looking so hot."

"I'm not *feeling* so hot," I replied, shivering from cold and blood loss both. "Now hand me that belt."

She did, and I wrapped it around the towel, cinching it down until it hurt too much to keep going. There wasn't a hole that small on the belt, so I had to force the tine through the leather to get it to stay, but it'd do the trick.

Next, with Kate's help, I slid on the gym shirt. Lime green, and emblazoned with a faded silkscreen for a charity 10K, it was both hideous and two sizes too small for Flynn's muscular frame, but still worlds less conspicuous than the blood-soaked undershirt I'd just removed.

"There," I said, "now help me up."

"This is nuts – you need to rest."

"Look, what went down back there was certain to attract some serious attention, and sooner or later, the cops are gonna talk to *somebody* who saw us leave. When that happens, they're gonna start looking for us, and we can't be here when they do. If they arrest us, you're as good as dead. I am *not* going to let that happen."

"Then let's take this thing," she said, eyeing the Taurus. "You got that piece of shit van started; you could get this going, too – right?"

I shook my head. "I'm in no shape to drive."

"Then let me. You could ride shotgun and rest up while I get us out of here."

"Do you even drive?"

"I've got my learner's permit," she replied, defiant and sheepish in equal measure.

Learner's permit. Jesus. "Kate, you saw how bad shit got back there once Bishop caught our scent. I'm not going to run the risk of having you behind the wheel when he catches up to us again. It's just too dangerous," I said, realizing as the words came out of my mouth how unintentionally parental they sounded. "We've just got to find someplace safe and hole up a while until things cool down," I added.

She fell silent a moment, and made no move to help me up. "Sam, can I ask you something?"

I rested my head on the side of the Taurus, and closed my eyes. "Sure, kid. Ask away."

"Back there, when Bishop was coming after us, he jumped from the cop to the woman, right? I mean, just like that," she said, snapping her fingers.

"Yeah? And?"

"Well, *look* at you. You look like shit. Why not just leave this guy here and hitch another ride?"

"Kate, I can't. He might tell them where we're going, and then we're fucked."

"But *you* don't even know where we're going – how the hell could *he*? Besides, that cop Bishop ditched back there, he knew the score, and I'm betting with all he's seen, this guy'd be no different. So why, then? Why, when this guy's doing nothing but hold us up?"

I sighed. "It's complicated, Kate."

"Yeah? Well, we need to get out of here fast, so uncomplicate it quick."

If I could have gotten to my feet then, I would have.

If I could have lied, or deflected, or thought of anything that might've gotten us out of there without having this discussion, I would have. Truth was, I just didn't have the energy. I was out of fight, and she knew it.

Some protector I was.

"Kate, when we met… that vessel was not like this one. He was different."

"Different? Different how?"

"Well, for starters, he was dead."

"Dead? I don't understand."

"You understand fine. See, most of my kind, they possess the living – after all, they're plentiful enough, and they can get you wherever you need to go. Chasing down a prisoner? Just hop a ride in a guard, or better yet a cellmate. Paranoid lunatic holed up in a bunker? If he's got himself a hostage, you're good to go. The problem is, the living are noisy. They're gonna claw and scratch and fight to regain control; it takes a while and no small amount of effort to get them to quiet down. That eventual subjugation doesn't come without a cost. It chips away at whatever it is that makes us human, and forces us to act as a demon would act – to cast aside our empathy, our humanity, and treat them as nothing but a nameless *other* to be used and discarded. A means to an end. Every time we take a living vessel, we lose touch of who we are. And with each vessel we discard, we leave a little bit of what makes us who we are behind."

"But if you only possess the dead, you get to stay human?"

I shook my head. "Kate, you don't understand. There is no getting to *stay* anything. See, the folks who end up like me, there's always a reason. Maybe in life they stripped someone of the life that was rightfully theirs – by murder or betrayal or whatever – and it ate them up inside. Maybe they made themselves a bargain, and took what wasn't theirs to take. Problem is, there's always a price. See, fate's sort of a zero-sum game: you take what isn't yours to take, and it's gotta come from somewhere else. Which means, you make yourself a bargain, and you're stealing someone else's luck, someone else's fate."

"So which were you? Did you strike yourself a deal? Or were your actions to blame?"

I laughed – a cold, humorless laugh. "A bit of both, I suppose. Truth to tell, it ain't the act that's important – it's the guilt. The remorse. The way it eats you up inside. That's the one thing most Collectors have in common – at least, at first."

"What do you mean, at first?"

I paused a moment, unsure as to how to continue. Eventually, though, the words came. "This job – this curse – it feeds on that remorse, forcing you to relive the choices that delivered you to this fate every time you snuff out a life. Every time you tear free a soul, you see every joy, every disappointment, everything that brought that person to where you yourself once were. Every time, some small part of you relives that moment of collection, again and again, in perfect, agonizing detail. With every soul you take, you're reminded of how beautiful life once was, and how you let it slip away. Every time you steal a victim's breath, you remember that first fateful choice you made that brought you to that point, only now, you have no defenses to fall back on. Not ignorance, nor arrogance – no justifications or excuses. It's just you and your actions, stripped bare, and eventually, it's just too much to take."

"So what happens then?" she asked. "What happens when a Collector reaches the breaking point?"

"They go mad. They begin to enjoy the work. They delight in their role. They bury their humanity so deep, they can't even hear its screams. And eventually, their soul just withers and dies. You wanna know what's worse than being damned? Allowing your soul to be snuffed out, just erased from the record books like it never *was*. There's no greater punishment in existence, and no greater crime, than being party to your own eradication. It's as if you're admitting that all you've touched, all you've done, everything you've seen, is for nothing. To choose oblivion is to turn your back on God. There is no greater betrayal. And once you do that, all that's left of you is a monster."

"Is that what happened to Bishop?"

"I guess so. I don't know. If the stories they tell of him are true, he was plenty corrupted in life. In his case, his appointment as Collector may have been more compliment than punishment. Perhaps his patron demon was amused by him, and chose to take him as a pet. But either way, whatever little of him was human when he died is long gone now, warped by centuries of possession and subjugation."

"But Sam, you're not like that! If anybody can find a way around it, it's you."

"Kate, it doesn't work like that. Whether it takes a dozen years or a thousand, this job isn't going anywhere, and not a Collector in existence has ever avoided their fate. All I'm doing every time I hitch a ride with a corpse is forestalling the inevitable. There's simply nothing I can do to stop it."

She replied, "I refuse to believe that."

"Do you? You saw what I did to Pinch back there. Do you think a decent person could have done that?"

"You said yourself: that wasn't Pinch."

"And you said yourself that I was a monster for doing what I'd done. That I was no better than the rest of them."

"Sam, I was upset. I didn't understand–"

"There's nothing to understand, Kate. No excuses to make. I did what had to be done. But what had to be done was just another mile down the road to where I'm going. That's the bitch about fate – there's just no getting around it."

Sirens echoed in the distance. Sounded like half the cops on Staten Island were converging on the hospital. "C'mon," I said to Kate, "it's time to go."

She helped me to my feet. My *foot*, really, since I was keeping my weight on my good leg, for fear of toppling to the pavement all over again. Gingerly, I shifted some weight onto my injured leg. My vision swam, but I didn't black out, and I managed to stay up. I took a step, and then another, one steadying hand never leaving the roof of the Taurus beside me.

Kate watched this process with concern, and when I'd gotten as far as the Taurus' roof would take me, she slid in under my armpit and put an arm around my waist. "All right," she said, "if you can't hop yourself another ride, let's see if we can't patch up this one, OK?"

I nodded, once, my jaw clenched tight against the pain of walking. Sirens approaching, we fled arm in arm across the parking lot.

23.

The house was a shabby old duplex, white with blue trim. A length of narrow pipe, painted white, jutted from the concrete of the lowest porch-step and led upward to the covered porch above. The porch itself was chipped and weathered and littered with cigarette butts and empty beer cans. Two doors, side by side, allowed entry to the house, and they were flanked by two mailboxes, each numbered by hand in black marker. One screen door sat crooked across its frame, its top hinge torn free of the jamb. It swayed lazily in the early morning breeze, creaking all the while.

Just above the rooftop hung a sky of navy blue, streaked with the dusky hues of an overripe peach – the beginnings of a beautiful sunrise. Truth be told, I barely noticed. I was mostly focused on the house – well, that and staying conscious – while the knife wound in my leg seemed content to spend its time bleeding through the towel I'd wrapped around it, throbbing like a son of a bitch all the while.

We were sitting on the darkened stoop of a pawnshop

across the street, its barred windows chock-full of guitars, electronics, and the sundry other crap people'd seen fit to part with for a little quick cash. No gold, though, I noticed – just a patch of black velvet where I supposed it ought to go. I guess they kept that stuff in back. Made sense. Any neighborhood with a pawnshop probably ain't the kind of place you want to leave your jewelry unattended.

I'd been resting my head against the pawnshop door, and I suppose I must've dozed off, because my eyes flew open at the sound of Kate's voice. Startled, I jerked upright. The sudden muscle tension sent waves of searing pain down my leg, and up into my gut. A cold sweat broke out across my face, and I thought I was gonna puke. At least it did a number on the cobwebs.

"Jesus, Sam, are you all right? I thought I might've lost you there."

"I'm fine," I replied. "What'd you say?"

"I said we've got movement," Kate replied. "Second floor. Bedroom, it looks like."

"Left side or right?"

"Left," she said.

"Huh. Looks like I owe you a buck."

We sat in silence for a while as lights came on and off inside. After maybe fifteen minutes, the lights went out, and the left-hand door clanged open. A heavyset dude in a pair of dusky blue coveralls and a good week's worth of scruff stepped out onto the porch, shuffled down the stairs, and hopped into the rusted-out Chevy pickup that sat in the driveway. It was the pickup that had tipped me off, or rather the Department of Sanitation sticker that adorned its rear window. Good thing I'd spotted it, too – I'd barely managed the six or so blocks from the hospital parking lot on this bum leg of mine, and it was only a matter of time before the cops fanned out looking for us. All of which meant we needed to get the hell off the street, and fast. The way I figured it, a garbage man is the first guy out the door in the morning, which meant we'd just scored ourselves an empty apartment, and the luxury of busting in while the rest of the neighborhood was fast asleep. Hell, it was practically Christmas. All we had to do was wait, and cross our fingers it wasn't our guy's day off.

Lucky for us, it wasn't. We watched him pull away, and as soon as his tail lights disappeared around the corner, we made our move. It was a slow, gimpy move, I'll admit – Kate helping me to my feet and supporting my weight as we crossed the street and scaled the porch steps – but it was the best that we could manage under the circumstances. Near as I could tell, there wasn't anyone awake for blocks to see us, anyway.

When we reached the door, I grabbed the jamb for support, and took a long, hard look at the lock. Just your garden-variety deal, damn near as old as the house itself, and no deadbolt, which was a relief. Still, I didn't have anything to pick it with, which meant we were gonna have to do this the hard way. I'm not sure which I relished less: the idea of trying to kick this thing in with a bum leg, or the attention the racket of doing so would attract. Still, it's not like we had a lot of options.

"Listen, Kate – here's what's gonna happen. I need you to grab hold of my left arm. I'm gonna give the door a swift kick with my good leg, and you've got to support my weight, you got me? It might take a couple kicks, so you've got to keep me up, OK? If I don't get the thing down quick, we're gonna wake half the neighborhood, and somebody's bound to call the cops. C'mon – we go on three."

But she just stood there, grinning at me. "What?" I snapped.

"You're really all about the hard way, aren't you?" Kate lifted the lid on the mailbox and reached a hand inside. After a moment of fishing, she pulled out a key. "I mean, seriously, were you even going to *look*?"

I mentally scrolled through a couple dozen witty rejoinders before settling on: "Just open the damned door."

She did, and once we were inside, she locked it behind us, setting the chain as well. The inside was at least as shabby as the outside. We were standing in a cramped living room, made all the more so by the oppressive green-brown of the carpet, and wood-paneled walls that seemed to press inward from all sides. The stench of spent cigarettes hung in the air. A thrift-store couch and easy chair were arranged around a TV that would've looked old when the Nixon hearings aired.

Kate dropped me into the easy chair and disappeared from sight, returning a moment later with an armful of supplies and a chipped glass half full of water. She dropped her payload on the couch, and handed me the glass. "Here," she said, shaking loose a handful of ibuprofen from the bottle she'd scored, "take these." I complied. "This place is a dump, by the way."

I said, "I've seen worse."

"Yeah? You may wanna check out the bathroom before you go making any claims like that. How long you figure we got here, anyway?"

"I dunno – eight hours, maybe nine?"

"We'd best get to it, then," she said. "C'mon, we've got to get you out of these pants."

I made no move to take them off. Kate just laughed. "Don't go all modest on me now, Sam. We've got to dress that wound, or you won't be going anywhere, and besides, this body isn't even *yours*."

Eventually, I acquiesced, undoing the belt I'd wrapped around my leg, and tossing the bloodied towel on the floor. I nearly dropped the belt as well, but Kate shook her head. "Unh-uh – you're gonna need that in a sec."

A few moments' struggle, and my tattered, blood-soaked pants were just a crumpled mess on the threadbare carpet. The meat-suit, as it happens, was a briefs guy. Can't say at that moment I was psyched with his choice, but Kate was polite enough to pay it no mind.

"Looks like that Bishop dude got you pretty good, but the bleeding's slowed at least. God knows where

Anders' knife has been, though – I'm gonna have to disinfect the wound if you want this guy to last the week." I nodded. She snatched up a bottle of rubbing alcohol from her pile of supplies, and twisted free the cap. "You might want to bite down on that belt of yours – this is gonna sting a bit."

That, as it turns out, was a bit of an understatement. I've been kicking around this world for going on ninety years – most of those damned – and I've gotta say, the ten or so seconds after the alcohol hit and before I blacked out were perhaps the most excruciating moments of my life. Every fucking muscle tensed at once, and I thrashed so hard, I thought this body might just tear itself apart. I clenched my eyes so tight I thought I was gonna pop 'em, and my teeth bit clean through the belt, even doubled over on itself as it was. Leather and blood mingled with the prickling scent of alcohol, and the roar of my pulse in my ears nearly drowned out my own tortured screams. And then, for a while, there was nothing.

When I awoke, I was on the couch, my leg bound tight with gauze and duct tape and propped up on a mound of pillows, the wound throbbing dully in time with my pulse. Kate sat on the floor, eating a bowl of cereal by the pale glow of the television. The easy chair was gone; in its place sat a tangled mess of splintered wood and rent fabric, littered with tape and gauze and paper towels, the whole of which was streaked with blood.

"Oh, good – you're up. You had me worried for a while, there."

"What..." My tongue felt like it was filled with sand. "What happened to the chair?"

"You sort of broke it when you started shaking. You're lucky you didn't hurt yourself any further. It wasn't easy dragging your ass to the couch, by the way – but I figured we had to get that leg elevated or it'd just keep on bleeding."

"What time is it?"

"Almost noon," she replied. "Speaking of, are you hungry? This guy doesn't have much that isn't growing fuzz, but there's cereal, and the milk's still good."

"I'm not really very hungry," I replied. As I said it, though, I realized I was lying – my stomach was an empty, gnawing pit, and I couldn't remember the last time I had anything to eat. "On second thought, I think I *will* take some."

Kate headed for the kitchen, returning with a heaping bowl of some God-awful looking pink-and-red marshmellowy concoction, floating atop a sloshing bit of milk. "What the hell is this?" I asked.

"Franken Berry!"

"I thought you said there was *food*."

"Just eat it, it's good."

I took one hesitant bite. I had to admit, it was pretty damn tasty. The second bite was a lot less hesitant. Before long, the bowl was empty, and I was feeling a whole lot better. My leg still ached like crazy, but the pain was of a more manageable sort, and thanks to the food, my head was clearing, and I could feel the strength returning to my abused limbs.

For the first time since coming to, the television caught my attention. It was tuned to CNN, and the sound was down so low, I couldn't make out what they were saying. The image, though, was clear enough: a well-dressed woman, mic in hand, standing at the corner of Park and Forty-second, the massive pillars of Grand Central Terminal jutting skyward behind her. The street around her was littered with shards of glass and bits of debris, and behind her was a massive, open-sided tent overflowing with injured men, women, and children, all being tended to by uniformed EMTs. The great arched windows of the main concourse had been shattered, and the columns streaked with soot. Blackened bits of window frame twisted outward from the building like some horrible, creeping vine. Yellow police barriers set a perimeter around the station, and cops manned them at regular intervals, trying in vain to keep the throng of onlookers at bay. Nearest the building, three fire engines and a handful of smaller fire-and-rescue vehicles sat crookedly, half on, half off of the sidewalk. Scraps of singed paper tumbled through the frame like autumn leaves.

"What happened there?" I asked.

"Some kind of explosion," Kate said. "Terrorists, they think. All the networks are covering it."

"Turn it up."

"At least they've stopped showing my picture every five minutes, right?"

"Kate, *turn the TV up.*"

The woman's voice filled the apartment. "*... authorities still have no idea what motivated the attack – which* *has left twenty dead so far, and dozens more injured – but they believe that this man, seen entering the area moments before the blast, may have been involved."* The image of the reporter was replaced with a still from a security camera of a trench-coated man of average height and weight, his features obscured as if by some odd, internal light. *"Despite his apparent proximity to the detonation site, it appears the man may not have perished in the blast, as several eyewitnesses claim they saw him fleeing the terminal in the ensuing confusion. Authorities declined to comment at this time, pending further review of the security footage, but anyone who recognizes this man is urged to call…"*

But her words were lost to me. Instead, I was focused on the medical tent at the edge of the screen. A man, clearly dazed, had been stretchered into the tent, and was being examined by a doc at the scene. His tattered left arm draped awkwardly off the side of the stretcher, and his clothes were singed black, but otherwise he appeared intact.

As his head lolled toward the camera, I had a flicker of recognition that confirmed what I'd been worried about since the scene first caught my eye.

"Christ," I said, "it's already begun."

"What, Sam?" Confusion twisted Kate's features into a scowl. "*What's* begun?"

"War."

24.

"Get your things," I said. "We're going."

"Sam, what the hell are you talking about? Where, exactly, are we going?"

"There," I said, nodding toward the TV.

"Are you out of your *mind*? Set aside the fact that you just lost a lot of blood, and shouldn't be going *anywhere* but to bed – half the cops in the city are there!"

"Half the cops, sure, and every looky-loo in town. You really think they're gonna notice two more?"

I dragged my ass off of the couch and limped over to the TV set, clicking it off. My leg hurt like a motherfucker, and set my teeth on edge, but the bandages held. It'd get me where I needed to go.

"C'mon, Sam, you're in no shape–"

"This isn't a debate, Kate. We're going."

"But why?"

"Because we need answers, and there's someone there who just might be able to give them to us. Besides, it's not like we've got any other leads. It's this or nothing, Kate, and if we do nothing, it's just a matter of time before they catch up with us."

She nodded, and snatched her leather jacket up off of the floor. "You know you can't go out looking like that, right? I mean, you're gonna need some clothes."

She was right, of course. Thanks to the mess Kate made dressing my wound, my shirt was once more bloodied, and my pants I'd left in tatters on the floor. I hobbled toward the staircase in search of our unwitting host's bedroom. Kate ran to my side, a steadying hand on my elbow, but I shrugged her off. She retreated, just a step or two, and watched with trepidation as I gingerly scaled the stairs.

The bedroom wasn't any nicer than the living room, and a quarter the size – just enough room for the musty, unmade bed and a small dresser. A door on one wall opened to a small bath. I peeled my soiled shirt off and headed to the bathroom, splashing some water on my face and drinking from cupped hands, before returning to the bedroom in search of fresh clothes. In the middle drawer of the dresser I found a rumpled flannel shirt, and in the bottom drawer, a pair of baggy, paint-stained jeans. I dressed quickly, cinching the jeans tight with a belt left atop the dresser. I tucked the lone ceramic cat-shard into my shirt pocket, and then it was back down the stairs, toward Manhattan, and toward our fates.

I had to admit, she looked fantastic. The nausea that had plagued her in the early weeks of the trial had abated, and the color had returned to her cheeks. No longer just the pricks of red over a backdrop of gray that screamed "lunger" to anyone who saw them – they were now a warm golden hue that highlighted the dusting of freckles across her nose and reminded me why I'd fallen in love with her to begin with. And her appetite had improved as well; I watched with amazement as she plowed her way through a plate of ham and eggs, delivered to her bedside by one of the team of nurses that tended to the thirty-odd patients in the study. I had to hand it to Dumas – whatever they were giving her was working.

"Strep-toe-my-sin," she said when I had asked, enunciating each syllable as though she'd memorized them individually. "Not terribly catchy, is it? I mean, you think they'd call it Tubercu-Cure or some such, wouldn't you? But anyway, they seem to think it's working – they say another month of treatment, and I'll be cured, can you believe it? Cured!"

"That's fantastic, love," I said, but my thoughts were elsewhere, a fact that wasn't lost on Elizabeth.

"They did warn me, though, that there are side effects," she said.

"Yeah?" I said, barely hearing her.

"They say I may grow a trunk and hooves."

"Huh."

"Seriously, Sam, where are *you today?"*

"Nowhere – forget it."

"It's this new job of yours, isn't it?"

"What? No, of course not."

I was lying, of course. This past month, Dumas had run me ragged, calling at all hours of the night to tell me he had a package to deliver, a client to entertain, a customs agent who needed a little paying off. Between the insane hours and the knowledge of what I was doing, I couldn't eat, couldn't sleep, and there was no doubt the job was taking its toll on my marriage, as well – I'd been nothing but short-tempered and distant for weeks.

"Sure," Elizabeth said. "Fine. When's the last time you had something to eat? I could talk to the nurse, have her grab a plate for you as well."

"I'm not hungry."

"You've been saying that for weeks. Have you seen a mirror recently? You're skin and bones, Sam. You need to start taking better care of yourself; after all, I've got to have a husband left to come home to, don't I?"

"Just leave it be, would you? I said I wasn't hungry."

Elizabeth fell silent for a moment, surprised by the sudden venom in my tone. Then she put a hand on my forearm and gave it a squeeze. "You know, I've got half a mind to give this Dumas a call and quit for you right now."

"You'll do no such thing," I said, anger once more creeping into my voice.

"I know we need the money, Sam, but honestly, no job is worth this. I never see you anymore, and when I do, we always bicker. I just want you to be happy is all. I just want to have my husband back."

"You want your *husband* back? Damn it, Liz, can't you see I'm doing this for *you*? For *us*?"

"But what's the point, if there's barely an us left to do it for?"

"You don't know what you're talking about," I said.

"Maybe not," she said, "but I do know you. *And I know that whatever's going on, it's eating you alive. Don't try to argue – it's written all over your face. So push me away all you like. I'm your wife – it's my job to worry about you. And right now, it's your job I'm worried about."*

"Look, I just got to stick with it a little while longer, OK? When you come home, I promise I'll quit, and then maybe we'll start over someplace new."

"I wish I understood the hold this job has over you," she said. I said nothing.

Just then, a nurse came trotting over from the nurses' station, her flats clattering against the institutional tile floor. "Mr Thornton?" she asked. "I'm so sorry to interrupt your visit, but there's a Mr Dumas on the phone for you. He says it's urgent."

Elizabeth shot me a look I chose to ignore. "You should let him wait," she said.

"Damn it, Liz, you know I can't."

"I don't know any such thing," she said. And then, with a sigh: "Fine. Go. But first, a kiss."

She leaned toward me, expectant. I pecked her absently on the forehead and made for the nurses' station.

"Hey!" Elizabeth called.

"Yeah?"

"I love you!"

"Yeah. Me too. Listen, Liz, I gotta go – I really shouldn't keep him waiting."

I turned and left, then, leaving nothing but silence behind.

The trip from the apartment to Grand Central took us damn near three hours. The ferry terminal was a mess

– National Guardsmen in full camo manned security checkpoints, frisking every passenger before boarding, and slowing the line to a crawl. What's worse, the city'd suspended all subway service north of Thirty-third, which meant a nine-block hike against a bitter northern wind. By the time we arrived, my leg wound had begun to seep, and a cold, acrid sweat had broken out across my face and chest.

The scene itself was one of utter panic. Nothing I'd seen on TV had prepared me for its scope. The streets were flush with people – many fleeing, although most, like us, pushed ever closer to the terminal. News choppers thudded overhead, and over their incessant din I heard a woman shrieking for her child, while behind her, a street preacher atop a milk crate shouted that the end was near. Since we'd left the apartment, a portion of the terminal's roof had collapsed, sealing shut the southern entrance to the station. Rescue workers struggled to clear the debris and reach those still trapped inside, while just outside the perimeter, the city pressed close – watching, waiting. The sheer volume of people had halted traffic for blocks before we'd even reached the barricades, and dozens of car horns sounded again and again in a futile attempt to break the jam.

We shoved our way through the crowd, me in the lead, and Kate trailing behind, her left hand gripped tightly in my right. Though the fire had long been out, thick dark smoke still poured out of the ruined windows of the terminal and hung over the crowd like an impending storm. The afternoon light was reduced to a trickle, and the acrid smoke burned my eyes, my nose, my throat. With every face that passed, I felt a flutter of anticipation, and I scanned them all in turn – each time dreading that flicker of recognition that would mean that we'd been made. I kept telling myself that there was no way for Bishop to know where we'd gone, that he was probably half a city away, but it did nothing to stop my heart from thudding in my chest, nor to quell the anxious tremors in my hands.

A knee connected with my injured thigh, and I stumbled. Pain radiated outward from the wound in nauseating waves, and my vision went dim. Eventually, I got my feet back under me, and we continued through the crowd, but my leg was once more slick with blood, and my head grew foggier with each mutinous heartbeat.

The barriers were a surprise. One moment, the crowd seemed to go on forever, and the next, I was expelled into a sawhorse with enough force that I nearly toppled over it. Kate's hand slipped free of my sweat-slick grasp, and I teetered for a moment, doubled over the grimy, yellow thing – my feet no longer touching street, my fingertips just inches from the pavement on the other side. A uniformed hand grabbed a fistful of my shirt, none too gently, and hoisted me upright.

"Easy, mac," said the cop. "Where the hell you think you're going?"

I confess that in my dazed and injured state, I didn't really have an adequate reply. Turns out, I didn't need one.

"My uncle, he's hurt. From the blast, I mean. He was walking past when it happened, and I think he mighta caught some shrapnel or whatever. It won't stop bleeding."

I stared at Kate for a moment like she had a second head. Then I broke into a smile when I realized what she was doing. Kate nudged me, her face set in a scowl. I followed suit, replacing my smile with a grimace of pain that wasn't just for show. The cop didn't see any of that, though – he was staring at my blood-soaked jeans.

"All right, come on," he said, yanking the barrier aside enough to admit both Kate and me.

Between the two of them, they managed to wrestle me to the medical tent, one under each arm, with my bum leg trailing out behind. For a while, I tried to hop along, but by then even my good leg was pretty shaky, and I think I was more hindrance than help. They dropped me onto a stretcher, soot-smudged and flecked with blood, and the cop disappeared into the fray to find a medic.

"That was some good thinking back there," I said, once the cop was out of earshot. I couldn't help but notice I was slurring my speech.

Kate replied, "Thanks."

I tried to swing my legs off of the stretcher, but I wasn't having much luck. "Help me get off of this thing, would you?"

"Sam, I'm not sure that's the best idea. I mean, your leg's in lousy shape – you might want to let them take a look at it."

"Jesus, Kate, listen to yourself! Do you even realize where we are? The last thing we need right now is attention! Now for God's sake, help me up!"

Just then, a woman emerged from the crowd, clad in dirty scrubs, a stethoscope draped around her neck. She carried with her a tray stacked with medical implements – gauze, needles, surgical thread, and the like. She couldn't have been more than thirty, and she was thin as a rail, her mouse-brown hair pulled into a no-nonsense bun above a face that looked as though it hadn't seen sunlight in weeks.

"I understand we've got a leg injury? A puncture of some kind?" said the doctor.

I tried to protest, but Kate cut me off. "That's right. He got it walking past. I tried to dress it myself, but it won't stop bleeding."

"Let's have a look, then, shall we?"

I watched as the doctor cut through my second pair of pants in a day, this time following upward along the inseam and peeling back the fabric like a denim banana. Her brow furrowed. "You got this here?"

"Yes," both Kate and I replied, doubtless a little more forcefully than was required.

"You're sure."

"Yes," I repeated, more casually this time. "I was in line for a pretzel when it happened. Next thing I knew, I was flat on my back, a hunk of metal sticking outta my thigh. I know I probably shoulda stuck around, but I was scared. I hobbled home, and my niece here patched me up, only it didn't take."

The doctor jabbed a needle into my thigh, and soon the wound went blissfully, disconcertingly numb. "No, I wouldn't expect it would have. Probably the worst thing you could have done was removed the shrapnel on your own – as it stands, you've lost a lot of blood. Speaking of, where is it?"

"Where is what?"

"The metal fragment," she said, her hands expertly drawing the nylon thread through the meat of my thigh and closing the wound tightly. "The police have requested that any shrapnel be saved and cataloged, so they can better reconstruct what happened."

Kate and I shared a glance. No doubt the doctor noticed. It was Kate that answered. "We, ah, left it at home."

"That's fine," the doctor replied, though her expression was not as light as her tone implied. "An officer will be by to take your statement, and I'm sure they can send someone along to collect it." The stitching done, she began wrapping my leg in layer upon layer of gauze.

"Our statement?" asked Kate.

Her wrapping stopped. The doctor sat there, roll of gauze in hand, and met both our gazes in turn. "Yes, your statement. Like it or not, you are both material witnesses to a federal crime – the police are going to want to know where you were when the blast went off, as well as what you saw. If I were you, I'd cooperate, and that means you'd better get your facts straight."

"Meaning what?" I asked, feigning offense.

"Meaning there's no way that wound was made by a flying hunk of twisted metal. The surrounding flesh is too clean, the borders too discrete."

"I don't know what you're talking about."

"I'm pretty sure you do." The doctor finished wrapping the wound and taped the gauze in place. "This," she said, "is a knife wound."

25.

I said, "Listen, lady, I think you've got this all wrong."

The doctor raised her hands, a placating gesture. "I'm not the one you should be talking to," she said. "I'm just here to patch you up – I don't much care *what* happened. But if you know something about what happened here, they will find out. You two don't look like terrorists to me – make things easier on yourself and cooperate."

Kate opened her mouth to protest, but I silenced her with a glance. "You're right," I said. "Of course you're right. About the wound. About everything."

The doctor said, "So you *were* stabbed."

Kate looked at me – puzzled, frightened. "Sam, don't–"

I shot Kate a silencing look and said, "It's all right, Mary – we have to tell her."

"Tell me *what*?"

"About the bomb. See, my brother – her father – he's always talking crazy, like one day, he'll have his revenge – that sort of thing, you know? He's been that way forever, and didn't nobody think he'd ever *do* anything about it. Only last week, when the city laid him off, he started gettin' twitchy – leavin' at all hours of the night, holing up in the basement for hours on end working on God knows what. I mean, I got worried. We *both* got worried. I took to snooping around, trying to figure what he was up to. That's when I found the book."

"The book?" the doctor asked, rapt.

"That's right. Some sort of anarchist's handbook. It was full of crazy crap about napalm and explosives and stuff. Truth is, it scared the shit out of me. So this morning, I followed him to the basement and confronted him – least, that was the plan. When I got there, there was one o' them bombs, I mean right out of the pictures, and when he saw I saw it, he freaked. Stabbed me in the leg, and just left me there. I musta passed out, because by the time this one brought me to, it was too late."

"And your brother?"

"I can't say for sure, but if I had to guess, I'd say he died in the blast."

"And you'd be willing to cooperate with the police on this?"

I nodded solemnly. "I guess at first I figured you got to stick up for your family, no matter what, but you're right – we owe it to everybody here to tell the truth."

She put a hand on my shoulder and gave it a comforting squeeze. "Stay here," the doctor said. "I'll be right back." Then she ducked out of the tent, setting off toward the makeshift command center the cops had established on the other side of the street.

"Sam, what the hell was *that* about?" Kate demanded.

"I was buying us some time," I said. I swung my legs down off the bed – a little easier, now that the wound was good and numb – and, with a little help from Kate, managed to find my feet. There was a pair of crutches lying across the empty bunk beside me, and I grabbed one of them, wedging it in my left armpit to take the pressure off my injured leg. I took a couple cautious, hobbling steps, and found that with the crutch's help, I got along just fine.

"By implicating us in the bombing? No way they let us walk out of here now!"

"You saw the way that that doctor was looking at us, Kate – she wasn't letting us out of here regardless. The only difference is, now she thinks that we're co-operating, which means instead of flagging some beat cop down from my bedside, she's gonna give us a little breathing room while she goes and fetches us a bigwig."

"Yeah, but when she realizes we're gone, this place is gonna snap shut so tight, *no one's* gonna get to leave."

"You're right," I said. "Which is why you've got to get out of here now, before they realize what's happened. I can ditch this body and follow, once I get what I came for."

She shook her head. "No."

"Kate, you *have* to."

"I'm not leaving you, Sam – not this time. Last time, it nearly got you killed, and it turned out Pinch wasn't quite so lucky. I won't make that mistake again."

"Damn it, Kate, you don't have a choice!"

"The hell I don't."

I said, "Look – the man I saw on the news, his name is Mu'an. He's a messenger-demon – sort of an emissary between the demon-world and their angelic counterparts. He's also a snitch. An information broker, to hear him tell it, but whatever you call it, the job's the same. I understand he's profited quite heartily in this détente, selling whatever it is he knows to whoever'd like to know it. It seems in war he's not so lucky. Now, there's a chance that Lilith was right, and this was nothing more than a random act of violence, one of a thousand such skirmishes to come. Then again, maybe Mu'an wasn't just a target of convenience. Maybe he knew something – something worth killing to keep quiet."

"But the man on the security tape – he was no demon, was he?"

"No – he was an angel. But an angel of the lowest order – a foot soldier. He'd just be carrying out orders, which means the call came from somewhere else. Relations being what they are, I suspect a whispered lie in the right ear would be enough to get the job done."

Kate asked, "You think this Mu'an was set up by his own kind?"

"I don't know. It's a pretty big leap, but right now, it's all I've got. That's why I'm not leaving until I talk to him."

"Well, then," she said, a wan smile flickering across her weary features, "I guess we'd better find him fast."

Turns out, he wasn't hard to find. Though the medical tent was a crowded, sprawling affair, the patients had been triaged according to the severity of their injuries. The end we'd been deposited in was full of scrapes and cuts and broken bones. At the far end of the tent was a makeshift ICU, a roiling mass of sound and fury as medical personnel struggled to stabilize the worst-hit so they could be loaded into one of the endless parade of ambulances that waited to whisk them away. Mu'an was somewhere in the middle. He lay uncovered atop a stretcher, eyes closed, in a navy suit of worsted wool. His tie they'd cast aside, and his shirt was unbuttoned to the waist, revealing a bloodied undershirt beneath. A coarse white hospital blanket lay tossed off on the ground beside him. His suit and hair – the latter pitch black, and tied into a loose ponytail at the base of his neck – were badly singed and reeked of smoke. His lips were dusky and cracked, his eyebrows gone; his broad cheekbones, normally so deeply tanned, were streaked with a raw, angry red that glistened beneath a thin layer of ointment. One arm was draped across his chest, his shirtsleeve cut away. The little of his arm that was visible around the gauze was blackened like an overcooked ham.

I approached his bedside. Mu'an didn't stir. But as I reached out to shake him awake, his eyes flew open, and his hand clamped down on my wrist. It was wrapped in bandages and crackled sickeningly as he tightened his grasp around me. Then he recognized me, and his grip slackened. His head, raised suddenly when I'd disturbed him, collapsed back onto the flimsy hospital pillow.

"Well, look at this – the man himself. I confess, I didn't expect I'd see *you* here." Mu'an's speech had the odd, musical cadence of some long-forgotten language, as though despite his easy fluency, he would not deign to think in a human tongue. He attempted a smile, but all he got for his effort was the slightest of upturns at the corners of his mouth, and the glisten of fresh blood in the cracks of his desiccated lips. "To what do I owe this pleasure?"

I said, "I need answers. I think you have them."

Mu'an blinked at me a moment, his eyes glistening and unfocused. A cough escaped his lips, spraying his lips and teeth with blood. He dabbed at his mouth with the back of his one good hand and frowned. "And I'm to just supply them, then, is that it? On account of we're such good friends, I suppose."

"Something like that."

He raised his head again, looked Kate up and down. Though his expression was defiant, the strain the movement placed on him was evident.

"Is *this* the girl? The one all the fuss is about?"

"The girl is no concern of yours."

"The hell she isn't!" Mu'an spat, his voice scarcely louder than a whisper. "You and her, you *put* me here. Believe me when I tell you that's not something I'll forget. If you ask me, I should kill you where you stand,

and bring her in myself – save us all a world of trouble. You have no idea the wrath that you've unleashed upon us – you and that little monkey bitch of yours."

I let my crutch fall away as my fingers found his hair. I yanked back his head, while my other hand drew the cat-shard from my pocket and held it to his exposed throat.

"I don't think you're in a position to be making threats, now, do you? Now, I hoped we could do this all friendly-like, but you just wouldn't play nice, now would you? So here's how it's gonna be: you're gonna tell me what it is I want to know, and maybe – just maybe – you cheat death a second time today. You get me?" I said.

Mu'an's Adam's apple bobbed as he swallowed hard, the shard digging into the tender skin of his neck. Ever so slightly, he nodded.

"Good. Now, why don't you tell me what happened in there?"

"There isn't much to tell," he said. "I was grabbing a cup of coffee at the Market when I spotted them: three, maybe four foot soldiers, cutting through the crowd toward me. I tried to duck out through the concourse – I thought perhaps if I could get out to the street, I could shake them – but they were too fast." He laughed, just a single, barking note. "These fucking meat-sacks, they all think our noble cousins are the good guys, but you know what? Their precious angels are worse than *we* are. I mean, when they thought I might evade them, they damn near leveled the place, without a thought in the world as to the consequences. Fucking animals, they are."

"Why were they after you?"

"How should *I* know? You should have seen it, friend – after all, it was all your doing. I mean, when that bastard let loose, there was just this thundering, heavenly note – and then chaos. I mean, the wrath of fucking God. Can you even imagine what that's like?"

"Like a chorus of children," Kate said. Her voice was small and tremulous, and her eyes had a strange, faraway quality to them, as though she was somehow no longer here with us. "Sweet. Innocent. Painful in its beauty. Or so you think, until the real pain comes."

"That's right," Mu'an said, eyeing Kate a moment with sudden suspicion before continuing. "But then the light – the heat – it stripped bone from flesh, and the closest to the blast were just... erased. Gone. Had I been a moment slower, I would have been as well. But I managed to take shelter behind a pillar, which worked out well for me, because now I get the pleasure of talking to you lovely people."

At the end of the tent from which we'd entered, there was a flurry of activity. Shouting, clanking, the crackle of static. If I had to guess, I'd say our time was running short.

"But why? Why are they after you?" I asked.

"What's it matter why? The body count's the same. All these lives lost, and all because you wouldn't do your fucking job." I pressed the shard tighter to his neck, dimpling his skin. "OK – all right. Let's not be too hasty. Truth is, I don't know why they chose to target me. In my capacity as courier, I'm often in possession of information of some import, and there are always parties that would be interested in obtaining that information, or ensuring no one else can. It's been some time since anyone has resorted to violence to that end. I rather thought that we were past all that."

"So what is it you've been tasked to convey?"

"I'm telling you, Collector, I have no idea. It's somewhat of an open secret that the appropriate compensation does wonders to loosen my tongue, which is why some of my clients choose to upgrade to a more *secure* method of communication. Once the rite of suppression is performed, I haven't the faintest idea what it is I'm carrying – or even who I'm carrying it to. You want to know what I know? I suggest you ask your lady friend."

I shot Kate a glance, but she looked as puzzled as I felt.

Mu'an laughed. "You're even dumber than you look, you know that? You should know better than to assume I'd condescend to trafficking the secrets of this monkey."

Then Mu'an let out a horrible, rasping cough, followed by another, and then another. I released his hair from my grip, and withdrew the shard – but not far. His face reddened, and he doubled over. A thin thread of blood trailed downward from his lower lip, and this time, he made no move to wipe it away. From the corner of my eye, I saw a crash team approaching, worried by his sudden fit, while behind them, a mass of uniformed security personnel were going bed-to-bed, looking for us. I turned my head to watch as a patient – just a few beds from the one I'd occupied – raised a shaky hand at whatever he'd been asked and pointed directly toward me. The cop's gaze was close behind, and our eyes locked across the massive, crowded tent.

"Well, Collector," said Mu'an, sucking breath after labored breath, "I've told you all I can. Kill me if you must – I only ask you make it quick."

"You aren't going to die today, Mu'an – at least, not by my hand. C'mon, Kate, it's time to go." I stuffed the shard into my pocket and grabbed Kate by the wrist, dragging her deeper into the teeming medical tent.

"You're just forestalling the inevitable!" called Mu'an, though his huddled form was already lost to the crowd. "She *will* be taken, and when she is, you'll pay!"

As we pressed through the crowd, Kate leaned close. Her voice was nearly swallowed by the din – the patients around us now were the worst-hit, and between the flurry of medical personnel, and the nightmarish arcade cacophony of their monitors, I could barely hear myself think.

"You think he's right? That my collection is inevitable?"

"Eh, you know demons – they just can't help but indulge in a bit of apocalyptic bluster every now and again." I flashed her a smile. It felt tight and awkward on my face.

Kate looked over her shoulder, and I followed suit. A half-dozen of New York's Finest were pushing toward us through the crowd, maybe thirty feet away and closing fast.

"You got a plan to get us out of here?" Kate asked.

"I'm working on it," I replied. I figured it sounded more encouraging than *no*.

The tent roof sloped steadily downward toward us, and through the crowd I caught a glimpse of open street and pale gray sky. I pushed aside a nurse in blood-spattered scrubs and broke for the edge of the tent. It wasn't till I could feel the kiss of fresh air across my face that I saw him.

He was a mountain of a cop, with dark deep-set eyes peering outward from a fleshy face, the features of which were twisted into an angry frown around a mustache the size of a small woodland creature. His barrel chest strained the buttons of his uniform blues as he approached, nightstick in hand. I sized him up as he approached, wondering if I could take him down. I was pretty sure the answer was no. A shame, that – he didn't look like one for talking.

I released Kate's hand and stepped clear of the tent, my hands raised in surrender. My crutch clattered to the ground, and I had the sudden, queasy realization that if this didn't work, I couldn't exactly make a run for it. Kate, for her part, had the good sense to stay a few steps behind me, hidden in the bustle of the tent, although if I didn't deal with this guy quick, it wouldn't matter – there were a bunch more just like him bringing up the rear.

"Stay where you are!" he shouted, his sandpaper growl slathered with a goodly helping of Bronx.

"I'm unarmed!" I replied. His eyes narrowed in suspicion, nearly disappearing between the flesh of his cheeks and meaty brow. If he hadn't planned on frisking me before, he sure as shit was gonna now.

Fine by me. The closer I could get to him, the better chance we had.

"Put your hands on your head." I complied. The cop holstered his nightstick and approached. "Now turn around." Again, I did as he asked.

His hands were the size of hams, and he was none too gentle patting me down. My muscles tensed in anticipation. When he gave my bum leg a good thwack, I made my move. And by *made my move*, I mean *fell down*.

Well, mostly, at least. Mr Suspicious here made my job easy by not skipping over the pound of gauze I had wrapped around my wound, and who could blame him? After all, the bandage gave me ample room to stash a weapon, and I was plenty shifty. His only mistake was in not knowing it was my hands he had to be afraid of.

When his hand connected with the bandaged meat of my thigh, I let out a wail. My leg buckled. That part wasn't just for show, but I'd expected it – in fact, I was counting on it. I twisted as I fell, so that we were chest to chest when he did his cop-ly duty and caught me. Or, rather, we would have been chest to chest, had my hands not been between us.

I plunged them both deep into his chest, grabbing hold of his soul with all I had. His eyes went wide, his features slack. The medical tent, the station, the pavement beneath our feet – all of it disappeared, replaced with a swirling morass of grays and blues and the occasional shining points of light, sparkling like stars as they orbited breakneck all around us. This was a good man, I realized – touched by darkness, but not consumed by it. It was then that I resolved not to kill him.

Soul in hand, I yanked, and now it was the cop who wailed. His pained cry brought tears to my borrowed eyes, but I had no time for such sympathies. His wails died suddenly as he collapsed, shuddering, to the ground – in shock, no doubt. But my work was not yet finished. I took care to reseat his soul just as I had found it, hoping that when he regained consciousness, all would be right in his world. Somehow, though, I doubted it. I only hoped I hadn't changed him for the worse.

When I released my grip on his soul, the world lurched back into focus. I found I was sprawled out on Park Avenue, lying half on and half off of my new cop-friend. Our tussle, which lasted a second at most, had drawn a small audience – two EMTs and a nurse on their way into the medical tent stood frozen in their tracks, staring. All looked puzzled by what had just happened, and at least one of them – a lean, angular Latina EMT – was clearly measuring the odds that I was dangerous against the odds the cop needed her help.

I took pity on her and clarified the matter: I popped the snap on the cop's holster and slid free his piece – a sleek black Glock 9mm, lighter than I'd anticipated. Then I hobbled back to the tent and grabbed Kate by the wrist, yanking her out into the street. I couldn't help but notice the cops in the tent were closing fast. In seconds, they'd be upon us.

"What are you waiting for?" I brandished the gun at our trio of onlookers. "The man needs help!"

Without a word, they sprung into action, racing to the felled cop's side and checking for vitals. Now it was Kate who stood frozen in obvious puzzlement, watching as they loosened his uniform collar and tried in vain to rouse him.

"Kate, come on!"

But she didn't respond – she just stood there, watching. "Did you..." she asked, the question trailing off to nothing. "I mean, is he–"

"He's unconscious," I replied. "With luck, he'll be just fine. *You* won't though, unless we get moving."

That seemed to shake off her preoccupation with the unconscious cop. She followed my lead as I hobbled north-west toward Vanderbilt. My leg was throbbing again, but I ignored it, gritting my teeth against the pain and forcing this meat-suit into a jog. Even Kate, uninjured, struggled to keep up.

"Sam, where the hell are we going?"

But as we rounded the corner onto Vanderbilt, her eyes went wide. Just fifty yards away sat a medevac chopper, idling in a makeshift pen of police barriers at the intersection of Vanderbilt and Forty-third.

"I'm not exactly sure," I replied. "But I know how we're gonna get there."

26.

"Sam, you can't be serious." Kate stopped dead in the street, looking first at me, and then at the helicopter that sat idling in the center of the intersection – its upper rotor still, but its engines emitting a high, keening whine.

"The way I see it, Kate, we don't have a lot of options."

"But we can't just steal a helicopter."

"We're not *stealing* a helicopter – we're hijacking one. And of *course* we can; I'm one of the bad guys, remember?"

"It's not that – it's just, I mean, they're not going to let us get away with it."

"Kate, they're not going to let us get away *period*, if they have their way. This is the only shot we've got."

From behind us, shouting. Our pursuers had cleared the tent, and it was clear now they weren't the only ones on our tail: two parties of six or so uniformed men had just finished flanking the tent on either side, and onlookers pressed ever tighter to the police barriers that cordoned off the station as officers on all sides of us abandoned their posts to join the chase. Standing in the empty stretch of street between the tent and the makeshift landing pad, Kate and I had nowhere to hide. As the men approached, guns drawn, I grabbed Kate by the arm and together we ran for the chopper. This time, she didn't argue.

The helicopter was facing north-east toward Forty-third, away from us, and the cabin door was open, though we could not see inside. Kate and I approached the door cautiously, creeping toward it along the tail. A glance behind us told me our pursuers weren't so psyched about our exit plan – the whole lot of 'em were sprinting toward us, shouting and waving like madmen in an attempt to alert the flight crew to our presence. Doubtless there were at least that many more approaching from the other side of the chopper, and it was only a matter of time before every cop, National Guardsman, and SWAT unit in the city descended upon our location. The time for caution had passed.

I wheeled toward the door, gun at ready. Inside the cabin were two flight nurses, both lean and efficient and rendered genderless by their flight suits and helmets as they busied themselves stowing gear and inventorying supplies. When they saw me, they froze. With a twitch of my gun barrel, I suggested they vacate the vehicle. They caught my drift just fine, and climbed out of the chopper, hands held high.

I gestured for them to back away, and reluctantly, they complied. One of them spoke, though the words were lost in the wail of the engine. Then I caught movement out of the corner of my eye, and I realized the words were not for me, but for whoever was on the other end of that helmet mic.

The pilot had climbed from his perch behind the controls and was sneaking through the cabin – toward the open cabin door, and toward me. In his hand, he held a flare gun. I spun, leveling my piece at his face, and he stopped short, my barrel a scant inch from the bridge of his nose. The flare gun clattered to the cabin floor, forgotten, and he, too, raised his hands. I liked this one, I decided. He was brave, but not stupid. He was also the only one of the two of us who could fly this fucking thing, so by my count that was two reasons I was glad he hadn't made me pull the trigger.

My pilot-friend again made for the cabin door, though slowly this time, as though anticipating my demand that he follow his crew. I shook my head and waved him back inside. Though his eyes were hidden behind the reflective visor of his flight helmet, I saw his features slacken as realization dawned. He climbed back into the pilot's seat, while behind him, Kate and I clambered aboard.

"Get this thing in the air!" I shouted, but this time, it was he who shook his head. He tapped the side of his helmet, twice, and gestured toward a headset hanging from the console before him.

I slipped on the headset, which looked to me like an old pair of headphones, and adjusted the microphone before repeating my command. "It'll take a minute," came the crackling reply.

"It takes any longer, and you and I have got a problem – you get me?" I pressed my gun tight to the base of his neck, and he nodded – a jerky, frightened gesture. "Just fly us out of here, and you have my word you won't be harmed." Again, he nodded, though if I were him, I probably wouldn't have believed me.

There was a tap on my shoulder, and I damn near jumped out of my skin. It was Kate, and she looked worried. I lifted one earpiece, and she leaned close, shouting: "Sam, we've got company!"

A glance out the open door proved her right: the cops had set up a perimeter around the chopper, just outside the barriers that marked off the landing area. Two men, crouched behind riot shields, crept across the landing area toward us, buffeted by the breeze kicked up by our rotor, which now swung lazily overhead.

I nodded toward the flare gun that lay on the floor of the cabin. "See if you can't slow 'em down a bit – and get that door closed!"

She nodded, retreating to the back of the cabin. Over the *whump, whump* of the rotor above, and the chatter of the police in my headset, I didn't even hear the flare go off. But the gray of the afternoon was shattered by a sudden orange-red burst that sent the uniforms surrounding us diving to the pavement, and forced their advance team to scamper backward toward the barriers. The pilot did his damnedest to ignore the spectacle outside, instead focusing his attention on the confusion of dials and switches that comprised the helicopter's control panel. I allowed myself a thin smile as I realized we might actually make it out of there alive.

The chopper rocked on its skids as Kate slid shut the cabin door. Then the rock became a lurch as we leapt skyward. We hovered just a few feet above the street, motionless but for the gentle pitch and yaw of the chopper as she was buffeted by the wind.

"What now?" asked our pilot.

"Just fly."

"Where?"

"Anywhere." He nodded, and we began to climb.

Below us was a flurry of activity as our pursuers swarmed the landing pad. Too late, the order came to take us down – shot after shot rang out, audible even over the racket of the chopper. As we rose, I heard a dull thud, and the helicopter shuddered.

"Are we hit?" I asked, a little more panicked than I would have liked.

The pilot nodded. "Feels like they dinged our elevator. Long as we don't lose it, we'll stay up all right – but it's gonna be a bumpy ride."

We continued upward, the helicopter hitching and shaking like a carnival ride too long past inspection. The pop of gunfire beneath us faded to nothing as we cleared the rooftops, pitching southward in slow, jerky arc that eventually brought our bearing east.

"No," I said, again pressing the gun to his neck, "keep us over the city. You think I'm gonna let you take us out to sea and ditch this thing?"

He maintained his heading. "If you expect to crash

us into a populated area, you'd better pull the trigger – alive, I won't let it happen."

"I told you, I've got no intention of killing you – or anyone else for that matter. Just keep us over the city, and soon enough, we all go our separate ways. Or you could keep on heading east and see what happens when I get angry."

The pilot hesitated, but only for a moment. Then, without a word, he turned the bird around. He was going easy on the throttle, but whether it was because of the chopper's ever-worsening tremors or to give the authorities on the ground a chance to keep up, I didn't know. Besides, I couldn't exactly tell him to hurry if I had no idea where we were headed, and right now, our speed was the least of my concerns.

No, what worried me was the radio.

"Hey," I said, gesturing toward my headphones, "these things got a volume knob?" He looked confused for a moment, and then pointed at the console. I fiddled with the knobs he'd indicated, flinching as I inadvertently changed the frequency, and my headset filled with static. Eventually, though, I found what I was looking for, and the police band rang loud and clear in my ears.

"... two suspects – one a teenaged girl, possibly a hostage..."

"... chopper headed northeast along Park, approximately forty miles per hour..."

"... flight nurses were evacuated – only the pilot remains..."

I sat lost in the radio transmissions for God knows how long, only snapped back to reality by a tug on my sleeve. It was Kate. Her brow was furrowed with worry, and she tapped at her ear with frustrated urgency as I stared, puzzled, back at her. Finally, she yanked the headset off of my ears, and I heard what it was she wanted me to hear.

It was a low, rhythmic whumping, out of sync with the thudding of our own blades. I looked from window to window to find the source of the noise, and soon enough, I spotted it: a news chopper, keeping pace with us maybe fifty yards to our left. Mounted on their nose was a camera, on a sort of swivelling rig that allowed it to pan from side to side. Right now, though, it wasn't panning anywhere – it was pointed right toward us.

It looked like our days of staying off the radar were over.

All right, I thought. No need to panic. All we needed was a plan, and we'd get out of this just fine.

And that's when everything went to shit.

There was a screech of rending metal as our damaged elevator tore free of the chopper's tail, and then a horrible racket like a golf ball caught in a box fan as it got chewed up by the tail rotor. The world outside the cabin lurched sideways and began to spin. Our pilot doubled over, and the cabin filled with the acrid reek of sick. As our pilot slumped across the control panel retching, his task forgotten, the chopper dipped precariously. Kate slammed head-first into the cabin ceiling, collapsing in a heap onto the floor. And then a hand, strong as iron, closed around my neck.

I struggled against the pilot's grasp, so impossible in its strength, my arms flailing wildly as I struggled for breath. His face split into a grin, and he pulled me close, breathing two words into my ear, somehow audible even over the roar of the chopper: "Hello, Samuel."

Fuck. Bishop. Apparently the bastard had nothing better to do than sit around and watch the news.

The world around us continued to spin, and I felt curiously light, as though I were barely even there. I thought then that it was just the lack of oxygen, playing tricks on my brain. It hadn't occurred to me that the chopper was going down.

I clenched shut my eyes and forced myself to focus. It wasn't easy, what with Bishop squeezing the life out of me while my overwhelming dizziness made my limbs heavy and uncooperative. If I didn't do something fast, I was gonna lose consciousness, and Kate was as good as gone. It was then that I realized I still had the gun.

I tried to bring the gun to bear on the pilot/Bishop's face, but he just slapped it away with his free hand, cackling with delight. A second try, the same result. I realized that as long as he had my neck in a vise, I was at a disadvantage. That's when I decided to shoot him in the wrist.

I pressed the barrel to his arm and pulled the trigger. The sound was deafening, and my face was spattered with blood and gunshot residue in equal measure. Still, it did the trick – Bishop's hand withdrew, his borrowed face twisted in pain. Thanks to the lurching of the chopper, the shot had been a graze – a diagonal furrow maybe two inches long, halfway up the forearm. In truth, I was grateful – if I'd shot the pilot's wrist clean through, he'd have bled out in no time flat. Least this way, I had a shot at saving him – but that meant I had to knock him out, and quick.

Bishop struggled to climb from his seat, his wounded arm clutched to his chest, but he was just as off-balance as I was, and he staggered backward into the chopper's control panel. I braced myself against my seat and kicked him in the face. His head snapped backward, his nose spouting blood. I kicked him again for good measure, and he tumbled to the cabin floor.

It was only then that I turned my attention outside. The horizon wobbled wildly, the Manhattan skyline racing by. I kicked Bishop aside, as much out of anger as necessity, and then climbed into the pilot's seat. Before me was a whole mess of stuff I didn't have the first idea how to use. I started with the joystick-looking thingy between my knees, yanking it upward in an attempt to halt our descent – after all, it always worked in the movies.

In real life, not so much. The helicopter skittered backward, still plummeting, and the cant of the cabin was so bad that if the door had been open, Kate would've rolled clean out. Sheer instinct made me slam on the left-hand pedal at my feet, but this was a chopper, not a Buick, and the spinning worsened. I tried the other pedal, and our rotation slowed – not much, but it was encouraging nonetheless. Not so encouraging were the rooftops we were fast approaching.

The only option left was the emergency brake – at least, that's what it looked like to me. We were maybe twenty feet above the high-rises of Midtown when I closed my eyes and yanked the lever. I waited for our imminent collision, and when it didn't come, I cautiously opened one eye. The bird was still spinning like a top, and she shook like she was six shots into an espresso binge, but I'll be damned if we weren't holding altitude. For the first time since we'd started falling, I had the feeling we might just get out of this alive.

That's when Bishop hit me.

I later realized that it had been a fire extinguisher. At the time, I thought it was a freight train. Whatever it was, it bounced off the crown of my skull and knocked me out of my seat. The chopper jerked, and once more began to descend. I shook the cobwebs from my head and made for the up-lever. Bishop leapt atop me, hands scrabbling to find purchase around my neck. His hand pressed against my face, and I shook free, biting down hard on the meat of his thumb. Then I dug my nails into the furrowed flesh of his forearm, and he shrieked in pain and rage.

I tossed him off of me, and scrambled to the lever. Buildings whooshed past us just inches from our blades as we descended below the skyline, Sixth Avenue sixty yards beneath us. I felt a hand on my leg, pulling me backward – away from the lever. I held fast for a moment, but it slipped from my grasp, and I tumbled backward.

Bishop, surprised by the sudden lack of resistance, released my leg and slid backward toward the rear of the cabin. For a moment, he eyed Kate's unconscious form, and then I was on him, grabbing his helmet by the sides and slamming it into the cabin floor, again and again until he moved no more. I hoped that this time, he'd stay down – I'd had quite enough of killing innocent vessels. Their lives were a mighty steep price, no matter the stakes.

Of course, if I couldn't stop us from crashing, any debate over killing the pilot was gonna be kind of moot.

I scampered back to the pilot's seat while the street rushed upward to meet us. Forty yards, thirty. The chopper spun still, and I watched horrified as, beneath us, Sixth Avenue erupted into chaos: cars were abandoned as their drivers fled, pedestrians trampled one another in a desperate attempt to get away; a cab leapt the curb and launched headlong into a sausage cart. Twenty yards, ten. Behind me, Kate raised her head, her mutter of confusion becoming a frightened wail as she realized we were going down. I gripped the up-lever with all I had and yanked it backward, just moments from impact.

The chopper began to rise.

The street receded beneath us, but we weren't out of the woods yet. Still we hurtled forward, the helicopter spinning wildly, and no amount of my slamming on the pedals at my feet seemed to change that. Sixth Avenue, so broad and impressive in my youth, was suddenly the eye of a needle – it was all I could do not to slam into the massive buildings that jutted skyward to either side. To make matters worse, thick black smoke billowed from our tail, blanketing the street, while on the control panel, a dozen alarms flashed and chimed. I didn't know exactly what they meant, but I was pretty sure I caught the gist: no matter what I did, we weren't long for the sky.

One of our skids caught on a street light, and the helicopter shuddered. I jerked the joystick aside, nearly careening into one of the buildings that whizzed past on my right. The skid clattered, useless, to the street below. A moment later, the street light followed, slamming down atop an abandoned Lincoln Town Car in a flurry of sparks and broken glass.

At the far end of the cabin, Bishop or our pilot stirred. Kate didn't wait to find out which of them was driving – she clocked him full-swing with the same fire extinguisher he'd used to hit me. He went down in a tangle of limbs, out this time for sure.

The chopper swung wildly now from right to left, and there was only so much I could do to correct. We were maybe twenty feet above the street, but we were barreling along too fast to simply jump – and besides, if we abandoned the bird now, she was gonna wind up rearranging some real estate, not to mention killing dozens. But as the familiar Art Deco façade of the Ritz-Carlton loomed large over us and I caught a glimpse of the sea of greenery beyond, I had me an idea.

We were gonna land in the park.

OK, *land* might've been too generous a term, what with a non-pilot at the stick and one of our skids a few hundred yards behind us, but still, if I could slow her down enough and drop her somewhere soft, maybe we could walk away from this OK. At least, that's what I *would* have been thinking had my thoughts not been preoccupied by a silent mantra of *oh shit oh shit oh shit*. With the chopper threatening to shake itself apart, and the joystick unresponsive, that last block and a half was one tough needle to thread.

Without warning, we kicked sideways. Behind us, a latticework of scaffolding buckled where our blades had torn through it, and collapsed to the pavement beneath. The helicopter pitched and tumbled like a rowboat in a hurricane, and there was nothing left for me to do.

One way or another, this bird was going down.

27.

The chopper shook so badly that my vision blurred and the horizon was rendered indistinct, but still I gripped the joystick between my knees, struggling with all I had to keep the chopper on course. Even in the best of circumstances, there was no way in hell I was gonna land this thing smoothly, but minus one skid, and with the controls unresponsive, I figured my only shot was to drop us in some water. Even then, I didn't know if we'd survive.

We rocketed over the intersection of Sixth and Central Park South, and the buildings of Midtown dropped away. The treetops of the park scraped against the underside of the helicopter like the scrabbling of some unholy scavengers, eager to partake of the tasty morsels within. I tried my damnedest to gain a little altitude, but the scrabbling continued. It looked like we were out of up.

I considered my options. The reservoir was damn near two miles away – no way were we gonna stay up that long. Besides, the reservoir is *huge* – even if I brought her down OK, we'd likely drown before we reached the shore. The lake was a better bet – a little closer, a little shallower – but still, I didn't see this bucket getting that far. That left the pond. Plenty close, if a bit shallow for my liking. Would a few feet of water be enough to cushion our impact? I suddenly found myself wishing I'd done a little better in physics as a kid – or, failing that, that I'd taken it more recently than seventy-odd years ago.

Oh, well, I thought – only one way to find out.

I yanked the joystick to the right. The chopper banked. She lost a little altitude as well, and a maelstrom of leaves and branches raged around us. I caught a glimpse of shimmering water just ahead before the chopper plunged entirely below the tree line, and then I saw nothing but green.

There was nothing left to do but pray.

We emerged from the canopy like a slug from a barrel, our rotor twisted and unmoving above us, our landing skids both certainly gone. The cabin tilted, and I fell from the pilot's seat, slamming hard into the window beside me. Through it, I saw the water rise to meet us, and then a murky nothing as it engulfed us in a roar of surf and a screech of rending metal. And then my forehead met the windshield, and the world went dark.

The gun was a dull, ugly affair, all scuffed and gray and worn. A tiny little revolver with a nasty snub nose and a peeling leather grip, it had the look of a featherweight boxer gone to seed. I hefted it in my hand, marveling at its weight. Then I extended my arm outward, lining the sight up with the clock that sat behind a wire cage just a few feet above the countertop.

"Whoa, pal, that iron's hot! Do me favor and maybe don't go ventilating my shop, huh?"

I looked at him and set the gun down on the counter. He was a wiry guy of maybe forty, with beady close-set eyes and nervous hands, which at the moment were tapping out a jaunty number on the countertop. He wore a pair of baggy wool trousers, held up by a set of suspenders over a grease-stained T-shirt. Except for me and him, the hock shop was empty. I looked him up and down, and wondered was he always this nervous, or was it my sparkling personality that had him on edge. Then again, I guess it coulda been the gun.

"You always keep 'em loaded?" I asked.

"No, not always. But guys like you, they come in wantin' a piece, I've found it ain't wise to keep 'em waiting."

"What do you mean, guys like me?"

"You know," he said, looking suddenly uncomfortable, "guys like you. Made guys."

So that's what I'd become? A made guy? My friend here said it with such reverence it made me want to puke.

"So how much?"

"For you? Twenty-five bucks."

"That seems a little steep."

The drumming on the counter sped up a bit. The guy looked a little green. "Hey, that thing's got no serial, no history. That's a good deal I'm giving you – Scout's honor."

I looked him up and down. "You were a Boy Scout?"

"Hey, we've all been something we ain't anymore, you know what I mean?"

Yeah, I knew what he meant. I tossed some bills down on the counter and stuffed the gun into my pants pocket.

"There's thirty here," he said.

"Keep it," I replied. I left him grinning like an idiot behind the counter as I left the shop and stepped out into the cool September night.

On the street, I hailed a cab, and told the cabbie the corner of Whitehall and Bridge. I was headed to the Alexander Hamilton U.S. Custom House, where I was to exchange the envelope in my pocket for another that I'd deliver to Dumas later tonight. The envelope in my pocket was full of cash. God knows what was in the other one. Documents, I'd guess – the kind of documents that could slap a veneer of legitimacy on whatever illegitimate shit Dumas was bringing in through the harbor. Or maybe they were raffle tickets. Truth be told, I didn't care.

This wasn't the first time I'd made the customs run for Dumas, or even the fifth, and every time it was the same. This time of night, the building was pretty quiet. My contact would meet me at the service entrance around back. We'd make the exchange and go our separate ways – no fuss, no mess, no complications.

So if everything was roses, why'd I need the heater? Because like I said, every time it was the same. Make the swap, bring the papers to Dumas. Always a spot of his choosing, always far from prying eyes. The only difference was, this time he was gonna get a little lead along with his envelope.

I wasn't happy with the thought of it, but I'd gone over it a thousand times, and every time, the outcome was the same. Elizabeth's program ended in just under a week, but she'd been off the drugs for days – the docs just wanted to keep an eye on her, make sure she didn't relapse. Once she was out, Dumas and I were done, at least to my mind. But when I'd broached the topic to him, he just laughed and shook his head. "Hate to have you get her home all healthy, just to have her take a nasty spill," he'd say, eyes dancing with mischief all the while. Always friendly, jovial – like he thought that it was cute. But I meant to get out, and if he didn't mean to let me, then I was gonna have to find another way.

The Custom House was an imposing Federal structure, six stories of cold granite overlooking Battery Park, and New York Harbor beyond. I set fire to a cigarette and made my way to the service entrance. Three cigarettes' wait, and the exchange went off without a hitch. My hands trembled with anticipation as I handed over the envelope, but if my contact noticed it, he didn't let on. The envelope he handed me, I folded, and stuffed into my pocket. For maybe the hundredth time, I thought myself a fool for going through with the swap, when I could've just taken the money and used it to help us disappear once the deed was done. But even if I could stomach taking it, the people it belonged to weren't likely to let its disappearance slide, and that'd result in a whole lot of the wrong kind of attention for me. No, it was best for me if they thought the hit and this transaction had nothing to do with one another. If that meant Elizabeth and I fled broke, then that was just how it had to be.

The walk across Battery Park seemed to take forever. My nerves were jangling, my knee was killing me, and despite the chill breeze that blew in across the harbor, my hands and neck were slick with sweat. Dumas and I were to meet at the entrance of the old fort. Designed to protect the harbor from the British navy in the War of 1812 but never once seeing battle, it now sat squat and lifeless beneath a starless sky. A little more exposed than I'd have liked to be, but I've since learned these things rarely go as smoothly as I'd like.

Dumas was chomping on an unlit cigar when I arrived. "Evening, Sammy," he said, though the words were garbled by the fact that he never removed the cigar from his mouth. "I trust you got something for me?"

"Yeah," I said. I thrust my hands into my pockets, producing the envelope from my left and handing it to him. My right hand stayed in my pocket, wrapping tight around the gun grip.

"You all right? You don't look so hot."

I laughed, cold and bitter. "Truth is, I don't feel so hot," I said. "But I think things are looking up."

"Yeah? Why's that?"

I wanted to have something cool to say to that. Something bad-ass. Something that let Dumas know that I was done playing the patsy for him. But when I opened my mouth, the words just wouldn't come.

Dumas cocked his head, eyeing me with sudden suspicion. "Sammy, what the hell is going on?" Then I pulled the gun, and he knew exactly what was going on.

I stepped in close. Grabbed him by the collar, shoved the gun into his gut. One, two, three, and it was done. His body muffled the reports, but still my ears rang. I didn't have long before the bulls arrived. I let go of him, then, watched him slump to the ground, eyes wide and blank and dead. Three blooms of red spread out across his chest. So much blood. I looked down at my hands, and they were spattered with it – that and gunpowder burns. The gun fell, forgotten from my hands. I stood trembling in the chill night air, tears stinging my cheeks. I thought that once the deed was done, I'd feel relief, but I didn't – I just felt sick. Sick and hollowed-out.

It felt like an eternity, standing there, looking down at the body at my feet, but really, it couldn't have been more than a few moments. I was shaken from my reverie by the sound of sirens, distant but approaching. I should have thought to take the gun. I should have thought a lot of things. But the truth is, I didn't think anything at all. I just ran.

Problem is, some things, you just can't run from.

When I came to, my head was throbbing. By the digital readout on the console, I'd been out less than a minute, but it felt more like a week. For a moment, I didn't move, didn't *blink* – I just lay there, still as death, so spent was I by our mad flight across Manhattan, not to mention our sudden descent. My everything hurt, but the way I figured it, that meant my everything was still *attached*, so that wasn't all bad news. In the sudden absence of the helicopter's droning wail, the cabin was so quiet I wondered briefly if I'd been struck deaf. Then I heard a low groan from the back of the cabin, and I realized my ears, at least, were fine.

The groaning was coming from Kate, who lay prostrate atop our pilot. It seemed he'd cushioned her impact, because she looked pretty much in one piece, if a bit dazed. There was a welt above her right eye from when she'd slammed into the ceiling, and blood ran freely from a scrape on her chin, but when my eyes met hers, she smiled.

Our pilot had not fared so well. He was still out, and his leg was bent beneath him in a manner not possible given the usual number of joints and bones. His face was a swollen, bloody mess, and his bullet-grazed forearm had soaked through the fabric of his flight suit. Looking at him, I wanted to feel anger at Bishop for forcing me to hurt that man, or horror at what I'd done; I wanted to feel regret for having put the pilot in this position in the first place. I wanted to feel those things because they would have given me something of my past life to hold on to, something human and decent and kind. Mostly, though, I just felt tired.

"Ugh," Kate said, rolling off of the pilot and collapsing against the cabin wall that now served as the floor. "That *sucked*. Next time you steal a vehicle, make sure it's one you know how to drive, OK?"

"I didn't steal it – I *hijacked* it. There's a difference. And I don't think you 'drive' a helicopter."

"I think it's pretty clear *you* don't."

"Funny." I hauled myself up onto my knees. It felt like I was trying to lift a bus. "What about our pilot-friend? He still breathing?"

"Yeah," she said. "You think he's still a bad guy?"

"I don't know. If he's out, Bishop's out, so there's a chance Bishop's still around. But if I had to guess, I'd say Bishop bailed the last time our guy came to – *I* would have. The way that leg's bent, though, I don't think we've got to worry about him giving chase either way."

"So what now?"

"Now we run."

I lifted myself up off the chopper window, now buried in the thick, brown-green muck that lined the bottom of the pond. An earthy stench permeated the cabin, and as I rose, I was surprised to find my clothes were damp with muddy pond water. It bubbled upward from the cabin wall beneath us; it oozed from the control panels. I helped Kate to her feet, and looked down at our pilot-friend, the inky water pooling around him.

"We've got to take him with us," Kate said. "If we leave him here, he'll drown."

"The water's barely three feet deep, Kate, and coming in slow. He'll be all right till someone gets here."

"You can't know that."

"I *don't* know that – but it's the best we can do."

"No, it's not. You can help me get him out of here. I can't do it on my own."

"Kate, that's nuts – we don't have time."

"Yeah? Well, I say we do. You plan to sit and watch while I try, the cops approaching all the while? Or would you rather try and drag *me* off? Carry me or carry him – it's your choice. At least with him, you've got help, and unlike me, he won't be kicking the whole way."

The way that leg looked, he might not be kicking ever again, but I wasn't gonna tell her that. What I said instead was: "OK. But we'd better hurry."

First, though, we had to find a door. The one we'd boarded through now lay beneath our feet – not to mention a good inch of pond water. I scanned the cabin. If there was an emergency hatch, it sure as hell wasn't obvious. That left Plan C.

What was once the left-hand side of the cockpit window was submerged, the water thick with particles churned up in our landing, but the right-hand side was clear, slate sky hanging low above a canopy of leaves.

"Cover your eyes," I said. Kate complied.

The gun thundered in my hand, painfully loud in the small, quiet space of the cabin. I, too, had covered my eyes against the threat of spraying glass, burying my face in the crook of my elbow. Once the reverberations died down, I allowed myself a peek.

The glass had buckled outward, the pane a tangled web of cracks framing a hole the size of a quarter. I climbed atop the now-horizontal seat and braced my good leg against the window, my heel atop the hole, and my back pressed tight against the seatback. Then, with an animal cry, I pushed.

The pane snapped free of its frame, not in a thousand tiny pieces as I expected, but all at once. It smacked into the surface of the water with a *slap*. Cool air kissed my face, and carried with it the sound of distant sirens. Been hearing those too often lately, I thought.

"Grab his feet," I said, looping my arms under the pilot's arms and around his chest. "And mind that leg."

Together, we wrestled him to the window and tossed him out. He splashed into the water about as gracelessly as the window had, bobbing face-down as we scampered after. The water was bitterly cold. It came up to my waist, and seeped into the knife wound in my thigh, bringing with it a dull, woozy ache that set my head reeling. I pushed past it, dragging the pilot to the shore and collapsing to the grass as Kate emerged dripping beside me. Just a couple dozen yards away, the Fifth Avenue traffic roared and honked, but I barely noticed. I was shivering and exhausted, and all I wanted to do was lay on this bed of grass and sleep. But Kate was having none of it.

"Sam, c'mon, we've got to go." She grabbed my by the wrist and yanked. I stayed down. She tried again.

"Sam, those sirens are getting closer. And we've got an audience."

I raised my head and looked around. Dotting the park were a couple dozen onlookers, watching us with expressions of confusion and surprise. Then, one by one, their faces changed, each becoming a twisted mask of hatred. Black fire raged in their eyes. As one by one they began to approach, I found my feet, putting an arm around Kate and ushering her toward the low stone wall that marked the border of the park.

"Sam, what's going on? Who *are* those guys?"

"Demons – foot soldiers, I'd guess. Ever since I first failed to collect you, they've been watching me."

"It doesn't look like they're content to watch you *now*."

"No, it doesn't. Mu'an blamed me for the attack at Grand Central – for the war that's brewing now. I'm sure he's not the only one. I suspect they've tired of waiting for me to do my job."

"So what happens if they catch us?" Kate asked.

"Torture, death, an eternity of torment. You know, the usual."

"Let's make sure they don't catch us then, OK?"

"That's the plan."

We reached the wall, and I helped her up and over. When she reached the other side, she gasped.

"Oh, Jesus, Sam – they're gaining."

A glance over my shoulder told me she was right. There were maybe a dozen of them, approaching at a brisk walk. I noticed then that they were not alone – the park was dotted with figures in suits and trench coats, fedoras worn low over faces obscured as if by an inner light. Angels. They weren't pursuing us like the demons were; they just hung back. Watching. Waiting. For what, I didn't know – and I wasn't about to stick around to find out.

I vaulted over the wall, and hit the sidewalk at a run, dragging Kate along by the wrist. The pain in my leg wasn't so much forgotten as rendered unimportant. The promise of eternal torment does wonders in adjusting one's priorities.

We darted into traffic amidst a squeal of brakes and a blast of horns. A dozen shouted curses hurled our way. I paid them no mind. Behind us, the demons had broken into a run, and were one by one hopping the wall, as graceful and powerful as a pride of jungle cats. As traffic resumed behind us, I headed south-west along Fifth. Across the street, our pursuers followed suit. As a delivery truck rumbled past, obscuring us from view, I reversed directions, darting north-east with Kate in tow. She let out a yelp as I jerked her arm, and then got wise to the plan, sprinting beside me with all she had.

A roar of anger, guttural and animal, sounded from the other side of the street. The demons had spotted us, and once again followed. The truck had provided meager cover, and our head-start couldn't have been more than half a block. The demons ate into our lead with glee, scrabbling across the hoods and rooftops of the midtown traffic as easily as bricks on a walkway. As we reached the corner of Sixtieth, I felt a surge of adrenaline. Before us was a subway entrance, just two narrow sets of steps leading downward to the darkness below. If only we could catch a ride, I thought, we might just shake these guys. Together, Kate and I descended, our feet barely touching the steps, while behind us, the demons closed the gap.

We were greeted by the warm breath of subway exhaust, stale and sickly sweet. As we descended, we passed beneath a mural of birds in flight – once no doubt brightly hued, they'd been beaten a dull gray-brown by years and years of grime. They hovered like vultures, circling in anticipation of a meal soon coming. I hoped to God we'd disappoint them.

A snarl behind us, a frightened gasp. One of the demons had reached the entrance to the subway stairs. He wore the flesh of a bike messenger, though he no longer moved as if human – he scrabbled along, half walking, half prowling on all fours, his eyes so full of raging darkness that it spilled outward from them, flickering black across the tiles of the stairwell. He pushed aside a woman in a jogging suit – the one who gasped, no doubt – and she tumbled down the stairs, landing in an awkward heap at my feet. Two others joined him at the head of the stairs – a woman in a brown tartan business suit, now streaked with dirt and grime, and an overweight man in a hot-dog vendor's apron, his face sweaty and purple from the unnatural exertion, a set of greasy tongs dangling forgotten from the apron tie around his waist. The bike messenger spoke then – just one word, and in no language that I understood, but I recoiled nonetheless. Those two syllables seemed to rise from the pit of hell itself, rendering every curse, every epithet ever uttered by Man a mere shadow, a trifle, a charming colloquialism.

It was then that they came for us.

I would say they came like animals, but that's not exactly true. Animals must abide by basic laws of nature and physics, but these things hold no sway over a demon. No, they came at us like death, like damnation, like the devil himself. They clawed and scratched their way down the stairs, crawling and bounding along the floors, ceiling, and walls – as if all three surfaces were the same, as if all three had been put there for the express purpose of conveying them to us. Soon the stairwell was filled with the dust of broken tiles and the spatters of their vessels' blood, the vessels that were so much more fragile than the monsters they disguised. I'd like to say I fought, or schemed, or even ran, but the truth is, in the face of their imminent arrival, I did nothing – just stood there, stock-still, watching. Tears streamed down my cheeks as I surrendered to my fate. I'm not proud of it. I'm not even ashamed. At that moment, there was simply nothing else that I could do.

Lucky for me, Kate didn't feel the same. Maybe it's because, deep down, she still had hope to cling to, where I had nothing but regret. Maybe I was just a coward. Maybe it doesn't matter, because when she yanked on my arm, she shook me from my dazed and sorry state. We hopped the turnstiles and sprinted together across the platform, in that moment denying the inevitability of our fates. Whatever had come over me had passed. But that didn't mean we were out of it yet. We were cornered, and they were coming fast.

Scratch that – they were here.

The platform was crowded with afternoon commuters, serious folk in business suits jostling for position with uniformed wage slaves as they waited for their trains to arrive. At least, that was the scene when we arrived. What happened next was more of a nightmare.

As we shoved through the crowd, no goal in mind but to get away from the demons at our heels, we were greeted with muttered curses and the occasional elbow in return, so annoyed were they to be disrupted in their routine. But when the demons reached the platform, that annoyance became fear. A scream rang out, and then another, and soon, the entire crowd jostled to get away, pressing tight to the far end of the platform as if those precious few feet would save them from the monsters that stood before them.

It didn't. The three demons, that followed us down the stairs tore into the crowd with savage delight, rending limbs and gouging flesh before tossing them aside like so much litter. I watched in horror as they took to the walls again, climbing toward us with chilling ease. Others charged across the crowded platform, pausing only long enough to toss aside whoever stood in their way. Though they were clad in human clothes, their vessels no longer looked human in the slightest, so warped were they by the demons within. They were impossible, horrible; their shapes refusing to resolve themselves in my borrowed eyes, my borrowed mind.

A cry rang out in the center of the crowd, quickly silenced. What replaced it was a low, wet gurgle, and as I wheeled to see what had happened, I saw an older gentleman in a blue blazer holding a girl in a waitress uniform up by her neck. She scratched and kicked at him to no avail, while he cackled with delight, black flames dancing in his eyes. His eyes met mine, and he threw the girl aside, starting toward me through the quickly parting crowd.

Beside me, another cry – this one from Kate. I wheeled toward her in time to see the woman beside her writhe as a demon overtook her, spilling sick across the floor as her eyes filled with dark flames. She reached toward Kate, who stumbled backward into me, narrowly avoiding the demon's grasp. My hand went to my shirt pocket, fumbling for the last remaining cat-shard, but it had been pulverized in the crash – that, or my fight with Bishop – and nothing remained of it but dust. Instead I dragged Kate through the crowd, the demon trailing behind.

Screams reverberated off the station walls, and the yellowed tile was streaked with blood. One by one the commuters fell, or worse, were possessed as yet more demons joined the fray. One by one the lights went out, smashed by accident or design I didn't know. Soon, though, it would be black as pitch, as death, and there would be no one left alive but me and Kate. If that happened, we were as good as damned, and this world was damned as well. The problem was, I couldn't see any way around it.

My foot came down on something soft and round beneath me – a leg, limp and unmoving – and I pitched forward, dragging Kate with me as I fell. I braced myself for the impact against the concrete, for the sudden grasp of the demon just behind us, but neither came. Instead we just kept falling, eventually slamming to rest some six feet beneath the level of the platform. Something hard and uncomfortable jabbed into my ribs – a subway rail, I realized. Above, the slaughter raged, but down here, all was quiet, with nothing but the occasional discarded body to keep us company.

I climbed gingerly to my feet, and extended a hand to Kate. She took it, and I lifted her wobbily upward. She was filthy, and a little dinged up, but she looked mostly OK. I looked around. Two sets of tunnels extended outward to our left and to our right – a commuter rail nearest the platform, and beyond it the express. We stood atop the tracks of the first of them, closest to the platform, the tunnel's overhead lamps a string of Christmas lights, disappearing into the gloom on either side of us. For the first time since the demons had arrived on the platform, I allowed myself a ray of hope. If we could reach the tunnels unnoticed, we might just get out of there alive.

But as the demon on the platform spoke, I knew that we'd have no such luck. It was the messenger again, or what was left of him, now that the creature inside had had his way. Again, it said only one word, but this one I understood just fine.

"Collector."

My eyes met the demon's, but this time, I did not freeze. I wrapped my arms around Kate and pulled her close. Her jaw was set in fierce determination, but she was shaking like a leaf, and her heart fluttered in her chest.

The demon eyed the two of us and smiled. "Give us the girl, Collector, and you and I have no quarrel."

"Go fuck yourself," I said.

"Actually," the demon said, "I had a certain someone else in mind." It licked its lips, and a chill worked its way along my spine.

"You don't know what you're doing," I said. A cool breeze buffeted my face, and I realized the chill I'd felt was not from the creature's words alone.

"I rather think I do. The two of you have brought war upon us. I intend to set things right – to restore the natural balance. They shall sing my praises in heaven and hell both. And all for the pleasure of devouring this lovely little morsel."

"The girl is an innocent," I said. My eyes were filled with the grit of dust suddenly disturbed. I blinked it back, tried not to react. "These skirmishes you've seen are gonna seem like a holiday compared to the world of shit that'll rain down on you if you devour her soul."

"Do you dare attempt to deceive a deceiver? I know what the girl has done. Nothing you say can change her fate. The only hide you can save today is your own."

"Actually," I said, as the rush of air became a roar, and the glare of headlights kissed my face, "I think I'm gonna have to disagree with you, there."

I threw Kate backward with all I had, lunging after her as the train roared past the place where we'd just stood. It screeched to a halt at the platform, blocking the demon's path, and the walls shook with a wail of fury so pure that there was nothing Kate and I could do but cling to each other, trembling, as we lay sprawled across the second set of tracks, its darkened tunnels stretching off to either side around us.

But as the echoes of the demon's cry faded into nothingness, we found our feet, and sprinted hand in hand into the darkness.

28.

Keep running, I thought. Don't stop. Don't think. Just keep running.

The air in the tunnel was cold and dank, the tracks uneven beneath our feet. Above us, sickly yellow lights pushed back the darkness at regular intervals, and cast long shadows of the tangle of pipes across the filthy concrete walls. The space between the rails was narrow, forcing us to run single file – Kate in front, with me scant inches behind, my thigh twingeing with every step despite the doctor's numbing agent. The lights of the next station were lost in the gentle curve of the tunnel. It could be fifty yards from where we stood; it could be five hundred. I told myself it didn't matter where it was – we just had to keep running. But of course it mattered. That train wasn't going to block their way forever. They'd find their way around it, or *through* it if need be. And when they did, they'd be coming for us. If we didn't reach the next platform before they broke through, we'd be trapped in this concrete tube with a horde of pissed-off demons. If that happened, I didn't like our odds.

Kate let out a yelp, and tumbled to the ground. Something squeaked angrily in the darkness. A pair of beady rodent eyes looked up from where she'd just stood, and then disappeared into the gloom. I dropped to a knee, panting, beside her.

"You all right?" I asked. Though I spoke at just above a whisper, my voice echoed through the tunnel, advertising our position to anyone – or anything – that cared to listen. I could only hope the constant clatter of distant trains was enough to drown out my words before they reached the ears of our pursuers.

"I stepped on something," Kate replied. "Something *alive.*" She twisted one arm out away from her, examining her elbow. A scrape the size of a silver dollar glistened black under the dim overhead lights.

"Rat," I said. "He's gone now, though." I nodded to her right, where, beneath a thin protective canopy, the third rail stretched the length of the track, just inches from where she lay. It looked so harmless, so unremarkable, that you couldn't help but doubt the countless admonitions you hear growing up in the city not to touch it. But still, there it sat – a challenge, a dare, a trap for the unwary. As Kate spotted it, she recoiled.

"That thing's got enough juice in it to animate a train," I said. "I suspect it's got the opposite effect on a person. Be careful getting up."

I extended a hand, and she took it. With a little more trouble than I'd expected, I hauled her to her feet, doubling over afterward and sucking air as waves of nausea radiated outward from the stab wound in my leg and turned my insides into knots.

"Jesus, Sam, are you OK?"

"Yeah," I said, straightening. "Just popped a stitch is all. C'mon, we gotta get moving."

She looked doubtful. I couldn't blame her – I didn't much believe me myself. But staying here wasn't really an option. So instead she slung an arm around my waist, and we set out down the tunnel, straddling the dead left-hand rail of the track, staying as far away from the third rail as we could manage.

We'd only gone ten paces when we heard it: a shriek of rending metal, a crash of shattered glass. A horrid slavering filled the tunnel, and one by one behind us, the overhead lights flickered and died. The darkness marched forward, step by step, as light after light gave up the ghost, and what remained was more than a mere absence of light: the darkness was pulsing, malevolent, *alive*. There was no mistaking what that darkness contained; it was the black fire of pure torment, of a being forever occluded from the nourishing light of grace, and in the face of it, all hope of escape withered and died.

They were here.

Without a word, Kate and I released each other from our awkward embrace, and took off down the tunnel at a dead sprint. Blind panic coursed through my borrowed frame. It made me strong. It made me fast. It didn't make me fast enough.

There were three of them, the bike messenger in the lead, followed by two others. In all my time walking this Earth, I'd never seen a demon so thoroughly warp its host as these three had theirs – nothing human of them remained. The clothes of the bike messenger hung in tatters around his now-massive frame. He galloped just ahead of the darkness on all fours, his flesh as black as the fire that raged in his eyes, as black as the Depths from which he had sprung. In the naming of things, humans have never been so wrong as when they called the brown-skinned "black" – for brown skin is full of warmth, of life, and this creature, black as pitch, was anything but. Its skin glistened and rippled as muscles pushed beyond the breaking point heaved and flexed like the haunches of a prized steed. Gristly streaks of red where the skin had split in deference to the form it now contained marred every swollen joint and twisted limb. Bloodied fingers, more claw now than digit, tore at the ground, propelling the beast forward, while the joints of its hind legs now bent backward, folding under the creature in an awkward, inhuman motion, and then extending in leap after bounding leap. I'm amazed I managed to keep my feet, so transfixed was I with the view over my shoulder. But keep my feet I did, and as I tore my eyes from the horrible visage behind me, I saw something that caused my heart to leap: a glimmer of light maybe a hundred yards ahead, the next station on the line. If we could just make it, just shake these beasts for long enough to disappear into the crowd...

But it may as well have been a hundred miles. Hot breath prickled at my neck as the liquid darkness engulfed us, and by instinct I pitched forward, snagging Kate on the way down and dragging her to the ground. The one-time bike messenger sailed overhead, one clawed hand swiping diagonally across where I had just stood. But the demon just passed through empty space, and the creature tumbled to the ground, rolling twice before finding its feet. It stood hunched in the center of the tracks, facing us, its chest heaving with every labored breath. A corona of light, pointed like the rays of a star, splayed out around the beast as its massive form eclipsed the light of the station beyond. The creature made a terrible chuffing sound. I suppressed a shiver as I realized that it was laughing.

"You have fought well, Collector," the demon said. "You have done yourself proud. But you have also caused us a great deal of trouble. I'm afraid you shall receive no mercy this day. Your death will be a slow and painful one."

Again I heard that awful chuffing sound, this time from behind. I wheeled to find the other two creatures standing guard just behind us, somehow clearly visible despite the darkness that enveloped us, blanketing the walls and rendering indistinct the ground beneath our feet. Both stood on all fours, their sudden heft supported by arms that now rippled with thick ropes of muscle. One lazily stretched a set of leathery wings, which made a sound like rustling leaves in the darkness. The other's flesh had split the length of its back, revealing two rows of bony protuberances – black as its skin and slick with its vessel's blood – that ran the length of its spine before terminating in a ridge of small horns at the bridge of its nose.

"And you, my dear," the once-bike-messenger said, addressing Kate. "So young, so petite, and yet so very, very dangerous. I've no doubt you'll find your new accommodations... satisfying. But don't fret; I'm certain that once your Collector is dispatched, we four can find the time for a little entertainment before we consign you to your fate. After all, you are such a pretty little girl..."

It was then that the winged one leapt. Maybe it was the rustle of its wingbeat; maybe it was just dumb luck. I guess it doesn't matter what it was that tipped me off, but as the winged demon closed the gap between us, its mouth of misshapen teeth open wide, I drew the gun from my pocket, closed my eyes, and pulled the trigger.

The report was deafening, and even through closed lids, the flash of the barrel was painful in its sudden brightness. The winged demon collapsed to the ground, whimpering like a wounded dog, its head a mess of blood and brain and teeth. I fired again. It shuddered and lay still.

The remaining demons roared in anger, shaking the ground beneath us. Then, as one, they pounced. The bike messenger was fast, I knew, but the horned demon was the closer of the two. I wheeled toward the latter and shot. I was too late, though, too slow – he slapped my arm away, and the bullet zinged off the concrete wall beside me. I heard a howl of pain, and as I tumbled to the ground, I caught a glimpse of the bike messenger lying across the right-hand rail, one clawed hand pressed tight to its ruined ankle. Blood welled red from beneath its fingers, and I felt a surge of savage delight as I realized my shot had not been wasted after all. Then the horned demon was atop me, one crushing hand around my neck, and I saw nothing but the encroaching darkness.

I clawed at the horned demon's arm with my free hand, but its grip was too powerful to break. I swung wildly at his face with my gun hand, and my heart surged as I heard a crunch of bone, a tortured cry. But the beast just slapped the gun away – it clattered useless to the ground, just out of reach. Suddenly, there were two hands on my neck, and the world began to recede.

My reality shuddered for a moment, and clarity returned. I realized it wasn't the world that had shuddered – just the demon atop me. Kate had kicked the thing across the face, and was kicking it still. The horned demon raised its hands to avoid the blows, but she just kept on kicking. Blood sprayed from the creature's mouth, sizzling as it spattered against the third rail. The scent of pennies filled the air.

The demon grabbed her foot and twisted. Kate went down, hard. It was on her in a flash.

The horned demon straddled her chest and buried one clawed hand in her hair, yanking back her head and regarding her carefully. Hatred burned in pitch-black eyes within the ruined meat of its face. A tongue, red-black and forked, extended from the demon's mouth and dragged across Kate's cheek, as one clawed digit traced a line across her neck – playful, taunting. The beast applied a little pressure, and the tender flesh of her throat furrowed, red blood welling in its wake. Kate shook with evident fear, and clenched shut her eyes as she steeled herself for the inevitable. The creature threw his head back and laughed – a full, throaty, baleful laugh that shook the tunnel around us, and sent showers of dust cascading from the rusted pipes above.

The bastard never even saw me coming.

I threw myself at him with everything I had. Considering what he was planning to do to Kate, my everything was a lot. I hit him in a full-on Superman horizontal, wrapping my arms tight around the demon's as the top of my head slammed into its cheekbone. Something in its cheek snapped like tinder, and the fucker went down. Momentum's a bitch, though, and I went down with him. We landed in a heap of bloodied limbs in the center of the tracks – me on top, with the demon kicking and scratching as it struggled to get free.

I pinned its arms above its head, with no thought but to stop it from gouging. Yellow-gray teeth scraped the flesh of my cheek as it snapped its jaw at me. By instinct, I recoiled. The creature seized the moment, shifting its weight and rolling me over. Cold steel dug into my back as it forced me down onto the rail, and again, its hands closed around my neck. I struggled against its grasp, but only for a moment. Then I slackened, and went still.

The creature's grip lessened – just a touch. It was all the opening I needed. I threw myself upward like a sit-up from hell, slapping away its hands and shoving the creature backward with all the strength I could muster. It tumbled backward, extending a hand to steady itself as it fell.

I could've danced a fucking jig.

Its hand connected with the third rail with a satisfying *fwap*. The creature convulsed as the voltage racked its body and every muscle tried to clench at once. Thick oily smoke snaked skyward, and pooled beneath the tunnel lights. A stench of salt and meat, sickly sweet, filled my nose, and I knew in that moment I'd never again attend a pig roast, no matter how long I walked this Earth.

Something happened to its eyes that I'd rather not describe, and then the beast was still. The liquid darkness of the tunnel faded somewhat, its walls and ceiling now no longer obscured from view. And there was something else, as well: a subtle swirling, a shifting of the smoke, which continued to pour off the body but now no longer pooled, instead drifting perceptibly toward the distant light of the station.

"Kate," I said, as I watched the smoke drift past, "we gotta go."

She followed my gaze, and then glanced back the way we'd come, worry clouding her delicate features. "Yeah," she replied, "I think you're right."

I snatched the gun up from the ground and extended Kate a hand. She took it, and we trotted side by side down the tunnel, leaving the husks of the two demons in our wake. But then we paused, spotting the third, the former bike messenger, propped against the tunnel wall ahead, its head thrown back in obvious pain, its brow damp with sweat. It sat, eyes closed, with one knee tight to its chest, and its other leg extended, the latter terminating in a bloody, glistening stump. A few feet away lay a mangled foot, connected only to the leg by a trail of blood. The gunshot, so far as I had seen, hadn't taken off the foot. As I watched the stump pulse and split like some nightmare egg, loosing fresh claws that kneaded the chill, damp air gingerly as if testing it, I realized the demon had removed the ruined appendage itself.

The creature pushed backward with its good leg, and its back slid haltingly up the wall. It stood there a moment, its weight supported by its undamaged foot, the new one scratching tentatively at the concrete floor below, and then it took one lurching step forward, its eyes opening at last.

"I cannot let you pass," said the demon. Once more, the dark enveloped us, radiating outward from the demon, but that darkness was fragile, somehow, barely there – like tissue paper. The creature took another lurching step, wincing as it did. It stood in the center of the tunnel now, the smoke from its fallen comrade now streaming overhead. A sudden breeze ruffled its tattered clothes.

"You're in no shape for this," I said, not unkindly. "Just let us go."

"I fear we've come too far for that."

"I don't believe that's true. You got a name?"

"I am but a foot soldier. We have no use for names."

From far behind us shone the rheumy glare of a subway car's lights. Kate shot me a worried look. I ignored her.

"You were an angel once – before the Fall. You had a name then, didn't you?"

Another shuffling step, another wince. Behind us, the train pressed ever closer.

"Yes."

"Then tell me, angel, what is your name?" I asked.

The creature swallowed hard. Its eyes closed in pain and concentration, and when it opened them again, I saw that the black flames they contained had dwindled to a flicker. "Veloch," it said.

"Veloch, I need you to listen to me. This girl is an innocent; her soul is unmarred. She's been set up – by who, I don't know. Whoever it is, they clearly want a war. If you take her, a war's exactly what they'll get. You and I both know what we've seen so far is nothing compared to what would happen if the Adversary were to lay claim to a pure soul."

The demon took another limping step forward. "Even if what you say is true, I have my orders. What is it you want from me?" Its voice was hoarse and weak, his words nearly lost to the rumble of the coming train.

"I want you to trust me," I said. "I want you to trust me because I need your help."

The creature snorted. "You want me to *trust* you? Why would I, when your kind, unlike mine, is not bound by your word?" It took another step forward. It stood only paces away now, wrapped in a shroud of guttering darkness that did little to repel the lights of the approaching train. Kate stood panicked and sweating beside me, looking continually from the demon to the train and back again. She clearly realized, as I did, there was no way we could beat the train to the station, no matter how fast we ran, and the thought seemed to imbue her with a kind of twitchy desperation that radiated off of her just as surely as the demon radiated darkness. Still, I remained calm. Maybe I'd resigned myself to the fact that we might not get out of this tunnel alive. Maybe I just didn't care. Or maybe I'd just found a little faith.

"I'm not asking you to trust my kind," I said, tucking the gun into my jeans and stepping toward Veloch, arms raised. "I'm asking you to trust *me*." I grabbed it by the wrist and pressed its hand tight to my breastbone. Behind me, Kate quailed, and a thin cry escaped her lips, while further back, the train inexorably approached.

The creature flexed its hand, and thick claws pierced my chest, but still I didn't flinch. Its borrowed eyes searched my face for any sign of duplicity, its own features twisted with suspicion. And then, suddenly, Veloch released me, the suspicion draining from its face.

"You speak the truth, so far as you believe. I shall help you in your quest."

"Thank you," I said.

"If I find that I have trusted you in error, I assure you, you will pay – and the girl as well."

"Of course."

"Tell me, Collector, what made you so sure that I would choose this path? After all, my fate is sealed – redemption, for me, is forever out of reach."

"Maybe. Maybe not. But it's not too late to try. Besides, if you sealed your fate, you did so by choosing to rebel – by choosing freedom over the bonds of servitude. It was a choice that, by all rights, wasn't yours to make, but you made it nonetheless. I guess I had to hope you'd make another."

"Uh, guys?" Kate said, voice tight with tension. "If we're all on the same side now, could we maybe move this somewhere else?"

"The girl is right – you must go. You and I are well-met, Collector."

I grabbed Kate by the arm and dragged her toward the station. The lights of the train were nearly upon us. But when Kate saw Veloch standing fast, she stopped.

"Wait," Kate said. "You're not coming?"

"No, child – not today. Perhaps someday we will meet again. I hope for your sake we do not. Now, go!"

At that last, Veloch roared, and Kate was shaken into action. We sprinted toward the station, while behind us, the train hurtled ever closer.

The tunnel shook with the force of the train, now just thirty yards behind us. Veloch's roar only seemed to build, swallowing the noise of the train whole until nothing remained but the demon's cry.

Twenty yards, ten.

I sprinted with all I had, Kate keeping pace by my side. Though I dared not look back, the shadow of Veloch that the subway's lights cast through the tunnel seemed, impossibly, to grow, until only the faintest trickle of illumination slid past.

Far too late, a screech of brakes.

In the moment of impact, I glanced back. Though the force of it shook the tunnel like an earthquake, and the sheer volume of the crash rendered me momentarily deaf, it's what I saw over my shoulder that will stick with me always: Veloch, eyes closed, arms clutched to his chest, a beatific smile gracing his warped and twisted features. The demon was near as tall now as the train itself, which crumpled around Veloch like a hatchback around a maple. For a moment, the demon didn't move, didn't flinch – it just stood stock-still while, impossibly, the train cleaved to either side of it, raining sparks as it ground along the tunnel walls, the metal yielding to Veloch's flesh as if it were stone. Then the life drained out of Veloch's face, and its massive body fell beneath the train, which, now unimpeded, surged forward once more.

The train bore down on us again, this time as a shrieking, flaming mass of twisted metal. The lights of the station loomed large, just a dozen yards ahead of us. But the train was coming fast, and hard as we were sprinting, it was all I could do to keep my feet.

Kate reached the station first, hauling herself up onto the platform and rolling clear of the tunnel entrance. I was not so lucky. Blood streamed from the wound in my leg, and I was growing weaker by the second. The heat of the flames licked at my back, and I knew the train was close. What's worse, even if I reached the station in time, as battered as I was, there was no way I was gonna be able to haul my ass up onto the platform. As the train squealed ever closer, I realized this vessel would be swallowed beneath it just as Veloch's had been.

But then, a hand. It reached down toward the tracks from the platform, just at the entrance of the tunnel. As the heat of the flaming wreck singed my neck and back, eating holes through the flannel of my shirt, I leapt, grabbing Kate's proffered wrist as she clamped down on mine.

And as I swung weightless toward the platform, forty tons of twisted metal bearing down behind me, I closed my eyes and prayed.

29.

I tumbled onto the platform, the flaming wreck of the train clipping my ankle as it sailed past and sending me skittering across the tiles. I came to rest in a dingy yellow corner littered with gum and filth and smelling faintly of piss, and as I lay there, taking stock to see if I'd brought all my limbs with me, I thought it might just be the most beautiful place I'd ever seen.

Kate lay on her back just a few feet away, her chest heaving with exertion, her face beet-red and drenched in sweat. As her eyes briefly met mine, though, I saw they were wild with life, as I'm sure mine were as well. What a sight we must've been, although there was no one there to see us; out of the corner of my eye, I caught a glimpse of the last stragglers from the platform fleeing streetside up the stairs. I guess nobody wanted to stay to watch the train wreck. Thinking back to the station we'd just come from, I suspected the people in this one had no idea how lucky they just were.

"You OK?" Kate asked. I rolled onto my side, watching her as she rested her head against the station floor and lay staring at the ceiling, chest heaving with breath after gasping breath. She looked as exhausted as I felt.

"Yeah. You?"

"Yeah. But what about Veloch? Is he dead?"

I shook my head, and then realized that, facing the ceiling as she now was, there's no way she could have caught that. "No," I said. "Just his vessel."

"Is that why they didn't go all buggy and stuff like Merihem did? Because we didn't really *kill* them?"

"Yeah. Most higher order demons, like Merihem or Beleth, have the ability to walk among us unseen – to trick our eyes and minds into seeing them as human. They'll possess someone if it suits their purposes, but it's hardly a necessity. The foot soldiers don't have that kind of power. If they want to hide their demon natures, they're forced to take a human vessel. Of course, a human vessel is nowhere near as powerful as an actual demon, but the upside is, it makes the demon less vulnerable to attack – if they get bounced from their vessel, they just retreat to their physical selves. Merihem didn't have that luxury, and now he's gone for keeps – hence the big, messy exit."

Kate fell silent for a moment while she caught her breath. "So those people they – what's the word – inhabited?"

"Possessed."

"Right. Possessed. Those people they possessed – we killed *them*, though, right?"

"Sort of," I said. "I mean, it's complicated. See, when a demon takes a host, it's not like when I do. I was human once, so human is how I see myself. If I remake a vessel in my image, I'm just rearranging their thoughts, the occasional mannerism – and even then, it takes time. When a demon possesses someone, they have a tendency to warp that person in their image. To some extent, they can't help it, although many use it to their advantage, as our friends back there did – they bring with them their strength, their speed, their *everything,* until not much of the host being remains. Those guys back there warped those bodies faster than I'd ever seen. Even if we had the time for an exorcism – which we didn't – I doubt they would have survived."

"So what does that mean about me? How'd the demon that killed my family change *me*?"

I sat up and looked at her, unsure of how to respond. After a moment's reflection, I decided to tell Kate the truth. "I don't know."

She seemed to turn my answer over in her mind as if inspecting it, and then she nodded. "So where *are* their physical selves? Where do they go, when you expel them from their vessels?"

"I have no idea. Hell's a big place."

"I thought you said that *this* was hell."

"For me, it is. For others, as well. But hell's not just this island, this city, this *planet*; it's *everywhere*, just a hair's breadth away from the 'reality' you see. You ask me, that gives them plenty of latitude to hide."

"Can they come back?"

I nodded. "All we did was slow 'em down."

"Well, then," Kate said, climbing to her feet and extending a helping hand to me, "what do you say to not being here when they do?"

We emerged from the station at the corner of Lexington and Sixtieth. Overhead, the gray sky deepened toward black as evening settled over the city. Sirens wailed in the distance – in response to the train wreck, no doubt. The midtown traffic must've slowed them up, though, because so far, they were nowhere to be seen. My thoughts turned to the hulking mass of twisted metal that sat burning beneath our feet, and the people doubtless trapped within it. I pushed those thoughts aside. There was nothing I could do for them. And if I failed to keep Kate safe, there was nothing I could do for anybody.

"Come on, we've gotta get moving." I took Kate by the arm, and led her away from the station. But just a few steps later, I stopped cold.

The squat storefront of *Mulgheney's* sat huddled before me, spilling neon red across the sidewalk like the last sixty years had never happened. Actually, that wasn't quite true: you could see those years in the film of grime that coated the storefront windows, in the dulling of its chromed marquee; a few feet above the door, an ancient air conditioner – not yet present when I'd last laid eyes on the place – dripped rust down the transom below. But all of that was swept away by the wave of remembrance that washed over me. The reek of the place, all cigarettes and whiskey and cheap cologne. The heady mix of lust and greed, of sin, that I'd mistaken for good cheer, for the promise of a better life for me and Elizabeth both.

No. Looking back, that wasn't true. I hadn't mistaken it for anything. Even then, I'd known better. Somewhere, deep down, I'd known exactly what it was that I'd so blithely bargained away. After all, I know better than anyone that's the way these bargains work. If the mark doesn't understand the stakes, then the deal is null and void.

So yeah, I'd known. I'd known it all along. And if I had the chance to do it all again, I'd probably play it the same. Guess I'm not one for learning my lesson.

As I stood there, staring at the place, a shiver coursed down my spine. A bead of sweat trickled down my side. Kate peered at me with concern. "You OK?" she asked.

"Just a shiver," I told her. She put an arm around my shoulder and gave it a squeeze. Then, the sound of sirens growing ever louder, we set off down the street.

"Sam, what are you doing here?"

I looked at Elizabeth, clad in darkness, rubbing the sleep from her eyes with balled fists. But for the occasional snore from her fellow patients – each separated from one another by curtains that extended outward from the walls – the ward was quiet, and the nurses' station was empty and unlit. The only illumination came from the window at the end of the long shared room: city lights reflected cold and brittle off the walls, the linens, the floors. But even in the dark, her expression wasn't hard to read. Liz was frightened. Frightened and suspicious.

"I don't know. I – I just had to see you. To make sure you were OK."

"It must be three in the morning!" she whispered. "People here are trying to sleep!"

"I know," I said. "I'm sorry." Actually, it was closer to four. I'd been walking the streets of the city since Battery Park, since Dumas, trying to wrap my head around what I'd done, but it was no use. I'd never taken a life before – hadn't thought myself capable – and it was just too much for me to deal with on my own. I didn't know at the time that I was coming here, at least not consciously. But while my thoughts went round and round, my feet had other plans. So here I stood. Broken. Trembling. Wanting nothing more than for her to tell me everything would be OK.

But Liz was having none of that. She clicked on her bedside lamp, looked me up and down. My eyes were red and swollen, and my cheeks stung from the salt of drying tears. My clothes were peppered with blood. Gunpowder burns had seared the flesh of my right hand, although the damage was hard to see, because try though I might, I couldn't stop my hands from shaking. She said, "Jesus, Sam, what happened to you?"

"Nothing – it's not important."

"The hell it's not! I haven't heard from you in days, and now you show up in the dead of night, looking like some kind of crazy person. And what is that all over your shirt? It's blood, isn't it? Oh, God, what kind of work are you doing for that man, anyway?"

"Believe me, you don't have to worry about Dumas anymore," I said.

Elizabeth's eyes went wide. She recoiled, her hands to her stomach, retreating to the far end of the bed. "You didn't. Tell me you didn't."

"You don't understand – this guy was as rotten as they come."

"Tell me you didn't," she repeated, tears welling in her eyes.

"I had no choice, Liz."

"Just please tell me that you didn't," she said, pleading now, tears pouring down her cheeks.

"I did what I had to do," I said. "I did it for us."

Elizabeth buried her face in her hands, her body racked with sobs. In the darkness, patients stirred around us, their sleep disrupted.

"I'm sorry, Liz, but there was just no other way. It's over now, though, and we can start fresh, you and me – maybe head back to California, or get that little place in Maine you're always talking about. But we gotta go now, if we're going. It's like we always said, love: it's just me and you, and to hell with everything else. C'mon, baby, what do you say?" I rested a hand atop her shoulder – a comforting gesture, I told myself, and I was only half-lying. The comfort was real. I just had the who it was comforting part backwards.

"Don't touch me," she spat, shaking off my hand. Her eyes were fixed on a spot somewhere in the middle of the bed, as though she couldn't even look at me.

"Liz, please."

"I want you to leave," she said.

"What?"

"I SAID LEAVE!"

At that last, the lights came on. I heard the grumble of patients in nearby beds, angry at the sudden disturbance. I heard a clatter of footfalls from down the hall, and the officious tones of hospital security ringing off the walls. And last, I heard the thudding of my heart, which threatened to burst inside my chest. I looked at Liz, my face a silent plea, but she was having none of it. So, security drawing closer, I fled.

I headed away from the nurses' station and hit the stairwell at a run, tears streaming down my cheeks. Four stories' worth of stairs passed unnoticed beneath my feet, and I spilled out into the biting cold night. I was in a narrow alley, the street beyond hidden behind a heaping mound of trash. Pavement bit the tender flesh of my hands and knees as I collapsed, retching, to the ground, my body racked with sob after painful sob. I didn't know if they were coming for me. At that point, I didn't care. I thought I'd reached the bottom, then. The worst that it could get.

I had no idea how wrong I was.

"Shit, Sam, I always figured you were kinda gutless, but this? Crying like a little bitch in the street?"

At the sound of his voice, my stomach clenched, but there was nothing left to purge. I didn't want to look at him. I knew I couldn't not. Almost without volition, I lifted my head.

Walter Dumas stood beside me, smiling. Black fire raged in his eyes. He was wearing the same suit I'd seen him in this evening, now filthy and blood-soaked. Three jagged holes, red-brown with drying blood and scorched around the edges, graced his shirt in the center of his chest. Beneath them, his skin was knotted and discolored, like a horrible injury decades old. As I stared at him, disbelievingly, Dumas tugged a blood-spattered kerchief from his pocket, and extended it to me. When I didn't take it, he just shrugged and returned it to his pocket.

"So what's the matter, Sammy-boy – lady troubles? Eh, them dames are all the same. Always squeamish when the killing starts."

My head was reeling. This couldn't be happening. "You… I mean, I…"

"Killed me, yeah. Well, tried to, at least. Made a pretty good go of it, too, if you don't mind my saying. Most folks just snap and make for the nearest blunt object, but you had yourself a plan – you even bought yourself a gun and everything. Gotta say, I'm proud o' you, son. Or, rather, I was, till I saw this *pathetic little display."*

"You… you wanted *me to kill you?" I asked.*

"Hell, yes, I did" he replied, "that's why your pal Johnnie dragged me into this affair! After all, you can't consummate a contract without blood. It's a common misconception in deals of this kind that the blood you sign with has got to be your own. Truth is, blood taken with malicious intent is always far more binding. I gotta tell you, I was beginnin' to think you'd never seal the deal – I been runnin' you ragged for months now, and you just kept on takin' it."

"'Deals of this kind'? Deals of what *kind?"*

"You mean you still haven't figured it out? I guess you always were a little dense. We own your soul *now, boy. Or, rather, the Boss Man does, though credit goes to Merihem – 'scuse me,* Johnnie, *for puttin' the whole thing together. How's fire and brimstone for all eternity sound, kiddo? Cause that's where you're headed."*

"You can't be serious."

Dumas said, "OK, you got me on the fire and brimstone. I mean honestly, I don't know who came up with that shit, but it sure as hell wasn't us. You kids and your books. It's downright cute, really. About the owning your ass, though, I'm afraid I'm quite serious."

"So what, then? You're just gonna whisk me off to hell, now?"

"Aw, come on, Sam, where's the fun in that? Nah, we'd rather let you sweat a bit. Don't you worry, though – your day is coming soon enough."

"I don't believe you," I said.

"You know what? I think you do."

There was no point arguing, I realized. Dumas was right. I did *believe. "What do you mean, my day is coming?"*

"Oh, you'll find out soon enough. You wanna know the funniest part?"

"What's that?"

"If you had only guessed at what I am, you wouldn't be in this predicament."

"How's that?"

"Ain't no sin to kill a demon. But as far as you knew, it wasn't a demon you were killin'. In this-here game of ours, intent is everything, and your intentions were just as black as can be. Tell me that *ain't the bit that's gonna keep you up at night." Dumas laughed. "Anyways, this has been fun and all, but I got places I need to be. See you 'round, Sam."*

And just like that, I was alone.

"Do you think they saw us?"

I glanced back through the glass door through which we'd ducked. It was plastered with multicolored sheets of paper – ads for roommates, dog-walking services, and the like, all obscuring my view of the street beyond. "I don't know."

We were standing in the vestibule of a Vietnamese noodle joint, just a tiny patch of threadbare floor mat stacked high with free weeklies and wedged between two doors. The interior door was propped open, giving me a view of the restaurant's spartan dining room and teasing my empty stomach with the aroma of ginger and lime and simmering meats. What few patrons there were made no attempts to hide their puzzled stares, and I couldn't blame them. What a pair we must make: Kate, scraped and filthy beneath her blue-streaked hair and studded choker, looking for all the world like a punk-rock zombie. Me, pallor ashen from loss of blood, much of which had dried red-brown into my tattered clothes. I, too, looked like a dead man walking, which was funny, cause for a change, I wasn't.

"So what do we do?" Kate asked.

It was a fair question. We'd barely made it a couple of blocks from the station before we'd spotted them: a pair of demons, combing the street, the black fire that burned in their eyes belying the impassive expressions that graced their otherwise human faces. I had no doubt that there were more of them – dozens, maybe hundreds by now – fanning outward from the spot we'd last been seen, determined to put a stop to this war, to this *girl,* once and for all. I wasn't about to let that happen, but that meant we needed a plan. From the looks on the diners' faces, we sure couldn't stay there.

I looked into Kate's eyes, so trusting and innocent despite all they'd seen, and I wished I had something to tell her. Truth was, I was out. Out of gas, out of ideas. I had no fucking clue where to go, or what we'd do when we got there. I'd fucked this job up from the get-go, and now, the whole city was on our tail – humans and demons alike. We'd be lucky if we lasted the night.

But of course I didn't say any of that. No, what I said was this: "We've got to get off the streets, and quick. Find a place to hole up while things calm down. If we stay off the radar for a while, there's no way for Bishop or the demons to get a bead on us. That means first we've got to get out of here. It's probably a matter of minutes before someone here calls the cops, if they haven't already. Right now, I'm thinking kitchen."

Kate nodded, and we ducked out of the vestibule, darting through the dining room and pushing open the kitchen's swinging double doors. The kitchen was hot and narrow and cramped, with two apron-clad cooks barely visible behind stainless steel counters stacked high with pots and pans and bins piled high with fresh-cut veggies. They shouted at us in their mother tongue, but we were gone as quickly as we'd come, banging open the heavy metal door that led to the alley behind the place. It slammed shut behind us, and I leaned against it while I got my bearings.

But for a mangy cat asleep atop a dumpster, the alley was empty. The way my heart was pounding, I guess I was surprised. I half-expected the place to be crawling with demons, eager to tear us limb from limb. Guess the damned aren't much for optimism.

I slid the Glock from my waistband and popped out the clip. Empty. I pulled back the slide and checked the chamber, a wave of relief washing over me as I realized there was one round left. We weren't gonna have the option of shooting our way out, if it ever came to that, but at least I could stack the odds a bit, make that one shot count. I dredged the powdered remains of the cat-shard from my pocket and funneled them as best I could into the barrel of the gun. I had no idea if the damn thing would fire, full of dust like that, much less whether these last sad scraps of cat-shard still had enough juju left to kill a demon, but faint hope was better than no hope at all. I tore a scrap of fabric from my shirt and stuffed it into the barrel to keep the powder in, and then I tucked the gun back into my jeans.

Kate, who had watched the process without a word, gave a slight nod, and then spoke. "All right, now where to?"

"Got me, kid. Seems to me, these are more your stomping grounds than mine. You got any suggestions, I'd be happy to hear 'em."

"Well, there's one place I can think of," she said.

"Yeah? Where?"

"Home."

30.

"You sure you're ready to do this?"

Kate stood looking upward at the building across the street, her hands worrying at the hem of her shirt. "Yeah," she said, the faintest quaver casting doubt on her assertion. "Yeah, I'm sure."

I remember now, having peered into her eyes for any evidence of doubt, and finding none. Of course, now I know it wasn't her I should've worried about. Turns out, I'm the one who wasn't ready.

We stood hand in hand at the crosswalk, waiting for the signal to change, and when it did, we set out across Park Avenue. Kate's building was a stunning pre-war co-op, draped in an elegant limestone façade. Arched transoms framed windows near as tall as I was, and each floor was delineated by an elaborate garland-and-wreath cornice. A limestone balustrade sat atop the building like a crown.

As we approached the massive Gothic arch that denoted the main entrance of the building, Kate stopped short, casting glances to either side.

"Something's not right here," she said.

That seemed, to me, an understatement – standing on this block, by this building, covered as I was in blood and filth, I felt like a kid out of class without a hall pass. But I'm guessing that wasn't the something she was talking about. "All right, I'll bite – what's wrong?"

"No Murray."

"No Murray?"

"Murray's our doorman."

"Your doorman," I echoed.

"Yes."

"And he's not here."

"Yes."

"If he were here, you think he'd be inclined to let us in?"

"Of course not. There's a service entrance around back, leads downstairs to the boiler room. It gets hot down there, so most days, the super leaves the door propped open. That's how I figured we'd get in."

"I'm still not seeing the problem here. The doorman pops out to grab a bite, and instead of slinking around in the hot basement, we get to walk in through the front door. Seems win-win to me."

"Sure, except Murray never leaves his post."

"Maybe somebody upstairs needed something? Some luggage carried or whatever?"

She shook her head. "They've all got staff for that."

"What about the bathroom?"

"The man's a freaking camel."

"So no Murray is bad."

"Yeah," Kate said, "no Murray is bad."

"Then we run," I said. "Find somewhere else to go to ground while we come up with a plan."

"I'm tired of running, Sam. Tired of hiding. Besides, what's the use? If they're waiting in there for us, they knew that we would come here before *we* did. If that's true, then where the hell are we gonna go?"

"So what, then – we just waltz in there and surrender?"

"No. We go in there and face them."

"Kate, that's suicide."

"Is it? Sam, I just saw you throw yourself at the mercy of a *demon*. A demon who could've killed us both, but instead decided to save us. As far as I'm concerned, that means all bets are off. I'm not asking you to die for me. I'm just asking you to have a little faith."

I stared her down. She didn't blink. Finally, I dropped my gaze and nodded.

"OK, then," I said, slipping a hand under my shirt and wrapping it tight around the gun grip. "Let's do this thing."

The elevator was quiet.

There was no attendant, no faint strains of insipid music, just the soft clatter of machinery high above, and the ragged sound of our breathing. The elevator car was paneled with mirrors, trimmed in mahogany and brass and polished to a perfect shine. As we rode upward, I blinked at the stranger that stood before me, watching as he blinked in kind. I wondered if the man whose body I'd borrowed was peering outward too. I wondered if he still recognized the man in the reflection.

The elevator slowed to a stop, a bell chiming to announce our arrival. It may as well have been a cannon report. I pressed myself against the mirrored wall – the gun in one hand, and Kate held fast to the wall beside me with the other. As the doors slid open, I held my breath. A bead of sweat traced its way along my spine.

Kate's apartment was the penthouse, a lush two-story affair with a view of the park. The elevator opened directly into the apartment's vast marble entryway, provided you knew the code. Kate, of course, did.

The entryway was dark, with only the faint illumination of the elevator light splashing across the marble tiles to guide our way. There was no police tape, no seal to break; evidently, the private elevator was deterrent enough. Of course, it also meant we didn't know if we were the first to enter or the fiftieth. I put the thought out of my mind and stepped out of the elevator.

The clack of my shoes against marble echoed through the entryway. I froze, straining to hear a response in the darkness, but there was none. I looked around. To my right was a massive staircase that curved upward to the second floor. A crystal chandelier dangled in the center of the room, its chain disappearing into the gloom above. Beneath the chandelier sat a round antique table. The vase that once rested atop it now lay shattered on the floor amidst a muddle of flowers, now withered and dead. A bloody handprint, matte brown against the high gloss of the tabletop, now sat where the vase had once stood. As I approached, I noticed the fingers of the hand were impossibly long, extending outward toward the elevator, as though whoever had made them had been clawing their way toward the exit, only to be dragged backward toward their horrible end. By the size of the print, I'm guessing it was the mother. I glanced back at Kate, her own delicate fingers wrapped around the elevator jamb to prevent the door from closing, and I thought – not for the first time – we were crazy to have come.

I whispered for Kate to follow. Another clatter of footfalls as she darted to my side. Behind her, the elevator door slid shut, plunging us into darkness. Kate made for the light switch, but I stilled her with a touch. We stood that way for what seemed like forever – listening, waiting.

Eventually, our eyes adjusted, and shapes appeared in the darkness. The faint glow of the eggshell walls, broken here and there by squares of black: by daylight, art, no doubt, and originals at that. Ribbons of man-made starlight, extending from floor to ceiling: the city lights, peeking through half-drawn curtains. The bulk of furniture: a high-backed chair, a low-slung chaise, more felt than seen in the darkness.

Kate gestured toward one of the hallways that extended outward from the entryway. We crept its length in silence, Kate clinging to my side.

At the end of the hallway was the largest kitchen I've ever seen, all granite and stainless and cherry, the surfaces gleaming blue by the light of the microwave display. It wasn't until I saw the place that I realized how hungry I was, how long it had been since I'd last eaten. I could've spent all night in there, chopping, roasting, sautéing – the place was a cook's dream. Then Kate turned on the lights, and that dream became a nightmare.

It was my fault – she'd asked me with a glance, and I'd acquiesced, my reluctance evaporating at the thought of something hot to eat. But when the lights came on, I lost my appetite.

The place was a fucking mess – cupboards emptied, drawers upturned, their contents scattered across the floor. A set of stools were tossed haphazardly into the center of the room, their cushions slashed, their batting stained brown-red.

In fact, the whole place was covered with blood: the floors, the counters, the walls. Even the tray ceiling above, a tasteful buttercream trimmed in purest white, was spattered with flecks of blood.

I looked to Kate, expecting to see her recoiling in horror, but she wasn't. Instead, she stood stock-still, her eyes glazed and faraway, her face slack and emotionless.

"This is where I killed them," she said.

"No."

"But it is," she said. Kate gestured toward the piano across the room, a baby grand. A bowl of cereal was perched atop it, half-empty and moldering. "Connor was sitting over there in his cowboy pajamas, banging away on the piano. He was supposed to be eating his breakfast, but as always, he had other plans. Dad was in his study, calling Tokyo, and he kept shouting at Connor to keep it down. And Patricia – Mom – was in the kitchen, making lunches for the both of us. She knew she didn't have to – our school provides lunch daily for everybody – but she always insisted. 'There's no *food* in their food,' she'd say. 'It's all fat and sugar and preservatives.' And that's when it happened."

"Kate–" I said, but she just ignored me.

"Connor was the first. I picked him up like he was nothing, and I tossed him across the room. When the piano stopped, Mom looked up. When she'd seen what I'd done, she started screaming, and Dad came rushing in. That's when I found the knife."

"Kate," I pleaded, "don't do this."

"Dad tried to stop me, of course, but I just shrugged him off. Connor was crying, I remember, and screaming for his mother. Then all of a sudden he wasn't crying anymore."

She nodded toward the far wall, where a streak of brown led downward to a floor crusted thick with dried blood. "There was so much blood," Kate said, "in my hands, my hair, my mouth. And so much screaming. My mother, my father – me, too, maybe, although that may have only been in my head. When Dad tried to stop me, it was bad. What I did to him made Connor's death seem merciful.

"But it was Mom that got the worst. I tied her to a chair, and fetched some rubbing alcohol from the bathroom. A tiny cut, a splash of alcohol, over and over again. Do you have any idea how excruciating that is?" Kate glanced down at the stab wound on my leg, seeping red-black through my ruined jeans, and smiled: thin, humorless. "But of course you do. Although at least you had the benefit of blacking out. I allowed her no such luxury."

She clenched shut her eyes, fighting back tears. When Kate opened them again, that faraway look was gone, replaced with one of sadness and regret. "Mom screamed for hours, you know. Screamed until her throat bled, until she forgot her own name. Screamed in fury and in agony, and eventually, she even screamed for mercy. But in the end, it didn't matter. I just kept cutting and dousing, cutting and dousing, until finally the police arrived. Only then, when she was of no more use to me, did I end her pain."

"That wasn't *you,* Kate. None of what you're saying was *you.*"

"What does that matter? What does it matter when the three of them are dead, and all I'm left with is the memory of their blood on my hands?"

I pulled her close, and held her tight. Kate resisted at first, but then the tears came, and she buried her head in my chest, sobbing for what seemed like hours. There was nothing I could say, so I just let her cry.

Finally, her sobs diminished; she dried her eyes on my shirt and let me go.

"It was a mistake, coming here," I said.

"No," Kate replied, "this was something that I had to do."

"Still, we shouldn't stay for long. It's not healthy. It's not *safe*. I think we should try to get some sleep, and then head out in the morning. We can grab some clothes, some food, maybe a little money, and then we'll see about getting out of the city."

Kate nodded, folding her arms across her chest and suppressing a shiver. "Yeah," she said. "Maybe getting out of here is not the worst idea."

31.

The problem was, I couldn't sleep.

I mean, the bed was plenty comfortable, and probably cost more than the average car, and the pajamas I'd borrowed were cool and clean against my skin, but I just couldn't stop my thoughts from racing. Maybe it was this place keeping me awake, with its echoes of the recent dead reverberating through its halls. Maybe it was the fact that, despite what I'd said to Kate, I hadn't a single fucking clue what we were gonna do next. Maybe it was the lack of food, or the phase of the moon, or any of a thousand mundane things that hold sleep just out of reach, but I doubt it. No, I think that maybe, just maybe, I couldn't sleep because I had a sense that something wasn't right.

I wish I could claim I'd listened to that feeling, that I'd posted myself at Kate's door and kept watch throughout the night. I didn't, though. We'd set up camp in a couple of guest rooms on the second floor – Kate, of course, could've slept in her own bed, but she'd opted not to, and who could blame her? I'd given the apartment a once-over before we retired to our rooms, but rather than allaying my fears, it only served to amplify them. The place was too big, too labyrinthine, with too many closets, nooks, and hidey-holes in which a would-be assailant could hide. Even with Kate by my side, I probably couldn't have checked them all, and after the scene in the kitchen, I didn't want to put her through all that again; so like an idiot, I'd gone it alone. To keep my worries at bay, I'd resolved to stay alert, to keep my ear to the ground – and I would have, had exhaustion not gotten the better of me.

But it did. And not just your garden-variety weariness, either; this was an exhaustion born of running balls-out for going on a week without a moment's peace, not to mention a decent meal. So as I watched the hours go ticking by, lying sleepless in my bed, I made a dumb-ass move. As the clock struck 3am, I dragged my ass out of bed and walked right past Kate's guest room to the bathroom down the hall. Just off the master bedroom, this bathroom was clearly an oasis for Kate's mom – all soft and floral and littered with make-up, a ginormous jetted tub wedged into the corner beneath a bubbled skylight. Like any self-respecting Upper East Side socialite, her medicine cabinet was a veritable pharmacy. I shook a couple sleeping pills from their amber bottle and washed them down with water from the tap. Then I stumbled back to my bed, not even bothering to pull back the covers before collapsing onto it.

I guess the pills did the trick, because that's the last thing I remember – at least until I jerked awake, panicked and sweating. Something had roused me from my slumber, but my brain was fuzzy, dulled from sleep and pills, and I couldn't focus. What was it that I'd heard?

Nothing, said my pillow. Just forget it and come back to sleep. But that pillow was a liar. I'd heard something – I knew I had. If I could just focus...

There. Again. A frightened whimper. A muffled thud. The fog lifted – not much, but a little – and I sat upright in bed, sliding the gun out from beneath the pillow as my feet found the floor. The scrap of fabric I'd used to hold in the powdered remains of the cat-shard protruded comically from the gun barrel, like a kerchief from a magician's sleeve, as though mocking me for putting my faith in so ridiculous a weapon. But it was too late to worry about that now. I crept over to my open bedroom door and peered out into the hall, but it was dark, and there was nothing to see.

I approached Kate's bedroom, gun held ready. The lights were off, the curtains drawn, but by the faint illumination of the alarm clock, I could tell the bed was empty. I padded barefoot down the hall to the staircase. At the top, I stopped, straining to hear what might be going on below. There, faintly – the whisper of something heavy being dragged across the floor. Something like a body.

The time for waiting had passed. I bounded down the stairs, two at a time, making for the source of the noise. The problem was, the whole damn place was marble and hardwood, and sounds bounced off the walls like an echo chamber. I ducked into three empty rooms before I was forced to admit I had no idea where the sound had come from. It was then that I heard the voice.

"Hello, Samuel."

It echoed through the darkened apartment as if from everywhere, or from nowhere at all. The voice itself was unfamiliar, but there was no mistaking that smug tone, that knowing sneer.

"Bishop," I said.

His laughter reverberated off the penthouse walls. "Of course, that's not my name of choice, but for now it should suffice."

I listened closely to his words, not caring a damn what they meant. No, all I cared about was where he was. The problem was, I still had no idea. If I wanted to find out, I was gonna have to keep him talking.

"Where's the girl, Bishop?"

"The child is fine," he replied, "for now."

I ducked into the living room, where brocaded high-backed chairs and a silken chaise gleamed dully by the city lights that trickled through the half-drawn curtains. But Bishop and Kate were nowhere to be seen. I leaned heavily against the mantel, trying to shake the cobwebs from my drug- and sleep-addled brain. If I couldn't focus, Kate was good as dead.

Once I gathered my wits, I tried again. "Let her go!" I called. I always wondered why people in the movies always said that; it's not like it ever works. Turns out, it doesn't matter. You say it because it buys you time. You say it because that's all there is to say.

"I don't think so," he replied, oblivious to my game, or perhaps not caring. "I rather like her where she is."

Another open door, this one to a darkened office. But they weren't there. I wondered how long it would be before Bishop tired of this game and ended her. I prayed I wouldn't find out.

I returned to the foyer, and called out to him again. "How did you know where to find us?"

"It was simple, really. The violent are so predictable, you see – so eager to return to their killing grounds. They always return eventually, desperate to reclaim that thrill, that joy, that ecstatic rush that only comes from taking a life. Tell me, dear, how did it feel when you bled your brother dry? When you snapped your father's bones in two? How did it feel when your mother begged for mercy as you tortured her? She did beg, didn't she? They all do, eventually. Even the biggest and bravest among us cower before the altar of suffering."

Kate whimpered, but didn't speak. It sounded like Bishop had her gagged. But suddenly, I realized where they were. I should've known from the start. He'd brought her back to where it had all begun. He'd brought her back to make Kate face what she had done. He'd brought her to the kitchen.

I snuck toward the kitchen hall, my bare feet noiseless against the hardwood floor. Before I began my approach, I ducked my head into a bathroom and shouted, "Don't you talk to her, you son of a bitch!" It was better, after all, if I was something less than expected.

"Son of a *bitch*? Oh, no, Samuel – you could not be more wrong. It was God himself that plucked me from this mortal coil, so pleased was he at my cleansing of the unrighteous."

I paid his words no mind, creeping down the hall toward the kitchen with my finger on the trigger.

Unbidden, Bishop continued. "Those boys were destined for a life of sin, and had I not intervened, their souls would roast still in the fires of hell. But I *did* intervene, purifying them and sending them into the arms of their loving God. Of course, they were young and poor and had so little to give, so they paid their tithe in blood. I assure you, He understood, which is why He made me his chosen son, his emissary in this realm."

At the threshold of the kitchen, I stopped, willing my heart to slow. My borrowed flesh was full of twitchy energy, muscle memory eager to put a bullet in Bishop's brain. Or perhaps it was something more? I'd never felt such willingness in a meat-suit before. I wondered if maybe after all he'd seen riding shotgun with me, Flynn was on my side.

The support was welcome, even if it might've been imagined. Whether willingly or not, we wheeled together around the corner, my gun hand drawing a bead on Bishop's smiling face, illuminated softly by the dim light that shone from up above the kitchen range.

But he was already a step ahead of me, crouching behind Kate to ensure I didn't have a clean shot. Her hands and feet were bound with duct tape, which wound as well around the limbs of a dining-room chair, affixing her in place. A strip of tape stretched across her mouth. Bishop held Kate by her hair – her head tilted awkwardly backward, her eyes pleading and terrified. A kitchen knife glinted cruelly at her throat.

"Ah-ah-ahhh!" he said, yanking back her head and pressing blade against flesh. "I wouldn't do that if I were you – someone might just get hurt!"

I took him in now, this familiar creature in an unfamiliar vessel. This one was a large-framed man, thick and meaty, like an athlete gone to seed. A few sad wisps of graying hair swept from one ear to the other in a foolish attempt to hide his baldness. He wore pants of bluish-gray, and an elaborate shirt to match – a doorman's uniform, no doubt. The shirt's double row of brass buttons were undone, his undershirted gut protruding from within. Mischief danced in his eyes, and his face was twisted into a manic grin.

"What are you going to do, Bishop – slit her throat? That's not the job, and you know it."

"It wouldn't be the first time I've taken a soul from a corpse," he said.

"I'll kill you before you ever get the chance. You have to know that. You've failed, Bishop. Just let the girl go, and you'll be spared."

"You expect me to be frightened of that popgun? You know as well as I that it won't kill me; like Lazarus, I shall rise again, and when I do, you'll pay. You and your little whore both."

"I wouldn't count on it," I said, training the sight of my rag-stuffed gun barrel at the bridge of his broad, crooked nose. "I'm pretty sure Beleth is never coming back."

"*DON'T YOU SAY HIS NAME!* Only the righteous may know the true name of the Lord!" Bishop cried.

"And what would you know about righteous?" I shot back.

"I was His chosen son! For centuries, I was the hand of God, smiting the wicked and ensuring His will be done! How dare you question me, when it is I who must step in after what you've done! It is a mantle I do not wear lightly, being God in His stead, but you've left me no choice. That is why we're here. That is why it's come to this."

"So that's what this is about?" I asked. "I kill your god, and now you want to make me pay?"

His eyes danced with anger and spite, and something else as well. Madness, I realized. The madness of a zealot.

"There is no payment great enough to repair what you have done. You've robbed the world of its Heavenly Father. I should have sensed the stink of wickedness upon you all those years ago when I first took your soul; I should have realized my attempts to cleanse the stain of sin from you were for naught. But I didn't, and it cost our Lord His life. That will forever be my burden to bear. But this girl, this harlot, this vile creature – she means something to you, does she not? Perhaps you see your wickedness reflected back at you. Perhaps you were seduced by her comely features, or tempted by her feminine wiles. I care not what it was that drew you to her; all I care is that you care for her – that this girl, this sinner, this foul creature holds meaning for you, as the Heavenly Father does for me. You see, I cannot extract payment enough from you to change what you have done, but I can force you to suffer. I can force you to watch as I tear her soul from her body. I can cause you to suffer as I myself have suffered, for I am a vengeful God, and all must learn that it is dangerous to cross me."

"You're no God at all, you fucking freak. You're a scavenger at best – or even less, you're just a cog in a machine. Your only task is to collect the souls of the damned, and even in that you're deluded. This girl's an *innocent*, Bishop. She didn't do it. That's why I've been protecting her. That's why I can't let you collect her."

"*I'm* deluded? Listen to yourself! You're not making any sense! Why would you be sent, if this girl was not to be taken? Why would the Lord himself have dispatched me to collect her?"

"Because she's been set up," I said.

"By whom? Who but she had motive to do what she has done? Who would stand to gain by the collection of an innocent soul? Who could possibly wish for war to erupt between the ranks of the righteous and the wicked?"

And just like that, I had it. It was obvious, really. I couldn't believe I hadn't seen it before.

The answer to Bishop's question fell softly from my lips, just one word, so quiet even I could barely hear it. Just one word, but with that one word, everything changed.

"So'enel."

32.

There was no thunderclap, no flash of lightning – no trumpet's blare to announce his presence. One moment, there was nothing to my left but empty space, and the next, the angel was there. In my jail cell, he'd worn a suit of charcoal gray, but now he wore nothing at all, his tall, slender body suffused with light and impossibly bright after the dimness of the room. As before, his features were indistinct, and almost painful in their beauty, but this time, I refused to look away.

"Collector," So'enel said, his rich baritone both confident and soothing, compassionate and strong.

"Seraph," I replied.

The angel looked around, taking in the scene before him: Kate, duct-taped to the chair, her gaze averted; Bishop, cowering behind her, the knife lying forgotten at his feet; and me, my silly rag-stuffed gun still trained at the spot over Kate's shoulder where, until recently, Bishop had stood. Then So'enel returned his gaze to me, his bright eyes of neither blue nor brown nor green penetrating into the furthest reaches of my tattered soul. "Tell me, Collector, why is it that you've brought me here?"

"Because I've done it," I said, willing the quaver out of my voice, the tremor from my limbs. "I figured out who it was that set up the girl."

The angel shook his head. "I see you're still persisting in this fiction of yours. It is understandable, I'll grant you, to refuse to believe one so young, so seemingly sweet, could be capable of such a terrible act, but as you recall, I looked into the matter myself. I assure you, the child is guilty."

"Yeah, so you said. Here's the thing, though – I'm positive she's not."

The angel smiled: blinding, beautiful. "Are you accusing me of lying, Collector?"

I ignored his question. "Before, in my cell, you told me my name was from the Hebrew for 'heard by God'."

"So I did, and so it is."

"Tell me, what does So'enel mean?"

"I fear I fail to see the relevance of the question."

"Oh, I think you see the relevance just fine. It means that you're a warrior, does it not?"

"A warrior for God, yes."

"Right," I said. "Not much to do these past millennia, though, huh? I mean, what with the détente and all."

"I'm sorry; I must be misunderstanding you. Are you suggesting that *I* am somehow involved in orchestrating an elaborate ruse to frame a poor innocent little girl?"

"I'm not suggesting that you orchestrated a thing. No, what I'm suggesting is it was *you* who possessed this girl. That it was *you* who killed her family. That it was *you* who tortured her mother until the police arrived, just to ensure there'd be no mistake in determining who was responsible. And that it was *you* who made sure she was marked for collection, covering your tracks so well that both sides are convinced she's guilty."

"That is preposterous," the angel said. "I am an angel of the highest order; a servant of God. I've no interest in being insulted by a lowly Collector."

"My apologies," I said. "I mean, it's not like any *other* angels have ever gone off the reservation. So tell me, this God of yours, you think he was just gonna let this slide? I mean, you damn an innocent soul to hell and start yourself a war, just for a little something to do? Sounds a lot like free will to me, my friend, and that's strictly verboten in angel-land, is it not?"

"What you're saying is heresy. You know not of what you speak."

"Maybe I do, and maybe I don't. But it seems to me it's a fine line between an angel and a demon; just a hint of jealousy, or of doubt, and you're off to the races. Are you telling me you *couldn't* have possessed the girl – that you don't have that kind of power? Of course you're not. If a demon can take a human host, it stands to reason an angel can, too. And here's the thing: Kate here told me that when she killed her family, she did it with a sense of calm, of peace, the likes of which she'd never felt before. She told me she did it with a song in her heart. Does that sound like any demonic possession you've ever heard of?"

The angel shook his head. "Don't you see what she has done to you? She's blinded you to her true nature! She's convinced you of this impossible scheme to blind you to the fact that she's responsible for these horrible acts!"

As he spoke, the angel approached, his action lending urgency to his words. I backed away from So'enel, and trained the gun at his chest.

"That weapon will not harm me," he said gently. "You sure about that? You may wanna ask Beleth." I found myself wondering if it's a bluff if you don't know for sure you're bluffing.

The angel raised his hands in acquiescence, a bemused smile settling across his beautiful face.

"What's so funny?" I asked.

"Nothing whatsoever, I assure you. It is just that I underestimated you, Collector – you're far more compassionate a creature than am I. After all, it must be difficult to defend the life of the girl who so brutally slaughtered your own granddaughter."

The blood drained from my face. I felt suddenly dizzy and weak, and my gun hand dropped to my side, the Glock pointed uselessly toward the floor. "*What* did you just say?"

So'enel replied, "Don't tell me you didn't know! I mean, the resemblance to your Elizabeth is astonishing! In the mother, and the boy as well; why, he would have been your great-grandson, would he not?"

Though the summer of '44 had been sweltering, October brought with it a brutal cold front, blanketing the city in the kind of chill that settles in your bones and makes you think you'll never feel the kiss of warmth again.

"But... she couldn't be." I said. "That's impossible."

It had been a month since that night, since Dumas, and I'd spent that time living on the streets. No, not living – trying desperately to drink myself to death, wishing every night as I lay down in the gutters and the alleyways that I would simply drift away with the next hard frost, never to wake again. The way I saw it, without Elizabeth beside me I was dead already. Sometimes, though, it takes a while for the meat to get the message.

"Is it?" the angel asked. "But you'd been following her, those months after she bid you adieu. You must have seen."

Liz had left the apartment in New Brighton, shacked up with a young doc from her program. I spent most nights camped out in a park across the street from his place, so desperate was I to be near her.

"No," I said, not in answer, but out of sheer denial.

I wanted to tell her I'd been wrong. I wanted to tell her I was sorry. I must've tried a dozen times, but her eyes would pass right over me, in the way that people's do when confronted with those who have fallen through the cracks, and every time, my voice would fail.

"You must have seen your child growing within her."

Every time but one. It was early evening, and Liz was walking briskly down the street, a bag of groceries in her hand. Her face was downcast, her brow furrowed in worry, and in that moment, I wondered if she was thinking of me. As she passed, I called to her – just her name, just once.

I said, "You're lying."

She turned around then, the bag falling forgotten to the sidewalk. I saw Liz peering into the crowd, searching for my face, but with my ratty hair and my twisted scraggle of a beard, she didn't see me looking back at her. But I *saw. I saw too much. I saw the weight she carried in her cheeks – just a touch, rounding out her face and glowing pink in the chill fall air. I saw her swollen belly, protruding from beneath her woolen jacket.*

"Did you tell yourself it wasn't yours?" said So'enel. "I assure you that it was."

And in that moment, I understood.

"Shut up."

Why she had pushed me away. Why she'd been forced to let me go.

"And that child grew into a woman, who had a child of her own."

She'd been protecting her child.

"I said shut up."

Protecting our child.

But he didn't shut up. "A child that grew up strong and sweet and brave and beautiful, so like your fair Elizabeth."

She'd been protecting it from me.

"Shut up shut up *shut up*!"

It was then, as I stood staring at the woman that I loved and the daughter I'd never know, that Bishop struck.

"A child that this one killed, without mercy, and without remorse."

The pain was excruciating as Bishop gouged my soul out of my chest, cackling gleefully all the while. In truth, I didn't mind. I knew then that I deserved it. For the person I'd become. For the choice I'd forced Elizabeth to make. And as the world around me disappeared, replaced by the swirling gray-black of my soul, I thought I heard her call out – just one heartbreaking syllable, her voice tremulous and full of hope: "Sam?"

My entire body shook in rage and pain and sudden doubt. I looked from the seraph to Kate, who once more fought against her restraints. She was trying in vain to speak, but the gag prevented it, deadening her words into a frantic series of grunts. Her eyes, wide with shock and terror, found mine, and even without her words to guide me, I knew that she was beseeching me not to listen.

"It seems the girl has something she'd like to say," the angel said. "Well, then, by all means, let her speak." He gestured, and the duct tape unwound from Kate's mouth as if of its own accord. "But first, my dear, a question. Your half-brother: what was his name?"

Kate forgot her fear for a moment, so thrown was she by the question. "C-c-connor," she said. "Connor MacNeil."

"Yes," said So'enel, not unkindly, "but what was his *middle* name?"

At that last, Kate's eyes went wide with shock and horror. When she spoke, it was flat, uninflected, barely audible. To me, though, it was a fucking knife in the gut.

"Samuel," she said. A single tear tracked downward across her trembling cheek. Then, as if from somewhere far away: "Patricia said it was in honor of her grandfather. But Sam, I never thought–"

"Enough of this," the angel said. "You see, Collector, I've steered you true. You know what it is you have to do."

I felt sick. Tears poured down my face, and my breath came in ragged, hitching gasps.

"Collector," So'enel said, and then he stopped short, correcting himself. "*Samuel.* This violation of your blood cannot be allowed to stand – the girl *must* pay."

"No!" I said, clenching shut my eyes as though to shut out the world – as though to shut out the angel's words.

"Samuel, you have to realize you were sent here for a reason. God isn't through with you yet, my child, and perhaps redemption is not so far off as you would think. It's time for you to do your duty. It's time for you to do what's right."

As the angel spoke, a calm settled over me, quieting the trembling in my limbs, the fire in my heart.

"You're right," I said, smiling at So'enel through my tears. "Of course I know you're right."

And then I aimed my gun and fired.

33.

I suspect there aren't many who've had occasion to see an angel die. I'm pretty sure that's for the best. When the bullet struck, he staggered backward, his face a rictus of shock and sudden pain. His inner light dimmed a moment, and then surged outward, engulfing us all in its radiance. Ceramic dust from the cat-shard, which had filled the room when the gun discharged, sparkled like stars in the sudden glare. The very building beneath us trembled, and I felt as though the flesh was being stripped from my bones, though perhaps it was my soul that ached, as the cleansing light of grace illuminated every moment of doubt and anger, of sin. A cry escaped my lips, agony and ecstasy in equal measure. From somewhere in the blinding light, Bishop let out a piteous wail, his own corrupted soul no doubt in flames. Only Kate failed to cry out, and it was in that moment that, by luck or providence, I realized my choice had been the right one.

Then, suddenly, the light collapsed upon itself, and unsupported by its presence, I fell to the floor, weeping like a child. In the angel's place, there writhed a single segmented insectile creature, black as sin, its back on the floor, its legs kicking frantically to right itself. It shrieked in agony, and then lay still, collapsing inward on itself as if a thousand years had passed in just one heartbeat, reducing the creature to no more than an ashen husk. I thought back to the demons Kate and I had so recently dispatched, full themselves of such creatures, and I wondered how long it had taken Beleth and Merihem once they fell to be consumed by the creatures, filled from within until there was nothing left of the beings of grace and light they once were. Then I thought of Veloch's act of mercy, and decided perhaps not all of them were so far gone.

I can't tell you how long I lay there crying, only that the first faint rays of morning light glimmered off the rooftops of Manhattan by the time I finally managed to rise. A glance around the room told me Bishop still lay trembling in the corner. I staggered over to where he lay and rolled him over, grabbing handfuls of his shirt and pulling him toward me until we were nose to nose.

"You're finished, here, you hear me? You'll not come for her again," I said.

But Bishop said nothing – he just stared wide-eyed at me, his face red and wet with tears. A stream of snot poured from his nose, and as I held him, he tried to pull away from me like a frightened animal.

I slapped him, hard. He yelped in pain and fear, but I thought I saw the slightest hint of clarity returned to his borrowed eyes.

"I *said* you're done here."

His head bobbed up and down – suddenly obsequious, eager to please. Eager, perhaps, to avoid another slapping. Not that I cared either way. Bishop was broken, beaten down by what he'd seen. So long as he remained that way, he'd pose no threat to Kate or me or anybody.

"I suggest you leave this body at once. He's an innocent in this, and I'd hate to have to kill him."

Again he nodded, and then his body went slack in my hands. Improbably, his abandoned vessel started snoring, and so I lowered him to the floor, leaving him to sleep. No doubt the man had earned it.

Kate, of course, was still fastened to the kitchen chair. I knelt at her side and tore through her restraints; the skin beneath was red and abraded. When she was free, Kate threw her arms around me and held me tight, tears streaming from her cheeks. "I'm sorry," she said. "I didn't know."

I held her close, her head pressed tight to the crook of my neck, her tears soaking through the fabric of my pajama shirt. "It wasn't *you*, Kate. And you *couldn't* have known."

"Is it over?" she asked.

"Yeah, kid. I think it is."

34.

The hall of the hospital was bright and clean and bustled with activity, its staff too busy with their evening rounds to pay me any mind. Not that there was anything out of the ordinary for them to see if they did. I'd learned my lesson the last go round, and this time, I'd snagged a napping orderly from another floor – familiar enough to anyone who passed to not warrant a second glance, but with luck unknown enough to the staff on this floor to avoid any pesky conversations. That was the theory, at least, and so far, it had worked; I'd gotten to where I was going without bothering a soul, unless you counted my orderly-suit, and there was a chance I'd have him back before he ever woke up.

I stood leaning just inside the doorway for a while, listening to the soothing rhythm of the heart monitor in the room within. Anders lay sleeping on the room's sole occupied bed, his face slack with peaceful sleep. From what little I could understand of his chart, he was gonna be just fine, and that made the whole body-swap worthwhile. I'd left my last meat-suit tied to a radiator in Kate's apartment, and I assured him I'd give the cops a ring as soon as my errand was done. It's tough to say for sure, but I think he understood. The man was a warrior, after all, and if it were one of his who'd fallen, I'm sure he would've done the same.

"Hello, Collector."

When I heard her voice, I jumped. I hadn't heard her coming, but then with Lilith, you never do.

She strolled past me into the room, tracing the line of my goateed jaw with one delicate finger. My meat-suit thrilled at her touch, her playful smile, her intoxicating scent. She was barefoot, of course, and clad in the barest suggestion of a cotton dress. It clung tight to her supple curves and halted scant inches beneath her luscious hips. The fabric was so thin one couldn't help but catch a hint of silken skin beneath.

"Evening, Lily."

She frowned – a beautiful, pouting, playful frown. "I've asked you not to call me that," she said.

"So you have."

"Nice to see you've chosen a live one once more – your second in as many vessels. Dare I hope this is the beginning of a trend?"

"Not likely."

"A shame, that – they do suit you so. Perhaps to honor the occasion, then, you and I should avail ourselves of that empty bed, and make this little errand of yours worth this meat-suit's time."

"Thanks, but no thanks," I replied. Lilith just shrugged.

"You did a good thing back there," she said. "Clearing that girl. I was wrong about her – we *all* were. I suppose I owe you an apology."

"You don't owe me anything."

"You'll be happy to know that she's been taken care of: new name, new face, new life. Easier than bringing back her family, I suppose, and with an entire city looking for her, I guess it's best she disappear."

"I assume it wasn't you who made the arrangements."

"No," Lilith said, smiling, "when the white hats realized it was one of their own who set her up, they were quick to volunteer. Nothing like an angry Maker to whip them into action, I suppose."

"So you don't know where they put her, then?"

"No, why?"

"I'd just hate to see you hunt her down, is all."

She shot me an odd look, then – puzzled, guarded. "Now why on Earth would I do *that*?"

"Oh, come now, Lily – don't play coy."

"Really, Collector, I haven't the faintest idea what you're talking about."

"Of course you don't. Only here's the thing: I spoke to Mu'an."

"Yes? And?"

"And there was something he said that I couldn't make any sense of at the time. See, he'd told me he was carrying a message, but that he couldn't tell me what it was, on account of he'd been bound by a rite of suppression."

"That makes sense," she said. "I mean, So'enel couldn't have marked the girl for collection on his own – he had to have some help. And it was angels, you recall, who destroyed Grand Central in an attempt to silence Mu'an. Maybe they're who he was there to meet, only once he'd served his purpose, he was nothing but a liability to them."

I shook my head. "I don't know – it doesn't track. See, when I asked him who it was he was playing courier for, he told me to ask my lady friend."

"Maybe he figured the MacNeil girl would know – I mean, it was the fact of her possession they were covering up."

"Sure, but Mu'an didn't know anything about the message he was carrying, and besides, we both know it wasn't Kate he was talking about."

"I don't understand. What exactly are you saying?" Lilith said.

"I'm saying he was talking about *you*."

"That's ridiculous."

"Is it? You've always taken quite an interest in me, Lily – there's no denying that. I always figured you enjoyed the game: tempting the poor tortured Collector and watching him squirm. Only maybe it was more than that."

"You flatter yourself, Collector. My interest in you is strictly professional."

"Is it? So it's a coincidence, then, that I'm dispatched

to collect a girl it turns out is responsible for the murder of my own flesh and blood?"

"You think it's not?" she asked.

"Damn right I think it's not. In all my time as a Collector, I've never been sent on a job I would have taken any joy in, and why would I be? After all, this gig is punishment for a life misspent. But if this job had been legit, it would've been a gift. Except it wasn't legit, was it? And the fact that I had a personal stake in it made for a nice little ace in the hole – if I got out of line, all So'enel had to do was play the family card, and I'd do my job like a good little soldier, with a smile on my face and a song in my heart."

"So So'enel set *you* up as well – is that any surprise, given what he did to the girl?"

"I suppose not. But what *is* a surprise is that he would have chosen me to do the job, not knowing me personally, and therefore having no idea how I'd react. No, I think he had help picking me – picking *her*. And I think that help was you."

"I assure you, Collector, you're mistaken."

"Am I? Then tell me – where were you when I went off the reservation? You said yourself – when we met in the park – you ought to report me for what I'd done. Why didn't you? You ask me, you didn't say anything because you were sure I'd eventually collect the girl, and you didn't want to be tied too closely to the job when I did. After all, if they suspected it was you who was responsible for the war that would have certainly ensued, you'd have both sides gunning for you."

"Assuming for a moment you're right," Lilith said, "what could I possibly stand to gain by inciting a heavenly war?"

"Revenge, for a start. I mean, the story says you were cast out of the Garden of Eden for refusing to be subservient to Adam. My guess is, if anybody's got a reason to start a war against God, it's you."

"Those are bedtime stories, Collector, nothing more. You of all people should know that."

"I *do* know that, but I also know that most of them contain a kernel of truth as well. Tell me, how long did it take you to tempt So'enel to your cause? Years? Centuries? Millennia? And how long, now that he's failed, before you try again?"

"You can't expect me to answer that."

I smiled and shook my head. "I suppose not."

"Nor can you prove a single word of what you just said."

"No, I guess I can't."

"So where does that leave us, then?"

I thought a moment. "Right back where we started, I suppose."

"Yes," she said carefully, "I suppose it does."

She strolled over to me, rising on tiptoes and kissing me softly on the cheek.

"You should take some time here with your friend," she said. "This work of ours can wait. After all you've seen, you deserve some rest – and believe me, you're going to need it. I have a feeling there's a storm brewing."

I said nothing: I just stood there watching as she strolled toward the open door. As she reached the threshold, she called to me, not looking back.

"See you 'round, Collector." Her voice hung in the air for what seemed like forever, long after she'd disappeared from sight.

Yeah, I thought. *I bet you will.*

About the author

Chris F. Holm was born in Syracuse, New York, the grandson of a cop with a penchant for crime fiction. He wrote his first story at the age of six. It got him sent to the principal's office. Since then, his work has fared better, appearing in such publications as *Ellery Queen's Mystery Magazine, Alfred Hitchcock's Mystery Magazine, Needle Magazine, Beat to a Pulp*, and *Thuglit*.

He's been a Derringer Award finalist and a Spinetingler Award winner, and he's also written a novel or two. He lives on the coast of Maine with his lovely wife and a noisy, noisy cat.

chrisfholm.com

Acknowledgments

There was a time when I – then but a lonely writer clacking away at a keyboard in a dark corner of my basement apartment – looked upon acknowledgments with skepticism. Writing is, by its nature, a solitary task. So who were these people to whom authors claimed they were so indebted?

Now, of course, I know better. Because it turns out those people are the difference between a dusty, unread manuscript cranked out by some lonely writer in a dark corner of a basement apartment, and the book you're now holding in your hands.

To that end, I'd like to first thank my agent, Jennifer Jackson, for her tireless work on my behalf. My path to publication has been circuitous, but Jennifer's enthusiasm and faith have been unflagging.

Thanks also to Marc Gascoigne, Lee Harris, and the rest of the Angry Robot team, for giving Sam and company such a loving (er, angry and robotic) home. Marc is also responsible for my stellar cover design, which was rendered beautifully by Martin at Amazing 15. Gents, I am forever in your debt.

My deepest gratitude to Charles Ardai, Frank Bill, Stephen Blackmoore, Judy Bobalik, Hilary Davidson, Leighton Gage, Jon and Ruth Jordan, Sophie Littlefield, Stuart Neville, and Mike Shevdon for their kindness and generosity of spirit. I can't tell you all how much it means to me.

I'm fortunate be part of an online writing community whose members' friendship and support I value more than I've room here to express. I would, however, like to single out a few of them for championing my work these many years (with my apologies to anyone I've missed, as this list is certainly inadequate to so Herculean a task): Patti Abbott, Patrick Shawn Bagley, Nigel Bird, Paul D. Brazill, R. Thomas Brown and the fine folks at Crime Fiction Lover, Joelle Charbonneau, David Cranmer and his cohorts at Beat to a Pulp, Laura K. Curtis, Neliza Drew, David Dvorkin, Jacques Filippi, Allan Guthrie, Sally Janin, Fiona Johnson, Naomi Johnson, John Kenyon, Chris La Tray, Jennifer MacRostie, Erin Mitchell, Lauren O'Brien, Sabrina Ogden, Dan O'Shea, Keith Rawson and the guys at Crimefactory, Chris Rhatigan, Darren Sant, Kieran Shea, the whole Spinetingler Magazine crew, Julie Summerell, Steve Weddle and the rest of the Needle team, Chuck Wendig, and the inimitable Elizabeth A. White.

I'd be remiss if I did not include a shout-out to the Cressey clan, fierce cheerleaders one and all. Thanks also to my family, who've not only supported my writing from the get-go, but have also given me no shortage of issues to work out in what one hopes are many books to come. (Kidding, family, kidding. Mostly.)

And last, but certainly not least, thank you to my lovely wife Katrina: my best friend, my sounding board, my first editor and ideal reader. I never would have had the guts to put pen to paper had she not encouraged me to do so – and even if I'd somehow managed to, I guarantee the result would have been nowhere near as good. Any mistakes contained herein are no doubt my own, but if ever you find I've stumbled onto a fleeting moment of grace, you now know who to thank.

NEXT

The Wrong Goodbye

Here is an exclusive extract

Rain tore through the canopy of leaves, soaking my clothes until they hung wet and heavy on my limbs but doing little to dispel the fetid stench of decay that pervaded every inch of this godforsaken place.

Just keep moving, I told myself. It's not far now.

Mud sucked at my shoes as I pressed onward, swinging my machete at the knot of vegetation that barred my way. The roar of the rain against the leaves was deafening, swallowing the noises of the jungle until they were little more than a distant radio signal, half-heard beneath the waves of static. Heavy sheets of falling water obscured my vision, reducing my entire world to three square feet of vines and trees and rotting leaves. I swear, that dank jungle stink was enough to make me gag. Then again, that could've been the corpse that I was wearing.

See, I'm what they call a Collector. I collect the souls of the damned, and ensure they find their way to hell. Believe me when I tell you, it ain't the most glamorous of jobs, but it's not like I really have a choice. Back in '44, I was collected myself, after a bad bit of business with a demon and a dying wife. I didn't know it at the time, of course, but this gig of mine was my end of the bargain. Most folks think of hell as some far-off pit of fire and brimstone, but the truth is it's all around them, a hair's breadth from the world that they can see – always pressing, testing, threatening to break through. That hell is where I spend my days, collecting soul after corrupted soul, all in service of a debt I can never repay.

Which brings me to Colombia, and to the dead guy I was wearing.

One of the bitches about being a Collector is that even though you're stuck doing the devil's bidding for all eternity, your body's still six feet under, doing the ol' dust to dust routine. But a Collector can't exist outside a body, which leaves possession as our only option. Most Collectors choose to possess the living – after all, they're plentiful enough, and they come with all kinds of perks, like credit cards and cozy beds. You ask me, though, the living are more trouble than they're worth. They're always crying and pleading and yammering on – or even worse, trying to wrestle control of their bodies back – and the last thing I need when I'm on a job is a backseat driver mucking everything up for me. That's why I stick to the recently dead.

Take this guy, for example. I found him on a tip from my handler, Lilith, who handed me a clipping from a local paper when she gave me my assignment. "Honestly," she'd said, her beautiful face set in a frown, "I don't understand your morbid desire to inhabit the dead, when the living are so much more convenient and, ah, pleasant-smelling."

"A living meat-suit doesn't sit right with me. It's kind of like driving a stolen car."

"You're aware you're being sent there to *kill* someone, are you not?"

"Yeah, only the folks I'm sent to kill need killing." I waved the article at her. "The hell's this thing say, anyway? I barely speak enough Spanish to find the restroom."

"Says he's a fisherman. Died of natural causes – and just yesterday, at that. He's as fresh as can be," she added, smiling sweetly.

Fresh. Right. Just goes to show you, you should never trust a creature of the night.

Turned out, Lilith's idea of natural causes included drowning. This guy'd spent six hours in the drink before they'd found him, washed ashore in a tangle of kelp a good three miles from where he'd gone overboard. I'd cleaned up as best I could in the mortuary sink, but no amount of scrubbing could erase the reek of low tide that clung to his hair, his skin, his coarse thicket of stubble. Still, if Lilith thought this guy would be enough to make me cave and snatch myself a living vessel, she was sorely mistaken. I'm nothing if not stubborn.

But the hassle with the meat-suit was nothing compared to the job itself. His name was Pablo Varela. A major player in the local drug trade. Varela's brutality was a matter of public record. In the two decades he'd been involved in the trafficking of coca, he'd only once been brought to trial. That had been seven years back, and the Colombian government turned the trial into quite the spectacle – TV, radio, the whole nine. Their way, I guess, of demonstrating their newfound dedication to the War on Drugs. Varela declined counsel, and mounted no defense. After eight weeks of damning testimony from the prosecution, it took the jury only minutes to acquit. Some say Varela got to them – that he threatened their lives and the lives of their families if they failed to set him free. Others claim he didn't have to, that his reputation alone was enough to guarantee his release. Whatever it was, the jury made the right choice. Save for them, everyone who set foot in the courtroom over the course of his trial was murdered – every lawyer, every witness, *everyone*. Some, like the bailiff and the court reporter, got off easy: two bullets to the back of the head. The judge and chief prosecutor weren't so lucky. They were strung up by their entrails in the city square – their throats slit, their tongues yanked through the gash in the Colombian style. One week later, the courthouse burned to the ground.

Now a guy like Varela, I don't much mind dispatching. Problem was, the man was paranoid. As soon as he caught wind that I was looking for him, he sent a couple of his goons around to take care of me. That didn't go so well for them, so he sent a couple more. I'm afraid they didn't fare much better. That's when I slipped up. See, I'm not much for killing anyone I don't have to. You could call it mercy, I suppose, or whatever passes for a conscience among the denizens of hell. I call it stupidity, because the bastard that I spared spilled his story to Varela, who grabbed a handful of his most trusted men – not to mention enough firepower to topple your average government – and disappeared into the jungle. Not a bad play, I'll admit. Hell, the first day or so, I even thought it was kinda cute. But as the hours wore on, and the rain continued unabated, the whole affair sort of lost its shine.

Now it'd been four days since I left Cartagena – four grueling days of tracking Varela and his men through blistering heat and near-constant downpours, without so much a moment to eat or sleep or even catch my breath. Varela's men were well-trained and familiar with the terrain, but they were also laden with gear and would no doubt stop to rest, so I was certain I could catch them. Still, October is Colombia's rainy season, and during that rainy season, there's not a wetter place on Earth. All I wanted was to turn around – to find some nice, secluded spot on the beach and watch the waves roll in off the Caribbean through the bottom of a bottle of beer. Which is exactly what I intended to do, just as soon as Varela was dead.

Woody ropes of liana hung low over the forest floor – clawing, scratching, winding themselves around my weary limbs as though they might at any moment retreat with me into the canopy, the rare unwary traveler too delicious a morsel to pass up. It was ridiculous to think, I know, but even the plant-life in the Amazon has a vaguely predatory air – from the strangler figs that choke the life from the mighty kapok trees, to the thick mat of green moss that blankets every surface, always probing, searching, feeding. By the light of day, the jungle wasn't so bad. But as the last gray traces of sun dwindled in the western sky and the brush around me came alive with the rustling of unseen beasts, panic set in. My heart fluttered. My spine crawled. The bitter tang of adrenaline prickled on my tongue. My lips moved in silent prayer – a useless habit – and I quickened my pace, pressing onward through the darkness.

I never even saw the embankment coming.

One moment, I was slashing through the underbrush, the jungle pressing in against me, and the next, there was just a queasy, terrifying nothing. It was like scaling a flight of stairs in the dark only to realize there's one fewer than you remembered, except in this case, my lead foot never hit ground.

I pitched forward. My arms pinwheeled, and my blade clattered to the forest floor, forgotten. I fell for what seemed like forever. Then I slammed into the side of the embankment so hard it knocked the wind out of me, and snapped my jaw shut on my tongue. My mouth filled with blood. My lungs seared as they begged for breath that wouldn't come.

And still, I wasn't done falling.

I tumbled down the steep, muddy slope, clawing frantically at every fern and rain-slick root, but it wasn't any use. I tried to dig in my heels, but one of them caught on something hard, and instead of stopping I hinged forward, somersaulting. End over end I bounced, every inch of my borrowed frame erupting in white-hot pain.

Then, suddenly, all was dark and still and quiet. I was lying face-down in two feet of muddy water, its vegetal stink invading my nose, my mouth, my very pores. Arms shaking, I pushed myself upward, gasping as my face cleared the surface of the muck.

I was at the edge of a broad, shallow stream, which blurbled a delicate melody as it passed along its rocky bed. Behind me, the embankment jutted skyward maybe thirty feet, more cliff-face than hill. From the dense bramble of exposed roots and the relative lack of greenery, my guess was it was the result of a mudslide, and a recent one at that. Not that it mattered much to me either way. I mean, a fall's a fall – and besides, I was way more interested in the fire.

It couldn't have been more than fifty yards downstream, nestled in a rocky crook on the far bank of the riverbed. The fire itself was lined with river rocks, and a makeshift spit of branches stretched across it, upon which roasted a goodly hunk of meat. Whoever'd chosen the spot knew what they were doing – the canopy was heavy there, providing shelter from the rain, and the stream supplied ample drinking water; the natural depression of the land hid the fire from view of anyone passing by above. Were it not for my fall, I would've walked right on past and never been the wiser. I allowed myself a smile as I pondered my sudden turn of fortune.

Though it had been days since I'd last eaten, and the aroma of cooking meat had set my mouth watering, I forced myself to hold my ground, counting to one hundred as I listened for any indication that Varela's men had seen me. I heard nothing but the growling of my stomach, and there was no sentry in sight. Given what I knew of Varela, the lack of perimeter guards was surprising, but maybe he believed the jungle to be protection enough from me. He had no idea how wrong he was.

I approached the stream at a crouch, suddenly grateful for the deepening twilight and the thin layer of mud that together served to obscure my approach. Water leeched into my boots as I crossed to the far bank, mindful all the while for any whisper of movement that might indicate snake. With Varela finally within my grasp, the last thing I needed was to tangle with a deadly coral, or have this meat-suit squeezed to death by an anaconda. I might not be too fond of this job of mine, but I'd still rather be predator than prey.

Twenty yards out, I knew that something was wrong. There was no idle chatter, no rustle of fabric – no sound at all from Varela's camp, save for a low, persistent buzzing, like a dentist's distant drill. From behind a massive kapok trunk, I hazarded a glance. Several men, their backs to me, were silhouetted by the fire, but all were still as death. I watched them for a moment, wondering if this was perhaps some kind of trap – a dummy camp set up to lure me in. Then I realized where the buzzing was coming from, and I knew this was no trap.

I stepped clear of my hiding place and wandered into the camp. The buzzing here was deafening, and up close its source was clear. The entire place was swarming with insects – millions of them – all fighting for their share of the feast laid out before them. The corpses of Varela's men teemed with them – from tiny flies and gnats to massive, iridescent beetles the likes of which I'd never seen, all attracted by the scent of spilled blood and dead flesh, still too faint for my meat-suit's nose to recognize. I counted seven men around the fire. Five of them were riddled with bullet-holes, and abandoned among them was a Kalashnikov assault rifle, its action open, its clip spent. Each of the dead men carried a Kalashnikov of their own, strapped across their backs as if they had been at ease when they'd been attacked.

By the look of the other two, I'd say those first five got off light.

The first of them lay face-down a few feet from the fire. His rifle lay beneath him, as if he had been holding it at ready when he was attacked. No doubt this was the sentry I'd been listening for. It looked to me like he'd come running to help his buddies when the shooting started. An admirable reaction, to be sure, but apparently not the smartest play. I rolled him over with the toe of my boot. His neck flopped like a wet noodle, and his head lolled to one side. A crushing blow from a rectangular something-or-other had caved in his nose and made tartare of his face – all meat and teeth and glistening bone. A glance at the abandoned Kalashnikov confirmed the gunstock was to blame; it was caked with blood and bits of flesh. Whether the blow had been enough to snap his neck, or his assailant had done it afterward for good measure, I couldn't say.

"What the fuck *happened* here?" I asked of no one in particular. For a moment, I thought I might just get an answer – the sentry's ruined lips parted and emitted a faint, rustling whisper. Then a cockroach the size of my fist crawled out of his mouth, antennae twitching in the still night air. I eyed it for a moment, but if it knew what went down, it sure as hell wasn't talking.

The last of the bodies lay spread-eagled on the forest floor. His hands and feet were staked to the ground with knives no doubt scavenged from the belts of his dead companions. His shirt lay open at his sides, exposing his mutilated chest, now crawling with all manner of bugs. Unlike the sentry, his face had been spared, though I suspect that was more for my benefit than for his. His eyes were clouded and glassy, and his features were twisted into a rictus of pain, but still, there was no mistaking that face.

Varela.

I crouched beside him and lay a hand atop his bloodied chest. Insects scampered across the back of my hand and crawled up my sleeve. I ignored them, instead closing my eyes and extending my consciousness – probing, searching. But it was no use. There was nothing left to find.

Varela's soul was gone.

My meat-suit's heart thudded in its chest as the realization hit. Now, I don't know how the white-hats play it, but the souls of the damned don't just up and leave on their own. That means whoever attacked these men wasn't human – as far as I knew, there wasn't a man alive who had the means to steal a soul. That meant Collector.

Problem is, we Collectors ain't exactly the Three Musketeers. All for one and one for all sounds all well and good, but hell doesn't work that way. Varela's soul was *my* responsibility – no exceptions, no excuses – which meant if *I* wasn't the one to bring him in, then I had failed in my mission. And believe me when I tell you, my employers don't take kindly to failure.

I took a calming breath, and willed my racing heart to slow. The last thing I needed now was to freak. I forced myself to look over the scene, certain there was something I had missed.

Turns out, I was right.

It's embarrassing, really, because in retrospect, it was so damn obvious. But when I'd first approached the camp, I had no reason to assume Collector. I just figured one of Varela's competitors had beaten me to the punch, in which case Varela's massive chest-wound made sense – I mean, he had to die of *something*. But when you take a soul, the body dies. So, then: why the bloodied chest?

I retreated to the fire, toppling the spit and sending the hunk of now-charred meat into the flames. For the first time, I realized how recently this must've all gone down – the meat, though burned, had yet to cook off the spit, and though the air was hot and thick with moisture, the bodies weren't bloated, and showed no signs of rigor. Whoever'd done this had beaten me by a matter of minutes. Of course, that knowledge didn't help me much – a few minutes was plenty of time for any Collector worth his salt to disappear. I pushed aside all thought of pursuit, instead focusing on my immediate task. I shoved one of the support branches from the spit into the embers until it caught. Then I returned to Varela's body, torch in hand.

The flame danced in the sudden breeze as I swung the branch at the writhing mass of bugs that blanketed Varela's chest. Reluctantly, they parted, frightened by the fire but unwilling to relinquish their blood meal. As they shifted, I caught a glimpse of something odd – letters, three inches high, carved into the dead man's flesh.

I lost my patience with the flame and dropped to my knees, scattering the remaining insects with a sweep of my arm. Beneath them was a message, ragged and crusted brown with drying blood:

SAM –

WE NEED TO TALK.

YOU KNOW WHERE.

-D

That bastard, I thought. I should've known.

I must've spent a half an hour sitting there, marveling at the presumption, the sheer arrogance that pervaded every grisly slice. Eventually, though, I rose and left the camp behind, plunging once more into the jungle – this time heading south.

Toward Bogotá.

Toward Danny.

SIXTY-ONE NAILS
MIKE SHEVDON
THE ROAD TO BEDLAM
MIKE SHEVDON
Darkness Falling
Peter Crowther

TIM WAGGONER
Dead Streets
Matt Richter, Private Eye, Zombie
TIM WAGGONER
Matt Richter, Private Eye, Zombie

JAMIESON
CHUCK WENDIG
MATT FORBECK